VOLUME ONE HUNDRED AND SEVEN

ADVANCES IN VIRUS RESEARCH

Immunopathology

VOLUME ONE HUNDRED AND SEVEN

ADVANCES IN VIRUS RESEARCH

Immunopathology

Edited by

JOHN P. CARR

University of Cambridge, United Kingdom

MARILYN J. ROOSSINCK

Department of Plant Pathology and Environmental Microbiology, Center for Infectious Disease Dynamics, Penn State University, University Park, PA, United States

ELSEVIER

ACADEMIC PRESS

An imprint of Elsevier

Academic Press is an imprint of Elsevier
50 Hampshire Street, 5th Floor, Cambridge, MA 02139, United States
525 B Street, Suite 1650, San Diego, CA 92101, United States
The Boulevard, Langford Lane, Kidlington, Oxford OX5 1GB, United Kingdom
125 London Wall, London, EC2Y 5AS, United Kingdom

First edition 2020

ISBN: 978-0-12-821588-3
ISSN: 0065-3527

For information on all Academic Press publications
visit our website at https://www.elsevier.com/books-and-journals

Publisher: Zoe Kruze
Acquisitions Editor: Ashlie M. Jackman
Editorial Project Manager: Hilal Johnson
Production Project Manager: James Selvam
Cover Designer: Alan Studholme

Typeset by SPi Global, India

Contents

Contributors

Victoria K. Baxter
Department of Pathology and Laboratory Medicine, University of North Carolina at Chapel Hill, Chapel Hill, NC, United States

David F. Boyd
Department of Immunology, St. Jude Children's Research Hospital, Memphis, TN, United States

Linfang Dong
College of Veterinary Medicine, Northwest A&F University, Yangling, Shaanxi, China

Andrea González-González
Department of Biology, University of Florida, Gainesville, FL, United States

Mark T. Heise
Department of Genetics; Department of Microbiology and Immunology, University of North Carolina at Chapel Hill, Chapel Hill, NC, United States

Christiane Herden
Institute of Veterinary Pathology; Center for Brain, Mind and Behavior, Justus-Liebig-University Giessen, Giessen, Germany

Kiwamu Hyodo
Institute of Plant Science and Resources, Okayama University, Kurashiki, Okayama, Japan

Paola Leonetti
Department of Biology, Agricultural and Food Sciences, Institute for Sustainable Plant Protection, CNR, Bari, Italy

Zhiwei Li
College of Veterinary Medicine, Northwest A&F University, Yangling, Shaanxi, China

Zhiqian Ma
College of Veterinary Medicine, Northwest A&F University, Yangling, Shaanxi, China

Pascal Miesen
Department of Medical Microbiology, Radboud Institute for Molecular Life Sciences, Radboud University Medical Center, Nijmegen, The Netherlands

Jonathan J. Miner
Department of Medicine; Department of Molecular Microbiology; Department of Pathology and Immunology, Washington University in St. Louis School of Medicine, Saint Louis, MO, United States

Jana Müller
Institute of Veterinary Pathology, Justus-Liebig-University Giessen, Giessen, Germany

Peter D. Nagy
Department of Plant Pathology, University of Kentucky, Lexington, KY, United States

Daniel Nobach
Institute of Veterinary Pathology, Justus-Liebig-University Giessen, Giessen, Germany

Tetsuro Okuno
Department of Plant Life Science, Faculty of Agriculture, Ryukoku University, Otsu, Shiga, Japan

Vitantonio Pantaleo
Department of Biology, Agricultural and Food Sciences, Institute for Sustainable Plant Protection, CNR, Bari, Italy

Derek J. Platt
Department of Molecular Microbiology, Washington University in St. Louis School of Medicine, Saint Louis, MO, United States

Hélène Sanfaçon
Summerland Research and Development Centre, Agriculture and Agri-Food Canada, Summerland, BC, Canada

Dennis Tappe
Bernhard Nocht Institute for Tropical Medicine, Hamburg, Germany

Paul G. Thomas
Department of Immunology, St. Jude Children's Research Hospital; Department of Microbiology, Immunology, and Biochemistry, College of Graduate Health Sciences, University of Tennessee Health Science Center, Memphis, TN, United States

Ronald P. van Rij
Department of Medical Microbiology, Radboud Institute for Molecular Life Sciences, Radboud University Medical Center, Nijmegen, The Netherlands

Marta L. Wayne
Department of Biology, University of Florida, Gainesville, FL, United States

Taylor L. Wilson
Department of Immunology, St. Jude Children's Research Hospital; Department of Microbiology, Immunology, and Biochemistry, College of Graduate Health Sciences, University of Tennessee Health Science Center, Memphis, TN, United States

Shuqi Xiao
College of Veterinary Medicine, Northwest A&F University, Yangling, Shaanxi, China

Ting Yang
College of Veterinary Medicine, Northwest A&F University, Yangling, Shaanxi, China

Preface

All of cellular life can be infected by viruses, and although many viruses are commensal, or even mutualistic, pathogens elicit a response in their hosts that involve various aspects of the host immune system. In humans and other animals, the innate immune system is often the first line of defense, followed by the adaptive immune system. Plants, insects, fungi, and even archaea and bacteria also invoke both innate and adaptive immunity in response to virus infections. The mechanisms of viral pathology in many cases are the result of this immune response being too strong. Immunopathology is well studied in human viruses. The importance of immunopathology also has been recognized more recently in plant–virus interactions. In this volume we cover a number of aspects of plant viral immunity that are involved in pathology in Chapters 1–4, before focusing on the viral immunopathology of vertebrates and insects in Chapters 5–10.

Chapter 1 provides a comparative review of the exploitation of small RNA-mediated defensive and regulatory systems by viroids and satellite RNA molecules in plants and by viruses in plant and insect cells. Chapter 3 takes up this narrative, discussing how viruses modulate symptom severity through their effects on small RNA pathways. The chapter also uses a range of specific examples to demonstrate how viruses and their gene products interfere with other plant immune systems and hormonal regulatory pathways to balance their need to replicate while preventing excess damage to the host. Small RNA-mediated RNA silencing systems were originally regarded as defenses against viruses, and indeed they may have originally arisen for this purpose. However, as these two chapters show, this primitive immune system has now become an exploitable target for viruses and subviral parasitic molecules to regulate their own replication and spread in the host.

Chapters 2 and 4 discuss the commandeering of host cellular components by viruses, and how viruses convert these to proviral factors that, instead of supporting normal metabolic or defensive processes, support viral infection and spread in susceptible plant hosts. Chapter 2 focuses on how plant viruses integrate plant defensive factors into protein complexes that facilitate viral infection and discusses the hijack of plant homeostatic processes by viruses, using specific examples. Chapter 4 describes how replication of tombusvirus-derived constructs in a yeast-based model system is used to identify host gene products that support multiplication of positive sense RNA viruses. The review describes how protein chaperones, the ubiquitin-mediated protein degradation system, and cellular RNA helicases are co-opted for tombusvirus replication. Such studies have wider relevance for our understanding of positive sense RNA viruses not only of plants but also those of insect and vertebrate hosts.

Chapter 5 describes the current knowledge about bornaviruses and the role of T cell-based immunopathology that has been known for some time, and recent studies using animal models that are revealing many other aspects of the immune systems role in pathology.

In Chapter 6 we learn about the most recent findings of Zika virus immunopathology, as well as recent therapies and vaccine development.

Chapter 7 describes the immunopathology of influenza virus that has been studied for over 100 years, and how the virus and host must find a balance between clearing the virus and immunopathology.

Chapter 8 discusses what is known about the immune response to insect viruses and gives a perspective of areas of research that could address immunopathology in insects as models.

Chapter 9 covers the alphaviruses, including West Nile virus, the equine encephalitis viruses, and chikungunya, whose immunopathology can include long-lasting arthritic-like symptoms. Understanding the role of the immune system in these viruses is important for development of therapies.

In Chapter 10 we provide a discussion of the development of a vaccine for SARS-CoV-2 using reverse genetics, and the concerns of immunopathology in the use of vaccines for coronaviruses.

This volume brings together pathogenic viruses from very diverse hosts to demonstrate that the negative effects of viruses are often linked to the host response to the virus, rather than a direct effect of the virus on host cells.

John P. Carr
University of Cambridge, United Kingdom
Marilyn J. Roossinck
Department of Plant Pathology and Environmental Microbiology,
Center for Infectious Disease Dynamics, Penn State University,
University Park, PA, United States

CHAPTER ONE

Viral and subviral derived small RNAs as pathogenic determinants in plants and insects

Paola Leonetti[a], Pascal Miesen[b], Ronald P. van Rij[b,*], Vitantonio Pantaleo[a,*]

[a]Department of Biology, Agricultural and Food Sciences, Institute for Sustainable Plant Protection, CNR, Bari, Italy.

[b]Department of Medical Microbiology, Radboud Institute for Molecular Life Sciences, Radboud University Medical Center, Nijmegen, The Netherlands

[*]Corresponding authors: e-mail addresses: ronald.vanRij@radboudumc.nl; vitantonio.pantaleo@cnr.it

Contents

Abstract

The phenotypic manifestations of disease induced by viruses and subviral infectious entities are the result of complex molecular interactions between host and viral factors. The viral determinants of the diseased phenotype have traditionally been sought at the level of structural or non-structural proteins. However, the discovery of RNA silencing mechanisms has led to speculations that determinants of the diseased phenotype are caused by viral nucleic acid sequences in addition to proteins. RNA silencing is a gene regulation mechanism conserved within eukaryotic kingdoms (with the exception of some yeast species), and in plants and insects it also functions as an antiviral

Advances in Virus Research, Volume 107
ISSN 0065-3527
https://doi.org/10.1016/bs.aivir.2020.04.001

mechanism. Non-coding RNAs of viral origin, ranging in size from 21 to 24 nucleotides (viral small interfering RNAs, vsiRNAs) accumulate in virus-infected tissues and organs, in some cases to comparable levels as the entire complement of host-encoded small interfering RNAs. Upon incorporation into RNA-induced silencing complexes, vsiRNAs can mediate cleavage or induce translational inhibition of nucleic acid targets in a sequence-specific manner. This review focuses on recent findings that suggest an increased complexity of small RNA-based interactions between virus and host. We mainly address plant viruses, but where applicable discuss insect viruses as well. Prominence is given to studies that have indisputably demonstrated that vsiRNAs determine diseased phenotype by either carrying sequence determinants or, indirectly, by altering host-gene regulatory pathways. Results from these studies suggest biotechnological applications, which are also discussed.

1. Introduction

Viruses are obligate intracellular infectious entities that affect all living organisms. The viral particle (i.e., virion) is composed of two, sometimes three components: (i) the viral nucleic acid genome, (ii) a coat protein organized to form a capsid that protects the genome and confers the shape to the virion and, only in the case of some families, (iii) an envelope that surrounds the virus particles (Hull, 2002). Viral genomes are considerably diverse in composition and organization and include DNA, RNA, linear, circular or segmented, single- or double-stranded, positive (+), negative (−) or ambisense (+/−) configurations. The type of genome determines viral replication strategies among groups of viruses. In addition to structural proteins, viruses express non-structural proteins, which perform other functions in the infectious cycle. RNA viruses encode their own RNA-dependent RNA polymerase (RDRp), DNA viruses use host and/or viral DNA polymerases for replication, whereas viroids use host RNA-dependent RNA polymerases and replicate via RNA intermediates (Hull, 2002).

Satellites (satRNAs or satDNAs), defective RNAs or defective-interfering RNAs are three different molecules associated with plant viruses (helper viruses) and generally referred to as subviral entities. Subviral RNAs are non-coding (except for cases of coding satRNAs) and can be linear or circular. They are indispensable for viral replication and propagation and are fully dependent on the helper virus (Simon et al., 2004). Another class of plant subviral agents is viroids, which are minimal circular non-coding RNAs that usurp host resources for replication, and complete the infectious cycle without any need of structural or non-structural proteins, or of a helper virus (Tabler and Tsagris, 2004).

The presence and replication of viruses in the host induce diverse mechanisms of defense at the single cell level as well as at the organismal level. These mechanisms range from the RNA silencing mechanism mainly found in plants, insects, nematodes and fungi to the sophisticated interferon-regulated gene response of higher animals. In plants, viral replication and other stages of the infectious cycle, such as single-stranded RNA (ssRNA) or double-stranded RNA (dsRNA) intermediates, can actively provoke RNA silencing-based host defense responses. RNA silencing is an evolutionarily conserved gene regulation mechanism in eukaryotes mediated by 21–24 nt small interfering RNAs (siRNAs) (Fire, 1999). siRNAs are processed from dsRNAs, hairpin RNA or highly structured RNAs (dsRNA-like) by the RNase III-like enzymes Dicer (in animals) or Dicer-like proteins (DCL, in plants) (see for review, Ding, 2010). To induce silencing, siRNA duplexes are loaded into an Argonaute (AGO) protein to form the RNA-induced silencing complex (RISC), the so-called passenger strand of siRNA is removed and the guide strand directs RISC to bind to complementary ssRNA and cleave the cognate RNA between nt 10 and 11, counting from the 5′ end of the small RNA. Plants have three basic types of small RNAs: 20–24 nt microRNA (miRNA), 21–22 nt siRNA, and 24 nt repeat-associated siRNA (rasiRNA). miRNAs induce post-transcriptional degradation or translational repression mainly of messenger RNAs (mRNAs) that encode regulatory proteins, such as transcription factors, and play a key role in plant and animal growth. The 21–22 nt siRNAs direct degradation of viral RNA and some endogenous mRNA, and are important in defense against viruses (Guo et al., 2019). The 24 nt rasiRNAs are unique to plants, and are involved in RNA-directed DNA methylation, which is important for maintaining genome stability by silencing transposons and repetitive DNA sequences (Lewsey et al., 2016).

RNA silencing pathways in insects follow the same principles as those in plants. Insects such as the fruit fly, *Drosophila melanogaster*, and vector mosquitoes encode two *Dicer* genes and two *AGO* genes that seem mostly dedicated to either the miRNA (*Dicer-1* and *AGO1*) or the siRNA (*Dicer-2* and *AGO2*) pathways (Lee et al., 2004; Okamura et al., 2004). In addition, insects have a third small RNA pathway, the PIWI-interacting RNA (piRNA) pathway, that was originally described as a defense mechanism against transposable elements in the germline (Brennecke et al., 2007), but is increasingly recognized to have additional functions in both germline and somatic tissues in a range of insects (Barckmann et al., 2015; Halbach et al., 2020; Lewis et al., 2018; Miesen et al., 2016; Rouget et al., 2010; Vourekas et al., 2016). Unlike the siRNA and miRNA pathways that process dsRNA structures, piRNAs are processed from ssRNA precursors encoded in the genome in so-called piRNA clusters. Precursors are processed in a series of steps into 25–30 nt small RNAs that associate with members of the PIWI clade of AGO proteins (Czech et al., 2018; Iwasaki et al., 2015). Primary piRNAs derived from piRNA clusters tend to be antisense to transposable elements, and are thus poised to target mRNAs from active transposons. All three RNA silencing pathways of insects have been implicated in virus-host interactions, with the siRNA pathway being the main antiviral defense system (Bronkhorst and van Rij, 2014; Guo et al., 2019).

Viral infections result in a variety of disease symptoms that vary in character and severity depending on the type of infection, the host factors involved, as well as the immune response triggered. The fact that plant viruses and subviral agents are obligate parasites that exploit vital functions of the host suggests that the development of symptoms is a side effect of the infection itself. However, in most cases host plants do not show any pathological phenotype, and in some cases viral infections even confer a selective advantage to the host in terms of reproduction and adaptation to disadvantageous climatic conditions (Groen et al., 2016; Marquez et al., 2007; Roossinck, 2011; Xu et al., 2008). A balanced co-existence of viruses and plants is particularly evident in ecosystems in which human influence is minimal (ecosystems with a status close to the climax) and in agricultural and forestry systems rich in biodiversity. In contrast, croplands characterized by a general loss of biodiversity can facilitate the emergence of plant viruses (Roossinck and Garcia-Arenal, 2015).

Pathological phenotypes can be the result of specific molecular interactions between host and viral factors. Interactions of both the plant and pathogenic virus can be protein-protein, protein-nucleic acid and between nucleic acids (RNA or DNA). Among pathogenicity determinants, emblematic is the case of viral silencing suppressors (VSSs), structural or non-structural proteins that counteract the antiviral RNA silencing machinery. It is well established that VSSs interfere with the endogenous RNA silencing machinery by interacting with siRNA and miRNA biogenesis and functionality at different stages in plants (Csorba et al., 2015). The discovery of RNA silencing has also led to speculate that determinants of the diseased phenotype can be caused by viral nucleic acid sequences in addition to proteins. This review aims to highlight the studies conducted so far that have indisputably demonstrated that vsiRNAs determine diseased phenotype by either carrying sequence determinants or, indirectly, by altering the host-gene regulation pathways. Moreover, we will explore recent insights into mechanisms of RNA suppression and evasion (Fig. 1).

2. vsiRNAs containing sequence-specific phenotypic determinants

Plant tissues infected with viruses, satellites and viroids contain easily detectable and high amounts (comparable to endogenous small RNAs) of vsiRNAs, which guide RISC to target and inactivate the viral RNA and restrict viral invasion. Herein, we discuss cases of vsiRNAs from satRNAs, viroids and viruses that target host sequences. These vsiRNA sequences thus act as pathogenicity determinants that are directly involved in the appearance of the diseased phenotype.

2.1 Satellites

SatRNA genomes range in length from 220 to 1500 nt and can be linear or circular, coding or non-coding. satRNAs determine symptoms with mechanisms mainly attributed to satRNA-helper virus interaction. Indeed, satRNAs participate in helper virus-host interactions by modulating the accumulation of helper virus and severity of helper virus-induced symptoms (Collmer and Howell, 1992). One emblematic example comes from cucumber mosaic virus (CMV), which is often associated with linear satRNAs. A satRNA associated with a severe pathogenic Shan-Dong (SD) strain of CMV attenuates the yellowing caused by SD-CMV infection. Modulation of the disease phenotype is due to one satRNA-derived vsiRNA, which associates with the AGO-containing RISC in vivo and targets the upstream region of the CMV 3′-untranslated region (UTR) for slicing (Zhu et al., 2011). In turn, the helper virus can harness host antiviral silencing to control the accumulation of satRNA. This is the case for the tombusvirus Cymbidium ringspot virus and its satRNA (Pantaleo and Burgyan, 2008).

Some satRNAs may originate from repetitive elements in the plant genome and, therefore, satRNA-derived vsiRNAs may be found in the sRNA transcriptome of healthy plants (Zahid et al., 2015). Yet, large amounts of vsiRNAs from satRNAs are usually detected in infected but

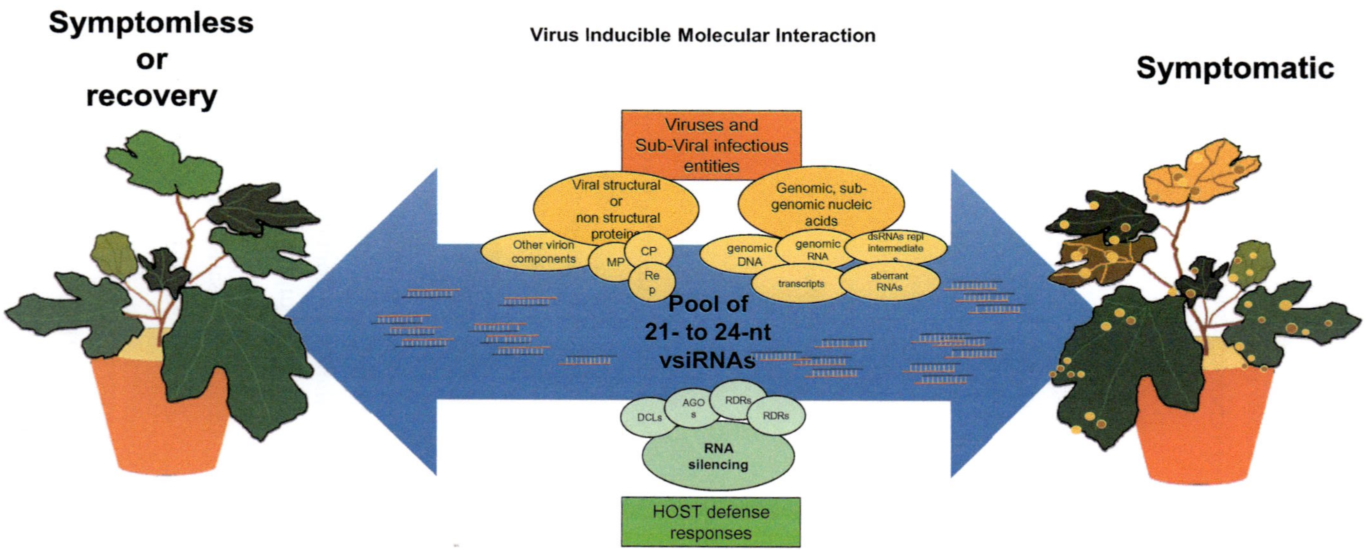

Fig. 1 Symptomatic or recovery plants are the showcase of RNA silencing-based host-pathogen interactions. Structural or non-structural viral proteins and nucleic acids of viral origin can induce molecular interactions and host defense responses. Pools of 21–24 nt viral siRNAs (vsiRNAs) accumulate in infected plants and may be involved in both the antiviral response and in the disease phenotype, including recovery. Dicer-like protein (DCLs), RNA-dependent RNA polymerase (RDRs), movement protein (MP), capsid protein (CP) and replication protein (Rep).

healthy plant tissues (for a review, see Pooggin, 2018), indicating that vsiRNAs are not necessarily involved in symptom development.

2.1.1 CMV Y-sat targeting ChlI transcripts

The first evidence that foliar symptoms are specifically induced by vsiRNAs was reported in the case of CMV-associated satRNAs (Shimura et al., 2011; Smith et al., 2011). Y-satRNA is one of several CMV satRNAs that confer a bright yellow mosaic in *Nicotiana tabacum* (Takanami, 1981). The determinant for these symptoms has been mapped to a restricted region of the Y-satRNA sequence (Masuta and Takanami, 1989), denoted as Satellite Yellow Region (SYR). By conducting microarray analysis in *N. benthamiana* leaves expressing a Y-sat inverted repeat (i.e., the Y-sat RNA is followed downstream by its reverse complement sequence), Shimura and collaborators identified genes that were down-regulated to levels lower than 40%. Of these genes, 31 were involved in the biosynthesis pathways of chloroplastic pigments; among the pigments, the magnesium-chelatase-subunit I (chlI) gene contained a 22 nt-long sequence that was highly complementary to the SYR, denoted as Yellow Region (YR). Importantly, small RNA sequencing (sRNAseq) of CMV/Y-sat-infected *N. benthamiana* revealed that plant tissues contained vsiRNAs, derived from both the virus and Y-sat. About 90% of Y-sat vsiRNAs were 21- or 22 nt long and predominantly from the (+) RNA strand with an average ratio of about 10 ([+] RNA strand) to 1 ([−] RNA strand). Many Y-sat vsiRNAs were from SYR, explaining the presence of uncapped 5′-end of the chlI mRNA in the middle of the YR, suggesting vsiRNA-mediated cleavage (Shimura et al., 2011) (Fig. 2).

The chlI gene sequence is highly conserved among plant species including *Arabidopsis thaliana* (hereafter, *Arabidopsis*), pepper, tomato and *Nicotiana* spp. However, of all the *Nicotiana* spp., which are highly permissive for CMV and Y-sat replication and invasion, *N. clevelandii* and *N. debneyi* do not reveal yellowing symptoms, and chlI mRNA remains unaltered in infected tissues. *N. clevelandii* expresses two chlI mRNA species, with the predominant species containing an A→G substitution in the YR sequence. Similarly, *N. debneyi* expresses two chlI mRNA species, with the predominant form containing a G→U substitution in the YR sequence. These conversions generate critical mismatches in the YR and abolish the capacity of Y-sat-derived vsiRNAs to target chlI mRNA (Smith et al., 2011), lending additional support to the relevance of Y-sat vsiRNAs in generating yellowing symptoms.

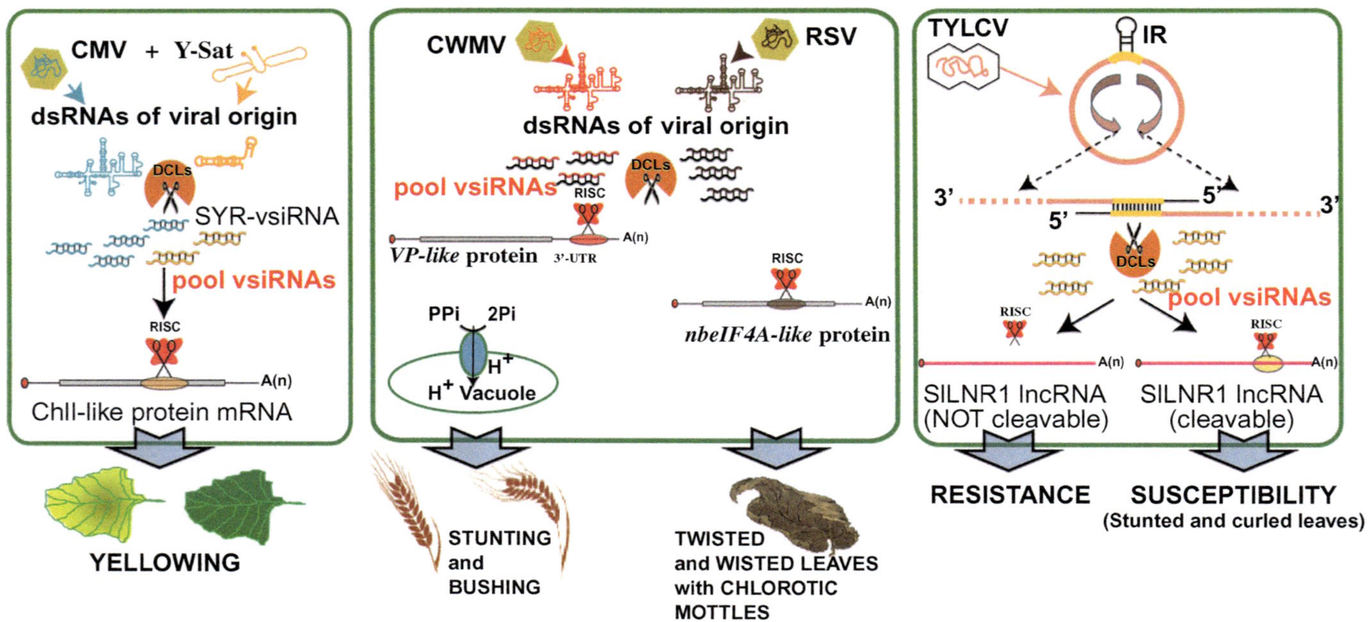

Fig. 2 Cucumber mosaic virus (CMV) Y-satellite RNA (Y-Sat) (left panel), Chinese wheat mosaic virus (CWMV) and rice stripe virus (RSV) (middle panel), and tomato yellow leaf curl virus (TYLCV) (left panel) deliver vsiRNAs in infected cells. These vsiRNAs target specific host mRNAs, resulting in the development of specific symptoms and plant resistance/susceptibility phenotypes. SYR, satellite yellow region.

2.2 Viroids

The importance of pathogenicity determinants in viroids, also termed virulence modulating regions (VMRs), in disease symptom severity is well established (Keese and Symons, 1985). The mechanism responsible for the development of symptoms in infected plants had remained elusive for decades, but now seems to be mediated, at least partly, by small RNAs.

Viroidal RNAs trigger antiviral RNA silencing in the plant and generate significant amounts of vsiRNAs (Adkar-Purushothama et al., 2015b; Di Serio et al., 2009; Diermann et al., 2010; Itaya et al., 2007; Li et al., 2012; Navarro et al., 2009). The observations that viroids produce easily detectable of vsiRNAs and that viroids (and thus VMRs) are non-coding have stimulated researchers to study pathogenetic determinants in vsiRNAs.

2.2.1 PLMVd vsiRNAs targeting cHSP90 transcripts

Peach latent mosaic viroid (PLMVd), a viroid of the Avsunviroidae family which replicate in plastids, produces a wide range of symptoms in peach trees, including strong albinism (known as peach calico, PC) in leaves, stems, and fruits (Hernandez and Flores, 1992). Navarro et al. (2012) dissected the VMRs within the PLMVd sequence that induce the PC phenotype. Small RNAseq analysis revealed that PLMVd is a source of vsiRNAs in tissues showing the PC phenotype. A selection of these vsiRNAs was used to query the peach transcriptome to identify possible mRNA targets. Two abundant species of 21 nt vsiRNAs aligned, with sufficient complementarity, with a nuclear mRNA encoding chloroplastic heat shock protein 90 (cHSP90). Accordingly, expression of the cHSP90 mRNA transcript was clearly reduced in tissues with the PC phenotype, and RNA ligase-mediated rapid amplification of cDNA ends (RLM-RACE) analysis detected a truncated cHSP90 mRNA transcript at a position corresponding to the 10–11th nucleotide of the vsiRNAs solely in symptomatic tissues (Navarro et al., 2012). The cHSP90 gene transcript was shown to be involved in the differentiation of plastids in *Arabidopsis* (Cao et al., 2003). Indeed, unlike the case of Y-sat-derived vsiRNA, the PLMVd vsiRNA determines the diseased phenotype by altering chloroplastic structures, not photosynthetic pigments (Fig. 3).

2.2.2 PSTVd vsiRNAs targeting FRL3 transcripts

In flowering plants the vegetative-to-reproductive transition is firmly determined by specific genetic factors (Wellmer and Riechmann, 2010) and induced by climatic conditions such as photoperiod and temperature

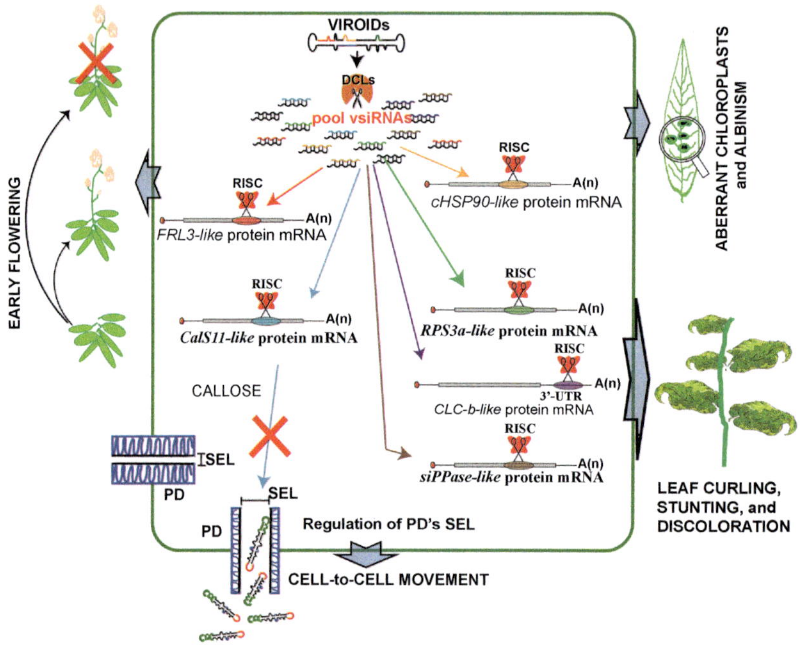

Fig. 3 Peach latent mosaic viroid (PLMVd) and Potato spindle tuber viroid (PSTVd) vsiRNAs contain sequence-specific phenotypic determinants. Strong albinism symptoms, early flowering and decrement of callose production and enlargement of the size exclusion limit (SEL) of plasmodesmata (PDs) are specific symptoms and can be accompanied by severe stunting, leaf curling and discoloration.

(Simpson and Dean, 2002). The timing of flowering is one of the most important developmental transitions in the life cycle of a plant and a basic requirement for reproductive success. In agro-ecosystems, the alteration of flowering time can lead to production losses and is therefore perceived by growers as a pathological manifestation. Oftentimes, viroid infection induces early flowering in plants, as previously observed in chrysanthemum infected with chrysanthemum stunt viroid and in tomato infected with potato spindle tuber viroid (PSTVd), two members of the *Pospiviroidae* family, which replicate in the nucleus (Adkar-Purushothama et al., 2018). To search for early flowering determinants, Perreault and colleagues (Adkar-Purushothama et al., 2018) dissected in silico the VMR sequence of PSTVd into 21 nt siRNAs. Pools of synthetic vsiRNAs were then used to interrogate the tomato transcriptome in an attempt to identify cognate target sequences. Two specific vsiRNAs deriving from two different PSTVd variants (PSTVd-I and PSTVd-M) were predicted to target the open reading frame (ORF) of FRIGIDA-like protein 3 (FRL3), an ortholog of *Arabidopsis FRIGIDA* that is required for transcription of *FLOWERING LOCUS C* (*FLC*), which in turn determines late flowering (Mouradov et al., 2002). FRL3-mediated cleavage by the selected vsiRNAs was confirmed by RLM-RACE coupled with high-throughput sequencing of uncapped mRNA ends (denoted as PARE) (German et al., 2008) in PSTVd-I-infected plants (but not in PSTVd-M-infected plants, likely due to lower viroidal titers, as discussed in Section 2.2.3). Adkar-Purushothama et al. have also shown that transient expression of the selected vsiRNAs in tomato could silence FRL3 transcripts and induce early flowering (Adkar-Purushothama et al., 2018) (Fig. 3).

2.2.3 PSTVd vsiRNAs targeting CalS11-like transcripts

Often, if not always, the extent and intensity of the disease phenotype associated with viruses and viroids depend on the accumulation of the infectious entity in infected tissues and their capillary spread throughout tissues and organs. The size exclusion limit (SEL) of plasmodesmata (PDs) regulates the passive diffusion of small, soluble molecules and the passage of macromolecular structures, such as viruses and viroids (Maule, 1991). Callose is a polysaccharide (β-1,3-glucan) that is deposited at PDs to regulate the SEL (Chen and Kim, 2009). Intuitively, the objective of viral invasion of the entire plant can be pursued by direct or indirect modification of the SEL of PDs. Initiation of systemic trafficking and cell-to-cell movement through PDs by PSTVd is influenced by two VMRs (Zhong et al., 2008). These two VMRs produce easily detectable 21–24 nt vsiRNAs (Adkar-Purushothama et al., 2015b). Perreault's group (2015a) demonstrated that some vsiRNAs mediate cleavage of Callose synthase 11-like (CalS11-like) mRNA at the cognate target site and, accordingly, determine down-regulation of the mRNA in infected tomato plants (Adkar-Purushothama et al., 2015a). In the same study, it was found that PSTVd-I spreads more rapidly and to higher levels (up to 10-fold) than the PSTVd-M strain. The CalS11-like mRNA transcript is expressed at lower levels in PSTVd-I infection than in PSTVd-M infection, which would explain the capacity of PSTVd-I, but not of PSTVd-M, to spread more rapidly and induce early flowering by more effective down-regulation of the FRL3 transcript (see Section 2.2.2) (Adkar-Purushothama et al., 2018) (Fig. 3).

2.2.4 PSTVd vsiRNAs targeting CLC-b-like and RPS3a-like transcripts

The distribution of virus and viroid vsiRNAs along the genome is never uniform and displays so-called "hotspots." These hotspots correspond to regions with a more complex secondary dsRNA-like structure that are therefore particularly attractive substrates for processing by DCLs. On the other hand, the functionality of siRNAs (both endogenous small RNAs such as miRNAs, as well as exogenous siRNAs such as vsiRNAs) is mainly based on a high loading efficiency with RISC, rather than merely their abundance (Shimura and Pantaleo, 2011).

Further investigations have focused on the ability of PSTVd to generate pathogenic vsiRNAs in infected tomato tissues, with particular attention to vsiRNAs from "non-hotspot" regions of the viroid genome, both from the (+) and (−) RNA strands (Adkar-Purushothama et al., 2017). In silico analyses suggested that (+) vsiRNAs from non-hotspot regions on either the (+) or the (−) RNA strand may target the 3′ UTR of chloride channel protein (CLC-b-like) mRNA and 40S ribosomal protein S3a-like (RPS3a-like) mRNA transcript in the ORF, respectively. Tobacco rattle virus vectors were used for virus-induced gene silencing (VIGS) (Marton et al., 2010) of the CLC-b-like and RPS3a-like gene transcripts. As expected, plants showed mild stunting and leaf curling or severe stunting, respectively. Similarly, tomato plants inoculated with either PSTVd-I or PSTVd-RG1 (the latter strain accumulates at higher levels than the former) exhibited initial disease phenotype such as leaf curling and stunted growth at 10–12 days post-infection (dpi) and both the CLC-b-lik*e* and RPS3a-like transcripts were significantly down-regulated. In addition, artificial miRNAs (amiRNAs) corresponding to viroid-derived vsiRNAs from the selected non-hotspot regions expressed in *N. benthamiana* specifically silenced GFP reporters containing the cognate sequences, and RLM-RACE or PARE analyses in symptomatic and infected tissues identified uncapped and truncated forms of the CLC-b-like and RPS3a-like transcripts at the expected cleavage sites. Importantly, PSTVd-RG1 infection resulted in down-regulation of both the CLC-b-like and RPS3a-like mRNAs transcripts, whereas PSTVd-I infection led to down-regulation of the RPS3a-like mRNA alone (Fig. 3).

2.2.5 PSTVd vsiRNAs (amiRNAs) targeting siPPase transcripts

In addition to the alteration of flowering time, PSTVd induces abnormal vegetative phenotypes in tomato and other plants of the *Solanaceae* family, such as stunting, bushiness and leaf discoloration. To assess whether vsiRNAs from VMRs (and/or flanking sequences) could determine such an abnormal phenotype, Eamens and collaborators expressed a set of amiRNAs (Schwab et al., 2006) in either *N. tabacum* or *N. benthamiana* (Eamens et al., 2014) and observed stunting, bushiness and leaf discoloration symptoms resembling those previously associated with PSTVd tomato infection. The authors then identified in *Nicotiana* spp. a soluble inorganic pyrophosphatase (siPPase) transcript harboring an amiR46 target sequence (Fig. 3). Molecular analyses revealed that the severity of the expressed phenotype strongly correlated with amiR46 accumulation and RNA silencing of the siPPase gene transcript. The authors, however, did not find orthologs of siPPAse in tomato and potato, indicating that the disease phenotype is the result of other interactions in these species.

2.3 Viruses

2.3.1 CWMV vsiRNAs targeting PPase transcripts

wIn plants, siPPase and other vacuolar PPase (VP) transcripts cooperatively regulate the hydrolysis of pyrophosphates (PPi) and are involved in regulating intracellular ion and pH gradients (Segami et al., 2018). In addition, vacuolar PPase transcripts are implicated in the response of plants to various abiotic stresses, including low nutrient, drought, cold, heat, heavy metal and salinity stresses. Chinese wheat mosaic virus (CWMV, genus Furovirus, family *Virgaviridae*) infecting wheat releases one vsiRNA (denoted as vsiRNA-20) capable of down-regulating a VP transcript (Yang et al., 2020). Indeed, in virus-infected wheat significant down-regulation of VP transcripts occurs after 15 dpi, and vsiRNA-20 can suppress PPi hydrolysis activity by regulating the expression of VP. As a result, intracellular pH increases, creating a more favorable cellular environment for CWMV replication (Yang et al., 2020). Moreover, Yang et al. (2020) have shown that over-expression of VP increases PPi hydrolysis leading to plant cell death, thus restricting viral multiplication and spread throughout infected tissues. As expected, vsiRNA-20 accumulates in CWMV-infected tissue and recognizes and cuts the VP transcript at the cognate target sequence located in the 3′ UTR. These results establish a new emblematic case of how a virus can harness the RNA silencing machinery to modify plant physiological mechanisms for intracellular pH regulation. *Stricto sensu*, vsiRNA-20 has not been associated with a specific phenotype; however, regulation of PPase transcript is crucial for CWMV infection in wheat (Yang et al., 2020). As a final result, CWMV-infected plants show disease phenotypes of stunting and bushing similar to the phenotypes induced by PSTVd and amiR46 targeting siPPase transcript in *Nicotiana* spp. (see Section 2.2.5) (Fig. 2).

2.3.2 RSV targeting NbeIF4A transcripts

Rice stripe virus (RSV) is a species of the genus *Tenuivirus* possessing four ssRNA segments with either negative or ambisense polarity (Hamamatsu et al., 1993). In nature, RSV is transmitted exclusively by a brown planthopper, and the leaves of infected rice fail to unfold properly and grow long becoming twisted and droopy with chlorotic mottles. RSV can be transmitted to *N. benthamiana* by mechanical inoculation, and the infection causes typical twisted leaves with chlorotic mottles, as in rice. With an integrated approach that includes sRNAseq, microarrays, reverse transcriptase-polymerase chain reaction and RLM-RACE analyses, Shi and collaborators characterized the host genes and their regulatory vsiRNAs responsible for RSV symptoms (Shi et al., 2016). The authors found that elongation factor 4A (NbeIF4A) transcripts were significantly down-regulated in RSV-infected *N. benthamiana* plants. Interestingly, a 20 nt-long region of the (+) RNA 1 strand was complementary to part of the NbeIF4A transcript ORF and 19–24 nt vsiRNAs from that region accumulated with above-average frequency in infected tissues. Efforts to detect cleaved transcripts did not produce clear results: RLM-RACE analysis detected truncated *NbeIF4A* transcripts. However, RSV-associated leaf twisting was dramatically attenuated in infected plants overexpressing the NbeIF4A transcripts (Shi et al., 2016). Altogether, these results suggest an additional case of pathogenic determinants carried by vsiRNAs, although the evidence is indirect (Fig. 2).

2.3.3 TYLCV vsiRNAs targeting SlLNR1 long non-coding RNA

In addition to targeting protein-coding genes, it has recently been shown that a vsiRNA can cause abnormal phenotypes by targeting and regulating host-encoded long non-coding (lnc)RNAs (Yang et al., 2019). Tomato yellow leaf curl virus (TYLCV) is a member of the *Geminiviridae* family that can cause severe damage and loss of many important crops. TYLCV has a circular single-stranded DNA genome which comprises a short intergenic region (IR) and six overlapping ORFs (Moriones and Navas-Castillo, 2000). The TYLCV IR can be transcribed bi-directionally during viral infection, although to lower levels than the protein-coding genes. Expression of the IR by VIGS in which a viral vector carried the

TYLCS IR and triggered the production of IR vsiRNAs in tomato plants, recapitulated the stunted and curling disease phenotype.

Yang et al. (2019) identified a 25 nt short segment of the TYLCV IR with near-perfect complementarity to a tomato lncRNA, termed SlLNR1. Notably, the disease phenotype could be induced by solely expressing the 25 nt IR segment with VIGS in tomato plants. Using RLM-RACE, the authors demonstrated that IR-derived vsiRNAs induces silencing of lncRNA and, accordingly, SlLNR1 transcripts were down-regulated during infection. Importantly, natural variants exist of SlLNR1 transcripts containing a 14 nt deletion in the targeted region, which evade vsiRNAs-mediated cleavage and confer resistance to TYLCV in some tomato cultivars. At the moment, the mechanism that links SlLNR1 transcripts to viral infection resistance has not been clarified (Yang et al., 2019) (Fig. 2).

2.4 Host targeting by viral small RNAs in insects

Although there is compelling evidence that vsiRNAs silence host genes in plants, thereby contributing to disease symptoms, to our knowledge this has not been systematically analyzed in insect systems. Virus infection may, however, modify host-gene expression via the insect miRNA machinery in two distinct ways (Fig. 4): (i) viruses may encode their own miRNAs that are dependent on the miRNA biogenesis machinery of infected host cells and (ii) the host miRNA profile changes as a response to viral infections either because viruses actively modulate global or specific miRNA levels or because the host changes miRNA expression as part of its defense strategy (discussed in Section 5). Expression of virus-encoded miRNAs is a common strategy for a large number of viruses with DNA genomes, such as herpes-viruses in mammals (Kincaid and Sullivan, 2012), and large DNA viruses of insects (Hussain et al., 2008; Webster et al., 2015). In contrast, miRNA expression from RNA viruses has been an issue of debate and to date only few convincing examples have been described. For example, in mammals, the retrovirus bovine leukemia virus encodes a cluster of miRNAs that is transcribed in the nucleus by polymerase III from the integrated viral sequence (Kincaid et al., 2012). Therefore, miRNA expression does not lead to the destruction of viral genomic RNA. In contrast, miRNA processing from cytoplasmically replicating RNA viruses inevitably requires the sacrifice of viral RNA that will consequently be unavailable for replication, or for translation of viral proteins. In addition, miRNA biogenesis from

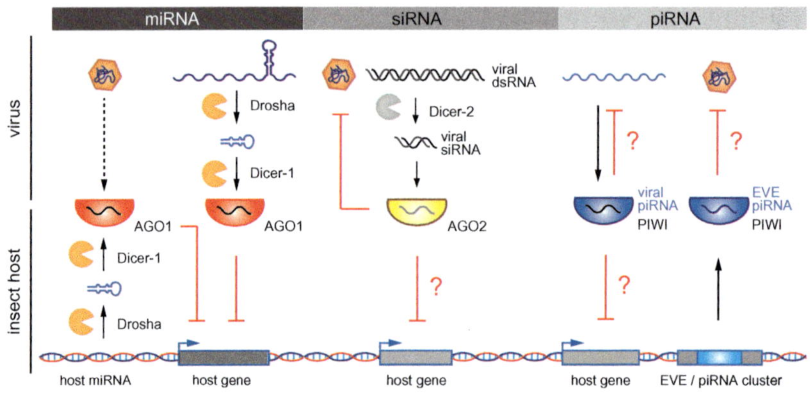

Fig. 4 RNA silencing pathways and viral infection in insects. Three RNA silencing pathways interact with viruses in insects. (Left) Viral infection may affect host microRNA (miRNA) expression, resulting in altered expression of host genes. Some viruses encode their own miRNAs to modulate expression of host and viral genes. (Middle) The small interfering RNA (siRNA) pathway controls replication of DNA and RNA viruses by processing viral double-stranded RNA (dsRNA) into siRNAs, which are used by Argonaute-2 (AGO2) to cleave viral RNAs and restrict replication. Whether viral siRNAs target host genes, like in the plant system, has not been studied thus far. (Right) The PIWI-interacting RNA (piRNA) pathway processes viral RNA into piRNAs in some insect vectors, including *Aedes* spp. The potential of viral piRNAs to target cellular transcripts has not been studied. Endogenous viral elements (EVEs) are segments of non-retroviral RNA viruses that are integrated into the host genome. These are enriched in piRNA clusters and give rise to piRNAs in some insects and may be involved in antiviral defense.

RNA viruses that do not replicate in the nucleus has been questioned, because cytoplasmic RNAs would not have access to the components required for releasing pre-miRNAs. However, Sindbis virus can be engineered to express artificial miRNAs in mammalian cells (Shapiro et al., 2010). Interestingly, the core component of the microprocessor, Drosha, was redistributed to the cytoplasm upon Sindbis infection without affecting cellular miRNA profiles (Shapiro et al., 2012). In addition, artificial introduction of a herpesvirus miRNA into tick-borne encephalitis virus (genus *Flavivirus*, family *Flaviviridae*) did not severely impact virus accumulation despite efficient production of mature miRNA molecules (Rouha et al., 2010). These studies indicate that miRNA processing from cytoplasmic RNA viruses is possible with no or little impairment of virus replication. Infection of insect cells with the West Nile virus Kunjin strain results in the production of a miRNA-like small RNA from the viral 3′ UTR, most likely from the flavivirus subgenomic RNA (Hussain et al., 2012).

Counter-intuitively, targeting of the cellular GATA4 transcript by this small RNA caused its upregulation and was proposed to enhance West Nile virus replication. Similarly, miRNA-like small RNAs were proposed to be generated from the UTR sequences of dengue virus, but the biological relevance of their production is an issue of debate (Finol, 2014; Hussain and Asgari, 2014; Skalsky et al., 2014).

Host-gene regulation by viral miRNAs is conceptually similar to host regulation by vsiRNAs in that in both cases viruses dedicate genomic space to modulate the host. Yet mechanistically these are quite distinct as processing of pre-miRNAs encoded in viral genomes is precise, with the pre-miRNA structure dictating the efficiency and accuracy of processing by the RNase III enzymes, Drosha and Dicer-1. This relative accuracy is important to ensure targeting of specific host transcripts, whereas a minor shift of the miRNA sequence results in a different seed sequence (nt 2–8) and hence a different repertoire of targeted host mRNAs. It is therefore somewhat surprising that, thus far, virus-encoded miRNAs have not been identified in plant viruses for reasons that are not well understood.

A wide variety of vsiRNAs are generated during RNA virus infection that seem to be distributed across the genomes of RNA viruses without prominent hotspots. Given the broad diversity of vsiRNAs, it is unclear how individual siRNA sequences are retained in plant viruses to target host genes and how the virus ensures that these are generated in an accurate manner and produced at sufficient levels. Interestingly, alphaviruses including Chikungunya virus, Semliki Forest virus, and Sindbis virus infections, give rise to virus-derived piRNAs that are abundant and produced in a reproducible manner (Miesen et al., 2015; Morazzani et al., 2012; Schnettler et al., 2013; Vodovar et al., 2012). However, whether these contribute to the silencing of host genes has not been studied (Fig. 4).

3. Revisiting the role of VSSs in infection and disease phenotype

Many viruses have evolved protein-based counter-strategies against host antiviral defense. Most VSSs in plants are viral structural or nonstructural proteins with important functions at different levels of the viral life cycle (e.g., coat protein, movement, insect transmission, replication). Thus, VSSs are generally multifunctional proteins that rarely possess motifs that can be said to be common to silencing suppression activity (Csorba et al., 2015).

The recovery phenotype of virus-infected plants is a sign of vsiRNA systemic movement, with viral silencing signals acting against viral invasion at distal regions. Initially, the plant manifests severe symptoms associated with a high titer of viral RNA in inoculated tissue or those initially systemically invaded. Later, during plant recovery the symptoms on new leaves are attenuated and the virus accumulates at lower levels in upper non-inoculated tissues. Moreover, the recovered tissues become resistant to re-inoculation (Pantaleo et al., 2007; Ratcliff et al., 1997).

3.1 VSSs in plant viruses

VSS-deficient model viral systems are critical to appreciate the importance of VSSs in determining both diseased and recovery phenotypes. It is now well accepted that VSSs may impair the movement and functionality of RNA silencing signals. Noteworthy in this regard are studies with p19-deficient Cymbidium ringspot virus and 2b-deficient CMV (Silhavy et al., 2002; Wang et al., 2010). On the contrary, in most cases VSSs are indispensable for viral replication and, thus, have mostly been studied in vitro, in VSS transgenic plants, or in plants transiently expressing VSSs in isolation. These studies have provided important insights into the mechanisms of VSS operation: VSSs are multifunctional and versatile proteins that act at various stages of RNA silencing in plants; thus, they influence many biochemical processes involved in fine-tuning gene regulation and virus-host interactions (Csorba et al., 2015).

Oftentimes, VSS-expressing plants exhibit disease-like phenotypes that include growth defects, which resemble those associated with alterations of cellular miRNA functions. Unfortunately, over-expression of VSSs in cellular compartments unrelated to the life cycle of viruses has in many cases overestimated the function of VSSs in determining symptoms. Important in this respect are the recent studies by Kontra et al. that have indisputably shown that p19 efficiently binds endogenous (*endo*-)siRNA duplexes only in the absence of the virus infection, suggesting that the impact of p19 on endo-siRNA pathways is restricted to specific cellular compartments under strict regulation by the virus (Kontra et al., 2016). Accordingly, p19 can impair endo-miRNA pathways (i.e., by binding miR403, a repressor of the antiviral AGO2) only at the very early stage of infections, fine-tuning initial RNA silencing-based antiviral responses (Pertermann et al., 2018) (Fig. 5). Where exactly in the cell and in the viral life cycle VSSs are expressed and how these elements are compatible with the side effects

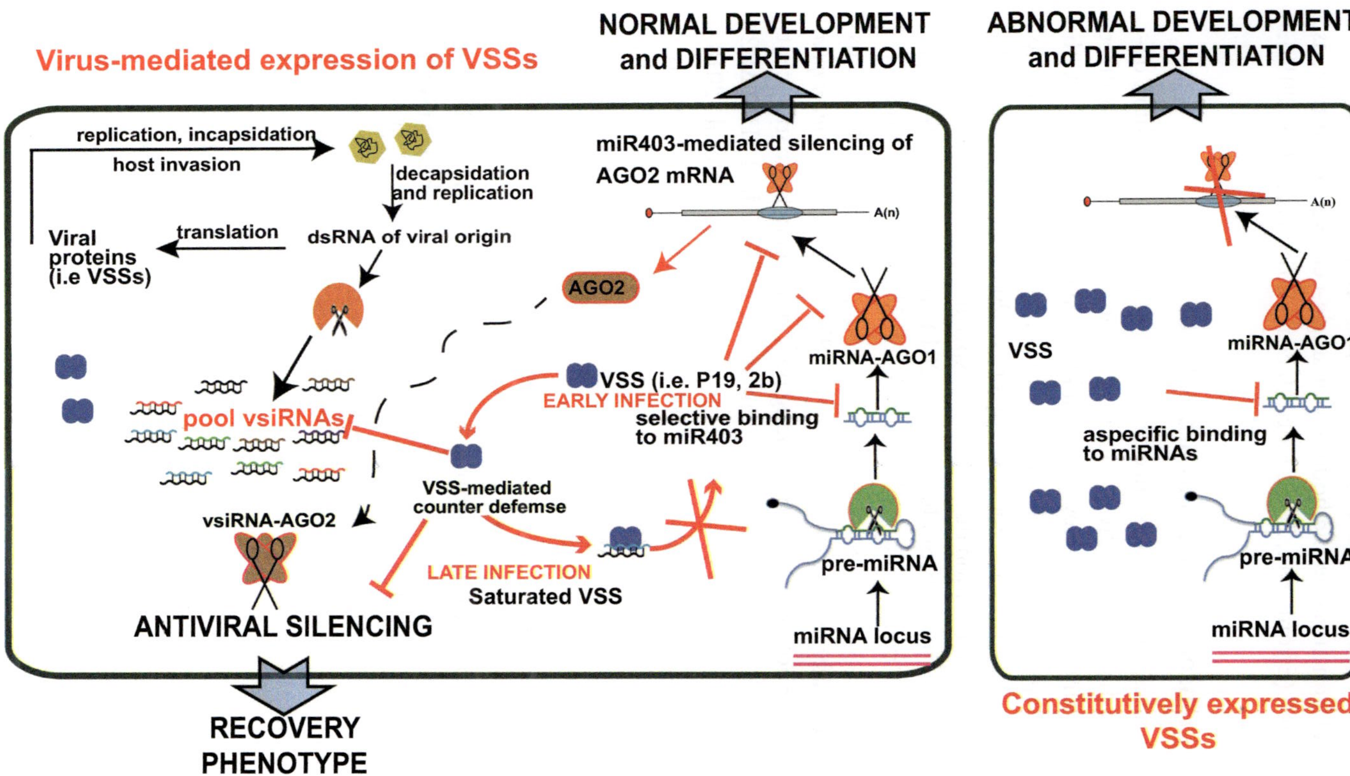

Fig. 5 Viral silencing suppressors VSSs in the context of viral replication have specific spatio-temporal functions promoting viral infection, but do not exert a direct role in symptom development as in the case of constitutive expression of VSSs.

described in artificial systems still remain open questions. The answers require novel non-invasive methodology at single-cell-type resolution, as recently applied (Efroni et al., 2015; Horvath et al., 2019; Ryu et al., 2019).

3.2 VSSs in insect viruses and arboviruses

Like plant viruses, insect viruses encode VSS proteins that may inhibit different stages of the siRNA pathway, but they generally do not seem to affect the miRNA pathway (for a review, see Bronkhorst and van Rij, 2014). With the notable exception of Flock House virus (FHV, *Nodaviridae* family) and its suppressor B2 (Aliyari et al., 2008; Wang et al., 2006), many insect VSSs have been studied in isolation due to lack of experimental systems to manipulate the viral genomes (Bronkhorst et al., 2014; Nayak et al., 2010; van Cleef et al., 2014; van Mierlo et al., 2012, 2014; Wang et al., 2006).

Interestingly, several recent studies have developed VSS-defective virus mutants, specifically cricket paralysis virus (CrPV) 1A mutants, *Autographa californica* nuclear polyhedrosis virus lacking p35, and invertebrate iridescent virus 6 (IIV6) Δ340R mutant, to complement the well-studied FHV ΔB2 mutant (Aliyari et al., 2008; Bronkhorst et al., 2019; Karamipour et al., 2018; Mehrabadi et al., 2015; Nayak et al., 2018; Wang et al., 2006). CrPV is a member of the RNA virus family *Dicistroviridae*, which encodes a VSS that inhibits slicing activity of AGO2 and recruits an E3 ubiquitin ligase to induce AGO2 degradation (Nayak et al., 2010; van Mierlo et al., 2012; Wang et al., 2006), IIV6 is a large DNA virus of the *Iridoviridae* family that encodes a dsRNA-binding domain protein, called 340R, which prevents small RNA loading into RISC (Bronkhorst et al., 2014), whereas FHV is a small RNA virus of the *Nodaviridae* family, which encodes the B2 protein that binds both long dsRNAs and siRNAs (Chao et al., 2005; Li et al., 2002; Lingel et al., 2005). Although these viruses differ dramatically in genome organization and in their mechanisms of RNA silencing suppression, a consistent phenotype emerges from studies in adult *Drosophila.* VSS-defective mutants show severe replication defects in RNA silencing-competent wild-type flies, which are rescued to levels of wild-type virus in Dicer-2 or AGO2 knockout flies (Aliyari et al., 2008; Bronkhorst et al., 2019; Nayak et al., 2018; Wang et al., 2006). These results suggest that RNA silencing suppression is the main, if not only, function of these proteins. Moreover, these results imply that the VSS proteins do not, or only to a small extent, affect global miRNA levels, in line with earlier reporter assays in which VSS proteins did not affect the function of individual miRNAs

(Nayak et al., 2010; van Mierlo et al., 2012; Van Rij et al., 2006). Yet, from quantitative analyses of small RNAs in IIV6 infection, a surprising phenotype emerged indicating that 340R specifically binds miR-305, as a 5p:3p duplex (Bronkhorst et al., 2019). The mechanism behind the selective binding of this miRNA duplex and its effect on disease phenotypes during IIV6 infection remain to be established.

A number of important arthropod-borne viruses (arboviruses) of vertebrates are transmitted between hosts by blood-feeding insects, in which they replicate. These viruses are also targeted by the RNA silencing response of the arthropod vector (Bronkhorst and van Rij, 2014; Samuel et al., 2018), and several studies have recently proposed VSS proteins in several arboviruses (Qian et al., 2020; Qiu et al., 2020; Samuel et al., 2016). For example, a dengue virus mutant defective in the RNA silencing suppressive activity of NS2 was rescued by knockdown of the RNA silencing genes Dicer-2 and AGO2 in insect cells, providing genetic support for its importance in RNA silencing suppression in insect vectors (Qiu et al., 2020).

Recently, an interesting twist to the counter-defense paradigm was proposed for the RNA silencing-based interaction between Drosophila C virus (DCV) and its host. The VSS of DCV, named 1A, contains a dsRNA-binding domain and was proposed to bind viral dsRNA and inhibit Dicer-2 function (Van Rij et al., 2006). Interestingly, even though DCV is a cytoplasmic virus, a large fraction of 1A resides in the nucleus where it binds and stabilizes a lncRNA, named VINR, thereby activating a non-canonical immune pathway that triggers the expression of a subset of antimicrobial peptides (Zhang et al., 2020). Whether this reflects a counter-counter-defense strategy, as proposed by the authors, and previously also observed in plants (Chen et al., 2017; Li et al., 1999), or a side effect of a putative proviral nuclear function of DCV 1A remains to be studied.

4. Small RNAs-mediated diversion of the plant RNA silencing machinery

RNA silencing-based interactions between virus and host are partly antagonistic, which often results in a molecular arms race. dsRNA of viral origin can be considered a virus-associated molecular pattern that is recognized and processed into vsiRNAs by DCLs, playing the role of pathogen recognition receptors. Once incorporated into RISCs, vsiRNAs guide RISC to viral nucleic acids for inactivation. Viruses actively suppress

RNA silencing by producing VSSs, which alleviate the antiviral defense of the host. As a consequence, plant viruses unable to suppress silencing fail to replicate or are severely restricted (Csorba et al., 2015).

As an alternative to the VSS-based strategy to suppress silencing, viruses may deploy vsiRNA-based strategies to evade silencing. The capacity of vsiRNAs to divert the antiviral RNA silencing machinery from viral transcripts might impact host *endo*-RNA silencing pathways and therefore induce disease phenotypes in a sequence-independent manner. In *Arabidopsis*, cauliflower mosaic virus (CaMV, a plant pararetrovirus) produces massive amounts of vsiRNAs, which are comparable in abundance to the entire complement of plant endo-siRNAs and endo-miRNAs (Blevins et al., 2011). DCL-dependent vsiRNAs are almost entirely produced from a 600-bp–long region of the 8-kb CaMV genome, corresponding to the 8S CaMV non-coding leader sequence. Counter-intuitively, DCL-defective *Arabidopsis* mutants accumulate minimal amounts of vsiRNAs, but do not exhibit hyper-susceptibility to CaMV infection. Moreover, in wild-type plants CaMV accumulates to high levels and robustly spreads throughout the plant, even though CaMV P6 is considered a weak VSS (Shivaprasad et al., 2008). These observations suggest that leader-derived vsiRNAs are ineffective in restricting virus replication, suggesting a decoy mechanism to evade RNA silencing (Blevins et al., 2011). Another situation in which vsiRNAs are produced but not effective occurs in viroid infection. PSTVd has a rod-shaped structure which gives rise to abundant vsiRNAs, but the viral genome itself is resistant to targeting by AGO owing to the compact structure of its covalently closed RNA genome (Denti et al., 2004; Itaya et al., 2007; Wu et al., 2012). These findings, together with the observation that large differences exist in the antiviral activity of individual vsiRNA sequences (Gago-Zachert et al., 2019), imply that vsiRNA production *per se* is not sufficient for antiviral activity, providing plants with ample opportunity to exploit antiviral RNA silencing to modulate host-gene expression. Moreover, they indicate that dicing of the dsRNA intermediate of viroidal and viral replication initiates but is not sufficient to limit invasions (Aliyari and Ding, 2009).

5. Small RNAs from viral and non-viral endogenous sequences

As recounted in the previous paragraphs, many studies indicate that vsiRNAs possess the ability to silence specific host genes with nearly perfect complementarity in plant virus infection. Nonetheless, in most cases specific disease phenotypes could not be fully and unambiguously ascribed to the characterized siRNAs. Indeed, genome-wide approaches have shown that vsiRNAs can target and cleave host mRNAs, but their direct involvement in the diseased phenotype could not always be demonstrated (Avina-Padilla et al., 2015; Marton et al., 2010; Miozzi et al., 2013; Qi et al., 2009; Thibaut and Claude, 2018). In addition to the production of vsiRNAs, viral infection may also modify small RNA profiles of the host, which in turn may affect the outcome of infection.

5.1 Virus-activated host small RNAs

Studies over the last 5 years have provided new insights into virus-induced and RNA silencing-dependent symptom determination, especially regarding generic yellowing symptoms due to alterations in photosynthetic pathways. In the model plant *Arabidopsis*, infections with two distinct RNA viruses—CMV and turnip mosaic virus (TuMV, *Potyviridae* family)—are accompanied by the generation of a new class of *endo*-siRNAs, called virus-activated (va)siRNA. vasiRNAs are massively produced from host-gene transcripts only upon viral infection, are DCL4 and RDR1-dependent and trigger widespread silencing of host genes upon incorporation into AGO1- or AGO2-containing RISC (Cao et al., 2014).

Endogenous vasiRNAs have also been observed in other viral infection models to act in *trans* in silencing photosystem components and determining disease phenotypes. Nucleotide-binding site-leucine-rich repeat (NBS-LRR)-containing proteins comprise the majority of resistance proteins: coiled-coil (CC)-NBS-LRR and Toll-interleukin-like receptor (TIR)-NBS-LRR (Collier and Moffett, 2009). In canola, the *braTIR1* gene, which lacks the NBS-LRR motif, produces siRNAs in response to TuMV infection, but not in non-infected plants (Cui et al., 2020). These siRNAs (bona fide vasiRNAs) can be considered DCL4-dependent. One of these vasiRNAs is able to guide the cleavage and silencing of BraCP24 transcripts, the homolog of which in *Arabidopsis*, chlorophyll protein 24 (CP24), is a subunit of the photosystem supercomplex that plays important roles in photomorphogenesis (de Bianchi et al., 2008; Wei et al., 2016). In turn, the accumulation of siRNAs derived from BraCP24 (siRBraCP24s) is remarkably increased after TuMV infection. These virus-activated *siRBraCP24s* may reinforce a cis-regulatory effect (Cao et al., 2014). Accordingly, silencing of *BraCP24* results in a change in the photosynthetic rate, leaf curling and multiple developmental defects, including early flowering and reduced growth (Cui et al., 2020). Similar observations have previously been made for tomato plants infected with tomato yellow leaf curl Sardinia virus (*Begomovirus* genus, *Geminiviridae* family). Indeed, a CC-NBS-LRR (Solyc008070.2.1) produces numerous siRNAs, some of which in phase, resulting in site-specific cleavage and dramatic down-regulation of chlorophyll *a*/b-binding proteins and of photosynthetic reaction center subunits of either Photosystem I or II (Chiumenti et al., 2018).

Besides encoding their own miRNAs, insect viruses may influence the expression pattern of cellular miRNAs in the infected host. For example, host miRNA expression changes have been reported both in *Aedes aegypti* and *Ae. albopictus* mosquitoes after infection with dengue virus and Zika virus (Campbell et al., 2014; Liu et al., 2015, 2016; Saldana et al., 2017). Similarly, infection of various insect cell lines with major human arboviruses results in the modulation of cellular miRNA levels (Shrinet et al., 2014; Slonchak et al., 2014; Yan et al., 2014). However, in most of these studies no cellular or viral targets have been identified, leaving the question open as to how differential miRNA expression influences virus replication. In adult *Drosophila* infected with DCV, several miRNAs are differentially expressed and consequently, their targets mis-expressed, which in turn alters DCV replication. Examples are regulatory circuits between miR-956-3p and ectoderm-expressed 4 and between miR-8-5p and the dJun transcription factor (Monsanto-Hearne et al., 2017a, b).

Besides modification of miRNA expression, *Ae. aegypti* infection with the alphavirus Sindbis virus has been reported to induce expression of a novel class of endo-siRNAs. Under normal conditions, insects may produce endo-siRNAs from convergent transcripts that overlap in the 3′ part of mRNA (Okamura and Lai, 2008). However, Sindbis virus infection additionally induced 29 novel loci that produced endo-siRNAs derived mostly from genes with long 3′ UTRs. From a mechanistic perspective, these endo-siRNAs seem to be different from plant vasiRNAs that depend on the activity of a DCL enzyme in combination with a cellular RDRp, which is lacking in insects (Adelman et al., 2012). The mechanism by which these small RNAs are induced, the importance for the viral life cycle, and whether other arboviruses likewise induce such endo-siRNAs remains to be established.

5.2 Endogenous viral elements and small RNAs

Metaviridae and *Pseudoviridae*, often referred to as Ty3-gypsy and Ty1-copia, respectively, are families of long terminal repeat retrotransposons that propagate through a copy-and-paste amplification mechanism via RNA intermediates. This property allows them to accumulate in genomic DNA, giving rise to the bulk of repeats in plant genomes. They are endogenous retroviruses and replicate via a virus-like intermediate referred to as virus-like particle (VLP). VLPs do not display infectivity according to the traditional virological definition. However, there is sound evidence that these particles are essential and direct intermediates in the life cycle of these transposable elements (Boeke et al., 2012; Eickbush et al., 2012). Retrotransposons in plants generate 21 and 22 nt epigenetically activated siRNAs (known as easiRNAs; Borges et al., 2018), which are de facto vsiRNAs and implicated in fertilization and the imprinting of developing seeds. Thus far, there are no reported disease determinants carried by vsiRNAs derived from these retroelements. Moreover, vsiRNA accumulation is unaffected by plant virus infection in plant vegetative tissues (Cao et al., 2014).

Other endogenous viral elements (EVEs) that are frequently found in the genomes of plants, animals, and fungi are fragments of viral genomes. Emblematic in plants is the case of DNA fragments from ancestral viruses of the *Caulimoviridae* family, which have been captured within the genomes of a wide range of angiosperms such as rice, sorghum, citrus, grape, apple, pear, strawberry, eucalyptus, poplar, tomato, potato, cucumber, and cotton (Geering et al., 2014). Endogenous fragments of viral genomes contribute to the evolution of plant genomes (Magidin et al., 2003) and viral genomes (at least when expressing a functional protein). Tromas et al. (2014) have indeed shown that a plant virus lacking a gene necessary for autonomous replication in wild-type plants has higher fitness in a transgenic plant expressing the viral gene than the full-length plant virus in transgenic plants. Furthermore, when the full-length plant virus evolved by serial passages in transgenic plants, viral genomic deletions within the viral gene were rapidly generated (Tromas et al., 2014). Many studies propose that EVEs in plants have a role in antiviral immunity (Han, 2019); and there is indeed evidence that vsiRNAs are produced from EVEs (Geering et al., 2014; Pooggin, 2018; Zahid et al., 2015). However, little is known regarding the possible roles of EVE-derived vsiRNAs in diseased phenotypes.

Although it is well established that retroelements make up large fractions of eukaryotic genomes (Bourque et al., 2018; Cosby et al., 2019), a wide range of host organisms also carry fragments of non-retroviral viruses in their genomes (Belyi et al., 2010; Horie et al., 2010; Katzourakis and Gifford, 2010). In the mammalian system, some EVEs have the potential to encode functional proteins. Indeed, an EVE derived from the bornavirus-like nucleoprotein sequence in the ground squirrel genome generates a protein that affects the replication of Borna disease virus (Fujino et al., 2014). It has recently become clear that vector mosquitoes of the *Aedes* genus contain an exceptionally large number of EVEs derived from non-retroviral RNA viruses (Aguiar et al., 2020; Crochu et al., 2004; Palatini et al., 2017; Suzuki et al., 2017; Ter Horst et al., 2019; Whitfield et al., 2017). Surprisingly, EVE sequences are enriched in piRNA clusters and, as a consequence, EVE-derived piRNA sequences are produced which tend to be antisense to viral-coding RNAs. These EVE-derived piRNAs thus have the potential to target incoming viral RNA for degradation, provided that sufficient sequence complementarity exists (Miesen et al., 2016; Ophinni et al., 2019; Parrish et al., 2015). To test this hypothesis, Tassetto et al. (2019) inserted sequences complementary to EVE-derived piRNAs into Sindbis virus and found that these viruses replicated to lower levels than viruses with control sequences (Tassetto et al., 2019). Excitingly, these results have now been corroborated in a natural virus-host system. Specifically, it was found that Cell fusing agent virus (CFAV, an insect-specific virus from the family *Flaviviridae*), replicated to lower levels in the ovaries of mosquitoes expressing a CFAV-derived EVE than in control mosquitoes lacking this EVE (Crava et al., 2020; Suzuki et al., 2020) (Fig. 4).

6. Concluding remarks

For more than 20 years it has been known that antagonistic interactions exist between viruses and RNA silencing mechanisms of the host. Yet, evidence has emerged suggesting even higher complexity than a simple defense-counter-defense paradigm, including unexpected activities of VSS proteins, the production of vasiRNAs, the discovery of EVEs that produce small RNAs, and the occurrence of counter-counter-defense systems. In addition, it is clear that vsiRNAs can act as sequence-specific pathogenesis determinants.

Work in several thoroughly investigated virus/viroid-host models has demonstrated how important vsiRNAs can be in determining specific and combined symptoms. These mechanisms can be expected to be present in other systems where co-evolution between virus and hosts has occurred, including crops. However, in such cases we may have visually imperceptible diseased phenotypes that may affect the general fitness of the plant and that obviously deserve careful investigation. Phenotyping and phenomic approaches, coupled with other yet to be applied "-omic" approaches, can finally disclose such unknown pathways and provide opportunities for better recognizing diseased phenotypes.

A crucial message emerges for plant breeding approaches: the disease phenotype caused by vsiRNAs can be prevented by the expression of either a naturally evolved, or artificially introduced, silencing-resistant sequence variant of host-targeted genes (see cases in Section 2). These observations reinforce the role of vsiRNA-mediated determination of symptoms and may also have important implications in developing virus-permissive but symptomless varieties through the use of marker-assisted breeding or gene (or genome) editing technologies. Greater knowledge of how pathogenic vsiRNAs can determine insect behavior and fitness can also significantly impact biotechnological applications in the field of agriculture. Insects that have important functions in pollination and insects as vehicles for the transmission of pathogens both have an unquestionable impact on global agricultural production. In-depth knowledge of the molecular mechanisms that cause disease-like phenotypes in plants and insects mediated by RNA silencing will likely open up new applications for sustainable agriculture.

Acknowledgments

This work was financially supported by a VICI grant from the Netherlands Organization for Scientific Research (grant number 016.VICI.170.090) to RPvR, and by funds from the Italian MIUR Progetto NUTR-AGE (FOE-2019, DSBAD004.271) to DiSBA of the National Research Council.

References

Adelman, Z.N., Anderson, M.A., Liu, M., Zhang, L., Myles, K.M., 2012. Sindbis virus induces the production of a novel class of endogenous siRNAs in Aedes aegypti mosquitoes. Insect Mol. Biol. 21, 357.

Adkar-Purushothama, C.R., Brosseau, C., Giguere, T., Sano, T., Moffett, P., Perreault, J.P., 2015a. Small RNA derived from the virulence modulating region of the potato spindle tuber viroid silences callose synthase genes of tomato plants. Plant Cell 27, 2178.

Adkar-Purushothama, C.R., Zhang, Z., Li, S., Sano, T., 2015b. Analysis and application of viroid-specific small RNAs generated by viroid-inducing RNA silencing. Methods Mol. Biol. 1236, 135.

Adkar-Purushothama, C.R., Iyer, P.S., Perreault, J.P., 2017. Potato spindle tuber viroid infection triggers degradation of chloride channel protein CLC-b-like and ribosomal protein S3a-like mRNAs in tomato plants. Sci. Rep. 7, 8341.

Adkar-Purushothama, C.R., Sano, T., Perreault, J.P., 2018. Viroid-derived small RNA induces early flowering in tomato plants by RNA silencing. Mol. Plant Pathol. 19, 2446.

Aguiar, E., Almeida, J.P.P., Queiroz, L.R., Oliveira, L.S., Olmo, R.P., Faria, I.J.S., Imler, J.L., Gruber, A., Matthews, B.J., Marques, J.T., 2020. A single unidirectional piRNA cluster similar to the flamenco locus is the major source of EVE-derived transcription and small RNAs in Aedes aegypti mosquitoes. RNA 26, 581.

Aliyari, R., Ding, S.W., 2009. RNA-based viral immunity initiated by the dicer family of host immune receptors. Immunol. Rev. 227, 176.

Aliyari, R., Wu, Q., Li, H.W., Wang, X.H., Li, F., Green, L.D., Han, C.S., Li, W.X., Ding, S.W., 2008. Mechanism of induction and suppression of antiviral immunity directed by virus-derived small RNAs in Drosophila. Cell Host Microbe 4, 387.

Avina-Padilla, K., Martinez de la Vega, O., Rivera-Bustamante, R., Martinez-Soriano, J.P., Owens, R.A., Hammond, R.W., Vielle-Calzada, J.P., 2015. In silico prediction and validation of potential gene targets for pospiviroid-derived small RNAs during tomato infection. Gene 564, 197.

Barckmann, B., Pierson, S., Dufourt, J., Papin, C., Armenise, C., Port, F., Grentzinger, T., Chambeyron, S., Baronian, G., Desvignes, J.P., Curk, T., Simonelig, M., 2015. Aubergine iCLIP reveals piRNA-dependent decay of mRNAs involved in germ cell development in the early embryo. Cell Rep. 12, 1205.

Belyi, V.A., Levine, A.J., Skalka, A.M., 2010. Unexpected inheritance: multiple integrations of ancient bornavirus and ebolavirus/marburgvirus sequences in vertebrate genomes. PLoS Pathog. 6, e1001030.

Blevins, T., Rajeswaran, R., Aregger, M., Borah, B.K., Schepetilnikov, M., Baerlocher, L., Farinelli, L., Meins Jr., F., Hohn, T., Pooggin, M.M., 2011. Massive production of small RNAs from a non-coding region of cauliflower mosaic virus in plant defense and viral counter-defense. Nucleic Acids Res. 39 (12), 5003–5014.

Boeke, J.D., Eickbush, T., Sandmeyer, S.B., Voytas, D.F., 2012. Pseudoviridae. In: King, A.M.Q., Adams, M.J., Carstens, E.B., Lefkowitz, E.J. (Eds.), Virus Taxonomy. In: Classification and Nomenclature of Viruses Elsevier, p. 467.

Borges, F., Parent, J.S., van Ex, F., Wolff, P., Martinez, G., Kohler, C., Martienssen, R.A., 2018. Transposon-derived small RNAs triggered by miR845 mediate genome dosage response in Arabidopsis. Nat. Genet. 50, 186.

Bourque, G., Burns, K.H., Gehring, M., Gorbunova, V., Seluanov, A., Hammell, M., Imbeault, M., Izsvak, Z., Levin, H.L., Macfarlan, T.S., Mager, D.L., Feschotte, C., 2018. Ten things you should know about transposable elements. Genome Biol. 19, 199.

Brennecke, J., Aravin, A.A., Stark, A., Dus, M., Kellis, M., Sachidanandam, R., Hannon, G.J., 2007. Discrete small RNA-generating loci as master regulators of transposon activity in Drosophila. Cell 128, 1089.

Bronkhorst, A.W., van Rij, R.P., 2014. The long and short of antiviral defense: small RNA-based immunity in insects. Curr. Opin. Virol. 7C, 19.

Bronkhorst, A.W., van Cleef, K.W., Venselaar, H., van Rij, R.P., 2014. A dsRNA-binding protein of a complex invertebrate DNA virus suppresses the Drosophila RNAi response. Nucleic Acids Res. 42, 12237.

Bronkhorst, A.W., Vogels, R., Overheul, G.J., Pennings, B., Gausson-Dorey, V., Miesen, P., van Rij, R.P., 2019. A DNA virus-encoded immune antagonist fully masks the potent antiviral activity of RNAi in Drosophila. Proc. Natl. Acad. Sci. U. S. A. 116, 24296.

Campbell, C.L., Harrison, T., Hess, A.M., Ebel, G.D., 2014. MicroRNA levels are modulated in Aedes aegypti after exposure to Dengue-2. Insect Mol. Biol. 23, 132.

Cao, D., Froehlich, J.E., Zhang, H., Cheng, C.L., 2003. The chlorate-resistant and photomorphogenesis-defective mutant cr88 encodes a chloroplast-targeted HSP90. Plant J. 33, 107.

Cao, M., Du, P., Wang, X., Yu, Y.Q., Qiu, Y.H., Li, W., Gal-On, A., Zhou, C., Li, Y., Ding, S.W., 2014. Virus infection triggers widespread silencing of host genes by a distinct class of endogenous siRNAs in Arabidopsis. Proc. Natl. Acad. Sci. U. S. A. 111, 14613.

Chao, J.A., Lee, J.H., Chapados, B.R., Debler, E.W., Schneemann, A., Williamson, J.R., 2005. Dual modes of RNA-silencing suppression by flock house virus protein B2. Nat. Struct. Mol. Biol. 12, 952.

Chen, X.Y., Kim, J.Y., 2009. Callose synthesis in higher plants. Plant Signal. Behav. 4, 489.

Chen, L., Yan, Z., Xia, Z., Cheng, Y., Jiao, Z., Sun, B., Zhou, T., Fan, Z., 2017. A violaxanthin deepoxidase interacts with a viral suppressor of RNA silencing to inhibit virus amplification. Plant Physiol. 175, 1774.

Chiumenti, M., Catacchio, C.R., Miozzi, L., Pirovano, W., Ventura, M., Pantaleo, V., 2018. A short Indel-lacking-resistance gene triggers silencing of the photosynthetic machinery components through TYLCSV-associated endogenous siRNAs in tomato. Front. Plant Sci. 9, 1470.

Collier, S.M., Moffett, P., 2009. NB-LRRs work a "bait and switch" on pathogens. Trends Plant Sci. 14, 521.

Collmer, C.W., Howell, S.H., 1992. Role of satellite RNA in the expression of symptoms caused by plant viruses. Annu. Rev. Phytopathol. 30, 419.

Cosby, R.L., Chang, N.C., Feschotte, C., 2019. Host-transposon interactions: conflict, cooperation, and cooption. Genes Dev. 33, 1098.

Crava, C., Varghese, F.S., Pischedda, E., Halbach, R., Palatini, U., Marconcini, M., Mattia, A., Redmond, S.N., Afrane, Y., Ayala, D., Paupy, C., Carballar-Lejarazu, R., Miesen, P., van Rij, R.P., Bonizzoni, M., 2020. Immunity to infections in arboviral vectors by integrated viral sequences: an evolutionary perspective. bioRxiv. https://doi.org/10.1101/2020.04.02.022509.

Crochu, S., Cook, S., Attoui, H., Charrel, R.N., De Chesse, R., Belhouchet, M., Lemasson, J.J., de Micco, P., de Lamballerie, X., 2004. Sequences of flavivirus-related RNA viruses persist in DNA form integrated in the genome of Aedes spp. mosquitoes. J. Gen. Virol. 85, 1971.

Csorba, T., Kontra, L., Burgyan, J., 2015. Viral silencing suppressors: tools forged to fine-tune host-pathogen coexistence. Virology 479-480, 85.

Cui, C., Wang, J.J., Zhao, J.H., Fang, Y.Y., He, X.F., Guo, H.S., Duan, C.G., 2020. A Brassica miRNA regulates plant growth and immunity through distinct modes of action. Mol. Plant 13, 231.

Czech, B., Munafo, M., Ciabrelli, F., Eastwood, E.L., Fabry, M.H., Kneuss, E., Hannon, G.J., 2018. piRNA-guided genome defense: from biogenesis to silencing. Annu. Rev. Genet. 52, 131.

de Bianchi, S., Dall'Osto, L., Tognon, G., Morosinotto, T., Bassi, R., 2008. Minor antenna proteins CP24 and CP26 affect the interactions between photosystem II subunits and the electron transport rate in grana membranes of Arabidopsis. Plant Cell 20, 1012.

Denti, M.A., Boutla, A., Tsagris, M., Tabler, M., 2004. Short interfering RNAs specific for potato spindle tuber viroid are found in the cytoplasm but not in the nucleus. Plant J. 37, 762.

Di Serio, F., Gisel, A., Navarro, B., Delgado, S., Martinez de Alba, A.E., Donvito, G., Flores, R., 2009. Deep sequencing of the small RNAs derived from two symptomatic variants of a chloroplastic viroid: implications for their genesis and for pathogenesis. PLoS One 4, e7539.

Diermann, N., Matousek, J., Junge, M., Riesner, D., Steger, G., 2010. Characterization of plant miRNAs and small RNAs derived from potato spindle tuber viroid (PSTVd) in infected tomato. Biol. Chem. 391, 1379.

Ding, S.W., 2010. RNA-based antiviral immunity. Nat. Rev. Immunol. 10, 632.

Eamens, A.L., Smith, N.A., Dennis, E.S., Wassenegger, M., Wang, M.B., 2014. In Nicotiana species, an artificial microRNA corresponding to the virulence modulating region of potato spindle tuber viroid directs RNA silencing of a soluble inorganic pyrophosphatase gene and the development of abnormal phenotypes. Virology 450–451, 266.

Efroni, I., Ip, P.L., Nawy, T., Mello, A., Birnbaum, K.D., 2015. Quantification of cell identity from single-cell gene expression profiles. Genome Biol. 16, 9.

Eickbush, T., Boeke, J.D., Sandmeyer, S.B., Voytas, D.F., 2012. Metaviridae. In: King, A.M.Q., Adams, M.J., Carstens, E.B., Lefkowitz, E.J. (Eds.), Virus Taxonomy. In: Classification and Nomenclature of Viruses Elsevier, p. 457.

Finol, E., 2014. Are viral small RNA regulating Dengue virus replication beyond serotype 2? Proc. Natl. Acad. Sci. U. S. A. 111, E2915.

Fire, A., 1999. RNA-triggered gene silencing. Trends Genet. 15, 358.

Fujino, K., Horie, M., Honda, T., Merriman, D.K., Tomonaga, K., 2014. Inhibition of Borna disease virus replication by an endogenous bornavirus-like element in the ground squirrel genome. Proc. Natl. Acad. Sci. U. S. A. 111, 13175.

Gago-Zachert, S., Schuck, J., Weinholdt, C., Knoblich, M., Pantaleo, V., Grosse, I., Gursinsky, T., Behrens, S.E., 2019. Highly efficacious antiviral protection of plants by small interfering RNAs identified in vitro. Nucleic Acids Res. 47, 9343.

Geering, A.D., Maumus, F., Copetti, D., Choisne, N., Zwickl, D.J., Zytnicki, M., McTaggart, A.R., Scalabrin, S., Vezzulli, S., Wing, R.A., Quesneville, H., Teycheney, P.Y., 2014. Endogenous florendoviruses are major components of plant genomes and hallmarks of virus evolution. Nat. Commun. 5, 5269.

German, M.A., Pillay, M., Jeong, D.H., Hetawal, A., Luo, S., Janardhanan, P., Kannan, V., Rymarquis, L.A., Nobuta, K., German, R., De Paoli, E., Lu, C., Schroth, G., Meyers, B.C., Green, P.J., 2008. Global identification of microRNA-target RNA pairs by parallel analysis of RNA ends. Nat. Biotechnol. 26, 941.

Groen, S.C., Jiang, S., Murphy, A.M., Cunniffe, N.J., Westwood, J.H., Davey, M.P., Bruce, T.J., Caulfield, J.C., Furzer, O.J., Reed, A., Robinson, S.I., Miller, E., Davis, C.N., Pickett, J.A., Whitney, H.M., Glover, B.J., Carr, J.P., 2016. Virus infection of plants alters pollinator preference: a payback for susceptible hosts? PLoS Pathog. 12, e1005790.

Guo, Z., Li, Y., Ding, S.W., 2019. Small RNA-based antimicrobial immunity. Nat. Rev. Immunol. 19, 31.

Halbach, R., Miesen, P., Joosten, J., Taşköprü, E., Rondeel, I., Pennings, B., Vogels, C.B.F., Merkling, S.H., Koenraadt, C.J., Lambrechts, L., Van Rij, R.P., 2020. A satellite repeat-derived piRNA controls embryonic development of *Aedes*. Nature 580, 274.

Hamamatsu, C., Toriyama, S., Toyoda, T., Ishihama, A., 1993. Ambisense coding strategy of the rice stripe virus genome: in vitro translation studies. J. Gen. Virol. 74 (Pt. 6), 1125.

Han, G.Z., 2019. Origin and evolution of the plant immune system. New Phytol. 222, 70.

Hernandez, C., Flores, R., 1992. Plus and minus RNAs of peach latent mosaic viroid self-cleave in vitro via hammerhead structures. Proc. Natl. Acad. Sci. U. S. A. 89, 3711.

Horie, M., Honda, T., Suzuki, Y., Kobayashi, Y., Daito, T., Oshida, T., Ikuta, K., Jern, P., Gojobori, T., Coffin, J.M., Tomonaga, K., 2010. Endogenous non-retroviral RNA virus elements in mammalian genomes. Nature 463, 84.

Horvath, R., Laenen, B., Takuno, S., Slotte, T., 2019. Single-cell expression noise and gene-body methylation in Arabidopsis thaliana. Heredity (Edinb) 123, 81.

Hull, R., 2002. Matthews' Plant Virology, Fourth ed. Academic Press, San Diego, CA.

Hussain, M., Asgari, S., 2014. MicroRNA-like viral small RNA from Dengue virus 2 autoregulates its replication in mosquito cells. Proc. Natl. Acad. Sci. U. S. A. 111, 2746.

Hussain, M., Taft, R.J., Asgari, S., 2008. An insect virus-encoded microRNA regulates viral replication. J. Virol. 82, 9164.

Hussain, M., Torres, S., Schnettler, E., Funk, A., Grundhoff, A., Pijlman, G.P., Khromykh, A.A., Asgari, S., 2012. West Nile virus encodes a microRNA-like small RNA in the 3′ untranslated region which up-regulates GATA4 mRNA and facilitates virus replication in mosquito cells. Nucleic Acids Res. 40, 2210.

Itaya, A., Zhong, X., Bundschuh, R., Qi, Y., Wang, Y., Takeda, R., Harris, A.R., Molina, C., Nelson, R.S., Ding, B., 2007. A structured viroid RNA serves as a substrate for dicer-like cleavage to produce biologically active small RNAs but is resistant to RNA-induced silencing complex-mediated degradation. J. Virol. 81, 2980.

Iwasaki, Y.W., Siomi, M.C., Siomi, H., 2015. PIWI-interacting RNA: its biogenesis and functions. Annu. Rev. Biochem. 84, 405.

Karamipour, N., Fathipour, Y., Talebi, A.A., Asgari, S., Mehrabadi, M., 2018. Small interfering RNA pathway contributes to antiviral immunity in Spodoptera frugiperda (Sf9) cells following Autographa californica multiple nucleopolyhedrovirus infection. Insect Biochem. Mol. Biol. 101, 24.

Katzourakis, A., Gifford, R.J., 2010. Endogenous viral elements in animal genomes. PLoS Genet. 6, e1001191.

Keese, P., Symons, R.H., 1985. Domains in viroids: evidence of intermolecular RNA rearrangements and their contribution to viroid evolution. Proc. Natl. Acad. Sci. U. S. A. 82, 4582.

Kincaid, R.P., Sullivan, C.S., 2012. Virus-encoded microRNAs: an overview and a look to the future. PLoS Pathog. 8, e1003018.

Kincaid, R.P., Burke, J.M., Sullivan, C.S., 2012. RNA virus microRNA that mimics a B-cell oncomiR. Proc. Natl. Acad. Sci. U. S. A. 109, 3077.

Kontra, L., Csorba, T., Tavazza, M., Lucioli, A., Tavazza, R., Moxon, S., Tisza, V., Medzihradszky, A., Turina, M., Burgyan, J., 2016. Distinct effects of p19 RNA silencing suppressor on small RNA mediated pathways in plants. PLoS Pathog. 12, e1005935.

Lee, Y.S., Nakahara, K., Pham, J.W., Kim, K., He, Z., Sontheimer, E.J., Carthew, R.W., 2004. Distinct roles for Drosophila Dicer-1 and Dicer-2 in the siRNA/miRNA silencing pathways. Cell 117, 69.

Lewis, S.H., Quarles, K.A., Yang, Y., Tanguy, M., Frezal, L., Smith, S.A., Sharma, P.P., Cordaux, R., Gilbert, C., Giraud, I., Collins, D.H., Zamore, P.D., Miska, E.A., Sarkies, P., Jiggins, F.M., 2018. Pan-arthropod analysis reveals somatic piRNAs as an ancestral defence against transposable elements. Nat. Ecol. Evol. 2, 174.

Lewsey, M.G., Hardcastle, T.J., Melnyk, C.W., Molnar, A., Valli, A., Urich, M.A., Nery, J.R., Baulcombe, D.C., Ecker, J.R., 2016. Mobile small RNAs regulate genome-wide DNA methylation. Proc. Natl. Acad. Sci. U. S. A. 113, E801.

Li, H.W., Lucy, A.P., Guo, H.S., Li, W.X., Ji, L.H., Wong, S.M., Ding, S.W., 1999. Strong host resistance targeted against a viral suppressor of the plant gene silencing defence mechanism. EMBO J. 18, 2683.

Li, H.W., Li, W.X., Ding, S.W., 2002. Induction and suppression of RNA silencing by an animal virus. Science 296, 1319.

Li, R., Gao, S., Hernandez, A.G., Wechter, W.P., Fei, Z., Ling, K.S., 2012. Deep sequencing of small RNAs in tomato for virus and viroid identification and strain differentiation. PLoS One 7, e37127.

Lingel, A., Simon, B., Izaurralde, E., Sattler, M., 2005. The structure of the flock house virus B2 protein, a viral suppressor of RNA interference, shows a novel mode of double-stranded RNA recognition. EMBO Rep. 6, 1149.

Liu, Y., Zhou, Y., Wu, J., Zheng, P., Li, Y., Zheng, X., Puthiyakunnon, S., Tu, Z., Chen, X.G., 2015. The expression profile of Aedes albopictus miRNAs is altered by dengue virus serotype-2 infection. Cell Biosci. 5, 16.

Liu, Y.X., Li, F.X., Liu, Z.Z., Jia, Z.R., Zhou, Y.H., Zhang, H., Yan, H., Zhou, X.Q., Chen, X.G., 2016. Integrated analysis of miRNAs and transcriptomes in Aedes albopictus midgut reveals the differential expression profiles of immune-related genes during dengue virus serotype-2 infection. Insect Sci. 23, 377.

Magidin, M., Pittman, J.K., Hirschi, K.D., Bartel, B., 2003. ILR2, a novel gene regulating IAA conjugate sensitivity and metal transport in Arabidopsis thaliana. Plant J. 35, 523.

Marquez, L.M., Redman, R.S., Rodriguez, R.J., Roossinck, M.J., 2007. A virus in a fungus in a plant: three-way symbiosis required for thermal tolerance. Science 315, 513.

Marton, I., Zuker, A., Shklarman, E., Zeevi, V., Tovkach, A., Roffe, S., Ovadis, M., Tzfira, T., Vainstein, A., 2010. Nontransgenic genome modification in plant cells. Plant Physiol. 154, 1079.

Masuta, C., Takanami, Y., 1989. Determination of sequence and structural requirements for pathogenicity of a cucumber mosaic virus satellite RNA (Y-satRNA). Plant Cell 1, 1165.

Maule, A., 1991. Virus movement in infected plants. Crit. Rev. Plant Sci. 9, 457.

Mehrabadi, M., Hussain, M., Matindoost, L., Asgari, S., 2015. The baculovirus antiapoptotic p35 protein functions as an inhibitor of the host RNA interference antiviral response. J. Virol. 89, 8182.

Miesen, P., Girardi, E., van Rij, R.P., 2015. Distinct sets of PIWI proteins produce arbovirus and transposon-derived piRNAs in Aedes aegypti mosquito cells. Nucleic Acids Res. 43, 6545.

Miesen, P., Joosten, J., van Rij, R.P., 2016. PIWIs go viral: arbovirus-derived piRNAs in vector mosquitoes. PLoS Pathog. 12, e1006017.

Miozzi, L., Gambino, G., Burgyan, J., Pantaleo, V., 2013. Genome-wide identification of viral and host transcripts targeted by viral siRNAs in Vitis vinifera. Mol. Plant Pathol. 14, 30.

Monsanto-Hearne, V., Asad, S., Asgari, S., Johnson, K.N., 2017a. Drosophila microRNA modulates viral replication by targeting a homologue of mammalian cJun. J. Gen. Virol. 98, 1904.

Monsanto-Hearne, V., Tham, A.L.Y., Wong, Z.S., Asgari, S., Johnson, K.N., 2017b. Drosophila miR-956 suppression modulates ectoderm-expressed 4 and inhibits viral replication. Virology 502, 20.

Morazzani, E.M., Wiley, M.R., Murreddu, M.G., Adelman, Z.N., Myles, K.M., 2012. Production of virus-derived ping-pong-dependent piRNA-like small RNAs in the mosquito soma. PLoS Pathog. 8, e1002470.

Moriones, E., Navas-Castillo, J., 2000. Tomato yellow leaf curl virus, an emerging virus complex causing epidemics worldwide. Virus Res. 71, 123.

Mouradov, A., Cremer, F., Coupland, G., 2002. Control of flowering time: interacting pathways as a basis for diversity. Plant Cell 14 (Suppl), S111.

Navarro, B., Pantaleo, V., Gisel, A., Moxon, S., Dalmay, T., Bisztray, G., Di Serio, F., Burgyan, J., 2009. Deep sequencing of viroid-derived small RNAs from grapevine provides new insights on the role of RNA silencing in plant-viroid interaction. PLoS One 4, e7686.

Navarro, B., Gisel, A., Rodio, M.E., Delgado, S., Flores, R., Di Serio, F., 2012. Small RNAs containing the pathogenic determinant of a chloroplast-replicating viroid guide the degradation of a host mRNA as predicted by RNA silencing. Plant J. 70, 991.

Nayak, A., Berry, B., Tassetto, M., Kunitomi, M., Acevedo, A., Deng, C., Krutchinsky, A., Gross, J., Antoniewski, C., Andino, R., 2010. Cricket paralysis virus antagonizes Argonaute 2 to modulate antiviral defense in Drosophila. Nat. Struct. Mol. Biol. 17, 547.

Nayak, A., Kim, D.Y., Trnka, M.J., Kerr, C.H., Lidsky, P.V., Stanley, D.J., Rivera, B.M., Li, K.H., Burlingame, A.L., Jan, E., Frydman, J., Gross, J.D., Andino, R., 2018. A viral protein restricts Drosophila RNAi immunity by regulating Argonaute activity and stability. Cell Host Microbe 24, 542.

Okamura, K., Lai, E.C., 2008. Endogenous small interfering RNAs in animals. Nat. Rev. Mol. Cell Biol. 9, 673.

Okamura, K., Ishizuka, A., Siomi, H., Siomi, M.C., 2004. Distinct roles for Argonaute proteins in small RNA-directed RNA cleavage pathways. Genes Dev. 18, 1655.

Ophinni, Y., Palatini, U., Hayashi, Y., Parrish, N.F., 2019. piRNA-guided CRISPR-like immunity in eukaryotes. Trends Immunol. 40, 998.

Palatini, U., Miesen, P., Carballar-Lejarazu, R., Ometto, L., Rizzo, E., Tu, Z., van Rij, R.P., Bonizzoni, M., 2017. Comparative genomics shows that viral integrations are abundant and express piRNAs in the arboviral vectors Aedes aegypti and Aedes albopictus. BMC Genomics 18, 512.

Pantaleo, V., Burgyan, J., 2008. Cymbidium ringspot virus harnesses RNA silencing to control the accumulation of virus parasite satellite RNA. J. Virol. 82, 11851.

Pantaleo, V., Szittya, G., Burgyan, J., 2007. Molecular bases of viral RNA targeting by viral small interfering RNA-programmed RISC. J. Virol. 81, 3797.

Parrish, N.F., Fujino, K., Shiromoto, Y., Iwasaki, Y.W., Ha, H., Xing, J., Makino, A., Kuramochi-Miyagawa, S., Nakano, T., Siomi, H., Honda, T., Tomonaga, K., 2015. piRNAs derived from ancient viral processed pseudogenes as transgenerational sequence-specific immune memory in mammals. RNA 21, 1691.

Pertermann, R., Tamilarasan, S., Gursinsky, T., Gambino, G., Schuck, J., Weinholdt, C., Lilie, H., Grosse, I., Golbik, R.P., Pantaleo, V., Behrens, S.-E., 2018. A viral suppressor modulates the plant immune response early in infection by regulating miRNA activity. mBio 9, e00419-18.

Pooggin, M.M., 2018. Small RNA-omics for plant virus identification, virome reconstruction, and antiviral defense characterization. Front. Microbiol. 9, 2779.

Qi, X., Bao, F.S., Xie, Z., 2009. Small RNA deep sequencing reveals role for Arabidopsis thaliana RNA-dependent RNA polymerases in viral siRNA biogenesis. PLoS One 4, e4971.

Qian, Q., Zhou, H., Shu, T., Mu, J., Fang, Y., Xu, J., Li, T., Kong, J., Qiu, Y., Zhou, X., 2020. The capsid protein of Semliki Forest virus antagonizes RNA interference in mammalian cells. J. Virol. 94 (3), pii: e01233-19.

Qiu, Y., Xu, Y.P., Wang, M., Miao, M., Zhou, H., Xu, J., Kong, J., Zheng, D., Li, R.T., Zhang, R.R., Guo, Y., Li, X.F., Cui, J., Qin, C.F., Zhou, X., 2020. Flavivirus induces and antagonizes antiviral RNA interference in both mammals and mosquitoes. Sci. Adv. 6, eaax7989.

Ratcliff, F., Harrison, B.D., Baulcombe, D.C., 1997. A similarity between viral defense and gene silencing in plants. Science 276, 1558.

Roossinck, M.J., 2011. The good viruses: viral mutualistic symbioses. Nat. Rev. Microbiol. 9, 99.

Roossinck, M.J., Garcia-Arenal, F., 2015. Ecosystem simplification, biodiversity loss and plant virus emergence. Curr. Opin. Virol. 10, 56.

Rouget, C., Papin, C., Boureux, A., Meunier, A.C., Franco, B., Robine, N., Lai, E.C., Pelisson, A., Simonelig, M., 2010. Maternal mRNA deadenylation and decay by the piRNA pathway in the early Drosophila embryo. Nature 467, 1128.

Rouha, H., Thurner, C., Mandl, C.W., 2010. Functional microRNA generated from a cytoplasmic RNA virus. Nucleic Acids Res. 38, 8328.

Ryu, K.H., Huang, L., Kang, H.M., Schiefelbein, J., 2019. Single-cell RNA sequencing resolves molecular relationships among individual plant cells. Plant Physiol. 179, 1444.

Saldana, M.A., Etebari, K., Hart, C.E., Widen, S.G., Wood, T.G., Thangamani, S., Asgari, S., Hughes, G.L., 2017. Zika virus alters the microRNA expression profile and elicits an RNAi response in Aedes aegypti mosquitoes. PLoS Negl. Trop. Dis. 11, e0005760.

Samuel, G.H., Wiley, M.R., Badawi, A., Adelman, Z.N., Myles, K.M., 2016. Yellow fever virus capsid protein is a potent suppressor of RNA silencing that binds double-stranded RNA. Proc. Natl. Acad. Sci. U. S. A. 113, 13863.

Samuel, G.H., Adelman, Z.N., Myles, K.M., 2018. Antiviral immunity and virus-mediated antagonism in disease vector mosquitoes. Trends Microbiol. 26, 447.

Schnettler, E., Donald, C.L., Human, S., Watson, M., Siu, R.W., McFarlane, M., Fazakerley, J.K., Kohl, A., Fragkoudis, R., 2013. Knockdown of piRNA pathway proteins results in enhanced Semliki Forest virus production in mosquito cells. J. Gen. Virol. 94, 1680.

Schwab, R., Ossowski, S., Riester, M., Warthmann, N., Weigel, D., 2006. Highly specific gene silencing by artificial microRNAs in Arabidopsis. Plant Cell 18, 1121.

Segami, S., Tomoyama, T., Sakamoto, S., Gunji, S., Fukuda, M., Kinoshita, S., Mitsuda, N., Ferjani, A., Maeshima, M., 2018. Vacuolar H(+)-pyrophosphatase and cytosolic soluble pyrophosphatases cooperatively regulate pyrophosphate levels in Arabidopsis thaliana. Plant Cell 30, 1040.

Shapiro, J.S., Varble, A., Pham, A.M., Tenoever, B.R., 2010. Noncanonical cytoplasmic processing of viral microRNAs. RNA 16, 2068.

Shapiro, J.S., Langlois, R.A., Pham, A.M., Tenoever, B.R., 2012. Evidence for a cytoplasmic microprocessor of pri-miRNAs. RNA 18, 1338.

Shi, B., Lin, L., Wang, S., Guo, Q., Zhou, H., Rong, L., Li, J., Peng, J., Lu, Y., Zheng, H., Yang, Y., Chen, Z., Zhao, J., Jiang, T., Song, B., Chen, J., Yan, F., 2016. Identification and regulation of host genes related to Rice stripe virus symptom production. New Phytol. 209, 1106.

Shimura, H., Pantaleo, V., 2011. Viral induction and suppression of RNA silencing in plants. Biochim. Biophys. Acta 1809, 601.

Shimura, H., Pantaleo, V., Ishihara, T., Myojo, N., Inaba, J., Sueda, K., Burgyan, J., Masuta, C., 2011. A viral satellite RNA induces yellow symptoms on tobacco by targeting a gene involved in chlorophyll biosynthesis using the RNA silencing machinery. PLoS Pathog. 7, e1002021.

Shivaprasad, P.V., Rajeswaran, R., Blevins, T., Schoelz, J., Meins Jr., F., Hohn, T., Pooggin, M.M., 2008. The CaMV transactivator/viroplasmin interferes with RDR6-dependent trans-acting and secondary siRNA pathways in Arabidopsis. Nucleic Acids Res. 36, 5896.

Shrinet, J., Jain, S., Jain, J., Bhatnagar, R.K., Sunil, S., 2014. Next generation sequencing reveals regulation of distinct Aedes microRNAs during chikungunya virus development. PLoS Negl. Trop. Dis. 8, e2616.

Silhavy, D., Molnar, A., Lucioli, A., Szittya, G., Hornyik, C., Tavazza, M., Burgyan, J., 2002. A viral protein suppresses RNA silencing and binds silencing-generated, 21- to 25-nucleotide double-stranded RNAs. EMBO J. 21, 3070.

Simon, A.E., Roossinck, M.J., Havelda, Z., 2004. Plant virus satellite and defective interfering RNAs: new paradigms for a new century. Annu. Rev. Phytopathol. 42, 415.

Simpson, G.G., Dean, C., 2002. Arabidopsis, the rosetta stone of flowering time? Science 296, 285.

Skalsky, R.L., Olson, K.E., Blair, C.D., Garcia-Blanco, M.A., Cullen, B.R., 2014. A "microRNA-like" small RNA expressed by Dengue virus? Proc. Natl. Acad. Sci. U. S. A. 111, E2359.

Slonchak, A., Hussain, M., Torres, S., Asgari, S., Khromykh, A.A., 2014. Expression of mosquito microRNA Aae-miR-2940-5p is downregulated in response to West Nile virus infection to restrict viral replication. J. Virol. 88, 8457.

Smith, N.A., Eamens, A.L., Wang, M.B., 2011. Viral small interfering RNAs target host genes to mediate disease symptoms in plants. PLoS Pathog. 7, e1002022.

Suzuki, Y., Frangeul, L., Dickson, L.B., Blanc, H., Verdier, Y., Vinh, J., Lambrechts, L., Saleh, M.C., 2017. Uncovering the repertoire of endogenous flaviviral elements in Aedes mosquito genomes. J. Virol. 91, e00571-17.

Suzuki, Y., Baidaliuk, A., Miesen, P., Frangeul, L., Crist, A.B., Merkling, S.H., Fontaine, A., Lequime, S., Moltini-Conclois, I., Blanc, H., van Rij, R.P., Lambrechts, L., Saleh, M.C., 2020. Non-retroviral endogenous viral element limits cognate virus replication in Aedes aegypti ovaries. bioRxiv. https://doi.org/10.1101/2020.03.28.013441.

Tabler, M., Tsagris, M., 2004. Viroids: petite RNA pathogens with distinguished talents. Trends Plant Sci. 9, 339.

Takanami, Y., 1981. A striking change in symptoms on cucumber mosaic virus-infected tobacco plants induced by a satellite RNA. Virology 109, 120.

Tassetto, M., Kunitomi, M., Whitfield, Z.J., Dolan, P.T., Sanchez-Vargas, I., Garcia-Knight, M., Ribiero, I., Chen, T., Olson, K.E., Andino, R., 2019. Control of RNA viruses in mosquito cells through the acquisition of vDNA and endogenous viral elements. eLife 8, e41244.

Ter Horst, A.M., Nigg, J.C., Dekker, F.M., Falk, B.W., 2019. Endogenous viral elements are widespread in arthropod genomes and commonly give rise to PIWI-interacting RNAs. J. Virol. 93, e02124-18.

Thibaut, O., Claude, B., 2018. Innate immunity activation and RNAi interplay in citrus exocortis viroid-tomato pathosystem. Viruses 10, 587.

Tromas, N., Zwart, M.P., Forment, J., Elena, S.F., 2014. Shrinkage of genome size in a plant RNA virus upon transfer of an essential viral gene into the host genome. Genome Biol. Evol. 6, 538.

van Cleef, K.W., van Mierlo, J.T., Miesen, P., Overheul, G.J., Fros, J.J., Schuster, S., Marklewitz, M., Pijlman, G.P., Junglen, S., van Rij, R.P., 2014. Mosquito and Drosophila entomobirnaviruses suppress dsRNA- and siRNA-induced RNAi. Nucleic Acids Res. 42, 8732.

van Mierlo, J.T., Bronkhorst, A.W., Overheul, G.J., Sadanandan, S.A., Ekstrom, J.O., Heestermans, M., Hultmark, D., Antoniewski, C., van Rij, R.P., 2012. Convergent evolution of argonaute-2 slicer antagonism in two distinct insect RNA viruses. PLoS Pathog. 8, e1002872.

van Mierlo, J.T., Overheul, G.J., Obadia, B., van Cleef, K.W., Webster, C.L., Saleh, M.C., Obbard, D.J., van Rij, R.P., 2014. Novel Drosophila viruses encode host-specific suppressors of RNAi. PLoS Pathog. 10, e1004256.

Van Rij, R.P., Saleh, M.C., Berry, B., Foo, C., Houk, A., Antoniewski, C., Andino, R., 2006. The RNA silencing endonuclease Argonaute 2 mediates specific antiviral immunity in Drosophila melanogaster. Genes Dev. 20, 2985.

Vodovar, N., Bronkhorst, A.W., van Cleef, K.W., Miesen, P., Blanc, H., van Rij, R.P., Saleh, M.C., 2012. Arbovirus-derived piRNAs exhibit a ping-pong signature in mosquito cells. PLoS One 7, e30861.

Vourekas, A., Alexiou, P., Vrettos, N., Maragkakis, M., Mourelatos, Z., 2016. Sequence-dependent but not sequence-specific piRNA adhesion traps mRNAs to the germ plasm. Nature 531, 390.

Wang, X.H., Aliyari, R., Li, W.X., Li, H.W., Kim, K., Carthew, R., Atkinson, P., Ding, S.W., 2006. RNA interference directs innate immunity against viruses in adult Drosophila. Science 312, 452.

Wang, X.B., Wu, Q., Ito, T., Cillo, F., Li, W.X., Chen, X., Yu, J.L., Ding, S.W., 2010. RNAi-mediated viral immunity requires amplification of virus-derived siRNAs in Arabidopsis thaliana. Proc. Natl. Acad. Sci. U. S. A. 107, 484.

Webster, C.L., Waldron, F.M., Robertson, S., Crowson, D., Ferrari, G., Quintana, J.F., Brouqui, J.M., Bayne, E.H., Longdon, B., Buck, A.H., Lazzaro, B.P., Akorli, J., Haddrill, P.R., Obbard, D.J., 2015. The discovery, distribution, and evolution of viruses associated with Drosophila melanogaster. PLoS Biol. 13, e1002210.

Wei, X., Su, X., Cao, P., Liu, X., Chang, W., Li, M., Zhang, X., Liu, Z., 2016. Structure of spinach photosystem II-LHCII supercomplex at 3.2 A resolution. Nature 534–569.

Wellmer, F., Riechmann, J.L., 2010. Gene networks controlling the initiation of flower development. Trends Genet. 26, 519.

Whitfield, Z.J., Dolan, P.T., Kunitomi, M., Tassetto, M., Seetin, M.G., Oh, S., Heiner, C., Paxinos, E., Andino, R., 2017. The diversity, structure, and function of heritable adaptive immunity sequences in the Aedes aegypti genome. Curr. Biol. 27, 3511.

Wu, Q., Wang, Y., Cao, M., Pantaleo, V., Burgyan, J., Li, W.X., Ding, S.W., 2012. Homology-independent discovery of replicating pathogenic circular RNAs by deep sequencing and a new computational algorithm. Proc. Natl. Acad. Sci. U. S. A. 109, 3938.

Xu, P., Chen, F., Mannas, J.P., Feldman, T., Sumner, L.W., Roossinck, M.J., 2008. Virus infection improves drought tolerance. New Phytol. 180, 911.

Yan, H., Zhou, Y., Liu, Y., Deng, Y., Chen, X., 2014. miR-252 of the Asian tiger mosquito Aedes albopictus regulates dengue virus replication by suppressing the expression of the dengue virus envelope protein. J. Med. Virol. 86, 1428.

Yang, Y., Liu, T., Shen, D., Wang, J., Ling, X., Hu, Z., Chen, T., Hu, J., Huang, J., Yu, W., Dou, D., Wang, M.B., Zhang, B., 2019. Tomato yellow leaf curl virus intergenic siRNAs target a host long noncoding RNA to modulate disease symptoms. PLoS Pathog. 15, e1007534.

Yang, J., Zhang, T., Li, J., Wu, N., Wu, G., Yang, J., Chen, X., He, L., Chen, J., 2020. Chinese wheat mosaic virus-derived vsiRNA-20 can regulate virus infection in wheat through inhibition of vacuolar- (H(+))-PPase induced cell death. New Phytol. 226, 205.

Zahid, K., Zhao, J.H., Smith, N.A., Schumann, U., Fang, Y.Y., Dennis, E.S., Zhang, R., Guo, H.S., Wang, M.B., 2015. Nicotiana small RNA sequences support a host genome origin of cucumber mosaic virus satellite RNA. PLoS Genet. 11, e1004906.

Zhang, L., Xu, W., Gao, X., Li, W., Qi, S., Guo, D., Ajayi, O.E., Ding, S.W., Wu, Q., 2020. lncRNA sensing of a viral suppressor of RNAi activates non-canonical innate immune signaling in Drosophila. Cell Host Microbe 27, 115.

Zhong, X., Archual, A.J., Amin, A.A., Ding, B., 2008. A genomic map of viroid RNA motifs critical for replication and systemic trafficking. Plant Cell 20, 35.

Zhu, H., Duan, C.G., Hou, W.N., Du, Q.S., Lv, D.Q., Fang, R.X., Guo, H.S., 2011. Satellite RNA-derived small interfering RNA satsiR-12 targeting the 3′ untranslated region of Cucumber mosaic virus triggers viral RNAs for degradation. J. Virol. 85, 13384.

CHAPTER TWO

Hijacking of host cellular components as proviral factors by plant-infecting viruses

Kiwamu Hyodo[a,*], Tetsuro Okuno[b]

[a]Institute of Plant Science and Resources, Okayama University, Kurashiki, Okayama, Japan
[b]Department of Plant Life Science, Faculty of Agriculture, Ryukoku University, Otsu, Shiga, Japan
[*]Corresponding author: e-mail address: khyodo@okayama-u.ac.jp

Contents

Abstract

Plant viruses are important pathogens that cause serious crop losses worldwide. They are obligate intracellular parasites that commandeer a wide array of proteins, as well as metabolic resources, from infected host cells. In the past two decades, our knowledge of plant–virus interactions at the molecular level has exploded, which provides insights into how plant-infecting viruses co-opt host cellular machineries to accomplish their infection. Here, we review recent advances in our understanding of how plant viruses divert cellular components from their original roles to proviral functions. One emerging theme is that plant viruses have versatile strategies that integrate a host factor that is normally engaged in plant defense against invading pathogens into a viral protein complex that facilitates viral infection. We also highlight viral manipulation of cellular key

Advances in Virus Research, Volume 107
ISSN 0065-3527
https://doi.org/10.1016/bs.aivir.2020.04.002

regulatory systems for successful virus infection: posttranslational protein modifications for fine control of viral and cellular protein dynamics; glycolysis and fermentation pathways to usurp host resources, and ion homeostasis to create a cellular environment that is beneficial for viral genome replication. A deeper understanding of viral-infection strategies will pave the way for the development of novel antiviral strategies.

1. Introduction

Plant-infecting viruses are responsible for many plant diseases that cause damage to crop yield and quality in all parts of the world; thus, they pose a serious threat to global food security (Jones and Naidu, 2019). Emerging and re-emerging plant viral diseases keep arising in part because of global warming, changes in agricultural environments, and/or worldwide trade in plant materials, which allow viruses to spread into new fields (Elena et al., 2014; Hanssen et al., 2010; Rojas and Gilbertson, 2008). The estimated economic impact of plant viruses is more than $30 billion annually (Sastry and Zitter, 2014).

Plant viruses are obligate biotrophic parasites with RNA or DNA genomes that replicate inside their host cells. Upon entry into a host cell and subsequent uncoating, viruses initiate their infection cycle; i.e., translation and replication of the viral genome, cell-to-cell and long-distance transport of virions or viral ribonucleoprotein complexes, and plant-to-plant dissemination mediated by vector organisms and/or direct physical contact. In general, virally encoded proteins have a multifunctional property with the potential to interact with several virus- and host-derived molecules, thus creating a complex web of molecular interactions in infected cells. In fact, plant viruses co-opt an array of host-derived factors (referred to as proviral host factors) to establish viral infection, in which viruses remodel their functions to facilitate every step of the viral life cycle (Cheng, 2017; Folimonova and Tilsner, 2018; Garcia-Ruiz, 2018; Hanley-Bowdoin et al., 2013; Heinlein, 2015; Huang et al., 2017; Hyodo and Okuno, 2014, 2016; Ishibashi and Ishikawa, 2016; Nagy, 2016; Nagy and Pogany, 2012; Noueiry and Ahlquist, 2003; Pitzalis and Heinlein, 2017; Sanfaçon, 2015; Wang, 2015; Zhang et al., 2019b).

During the co-evolutionary arms race between plants and viruses, plants have developed multiple defense systems against invading viruses that can recognize viral proteins or nucleic acids, resulting in the inhibition of critical step(s) of the viral life cycle (Garcia-Ruiz, 2019; Mandadi and Scholthof, 2013; Nicaise, 2014; Niehl and Heinlein, 2019). Among them,

RNA silencing is the best-characterized antiviral mechanism and mediates either the cleavage/translation repression of viral RNA (for both RNA and DNA viruses) or the inhibition of transcription from viral DNA (for DNA viruses) in a sequence-specific manner (Andika et al., 2016; Carbonell and Carrington, 2015; Yang and Li, 2018). Plants also possess an array of nucleotide-binding domain and leucine-rich repeat (NLR)-containing proteins that can confer dominant resistance to a limited range of viruses (De Ronde et al., 2014; Meier et al., 2019). Upon recognition of a specific feature of viral proteins, the NLR proteins are activated and trigger antiviral immune signaling, which is often associated with localized cell death and is referred to as hypersensitive response. Some plants also contain non-NLR-type antiviral proteins that attenuate viral infection by either inhibiting the functions of viral proteins, degrading viral nucleic acids, or acting via other unknown mechanisms (De Ronde et al., 2014; Ishibashi and Ishikawa, 2014; Ishibashi et al., 2019; Kloth and Kormelink, 2020; Liu et al., 2017; Sasvari et al., 2014; Yamaji et al., 2012; Yoshida et al., 2019). Moreover, it is becoming clear that cellular quality-control systems, including the autophagy and RNA decay pathways, are an integral part of the antiviral immunity in plants (Kushwaha et al., 2019; Li and Wang, 2019). As countermeasures for these multilayered defense systems, plant viruses have evolved to circumvent host antiviral defenses by interfering with key components involved in the aforementioned antiviral defense pathways (Csorba et al., 2015; Kushwaha et al., 2019; Li and Wang, 2019). Recent advancements in the field of plant–virus interactions at the molecular level have provided several key insights into how plant viruses usurp host cellular resources for the establishment of their infections. One such revelation is that plant viruses can even trick a host factor that is normally engaged in plant defense against invading pathogens into a viral protein complex (as a proviral factor), to facilitate their infection. Therefore, intricate molecular interactions between viruses and pro/antiviral host factors determine the fate of plant–virus interactions.

In this review, we summarize our understanding of the molecular interactions between plant viruses and multiple cellular components, focusing on recent advances covering specific topics: the hijacking of host defense-related factors as proviral components; the diversion of host posttranslational modification systems into viral protein complexes for successful viral infection and dissemination; and viral manipulation of cellular glycolysis and ion homeostasis to create a cellular environment that is beneficial for viral genome replication. Last, we discuss the potential control measures for viral diseases in plants that apply deeper knowledge by integrating recent advances in this field of research.

2. Diversion of host defense components into viral protein complexes

Plants have multilayered immune systems to ward off intruding viruses. An emerging theme from recent studies is that plant viruses have evolved diverse strategies not only to circumvent the hostile antiviral defense machineries of their hosts, but also to trick them into being a part of the viral infection strategy and support viral infection.

2.1 RNA silencing

In eukaryotes, RNA silencing, also referred to as RNA interference (RNAi), is a highly conserved, sequence-specific mechanism that regulates the fate of RNA molecules. In plants, as well as fungi and insects, RNA silencing plays a major role in restricting viral infection (Guo et al., 2019b). Upon viral infection, the double-stranded RNA (dsRNA)-specific RNaseIII-like endoribonuclease Dicer or Dicer-like (DCL) family of proteins detect and process viral dsRNAs (originated from highly structured regions of viral single-stranded RNAs (ssRNAs), replication intermediates of ssRNA viruses, overlapping bidirectional read-through transcripts from DNA virus genomes, or de novo synthesis by host-encoded RNA-dependent RNA polymerases (RDRs)) into 21–24 nucleotide-long virus-derived small interference RNA (vsiRNA) duplexes. Subsequently, vsiRNAs are incorporated into RNaseH-like Argonaute (AGO) proteins to form the RNA-induced silencing complex (RISC), and one strand of the vsiRNA duplex is used to guide RISC to perform cleavage or translational repression of homologous viral RNAs (Andika et al., 2016; Carbonell and Carrington, 2015; Yang and Li, 2018). Land plants typically encode multiple AGO proteins, each with specific functions, expression patterns, and small RNA preferences (Zhang et al., 2015). In *Arabidopsis thaliana*, multiple AGO isoforms participate in the antiviral defense against plant viruses, while in *Nicotiana benthamiana*, AGO1 and AGO2 are known to be involved in antiviral immunity (Carbonell and Carrington, 2015). To counteract host RNA silencing-based defenses, most, if not all, plant-infecting viruses encode an RNA silencing suppressor (RSS) protein (Csorba et al., 2015). The modes of action of viral RSS are diverse and include sequestration of virus-derived RNA molecules through ss-, ds-, and si-RNA-binding activity or inhibition of the activities of silencing effector proteins (e.g., DCLs, AGOs, or RDRs) through protein–protein interactions. Moreover, some viruses can hijack endogenous regulators of RNA silencing to attenuate the host antiviral defense. The calmodulin-like protein rgsCAM is an endogenous RSS that downregulates the RNA silencing pathway through the inhibition of the expression of the *RDR6* gene, which encodes a key player in antiviral RNA silencing, and the degradation of suppressor of gene silencing 3 (SGS3), which is a chaperone protein that is essential for RDR6 function (Anandalakshmi et al., 2000; Li et al., 2014, 2017). The βC1 protein, which is encoded by betasatellite DNA associated with a ssDNA virus tomato yellow leaf curl China virus (TYLCCNV; genus: *Geminivirus*; family: *Geminiviridae*), or the AL2 protein of tomato golden mosaic virus (TGMV; genus: *Geminivirus*; family: *Geminiviridae*) induces the expression of rgsCAM, leading to the attenuation of host antiviral RNA silencing through the downregulation of RDR6-mediated secondary vsiRNA amplification (Chung et al., 2014; Li et al., 2014). Another example comes from a study that used the turnip mosaic virus (TuMV; genus: *Potyvirus*; family: *Potyviridae*) and *A. thaliana* model system (Endres et al., 2010). Viruses in the family *Potyviridae* have a monopartite positive-sense (+)ssRNA genome of ~10 kb, which is linked covalently to viral protein genome-linked (VPg) at the 5′ terminus and is polyadenylated at the 3′ terminus. The genome of potyviruses contains a long open reading frame (ORF), which encodes a polyprotein of ~360 kDa that is processed by three viral proteinases into 10 mature proteins (Revers and García, 2015), and another cryptic ORF PIPO (pretty interesting *Potyviridae* ORF), which is expressed as a fusion with the N-terminal region of P3 (referred to as P3N-PIPO) that is translated from a viral RNA variant synthesized via programmed transcriptional slippage during viral (+)RNA synthesis (Olspert et al., 2015; Rodamilans et al., 2015). TuMV appears to exploit the *Arabidopsis* RAV2 protein, which is a transcription factor that regulates the expression of endogenous RSSs, including FIERY1 and CML38, a homolog of rgsCAM (Anandalakshmi et al., 2000; Gy et al., 2007), to attenuate host antiviral RNA silencing (Endres et al., 2010). The TuMV-encoded helper component-proteinase (HC-Pro) protein binds to RAV2 and mediates the impairment of the RNA silencing pathway in a RAV2-dependent manner. Interestingly, RAV2 is also required for the RSS activity of p38 from turnip crinkle virus (TCV; genus: *Betacarmovirus*; family: *Tombusviridae*) (Endres et al., 2010), suggesting that viral manipulation of RAV2 is an effective countermeasure to circumvent the host antiviral defense response.

A recent study using the bamboo mosaic virus (BaMV; genus: *Potexvirus*; family: *Alphaflexviridae*) and *N. benthamiana* system has provided an example of the viral hijacking of an AGO protein to facilitate viral infection (Huang et al., 2019c). Viruses in the genus *Potexvirus* have a monopartite 5.9–7 kb (+)ssRNA genome with a 5′ cap and a 3′ polyA tail that encodes five ORFs (Martelli et al., 2007). A viral RNA-dependent RNA polymerase (RdRP) encoded by the 5′ proximal ORF is translated from genomic RNA, while other ORFs are translated from subgenomic RNAs. Interestingly, the downregulation of AGO10a, which is the closest paralog of antiviral AGO1, attenuates BaMV replication and infection in *N. benthamiana*, indicating that AGO10a has proviral functions in BaMV infection. It has been shown that *Arabidopsis* AGO10 plays a key role in the fine control of plant developmental processes through the sequestration of cellular microRNAs from AGO1 (Zhu et al., 2011), and participates in the antiviral defense against TuMV in inflorescence tissues (Garcia-Ruiz et al., 2015). Interestingly, the BaMV p28 RSS protein promotes both the expression of the *AGO10a* gene and the stability of its protein product, leading to increased AGO10a abundance in BaMV-infected cells (Huang et al., 2019c). BaMV-derived vsiRNAs are sequestered into AGO10a and subsequently degraded by AGO10a-associated small RNA degrading nuclease 1 (SDN1). Thus, BaMV may take advantage of the host AGO10a–SDN1 module to hide vsiRNAs from other AGO proteins, to circumvent the formation of antiviral RISCs (Huang et al., 2019c). It is of interest to investigate whether other plant-infecting viruses adopt a similar strategy to attenuate host antiviral RNA silencing pathways.

2.2 RNA decay pathway

In addition to RNA silencing, RNA decay is an important mechanism that regulates cellular RNA metabolism (Zhang and Guo, 2017). In the RNA decay pathway, mRNAs can be degraded in either the 3′→5′, or the 5′→3′ direction, which is executed by two sets of exonucleases, a multisubunit complex called the exosome and exoribonucleases (XRNs) (Fig. 1). The degradation process is initiated by the shortening of the 3′ polyA tail, which is known as deadenylation, followed by either 3′→5′ decay or a decapping process and subsequent 5′→3′ exonucleolytic digestion. Prior to 5′→3′ mRNA decay, a collection of decapping factors are required to catalyze the removal of the 5′ cap from capped transcripts. These decapping factors include Decapping1 (DCP1), DCP2, the WD domain-containing protein VARICOSE, heptameric Sm-like (LSM) complexes (LSM1–7), and DHH1 DEAD-box RNA helicase (Golisz et al., 2013; Perea-Resa et al., 2012; Xu et al., 2006). XRN4 (Xrn1 in yeast and mammals) is an

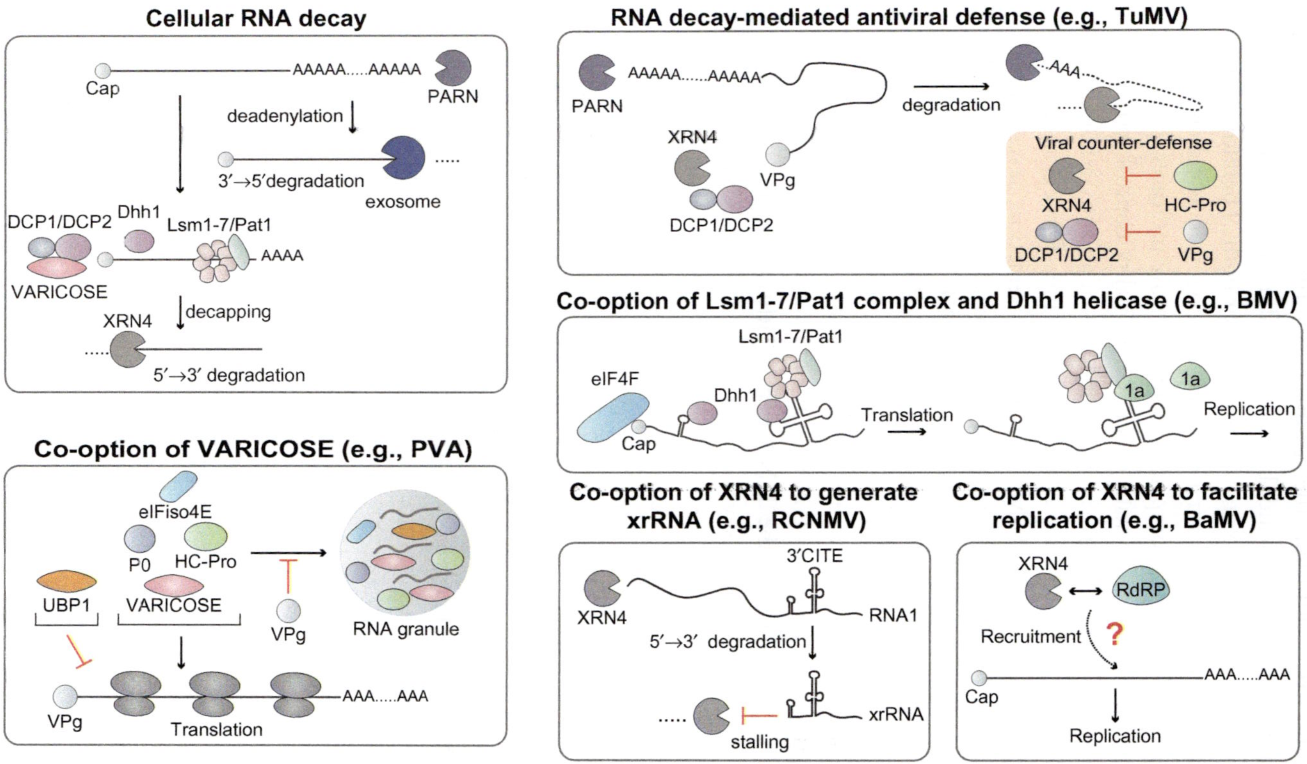

Fig. 1 Schematic representation of the interplays between plant-infecting viruses and the host RNA decay pathway. Red blunted lines represent negative regulation. PARN, polyA-specific ribonuclease; DCP, decapping; Lsm, Sm-like; XRN4, exoribonuclease 4; eIFiso4E, eukaryotic initiation factor isoform 4E; P0, ribosomal protein P0; UBP1, oligouridylate-binding protein 1; VPg, viral protein genome-linked; HC-Pro, helper component-proteinase; 3′CITE, 3′-cap-independent translational enhancer; xrRNA, Xrn1-resistant RNA; RdRP, RNA-dependent RNA polymerase; TuMV, turnip mosaic virus; PVA, potato virus A; BMV, brome mosaic virus; RCNMV, red clover necrotic mosaic virus; BaMV, bamboo mosaic virus.

exoribonuclease that is responsible for 5′→3′ degradation of cytoplasmic uncapped mRNAs. Although the primary determinants of cellular mRNA stabilities are the 5′ cap and the 3′ polyA tail, many plant-infecting RNA viruses lack either a 5′ cap or a 3′ polyA tail, or both. Therefore, it is reasonable to expect that these viral RNAs are recognized as "aberrant" RNAs in plant cells. In fact, compelling evidence now indicates that the RNA decay pathway, together with other RNA quality control systems, such as nonsense-mediated decay, is an integral part of an antiviral defense mechanism in plants (Li and Wang, 2018, 2019; Martínez-Pérez et al., 2017; May et al., 2018). During the TuMV–*A. thaliana* interaction, several host genes associated with the RNA decay pathway are significantly upregulated (Li and Wang, 2018). These include DCP1, DCP2, XRN4, and an mRNA deadenylase polyA-specific ribonuclease (PARN). Notably, loss of these RNA-decay-related factors facilitates TuMV infection in plants, while their overexpression inhibits TuMV infection (Li and Wang, 2018). These findings support the role of the host RNA decay pathway in antiviral defense. Accordingly, components of the RNA decay pathway, such as XRN4 and DCP2, limit infections by several plant-infecting (+)ssRNA viruses, including tomato bushy stunt virus (TBSV; genus: *Tombusvirus*; family: *Tombusviridae*) and the related cucumber necrosis virus (CNV), *Plantago asiatica* mosaic virus (PLAMV; genus: *Potexvirus*; family: *Alphaflexviridae*), tobacco mosaic virus (TMV; genus: *Tobamovirus*; family: *Virgaviridae*), tobacco rattle virus (TRV; genus: *Tobravirus*; family: *Virgaviridae*), and a (−)ssRNA virus tomato spotted wilt virus (TSWV; genus: *Orthotospovirus*; family: *Tospoviridae*) (Cheng et al., 2007; Jaag and Nagy, 2009; Jiang et al., 2018; Ma et al., 2019; Peng et al., 2011). However, viruses might have evolved to acquire countermeasure strategies to attenuate the host RNA decay pathway (Li and Wang, 2018, 2019; Lukhovitskaya and Ryabova, 2019; May et al., 2018, 2020). For example, the HC-Pro and VPg proteins, which are two well-known RSSs encoded by TuMV, interact with XRN4 and DCP2, respectively, to suppress their antiviral defense functions (Li and Wang, 2018).

Interestingly, studies using the yeast model system for brome mosaic virus (BMV; genus: *Bromovirus*; family: *Bromoviridae*) replication have revealed the proviral functions of the host decapping activator complex in the viral life cycle (Beckham et al., 2007; Díez et al., 2000; Kushner et al., 2003; Mas et al., 2006; Noueiry et al., 2003). Pat1 forms a stable complex with Lsm1–7 and acts as a scaffold protein to allow sequential recruitment of decapping factors onto mRNA. The Lsm1–7/Pat1 complex is recruited to viral RNAs through binding to a tRNA-like structure located in the 3′ untranslated regions (UTRs) of BMV RNAs (Galão et al., 2010). Dhh1 binds directly to *cis*-acting RNA elements that regulate the translation/replication of BMV genomic RNAs and form complexes with the eukaryotic translation initiation factors eIF4E, eIF4G, and eIF4A bound at the 5′ UTR (Jungfleisch et al., 2017). Loss of Lsm1–7, Pat1, or Dhh1 leads to inhibition of BMV translation at the initiation step and subsequent viral replication, indicating that these co-opted factors act as translation enhancers during BMV infection. After translation of viral replication proteins, the Lsm1–7/Pat1 complex is then recruited by the BMV 1a replication protein through a protein–protein interaction, leading to the redirection of the function of the Lsm1–7/Pat1 complex from the translation step toward facilitation of the recruitment of viral genomic RNAs into the replication step (Jungfleisch et al., 2015). These studies provide an example of the virus-mediated redirection of host RNA decay factors to proviral functions. The insect or human counterparts of Lsm1–7/Pat1 complex and Dhh1 DEAD-box helicase are also targeted by several insect- or human-infecting (+) ssRNA viruses, including Flock House virus (FHV; genus: *Alphanodavirus*; family: *Nodaviridae*), dengue virus (DENV; genus: *Flavivirus*; family: *Flaviviridae*), West Nile virus (WNV; genus: *Flavivirus*; family: *Flaviviridae*), and hepatitis C virus (HCV; genus: *Hepacivirus*; family: *Flaviviridae*), to promote viral translation and/or RNA replication (reviewed in Jungfleisch et al., 2016).

In addition to decapping activator complexes, plant (+)ssRNA viruses may subvert other host factors associated with the RNA decay pathway to facilitate viral infection. HC-Pro encoded by potato virus A (PVA; genus: *Potyvirus*; family: *Potyviridae*) induces RNA granule-like structures in infected cells that contain HC-Pro and viral RNA, as well as host RNA-binding proteins, including the ribosomal protein P0, eIFiso4E, VARICOSE, and oligouridylate-binding protein 1 (UBP1) (Hafrén et al., 2015). Downregulation of these host factors results in impairment of the granule formation and PVA infection, indicating that they have a proviral function in viral infection. Interestingly, P0, eIFiso4E, and VARICOSE facilitate VPg-mediated PVA translation, while UBP1 negatively regulates this process. As VPg negatively regulates the formation of RNA granule-like structures (Hafrén et al., 2015), these co-opted host factors, in concerted action with the PVA-encoded HC-Pro and VPg proteins, may participate in the fine tuning of viral translation and other step(s) of the viral life cycle through the formation/disruption of PVA-induced RNA granules.

It has been shown that BaMV hijacks another cellular component of the RNA decay pathway to facilitate viral replication (Lee et al., 2016). BaMV-encoded RdRP, termed REP_{BaMV}, interacts with a cytoplasmic exoribonuclease (XRN4) of *N. benthamiana*. Downregulation of XRN4 attenuates BaMV infection, while its overexpression increases viral replication in protoplasts in a catalytic-activity-dependent manner. XRN4 also promotes the replication of foxtail mosaic virus (FoMV; genus: *Potexvirus*; family: *Alphaflexiviridae*) and beet necrotic yellow vein virus (BNYVV; genus: *Benyvirus*; family: *Benyviridae*) (Flobinus et al., 2018; Lee et al., 2016), supporting a proviral role for XRN4 in plant–virus interactions, although the mechanism by which the co-opted $5' \rightarrow 3'$ exoribonuclease facilitates the RNA replication of these viruses is currently unknown. In addition to its function in RNA turnover, Xrn1 can regulate the transcription initiation, as well as the translation initiation, of a set of cellular mRNAs (Blasco-Moreno et al., 2019; Haimovich et al., 2013). It would be interesting to investigate whether these newly found functions of Xrn1 (and possibly XRN4) are associated with a proviral function of the co-opted $5' \rightarrow 3'$ exoribonuclease during plant virus infection.

Interestingly, accumulating evidence now indicates that some (+) ssRNA viruses produce noncoding subgenomic RNAs by co-opting a host $5' \rightarrow 3'$ exoribonuclease via highly structured regions of viral RNAs, which hinder the enzyme's progression, thereby leading to the production of a stable RNA decay intermediate (Clarke et al., 2015; Göertz et al., 2018; Kieft et al., 2015). The generated RNAs are referred to as xrRNAs (Xrn1-resistant RNAs) and play diverse roles during infection. The best-characterized examples include a 300–500-nucleotide-long xrRNA of insect-borne flaviviruses, such as DENV, WNV, and Zika virus, which promotes viral replication or infection by inhibiting the insect RNA silencing pathway or the mammalian interferon response (Clarke et al., 2015; Göertz et al., 2018; Kieft et al., 2015). Regarding plant (+)ssRNA viruses, red clover necrotic mosaic virus (RCNMV; genus: *Dianthovirus*; family: *Tombusviridae*) and other members of dianthovirus possess a 43-nucleotide sequence with an elaborated tertiary RNA structure that can stall Xrn1 progression at the beginning of the 3′ UTR of viral genomic RNA1, leading to the generation of a 400-nucleotide-long xrRNA that can antagonize the translation of cellular and viral RNAs, possibly through competition for host translation initiation factors (Iwakawa et al., 2008; Steckelberg et al., 2018a). Viruses in the family *Tombusviridae* have neither a 5′ cap nor a 3′ polyA tail; rather, they have developed a *cis*-acting RNA element referred to as 3′-cap-independent translational enhancer (3′CITE) in their 3′ UTRs that can recruit host translation initiation complexes (reviewed in Simon and Miller, 2013; Truniger et al., 2017). The 3′ UTR of RCNMV RNA1 possesses a canonical 3′CITE and a ∼25 nt adenine-rich sequence, which recruit the host translation initiation complex through binding to eIF4G and polyA-binding protein, respectively (Iwakawa et al., 2012; Kraft et al., 2013; Tajima et al., 2017). In addition to dianthoviruses, xrRNAs have been recently discovered in diverse members of plant-infecting (+)ssRNA viruses (Dilweg et al., 2019; Steckelberg et al., 2018b). Examples include the 212-nucleotide long xrRNA of tobacco necrosis virus-D (genus: *Betanecrovirus*; family: *Tombusviridae*) and the noncoding RNA3 of BNYVV, which play an important role in viral translation and systemic infection, respectively (Flobinus et al., 2016, 2018; Gunawardene et al., 2019; Peltier et al., 2012).

Collectively, these examples represent intricate molecular interactions between plant viruses and the host RNA decay pathway (Fig. 1), and provide new insights into how (+)ssRNA viruses take advantage of these cellular components to facilitate viral translation, replication, or other processes of the viral life cycle.

2.3 Autophagy

The role of autophagy in immunity against invading viruses in mammalian cells has been well established (Choi et al., 2018). Although our understanding of the roles of autophagy in plant–virus interactions is just starting to unravel, several recent studies have highlighted the importance of autophagy in the antiviral defense against plant viruses (reviewed in Kushwaha et al., 2019; Signorelli et al., 2019). Upon various unrelated RNA and DNA virus infections, autophagy is activated by largely unknown mechanism(s), leading to the engulfment of viral proteins and/or particles into double-membrane vesicles called autophagosomes that mediate their vacuolar degradation (Hafrén et al., 2017, 2018; Haxim et al., 2017; Li et al., 2018, 2020). ATG6/Beclin1, a key component of the phosphatidylinositol (PI) 3-kinase (PI3K) complex for autophagosome nucleation, recognizes TuMV RdRP through binding to the highly conserved GDD motif, which is a hallmark of (+)ssRNA viral RdRP (Li et al., 2018). Subsequently, ATG6/Beclin1 mediates the vacuolar degradation of viral RdRP and restricts viral replication. In addition to autophagic degradation, binding of ATG6/Beclin1 to TuMV RdRP also suppresses its replication activity in an autophagic-degradation-independent fashion. ATG6/Beclin1 may be a key player in plant antiviral immunity because it also interacts with the RdRPs of several other plant (+)ssRNA viruses, including cucumber green mottle mosaic virus (genus: *Tobamovirus*; family: *Virgaviridae*) and pepino mosaic virus (genus: *Potexvirus*; family: *Alphaflexiviridae*), via binding to GDD motifs, and inhibits their activities in plants (Li et al., 2018). To counteract host autophagy-mediated antiviral defense mechanisms, viruses have acquired the potent ability to inhibit and circumvent autophagic viral degradation by sequestering viral proteins and/or interfering with a key step of the autophagy pathway (Hafrén et al., 2017, 2018; Yang et al., 2018a; Zvereva et al., 2016).

In addition to nullifying autophagy, some plant-infecting viruses take advantage of the cellular autophagy pathway to eliminate host antiviral factors, such as Remorin and SGS3. Plant cells are connected with each other via membrane-lined nanoscale pore-like structures, called plasmodesmata (PD), that enable intercellular communication. To spread into neighboring cells, plant viruses generally encode one or more viral movement proteins (MPs), which target PD and increase their size exclusion limit to facilitate viral intercellular movement (Navarro et al., 2019). Remorin restricts PD opening through the reorganization of PD lipid nanodomain and/or sequestration of the plasma-membrane-associated Ca^{2+}-binding protein 1, which is a proviral host factor that is required for increasing PD permeability, thereby leading to impairment of the intercellular movement of a range of plant viruses (Cheng et al., 2020; Huang et al., 2019b). SGS3 is an essential component of antiviral RNA silencing, as mentioned previously. TuMV VPg interacts with these antiviral host factors and mediates their autophagic degradation to counteract Remorin- and SGS3-mediated antiviral activities (Cheng and Wang, 2017; Cheng et al., 2020). Similarly, the NSvc4 protein encoded by rice stripe virus (RSV; genus: *Tenuivirus*; family: *Phenuiviridae*) binds to host Remorin and mediates its degradation via both the autophagy and ubiquitin–proteasome pathways, to enhance intercellular virus spread in plants (Fu et al., 2018). Notably, the autophagic degradation of Remorin is regulated by *S*-acylation modification, and the binding of NSvc4 to Remorin inhibits this process (Fu et al., 2018). The P0 protein of polerovirus interacts with AGO1 at endoplasmic reticulum (ER) membranes by recognizing an 18-amino-acid-long degron motif, which is highly conserved among multiple AGO isoforms, resulting in the impairment of antiviral RNA silencing by inducing the degradation of AGOs through an ER-derived autophagy pathway (Derrien et al., 2018; Michaeli et al., 2019).

During the long-term virus–host arms race, viruses may have evolved various strategies to exploit the host autophagic pathway for their benefit (Choi et al., 2018). All eukaryotic (+)ssRNA viruses remodel host intracellular membranes to generate distinct membranous compartments, which are referred to as replication organelles (ROs), for robust multiplication of their genome. ROs originate from different cellular membranes, depending on the virus, including the ER, Golgi, peroxisomes, endosomes, mitochondria, chloroplasts, tonoplasts, and plasma membrane (Jin et al., 2018; Laliberté and Sanfaçon, 2010; Laliberté and Zheng, 2014; Miller and Krijnse-Locker, 2008). Poliovirus (genus: *Enterovirus*; family: *Picornaviridae*), a human-infecting (+)ssRNA virus, induces autophagy to accumulate autophagosome-like vesicles that may be utilized for membrane scaffolding for ROs, while DENV and Zika virus subvert the autophagy pathway to facilitate viral infection at post-replication step(s) (Abernathy et al., 2019). Poliovirus also exploits the host autophagy pathway for nonlytic viral release from an infected cell and subsequent cell-to-cell spread of virions (Bird et al., 2014). For plant-infecting viruses, whether they exploit autophagosomal membranes for their replication in a similar manner to poliovirus and other animal-infecting viruses remains largely unknown. Using the BaMV *N. benthamiana* system, the recent work reported by Huang and colleagues provides compelling evidence that a plant (+)ssRNA virus hijacks components of autophagy in a similar fashion to poliovirus (Huang et al., 2019a). BaMV infection induces the accumulation of autophagosomes, likely via the upregulation of the expression of autophagy-related molecules, including PI3K, Beclin1/ATG6, and the ubiquitin-like protein ATG8f in *N. benthamiana*. The ATG8 family of proteins are evolutionarily conserved proteins that play key roles in multiple steps of autophagosome biogenesis and selective cargo recruitment (Kellner et al., 2017). Pharmacological inhibition of PI3K or downregulation of several key autophagy-related genes leads to the attenuation of BaMV replication, while the overexpression of ATG8f stimulates BaMV infection, supporting the notion that autophagy plays a proviral role, rather than an antiviral role, in BaMV replication. Interestingly, NbATG8f-positive vesicles contain viral (+) and (−)RNAs, as well as REP_{BaMV}, in BaMV-infected cells (Huang et al., 2019a). These data support the conclusion that BaMV hijacks the host autophagy pathway to recruit ATG8f-positive double-membrane vesicles to the chloroplast outer membrane (BaMV replication site), thus providing a membrane platform for the biogenesis of ROs (Huang et al., 2019a). Rice gall dwarf virus (RGDV; genus: *Phytoreovirus*; family: *Reoviridae*) and the related rice dwarf virus induce the accumulation of the virion-containing autophagosomes during the persistent infection of the leafhopper vectors to facilitate nonlytic viral release from infected cells, leading to enhanced viral spread in vector organisms and viral transmission by the insect vector to rice plants (Chen et al., 2017a). Interestingly, RGDV does not activate the autophagy pathway in rice plants (Chen et al., 2017a), suggesting that RGDV has evolved to specifically manipulate the autophagy in the leafhopper vectors.

Phosphatidylinositol-3-phosphate, PI(3)P, which is generated by PI3K, plays key roles in multiple cellular physiological processes, including endosomal vesicle trafficking and autophagosome formation (Noack and Jaillais, 2017). TBSV induces the expression of the *PI3K* gene and increases PI(3)P abundance in infected cells, and the TBSV p33 replication protein binds to PI3K and diverts it into viral replication by facilitating the enrichment of PI(3)P at the vicinity of ROs (Feng et al., 2019; Inaba et al., 2019). PI(3)P is critical for TBSV replication, likely by affecting the stability of the p33 replication protein, which is degraded by the 26S proteasome if PI3K activity is blocked (Feng et al., 2019). PI3K and its product, PI(3)P, appear to be broadly involved in infections by plant (+)ssRNA viruses: pharmacological inhibition of PI3K activity in *N. benthamiana* attenuates the replication of several members of the family *Tombusviridae*, including carnation Italian ringspot virus (CIRV; genus: *Tombusvirus*), cucumber leaf spot virus (genus: *Aureusvirus*), TCV, and RCNMV (Feng et al., 2019). Although TBSV, and possibly other members of the family *Tombusviridae*, recruit the PI3K complex into ROs and induce its activation to increase PI3P abundance, as mentioned above, the proviral function of PI3K appears to be independent of the autophagy pathway, because the disruption of key autophagy-related genes, including *ATG1*, *ATG5*, *ATG7*, *ATG8*, and *ATG14* in yeast, has no obvious effect on TBSV replication (Feng et al., 2019). Also, it remains unaddressed if and how TBSV circumvents autophagy-mediated antiviral defense during infection. The recruitment of the PI3K complex into ROs might contribute to its sequestration from other autophagy components during TBSV infection.

2.4 Reactive oxygen species

In plants and other aerobic organisms, reactive oxygen species (ROS), including the superoxide anion (O_2^-), hydrogen peroxide (H_2O_2), and the hydroxyl radical (OH^-), exhibit high reactivity to various types of molecules and initiate multiple signaling pathways for acclimation to various environmental stimuli and cellular metabolic perturbations (Waszczak et al., 2018). ROS are produced in various cellular compartments, such as chloroplasts, mitochondria, peroxisomes, and the apoplast. Because of their high reactivity, the lifetime of ROS is likely to be short (nanoseconds for OH^-, and milliseconds to seconds for O_2^- and H_2O_2). Intracellular ROS production is often associated with photosynthesis and photorespiration, while apoplastic ROS are primarily produced by plasma-membrane-localized NADPH oxidases (respiratory burst oxidase homologs, RBOHs), which are responsible for O_2^- generation under various biotic and abiotic stresses (Kimura et al., 2017). RBOH-derived O_2^- can be converted to H_2O_2 through the action of superoxide dismutases or peroxidases. Plant cells can distinguish the type, concentration, and subcellular origin of ROS; therefore, the specificity of ROS signaling is likely to be determined by compartmentalization of the production/scavenging of ROS (Kimura et al., 2017; Waszczak et al., 2018). This may be true for plant–virus interactions, although whether and how the different subcellular origins of ROS affect plant virus infection needs future investigations. One of a hallmark feature of the plant immune response is the rapid production of ROS, which have antimicrobial activities, and defensive signaling functions (Sewelam et al., 2016). Although further research is needed to understand the roles of ROS during plant virus infection, accumulating evidence now indicates the double-edged action of ROS in compatible plant–virus interactions.

In the RSV–rice interaction, a higher ROS accumulation appears to be well correlated with antiviral defense activity (Wu et al., 2017). In a compatible TMV–*N. benthamiana* interaction, ROS act as antiviral signaling molecules during brassinosteroid-mediated systemic resistance (Deng et al., 2016a,b). The role in induction of antiviral mechanisms of ROS during plant virus infection is further supported by the finding that exogenous treatment with H_2O_2 attenuates barley stripe mosaic virus (BSMV; genus: *Hordeivirus*; family: *Virgaviridae*) infection in *N. benthamiana* (Yang et al., 2018b) (Fig. 2). A peroxisome-resident glycolate oxidase (GOX), which produces glyoxylate and H_2O_2 from glycolate, appears to be a primary source of ROS during BSMV invasion. As a counter-defense, the multifunctional BSMV γb protein binds to GOX at peroxisomes, resulting in the enhancement of BSMV infection via the inhibition of GOX-mediated H_2O_2 production. For citrus tristeza virus (CTV; genus: *Closterovirus*; family: *Closteroviridae*), which is a phloem-limited virus, the virally encoded protein p33 induces ROS production and cell death in both

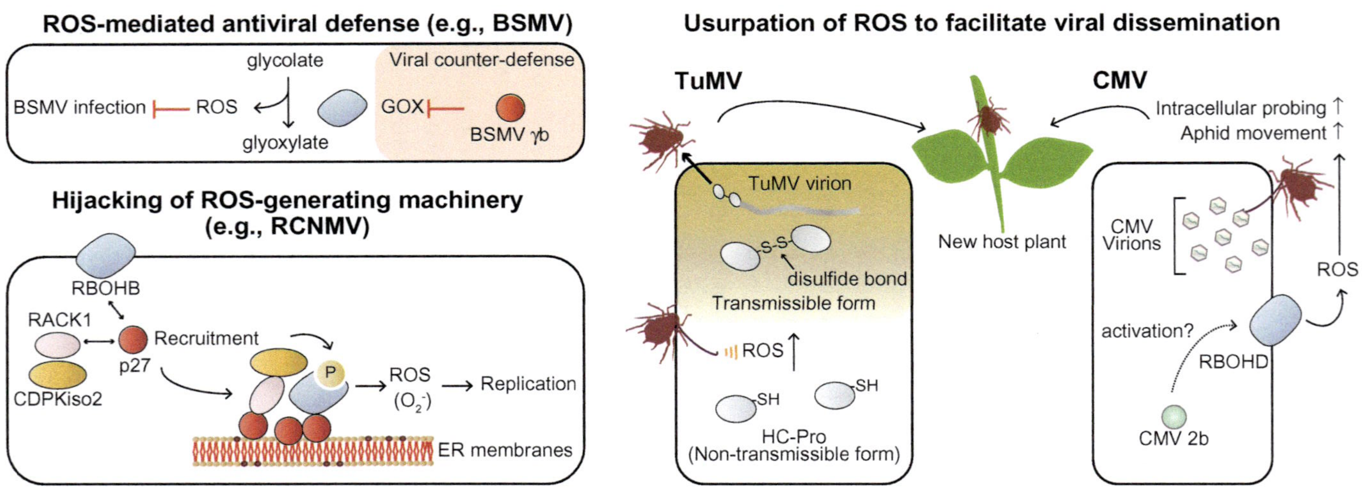

Fig. 2 Schematic representation of the interactions between plant-infecting viruses and the host reactive oxygen species (ROS)-generating machineries. Red blunted lines represent negative regulation. Double-headed arrows indicate protein–protein interactions. GOX, glycolate oxidase; RBOHB, respiratory burst oxidase homolog B; RACK1, receptor for activated C kinase 1; CDPKiso2, calcium-dependent protein kinase isoform 2; P, phosphate group; ER, endoplasmic reticulum; BSMV, barley stripe mosaic virus; RCNMV, red clover necrotic mosaic virus; TuMV, turnip mosaic virus; CMV, cucumber mosaic virus.

N. benthamiana and *Citrus macrophylla* host plants through a yet unknown mechanism (Sun and Folimonova, 2019). Deletion of p33 leads to reduced ROS production and increased viral accumulation in CTV infected tissues, resulting in invasion of the mutant virus into xylem tracheid cells, as well as perturbation of the proper differentiation of the vascular tissue of infected *C. macrophylla*. Thus, p33-mediated ROS production is likely to be important for the control of viral invasiveness and disease severity, which may be beneficial for CTV by allowing the long-term persistence of infection.

The proviral role of ROS in plant–virus interactions is supported by the finding that the downregulation or pharmacological inhibition of RBOHB in *N. benthamiana* attenuates the replication of RCNMV and distantly related BMV (Hyodo et al., 2017a) (Fig. 2). The RCNMV replication protein p27 binds to RBOHB and recruits it to the vicinity of ROs, leading to enhanced viral replication through the induction of RBOHB-dependent intracellular ROS production. To avoid unwanted ROS production, the enzymatic activity of RBOHs is tightly regulated by various mechanisms, including calcium ions, phosphatidate binding, and phosphorylation (Adachi and Yoshioka, 2015). RCNMV hijacks a host protein kinase, the calcium-dependent protein kinase isoform 2 (CDPKiso2), and a WD40-domain-containing multifunctional scaffold protein (receptor for activated C kinase 1, RACK1) to activate RBOHB (Hyodo et al., 2017a, 2019). p27 recruits RACK1 to ROs, where RACK1 mediates the bridging of p27 with CDPKiso2, resulting in RBOHB-mediated ROS production through the CDPKiso2-mediated phosphorylation of RBOHB. Although RCNMV and BMV commonly target RBOHB for their infections, they appear to require different types of ROS; RCNMV replication is sensitive to an O_2^- scavenger but insensitive to an H_2O_2 scavenger, whereas BMV replication is sensitive to both types of scavengers (Hyodo et al., 2017b). Although the molecular mechanism underlying the promotion of viral replication by RBOHB-derived ROS is not fully understood, the ROS-induced oxidative modification of viral proteins may contribute to this phenomenon. This is supported by the finding that the disulfide-linked oligomerization of 1a replication proteins of BMV or the 130K replication protein of tomato mosaic virus (genus: *Tobamovirus*; family: *Virgaviridae*) is critical for their functions to facilitate viral replication (Nishikiori et al., 2012). Similarly, the BaMV replication protein associates with the RBOHB protein, and downregulation of RBOHB attenuates BaMV infection (Lee et al., 2016). This finding suggests that RBOHB-derived ROS may have a proviral function in BaMV infection. However, BaMV replication relies on a host glutathione S-transferase (GST)

(Chen et al., 2013), which transfers the reduced glutathione (GSH) to a reactive electrophilic center of substrates for detoxification of toxic molecules or reduction of cellular oxidative stress (Gullner et al., 2018). BaMV recruits one isoform of GST of *N. benthamiana* (NbGSTU4) through GSH-dependent binding to the 3′ UTR of the viral genomic (+)RNA, to facilitate its replication (Chen et al., 2013). Exogenous supplementation with GSH or other reducing agents specifically stimulates a very early stage of BaMV replication (i.e., negative-strand RNA synthesis), supporting the conclusion that a reducing environment is favorable for viral (−)RNA synthesis. Alternatively, a dynamic reducing–oxidizing (redox) balance within BaMV ROs might exert some beneficial effects on BaMV replication, as proposed for Sindbis virus replication (Gullberg et al., 2015).

Most plant-infecting viruses depend on their vector organisms for interplant transmission (Whitfield et al., 2015). Viruses can attract insect vectors by manipulating plant defense signaling to facilitate virus transmission (Carr et al., 2018). Recent studies have provided evidence that some insect-transmissible plant-infecting viruses take advantage of host-derived ROS to facilitate viral plant-to-plant dissemination mediated by their vectors (Berthelot et al., 2019a,b; Guo et al., 2019a). TuMV appears to exploit the ROS generated during the plant immune signaling elicited by aphid infestation (Berthelot et al., 2019a) (Fig. 2). Aphid feeding triggers the generation of cellular ROS (Prince et al., 2014), which induce the cytosolic oxidization and oligomerization of TuMV HC-Pro via intermolecular disulfide bond. HC-Pro oligomers can bind to TuMV particles to form a viral transmissible complex that mediates vector binding (Berthelot et al., 2019a). The ROS-mediated oligomerization of HC-Pro may be conserved among potyviruses, because HC-Pro encoded by another potyvirus, lettuce mosaic virus, forms an intermolecular disulfide bond upon treatment with H_2O_2 (Berthelot et al., 2019a). ROS-induced HC-Pro oligomerization may represent an example of the functional switching of a multifunctional viral protein when and where the virus needs it. In fact, HC-Pro is a well-known multifunctional viral protein that is involved in the autoproteolytic cleavage of viral polyproteins, viral replication, viral long-distance movement, suppression of host RNA silencing, and vector transmission (Valli et al., 2018). ROS are also required for the aphid vector-mediated transmission of cauliflower mosaic virus (CaMV; genus: *Caulimovirus*; family: *Caulimoviridae*) (Berthelot et al., 2019b). However, the molecular mechanisms underlying the ROS-mediated enhancement of virus transmission may vary between TuMV and CaMV; the thiol-reactive compound *N*-ethylmaleimide, which prevents the formation of new disulfide bridges, inhibits TuMV HC-Pro oligomerization and TuMV transmission, while it has no effect on CaMV transmission (Berthelot et al., 2019a,b). The P2 protein of CaMV is the transmission helper component, which bridges CaMV particles to the aphid stylet, thereby mediating viral transmission to aphid vectors (Whitfield et al., 2015). P2 and virus particles are spatially separated: P2 is concentrated in specific cytoplasmic inclusions called transmission bodies, while most virus particles are localized within another type of viral inclusions (Bak et al., 2013; Martinière et al., 2013). Aphid infestation triggers the release of P2 and virus particles from the respective viral inclusions to microtubes, leading to the formation of the transmissible complex, which is the virus form that the aphid vector can acquire and transmit (Bak et al., 2013; Martinière et al., 2013). At present, the manner in which ROS stimulate CaMV vector transmission and whether ROS affect the formation of CaMV transmissible complex remain unknown. The SD strain of cucumber mosaic virus (CMV; genus: *Cucumovirus*; family: *Bromoviridae*) exploits *N. tabacum* RBOHD through the viral RSS 2b protein to induce ROS (Guo et al., 2019a) (Fig. 2). In turn, 2b-induced ROS lead to the impairment of aphid phloem feeding, which results in increased intracellular probing and aphid movement, thereby enhancing the probability of virus acquisition and transmission by aphid vectors (Guo et al., 2019a). RBOHD-mediated ROS production appears to be specifically involved in the vector-mediated transmission of CMV, because downregulation of RBOHD has no significant impact on CMV replication and infection within plants (Guo et al., 2019a). Although the manner in which 2b activates RBOHD warrants further investigation, this study provides a new insight into how plant viruses manipulate the behaviors of their insect vectors by exploiting ROS to facilitate viral transmission.

In addition to the examples mentioned above, the proviral action of ROS can be extended to other virus–host interactions. Several human-infecting RNA viruses, such as influenza A virus (IAV; genus: *Alphainfluenzavirus*; family: *Orthomyxoviridae*), human immunodeficiency virus 1 (genus: *Lentivirus*; family: *Retroviridae*), and DENV, induce endosomal superoxide production by activating NADPH oxidase NOX2, resulting in the inhibition of host antiviral signaling via the oxidative modification of the toll-like receptor 7, which recognizes invading viral RNAs (To et al., 2017). Members of the *Flavivirus* genus, including DENV2 and Kunjin virus, also induce ROS production in infected cells (Gullberg et al., 2015). H_2O_2 stimulates the DENV2 NS5 capping enzymatic activity and viral RNA replication by inducing oxidative modification of NS5 at Cys-140 and Met-219. In contrast, treatment with the antioxidant agent butylated hydroxyanisole (BHA) causes inhibition of the NS5 activity and, thus, viral RNA replication. Similarly, BHA treatment leads to impairment of the capping enzymatic activity of the nsP1 protein of Venezuelan equine encephalitis virus (genus: *Alphavirus*; family: *Togaviridae*), which results in the attenuation of viral replication (Gullberg et al., 2015). However, the replication of another alphavirus (Sindbis virus) is sensitive to both H_2O_2 and BHA, suggesting that the dynamic cellular redox balance during infection is critical for optimum viral replication (Gullberg et al., 2015). Regarding hepatitis B virus (HBV; genus: *Orthohepadnavirus*; family: *Hepadnaviridae*), the viral X protein triggers the generation of ROS originated from mitochondria, leading to the activation of c-Jun NH_2-terminal kinase signaling, which is beneficial for HBV DNA replication by enhancing autophagosome formation in infected human hepatocyte cells (Zhong et al., 2017).

3. Usurpation of cellular posttranslational dynamics during plant viral infections

A variety of cellular proteins undergo posttranslational modifications (PTMs), including phosphorylation, ubiquitination, methylation, acetylation, glycosylation, and SUMOylation, to be dynamically and properly regulated regarding their subcellular localization, activities, and/or abundance. PTMs are crucial for the rapid reprogramming of cellular signaling for a variety of biological processes, including immune response to invading bacterial and fungal pathogens; in turn, successful pathogens manipulate cellular PTMs to attenuate immunity (Withers and Dong, 2017). Viral proteins also undergo PTMs to increase their functional diversity during infection (Alcaide-Loridan and Jupin, 2012; Jakubiec and Jupin, 2007). In addition, similar to bacterial and fungal pathogens, plant viruses may target host PTM dynamics to circumvent the host antiviral immunity.

3.1 Fine control of viral protein dynamics using host kinases

Protein phosphorylation is a reversible posttranslational modification that plays a fundamental role in the regulation of many cellular processes. Plant viruses appear to hijack a variety of cellular kinases to fine-tune the functions of viral proteins and facilitate infection at multiple steps of their life cycles. For example, BSMV uses host casein kinase 2 (CK2) to phosphorylate the viral TGB1 movement protein at its Thr-401 residue, resulting in enhanced viral systemic infection (Hu et al., 2015). CK2-mediated phosphorylation appears to promote the assembly of the viral ribonucleoprotein complex by enhancing the TGB1–TGB3 interaction, which is essential for BSMV intercellular trafficking. Another BSMV-encoded protein, γb, undergoes phosphorylation at Ser-96 by a host cAMP-dependent protein kinase (PKA)-like kinase during viral infection (Zhang et al., 2018). This phosphorylation induces a two- to threefold increment in the siRNA-binding capacity of γb, which is required for efficient sequestration of BSMV-derived vsiRNAs from antiviral RISCs. In addition to RNA silencing, γb phosphorylation at Ser-96 appears to be crucial for attenuating salicylic acid (SA)-mediated host programmed cell death, which otherwise efficiently restricts viral spread before the establishment of infection. In the case of CMV, the tobacco (*N. tabacum*) calcineurin B-like calcium sensor-interacting protein kinase 12 phosphorylates viral 2a RdRP at the late stages of infection, leading to the inhibition/acceleration of the assembly/disassembly of the viral replicase complex, to switch from RNA replication to other steps, such as cell-to-cell movement and virion formation (Kang et al., 2012; Kim et al., 2002). Another CMV-encoded protein, 2b, also undergoes phosphorylation at Ser-40 and Ser-42 by an unidentified host kinase that regulates its RSS activity and virulence function by limiting its nucleus/cytoplasm shuttling (Nemes et al., 2017; Zhou et al., 2014). Tomato leaf curl Yunnan virus (TLCYnV; genus: *Begomovirus*; family: *Geminiviridae*) regulates the nucleus/cytoplasm shuttling of the viral C4 protein by hijacking glycogen synthase kinase 3/shaggy-like kinase η of *N. benthamiana* (SKη) (Mei et al., 2018a,b). C4 binds to SKη and the interaction induces the phosphorylation of the C4 protein at Thr-51 in the nucleus, which results in the noncanonical *N*-myristoylation of the viral protein. After these modifications, the C4–SKη complex binds to exportin-α and then translocates to the cytoplasm. The export of the C4–SKη complex from the nucleus causes the sequestration of host SKη from cyclin-dependent protein kinase D1;1 (CycD1;1), to prevent the SKη-triggered 26S proteasome-mediated degradation of CycD1;1 (Mei et al., 2018a,b). The increased accumulation of CycD1;1 induces an abnormal cell division in infected cells that is beneficial for TLCYnV. The induction of cell division promotes the expression of host DNA replication machineries in differentiated plant cells, which is a critical step for DNA virus replication (Hanley-Bowdoin et al., 2013). Multiple C4 proteins of other members of the family *Geminiviridae* also bind to host SK family members (Bi et al., 2017; Mills-Lujan et al., 2015; Piroux et al., 2007).

Additional examples highlight the role of phosphorylation in the assembly of viral particles. The CaMV capsid precursor protein undergoes phosphorylation by host CK2 kinase, which triggers its proteolytic processing to generate the mature form of CP (Champagne et al., 2007; Chapdelaine et al., 2002). In the case of beet black scorch virus (genus: *Betanecrovirus*; family: *Tombusviridae*), a host PKA-like kinase phosphorylates CP at Thr-41, which is a crucial step for the assembly of stable virions (Zhao et al., 2015). After virion assembly and intercellular movement, the regulation of viral particle stability is crucial for determining the timing of translation and replication in newly infected cells. PVX genomic RNA in virions appears to be tightly associated with CP and incompetence for translation. Phosphorylation of CP at the N-terminal residues by host PKA-like kinase or CK2 kinase reduces the RNA-binding activity of CP, which results in the release of the encapsidated viral RNA to initiate translation (Atabekov et al., 2001). Phosphorylation of BaMV CP by the host CK2 kinase likely occurs at PD, which promotes the translational activation of viral RNA in newly infected cells by reducing the interaction of CP with the encapsidated RNA (Hung et al., 2014). Similarly, the phosphorylation of BMV CP modulates the interaction between CP and viral RNAs within virions, and it regulates the timing of viral translation and replication by modulating virion stability (Hoover et al., 2016). In addition to being a building block of mature virions, CP has multiple functions and plays regulatory roles in various stages of the viral life cycle (Weber and Bujarski, 2015). PVA CP is important for switching from translation to replication through the co-translational recruitment of accumulated CP molecules (supplied in *trans*) to viral RNA, which allows the dissociation of translating ribosomes from viral RNA and the subsequent recruitment of viral RdRP (Besong-Ndika et al., 2015). CP appears to be removed from viral RNA before the onset of RNA synthesis (Lõhmus et al., 2017). PVA recruits host CK2 kinase to phosphorylate CP at Thr-243, which weakens the RNA-binding activity of CP and results in its dissociation from viral RNA (Ivanov et al., 2001, 2003; Lõhmus et al., 2017). The downregulation of host CK2 inhibits PVA infection, suggesting that the regulation of CP by co-opted CK2 appears to be critical for switching from viral translation to replication in an early stage of infection.

3.2 Viral manipulation of cellular phosphorylation dynamics

The target of rapamycin (TOR) protein is a highly conserved serine/threonine kinase that acts as a central hub to integrate nutrient and environmental signals (Xiong and Sheen, 2015). TOR regulates various cellular processes, including mRNA translation, energy production, metabolism, and autophagy. Several lines of evidence have suggested the link between plant virus infection and TOR. For example, the CaMV P6/TAV (transactivator/viroplasmin) protein binds to TOR and mediates its activation, leading to the phosphorylation of the ribosomal protein S6 kinase 1 (S6K1) (Schepetilnikov et al., 2011). The phosphorylated S6K1 then activates host translation factors, including the re-initiation supporting protein and subunit h of eIF3, to facilitate the translation re-initiation of viral pregenomic RNA (Schepetilnikov et al., 2011). P6-mediated activation of TOR also results in the attenuation of autophagy induced by the plant defense-related phytohormone, SA, which would be beneficial for CaMV (Zvereva et al., 2016). Downregulation or pharmacological inhibition of TOR attenuates infections by two unrelated plant viruses, CNV and watermelon mosaic virus (WMV; genus: *Potyvirus*; family: *Potyviridae*) (Inaba and Nagy, 2018; Ouibrahim et al., 2015), further supporting the proviral function of TOR in plant virus infection. Notably, CNV infection leads to partial impairment of cellular TOR activity, likely because of the reduced cytosolic ATP concentrations caused by the hijacking of host glycolytic enzymes during virus replication (see Section 4), although such mild inhibition of TOR activity appears to not be adverse for virus replication (Inaba and Nagy, 2018). As TOR plays versatile roles in various aspects of the life cycle of plants, it would be interesting to investigate whether virus-mediated alternations of the TOR activity exert some effects on the physiological changes in plant cells caused by plant virus infections.

Phosphorylation dynamics are crucial for the regulation of plant immune responses to bacterial and fungal pathogens (Withers and Dong, 2017). Among them, the mitogen-activated protein kinase (MAPK)-mediated phosphorylation cascades play a central part in the transduction of immune signaling that mediates transcriptional reprogramming for defense (Meng and Zhang, 2013). Although a previous report showed the activation of MAPKs in an incompatible TMV–tobacco interaction (Zhang and Klessig, 1998), our understanding of their roles in plant virus infection is just beginning to unravel. In the case of BaMV, MAPK phosphatase 1 (MKP1) has been detected in an affinity-purified fraction of REP_{BaMV} in *N. benthamiana* (Lee et al., 2016). The downregulation of MKP1, which dephosphorylates MAPK to downregulate its activity, enhances BaMV and related FoMV replication in *N. benthamiana*, while the overexpression of MKP1 inhibits viral infection in a phosphatase-activity-dependent manner (Lee et al., 2018). The proposed function of MKP1 in BaMV and

FoMV replication is the downregulation of viral RNA replication by attenuating MAPK activity; i.e., the activation of MAPK signaling might be beneficial for viral replication. The manner in which MKP1 (and possibly MAPK) is involved in BaMV replication and whether BaMV modulates MAPK activity through MKP1 warrant further investigation. Conversely, MAPK signaling appears to be important for antiviral defense responses to TYLCCNV (Hu et al., 2019). *A. thaliana* and *N. benthamiana* can trigger MAPK activation in response to TYLCCNV infection through an unknown mechanism, and loss of MPK4 (a MAPK isoform) or its upstream kinase, MKK2, leads to enhanced viral infection and severe symptoms, supporting the antiviral role of the MKK2–MPK4 branch. Importantly, the βC1 protein of the TYLCCNV betasatellite binds directly to both MKK2 and MPK4 and inhibits their kinase activities, to attenuate the MKK2–MPK4-mediated plant defense response. These two examples suggest a dual role for MAPK signaling in virus replication and plant defense, although further studies are needed to elucidate the exact role(s) of MAPK signaling in various plant–virus interactions.

3.3 Ubiquitination machineries

Ubiquitin is a 76-amino-acid, highly conserved polypeptide that becomes posttranslationally attached to other intracellular proteins via a dedicated conjugation cascade, ultimately resulting in the link between one or more ubiquitin molecules and accessible lysine residues within the target protein (referred to as ubiquitination) (Miricescu et al., 2018; Xu and Xue, 2019). Ubiquitin is transferred by the ubiquitin-activating enzyme (E1) to the ubiquitin-conjugating enzyme (E2), and is then conjugated to the target through the action of the E3 ubiquitin ligase. Once attached, the ubiquitin moiety(ies) can dramatically impact the target, including altering its conformation, function, activity, and/or subcellular localization. The conjugation of ubiquitin polymers with the substrate (polyubiquitination) leads to its rapid degradation by the 26S proteasome with the release of the ubiquitin and its recycling. Ubiquitylation is a highly dynamic process that can be reversed by enzymes known as deubiquitinases or deubiquitylating enzymes (DUBs). Most of these enzymes are Cys proteinases, which can either trim or remove poly-ubiquitin linkages from the substrate, thus contributing to the reversal of the cellular ubiquitin-dependent system. The ubiquitin system in plants is widely recognized as a crucial hub for adaptation to multiple environmental stresses, including viral infection (Miricescu et al., 2018;

Verchot, 2016; Xu and Xue, 2019). Therefore, it is not surprising that successful viruses have evolved that divert the host ubiquitin system and integrate it in the viral life cycle by multiple ways. Such mechanisms may be beneficial to the virus, by creating a more favorable cellular environment, by fine-tuning viral regulatory processes, or by attenuating the host defense mechanisms. For example, the TBSV p33 replication protein binds to Cdc34p and Rad6p, which are yeast E2 ubiquitin-conjugating enzymes, to recruit them into TBSV ROs (Imura et al., 2015; Li et al., 2008). UBC2, an *Arabidopsis* counterpart of yeast Rad6p, also interacts with p33 (Imura et al., 2015). Rad6p/UBC2 appears to be able to transfer ubiquitin to the substrate without the help of E3 ubiquitin ligase. The binding of p33 to Cdc34p and Rad6p/UBC2 mediates the mono- and bi-ubiquitination of p33. The ubiquitinated forms of p33 then recruit the host ESCRT complex to ROs to modify the structure of the peroxisomal boundary membrane, where TBSV replication takes place (Barajas and Nagy, 2010; Barajas et al., 2009, 2014; Imura et al., 2015; Kovalev et al., 2016, 2017). The replication initiator protein (Rep) of chili leaf curl virus (genus: *Begomovirus*; family: *Geminiviridae*) binds to host UBC2 and histone monoubiquitination 1 to promote H3K4me3 on viral chromatin, which results in the enhanced transcription of viral genes (Kushwaha et al., 2017). The 98K protein encoded by turnip yellow mosaic virus (TYMV; genus: *Tymovirus*; family: *Tymoviridae*) is a functional ovarian tumor-like DUB that mediates the removal of polyubiquitin moieties from TYMV 66K RdRP to prevent its degradation through the 26S proteasome (Camborde et al., 2010; Chenon et al., 2012). A structure-based mutagenesis analysis indicated that the DUB activity of the TYMV 98K protein plays a crucial role in TYMV replication and in systemic infection (Jupin et al., 2017), thus providing an example of a viral strategy that manipulates the host ubiquitin–proteasome degradation system via a virally encoded DUB, to facilitate viral RNA replication.

3.4 Viral manipulation of host ubiquitination machineries

Skp1/Cullin/F-box (SCF)-type E3 ubiquitin ligase complexes consist of a Skp1 adaptor protein, a Cullin scaffold protein, an F-box protein, and a Ring-box protein that binds to the E2 ubiquitin-conjugating enzyme (Miricescu et al., 2018). The F-box protein confers substrate specificity to the complex. Upon association with a substrate protein, the SCF complex attaches monoubiquitin or polyubiquitin chains to the substrate. As SCF complexes regulate several key cellular processes, including RNA silencing, the hijacking of the host ubiquitin systems by multiple ways would be helpful for plant viruses to facilitate their infection. The P0 protein of turnip yellows virus (TuYV; genus: *Polerovirus*; family: *Luteoviridae*) and the related Brassica yellows virus (BrYV) is a virus-encoded F-box protein that hijacks the host SCF E3 ubiquitin ligase complex through an interaction with SKP1, to promote the autophagic degradation of AGO1 (Baumberger et al., 2007; Bortolamiol et al., 2007; Csorba et al., 2010; Derrien et al., 2012, 2018; Michaeli et al., 2019; Pazhouhandeh et al., 2006). The BrYV P0–SKP1 interaction is important for the increased stability of the BrYV P0 protein, which results in strong RNA silencing suppression, thereby promoting robust virus infections (Li et al., 2019b). The BNYVV p25 protein appears to regulate the assembly of the host SCF complex via an interaction with an F-box family protein, to attenuate plant antiviral defenses (Thiel et al., 2012). The p22 RSS protein of the cucurbit chlorotic yellows virus (genus: *Crinivirus*; family: *Closteroviridae*) and the P7-2 protein of rice black-streaked dwarf virus (RBSDV; genus: *Fijivirus*; family: *Reoviridae*) is a potential F-box-like protein that can be incorporated into host SCF complexes via an interaction with the SKP1 protein (Chen et al., 2019; Tao et al., 2017; Wang et al., 2013). Very recently, it was shown that a (−)ssRNA virus, rice grassy stunt virus (RGSV; genus: *Tenuivirus*; family: *Phenuiviridae*), induces the expression of a rice E3 ubiquitin ligase-encoding gene via the action of the virally encoded P3 protein, resulting in enhanced virus infection and virulence through the 26S-dependent degradation of NRPD1, which is a core subunit of host RNA polymerase IV that mediates RNA-directed DNA methylation (RdDM) in rice (Zhang et al., 2020). The manner in which the suppression of RdDM contributes to the enhancement of RGSV infection and whether RdDM generally mediates antiviral activity against plant RNA viruses warrant further investigation.

Notably, multiple phytohormone signaling pathways, including those of auxin, gibberellic acid (GA), jasmonic acid (JA), and strigolactone, are regulated by SCF complexes (Blázquez et al., 2020). The perception of an active form of JA by an F-box protein, coronatine insensitive 1 (COI1), triggers the rapid polyubiquitination and degradation of transcription repressors, i.e., the Jasmonate ZIM domain (JAZ) family of proteins. Upon JAZ degradation, activation of transcriptional programs in response to JA can occur. The βC1 protein encoded by betasatellite of cotton leaf curl Multan virus (genus: *Begomovirus*; family: *Geminiviridae*) interacts with SKP1 to inhibit the function of SCF^{COI1} and SCF^{SLY1}, which results in perturbation of

JA- and GA-mediated antiviral signaling and enhanced viral infection (Jia et al., 2016). The SCF complexes are regulated by covalent attachment and removal of a ubiquitin-like protein (referred to as RUB), mediated by the COP9 signalosome (CSN) complex, which consists of eight subunits (CSN1 to CSN8) (Schwechheimer, 2018). The RBSDV-encoded P5-1 protein interacts with the CSN5A subunit and alters the deRUBylation patterns of the Cullin subunit of the SCF complexes (He et al., 2020). The action of P5-1 causes impairment of SCF^{COI1}, which leads to the suppression of the JA-mediated antiviral signaling pathway to facilitate virus infection. Similarly, the C2 proteins of tomato yellow leaf curl Sardinia virus (TYLCSV; genus: *Begomovirus*; family: *Geminiviridae*) and beet curly top virus (genus: *Curtovirus*; family: *Geminiviridae*) interfere with the JA- and GA-mediated signaling pathways by interacting with the deRUBylation activity of the CSN5 subunit (Lozano-Durán et al., 2011). In addition to interfering with deRUBylation, the C2 proteins of some members of monopartite begomovirus can bind to and sequester ubiquitin, which results in the reduced ubiquitin-dependent degradation of the JAZ1 transcription repressor, thereby preventing the JA-mediated induction of terpene biosynthesis (Li et al., 2019a). The reduced production of terpene and other JA-regulated metabolites causes increased whitefly performance, thereby helping whitefly-mediated viral transmission. The other examples of the viral manipulation of JA signaling in tripartite plant–virus–insect vector interactions can be found in recent reviews (Wu and Ye, 2020; Ziegler-Graff, 2020).

3.5 SUMOylation enzymes

Small ubiquitin-related modifier (SUMO) is an ~100-amino-acid polypeptide that is structurally related to ubiquitin. Like other PTMs, conjugation of SUMO to cellular proteins can dynamically regulate its function (Augustine and Vierstra, 2018). Plants typically express several isoforms of SUMO, and SUMO3 of *Arabidopsis* participates in plant immunity through the SUMOylation of the nonexpressor of PR1 (NPR1), an important immune regulator in plants (Saleh et al., 2015). SUMO3-mediated SUMOylation switches the function of NPR1 from transcription repressor to transcription activator, thereby inducing defense-related transcriptional reprogramming during pathogen invasion. TuMV RdRP NIb binds to the host E2 SUMO-conjugating enzyme SCE1, which can conjugate SUMO to target proteins (Xiong and Wang, 2013). The SCE1-binding domain of TuMV NIb has been mapped, and the amino acid sequence of this domain is highly conserved among members of the *Potyvirus* genus (Xiong and Wang, 2013). During TuMV infection, SCE1 induces NIb SUMOylation by SUMO3 to facilitate virus replication (Cheng et al., 2017; Xiong and Wang, 2013). SUMOylation appears to regulate the subcellular localization of the NIb protein. Similar to other (+)ssRNA viruses, potyviruses replicate in the cytoplasm. However, NIb contains a functional nuclear localization signal (NLS) and predominantly accumulates in the nucleus during infection (Restrepo et al., 1990). Disruption of the NIb NLS compromises viral accumulation, suggesting that the nuclear–cytoplasm trafficking of NIb functions in viral replication (Li et al., 1997). The SCE1-mediated SUMO3 conjugation appears to function in NIb export from the nucleus, leading to the formation of ROs in the cytoplasm (Cheng et al., 2017). NIb appears to compete with NPR1 for SUMO3-mediated SUMOylation in the nucleus, leading to the suppression of NPR1-mediated plant immunity (Cheng et al., 2017). The proviral function of SCE1 can be extended to other plant virus infections, as downregulation of *SCE1* decreases TMV accumulation in *N. benthamiana* (Xiong and Wang, 2013). In addition, several lines of evidence indicate that some members of the family *Geminiviridae* also take advantage of the host SUMOylation machinery by targeting the SCE1 protein. Geminiviruses do not encode their own DNA polymerases and, therefore, depend on the host DNA replication machinery for viral DNA replication (Hanley-Bowdoin et al., 2013). The viral replication initiator protein (Rep) is essential for the reprogramming of the cell cycle and subsequent initiation of viral DNA replication by interacting with a range of host components related to host DNA replication (Hanley-Bowdoin et al., 2013). One of the host proteins that facilitate viral DNA replication is the DNA clamp protein proliferating cell nuclear antigen (PCNA). The function of PCNA can be modulated by SUMO conjugation, which affects the interaction of PCNA with its binding partners (Choe and Moldovan, 2017). Rep proteins encoded by TGMV and TYLCSV have been shown to bind to SCE1 (Castillo et al., 2004; Sánchez-Durán et al., 2011). The interaction between Rep and SCE1 does not alter the global SUMOylation pattern in plants; rather, it interferes specifically with the SUMO1/2 conjugation of selected host proteins, such as PCNA (Arroyo-Mateos et al., 2018; Sánchez-Durán et al., 2011). TGMV and TYLCSV appear to perturb the SCE1-mediated SUMOylation of PCNA, thereby modulating its cellular function to facilitate viral DNA replication (Arroyo-Mateos et al., 2018).

4. Hijacking of host glycolytic enzymes for ATP production

As viruses lack a self-containing metabolism, they rely completely on their host cells to ensure energy and components for their replication and infection. To accomplish the energy costs of the replication of their genome and other processes, viruses have evolved to develop multiple strategies to induce cellular metabolic pathways. Many human-infecting viruses, such as DENV and human herpesvirus 5 (genus: *Cytomegalovirus*; family: *Herpesviridae*), upregulate cellular glucose uptake and aerobic glycolysis, which converts glucose to lactate (in animals) or ethanol (in plants/yeast) in the presence of plentiful glucose and oxygen, to produce a wealth of energy rapidly (in the form of ATP, NADH, and NADPH), as well as the core building blocks of biomolecules, including nucleotides, amino acids, lipids, and carbohydrates (Goodwin et al., 2015; Sanchez and Lagunoff, 2015). However, until recently, the manner in which plant-infecting viruses fulfill their energy demands for infection has remained largely unknown.

Recent elegant studies taking advantage of ATP biosensors have revealed the molecular mechanism via which plant-infecting (+)ssRNA viruses induce local ATP biosynthesis at the vicinity of ROs (reviewed in Nagy and Lin, 2020). TBSV p33 and p92^{pol} bind to two host glycolytic ATP-generating enzymes, phosphoglycerate kinase (PGK1) and pyruvate kinase (PK), leading to their recruitment into the site of viral replication (Chuang et al., 2017; Prasanth et al., 2017). Although PGK1 is released from TBSV replication proteins upon replicase complex assembly, PK seems to be retained in ROs. The co-opted PGK1 appears to supply ATP for heat-shock protein 70 (HSP70) activation during the assembly of the replicase complex. HSP70 plays multiple roles in the TBSV life cycle, including the stabilization and subcellular localization of viral replication proteins, the formation of the replicase complex, the activation of p92^{pol} RdRP, and the assembly and disassembly of viral particles (Alam and Rochon, 2015, 2017; Pogany and Nagy, 2015; Pogany et al., 2008; Serva and Nagy, 2006; Wang et al., 2009a,b). Conversely, the ATP generated by PK seems to be important for the activation of the co-opted DEAD-box helicases, including DED1, which unwind the local structure of the 3′ end of viral (−)RNA to facilitate (+)RNA synthesis (Chuang et al., 2015; Kovalev and Nagy, 2014;

Kovalev et al., 2012a,b). To recruit PK efficiently to the close vicinity of DED1, TBSV usurps the RPN11 deubiquitinase/metallocaspase, another proviral host factor for TBSV replication (Prasanth et al., 2015); PK, RPN11, and DED1 form a trimeric complex in yeast that is likely to be recruited by TBSV through a $p92^{pol}$–RPN11 interaction. In RPN11-deficient yeast, TBSV fails to recruit PK and DED1 into ROs, which results in decreased RNA synthesis (Chuang et al., 2017; Prasanth et al., 2015).

A nuclear-encoded chloroplast phosphoglycerate kinase (cpPGK) has been shown to be involved in BaMV replication (Cheng et al., 2013; Lin et al., 2007). cpPGK is translated in the cytoplasm and contains an N-terminal transit peptide that carries chloroplast-targeting information. BaMV recruits cpPGK to viral RNA in the cytoplasm via binding to the 3′ UTR and is transported by cpPGK to the chloroplast membrane, where it builds ROs (Cheng et al., 2013). In this case, the enzymatic activity of cpPGK (i.e., ATP production) is likely to be irrelevant to BaMV replication, because artificial chloroplast-targeting eEF1A, which is known to bind to the 3′ UTR of BaMV RNA, can near-fully complement cpPGK deficiency in BaMV replication (Cheng et al., 2013). PGK is also involved in infections of several plant viruses, including RSV, WMV, plum pox virus (PPV; genus: *Potyvirus*; family: *Potyviridae*), and sugarcane mosaic virus (genus: *Poacevirus*; family: *Potyviridae*) (Bi et al., 2019; Chen et al., 2017c; Ouibrahim et al., 2014; Poque et al., 2015). Whether these viruses require the ATP-generating activity of PGK remains unknown.

TBSV also exploits the two host fermentation pathway enzymes (pyruvate decarboxylase 1 (PDC1) and alcohol dehydrogenase 1 (ADH1)) to circumvent the glycolysis arrest caused by NAD^+ shortage via the regeneration of the NAD^+ pool from NADH (Lin et al., 2019). TBSV induces the expression of the *PDC1* and *ADH1* genes during infection and hijacks these enzymes for efficient replenishment of NAD^+ at ROs. This strategy is unlikely to be limited to TBSV, as downregulation of PDC1 and ADH1 attenuates the replication of several plant (+)ssRNA viruses, including CNV, CIRV, TCV, TMV, and BaMV (Lin et al., 2019). Similar to TBSV, these viruses induce the expression of PDC1 and ADH1 during infection; however, the manner in which they induce a fermentative transcriptional program remains unknown. It is worth mentioning here that the group VII ethylene responsive factor family of transcription factors have been shown recently to be involved in the expression of fermentative genes in response to herbicides in *Arabidopsis* grown under aerobic conditions (Gil-Monreal et al., 2019).

5. Viral manipulation of intracellular ion homeostasis

Regulation of ion homeostasis appears to be crucial for virus infection at various stages. For example, a low pH condition has been postulated to be favorable to the assembly of TYMV virions (Rohozinski and Hancock, 1996). Protons could affect the structure and function of cellular proteins, as well as the electrical potential across the membrane (Casey et al., 2010). In the case of BaMV, the efficiency of viral RNA synthesis can be modulated by the proton concentration at ROs, which may be optimized in part by a nuclear-encoded chloroplast-resident carbonic anhydrase (Chen et al., 2017b). The GEF1p Cl^-/H^+ antiporter affects TBSV replication by keeping the organelle membrane potential to maintain the activity of the Ccc2p copper pump, which mediates Cu^{2+} uptake into the organelle lumen (Sasvari et al., 2013). Increased cytoplasmic Cu^{2+} accumulation may be toxic to viruses that replicate in the cytoplasm, because Cu^{2+} ions strongly hinder viral RdRP activities (Sasvari et al., 2013). It has been shown that many human-infecting RNA viruses, such as IAV and HCV, as well as several DNA viruses, manipulate cellular ion homeostasis via virus-encoded pore-forming small transmembrane proteins, called viroporins, to render cells more favorable for virus infection (Nieto-Torres et al., 2015). These studies illustrate the importance of intracellular ion homeostasis in various virus–host interactions. In plant viruses, the BMV 1a protein has been shown recently to possess a viroporin-like activity to permeabilize membranes, which results in the release of oxidant agents from the ER lumen into ROs, to facilitate intermolecular disulfide bond formation between viral replication proteins (Nishikiori and Ahlquist, 2018). However, if and how plant-infecting viruses manipulate host ion homeostasis remain largely unknown.

A study reported recently by Sun and colleagues demonstrated the molecular interaction between a plant virus and a host ion transporter (Sun et al., 2018) (Fig. 3). PVY appears to manipulate intracellular pH through the hijacking of a host proton transporter, CLC-Nt1 (in *N. tabacum*) or CLC-Nb1a/b (in *N. benthamiana*), which is an ER-resident putative Cl^-/H^+ antiporter that functions in maintaining ER luminal alkalization via the transport of protons from the ER lumen to the cytoplasm. The PVY 6K2 protein induces an increment in ER luminal pH through an interaction with CLC-Nb1a/b without affecting its gene expression, suggesting that PVY boosts its transporter activity to alkalinize the ER lumen. CLC-Nb1a/b

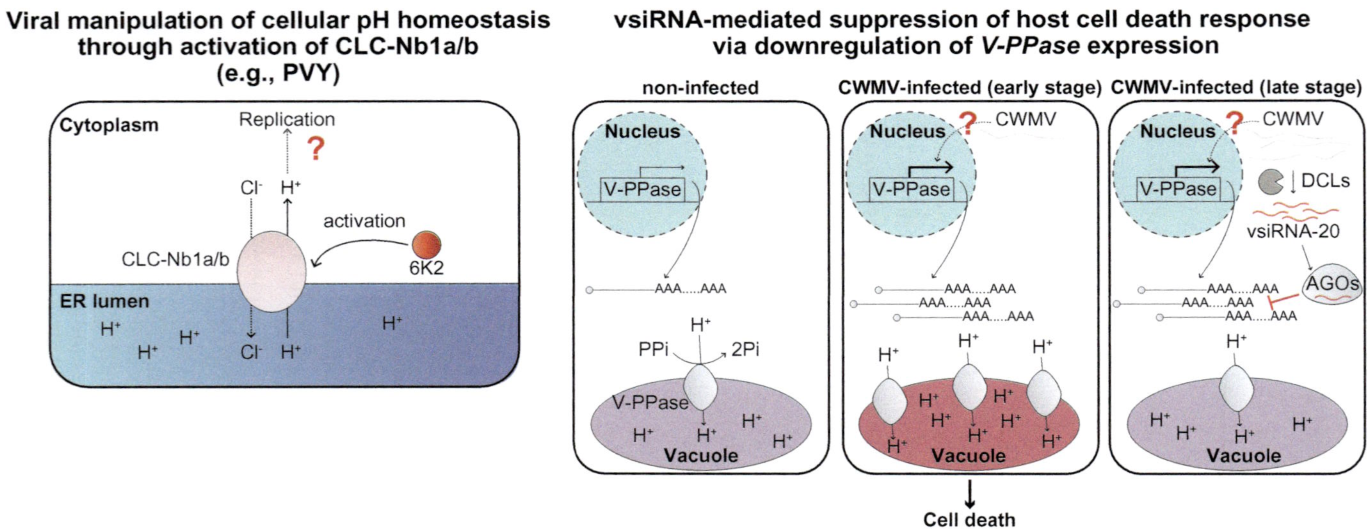

Fig. 3 Viral manipulation of cellular pH homeostasis. Red blunted lines represent negative regulation. The solid line with an arrow represents expression of the *V-PPase* gene. CLC, chloride channel; vsiRNA, virus-derived small interference RNA; V-PPase, vacuolar-type pyrophosphatase; DCL, Dicer-like; AGO, Argonaute; PVY, potato virus Y; CWMV, Chinese wheat mosaic virus.

promotes PVY replication and systemic infection in a proton-transport-activity-dependent manner. Although the molecular mechanism via which the ER luminal pH regulates PVY replication remains unclear, this work provides an example of the subversion of a host proton transporter to manipulate cellular ion homeostasis in a manner that is suitable for viral replication. Conversely, a vacuolar-type pyrophosphatase (V-PPase) has been shown recently to be an antiviral factor during infection with Chinese wheat mosaic virus (CWMV; genus: *Furovirus*; family: *Virgaviridae*) (Yang et al., 2020) (Fig. 3). V-PPase is a membrane-bound transporter that uses inorganic PPi as a driving force to transport protons from the cytoplasm to the vacuolar lumen (Graus et al., 2018). Overaccumulation of V-PPase disrupts the membrane potential of the plasma membrane, leading to cell death (Graus et al., 2018). Downregulation of V-PPase expression in wheat (*Triticum aestivum*) enhances CWMV genomic RNA accumulation, whereas ectopic expression of wheat V-PPase in *N. benthamiana* induces cell death, thus attenuating CWMV infection. V-PPase expression is highly induced at the early stage of CWMV infection, indicating that upregulation of V-PPase expression may be a plant defense response to restrict virus infection by inducing cell death. Interestingly, one of CWMV-derived vsiRNAs, i.e., vsiRNA-20, has a complementary sequence to the wheat V-PPase mRNA and accumulates abundantly at the late stage of infection. vsiRNA-20 is incorporated into host AGO proteins and guides the posttranscriptional cleavage of the wheat V-PPase mRNA, which results in the acidification of the cytoplasm and reduced PPi hydrolysis activity in infected cells (Fig. 3). This virus-induced host V-PPase silencing appears to be important for manipulating cytoplasmic pH and PPi homeostasis, to facilitate CWMV infection by preventing a host defense response controlled by V-PPase-mediated cellular ion homeostasis (Yang et al., 2020).

6. Conclusions

Recent advances in the field of molecular plant–virus interactions have provided fruitful insights into how plant-infecting viruses manipulate host cells. They provide compelling evidence showing that plant viruses can convert hostile cellular components to proviral forms. As discussed in this review, versatile viral hijacking strategies that usurp host components related to antiviral defense, such as RNA silencing, RNA decay, autophagy, and ROS production, represent remarkable examples. Moreover, we are beginning to understand how plant viruses manipulate host cells to create a beneficial environment for virus infection and dissemination. Emerging key cellular pathways include PTMs, glycolysis/fermentation, and ion homeostasis. Despite such progresses, our understanding of plant–virus interactions seems to be far from complete and is limited to several model pathosystems. Although PTMs appear to play a central part in diverse plant–virus interactions, the identification of key viral PTMs and the characterization of cellular modification enzymes remains challenging. Recently developed powerful techniques, such as proximity-labeling of protein complexes in planta (Mair et al., 2019; Zhang et al., 2019a), would be helpful to detect such transient or weak protein–protein interactions. We also rarely know how interactions of plant virus proteins with cellular proviral components affect the host proteome and metabolome, which ultimately affect plant fitness and survival. As plant defense responses are postulated to have a high metabolic requirement (Huot et al., 2014; Smakowska et al., 2016), viral hijacking of host glycolytic enzymes and the accompanying cytosolic ATP reduction might overtax the limited resources of plants. It will be of interest in the future to design experiments that elucidate the consequences of the viral manipulation of proviral host factors on plant growth/development under multiple environmental stress conditions. Further characterization and in-depth study of host metabolic enzymes across diverse plant–virus pathosystems is an interesting research direction that would be helpful for obtaining a complete picture of the plant–virus molecular arms race.

One major driving force of studies of proviral host factors comes from their potential applications for the creation of novel virus-resistant crops. Theoretically, mutations that lead specifically to the loss of the proviral function without any obvious adverse impact on crop yield and quality can confer genetic resistance traits to a susceptible cultivar that is governed by loss-of-susceptibility (a form of genetically recessive resistance) (Garcia-Ruiz, 2019; Hashimoto et al., 2016). However, at present, our genetic resources for recessive resistance remain limited. To make matters worse, recessive resistance based on single-gene mutations seems not to be durable because of the frequent emergence of resistance-breaking viral isolates (Kobayashi et al., 2014), which is caused in part by the fast-evolving feature of viral genomes. Superimposing multiple recessive resistance traits would be a valid and feasible way to overcome these challenges. The identified proviral host factors presented in this review may represent a valuable arsenal for the development of novel robust "pyramiding" resistance that would reduce the opportunity for the emergence of resistance-breaking isolates. However, the emerging viral strategies that hijack cellular components related to the antiviral defense response and/or cellular homeostasis would pose difficult challenges toward the development of down-to-earth virus-resistant cultivars in fields, where numerous biotic/abiotic stresses threaten crop yield and quality. Nevertheless, because of the indispensability of proviral host factors for the plant virus life cycle, they apparently represent a viral Achilles' heel and provide attractive breeding targets for the management of viral diseases. Untangling the complex web of molecular plant–virus interactions combined with an in-depth dissection of the roles of the identified proviral host factors in the plant life cycle under various conditions would definitely contribute to the increase of potential resources for a better management of plant virus diseases.

Acknowledgments

The authors would like to thank Drs. Nobuhiro Suzuki and Hideki Kondo (IPSR, Okayama University) for their invaluable comments and suggestions. The authors apologize to those colleagues whose work could not be discussed in this review because of space limitations. This work is supported by JSPS KAKENHI grant 19K15841 and Grants-in-Aid for Scientific Research on Innovative Areas from the Ministry of Education, Culture, Science, Sports and Technology (MEXT) of Japan 16H06429, 16K21723, and 19H04836.

References

Abernathy, E., Mateo, R., Majzoub, K., van Buuren, N., Bird, S.W., Carette, J.E., Kirkegaard, K., 2019. Differential and convergent utilization of autophagy components by positive-strand RNA viruses. PLoS Biol. 17 (1), e2006926.

Adachi, H., Yoshioka, H., 2015. Kinase-mediated orchestration of NADPH oxidase in plant immunity. Brief. Funct. Genomics 14 (4), 253–259.

Alam, S.B., Rochon, D., 2015. Cucumber necrosis virus recruits cellular heat shock protein 70 homologs at several stages of infection. J. Virol. 90 (7), 3302–3317.

Alam, S.B., Rochon, D., 2017. Evidence that Hsc70 is associated with cucumber necrosis virus particles and plays a role in particle disassembly. J. Virol. 91 (2), e01555-16.

Alcaide-Loridan, C., Jupin, I., 2012. Ubiquitin and plant viruses, let's play together!. Plant Physiol. 160 (1), 72–82.

Anandalakshmi, R., Marathe, R., Ge, X., Herr Jr., J.M., Mau, C., Mallory, A., Pruss, G., Bowman, L., Vance, V.B., 2000. A calmodulin-related protein that suppresses posttranscriptional gene silencing in plants. Science 290 (5489), 142–144.

Andika, I.B., Kondo, H., Sun, L., 2016. Interplays between soil-borne plant viruses and RNA silencing-mediated antiviral defense in roots. Front. Microbiol. 7, 1458.

Arroyo-Mateos, M., Sabarit, B., Maio, F., Sánchez-Durán, M.A., Rosas-Díaz, T., Prins, M., Ruiz-Albert, J., Luna, A.P., van den Burg, H.A., Bejarano, E.R., 2018. Geminivirus replication protein impairs SUMO conjugation of proliferating cellular nuclear antigen at two acceptor sites. J. Virol. 92 (18), e00611-18.

Atabekov, J.G., Rodionova, N.P., Karpova, O.V., Kozlovsky, S.V., Novikov, V.K., Arkhipenko, M.V., 2001. Translational activation of encapsidated potato virus X RNA by coat protein phosphorylation. Virology 286 (2), 466–474.

Augustine, R.C., Vierstra, R.D., 2018. SUMOylation: re-wiring the plant nucleus during stress and development. Curr. Opin. Plant Biol. 45 (Pt. A), 143–154.

Bak, A., Gargani, D., Macia, J.L., Malouvet, E., Vernerey, M.S., Blanc, S., Drucker, M., 2013. Virus factories of cauliflower mosaic virus are virion reservoirs that engage actively in vector transmission. J. Virol. 87 (22), 12207–12215.

Barajas, D., Nagy, P.D., 2010. Ubiquitination of tombusvirus p33 replication protein plays a role in virus replication and binding to the host Vps23p ESCRT protein. Virology 397 (2), 358–368.

Barajas, D., Jiang, Y., Nagy, P.D., 2009. A unique role for the host ESCRT proteins in replication of Tomato bushy stunt virus. PLoS Pathog. 5 (12), e1000705.

Barajas, D., Martín, I.F., Pogany, J., Risco, C., Nagy, P.D., 2014. Noncanonical role for the host Vps4 AAA+ ATPase ESCRT protein in the formation of Tomato bushy stunt virus replicase. PLoS Pathog. 10 (4), e1004087.

Baumberger, N., Tsai, C.H., Lie, M., Havecker, E., Baulcombe, D.C., 2007. The Polerovirus silencing suppressor P0 targets ARGONAUTE proteins for degradation. Curr. Biol. 17 (18), 1609–1614.

Beckham, C.J., Light, H.R., Nissan, T.A., Ahlquist, P., Parker, R., Noueiry, A., 2007. Interactions between brome mosaic virus RNAs and cytoplasmic processing bodies. J. Virol. 81 (18), 9759–9768.

Berthelot, E., Ducousso, M., Macia, J.L., Bogaert, F., Baecker, V., Thébaud, G., Gallet, R., Yvon, M., Blanc, S., Khelifa, M., Drucker, M., 2019a. Turnip mosaic virus is a second example of a virus using transmission activation for plant-to-plant propagation by aphids. J. Virol. 93 (9), e01822-18.

Berthelot, E., Macia, J.L., Martinière, A., Morisset, A., Gallet, R., Blanc, S., Khelifa, M., Drucker, M., 2019b. Pharmacological analysis of transmission activation of two aphid-vectored plant viruses, turnip mosaic virus and cauliflower mosaic virus. Sci. Rep. 9 (1), 9374.

Besong-Ndika, J., Ivanov, K.I., Hafrèn, A., Michon, T., Mäkinen, K., 2015. Cotranslational coat protein-mediated inhibition of potyviral RNA translation. J. Virol. 89 (8), 4237–4248.

Bi, H., Fan, W., Zhang, P., 2017. C4 protein of *sweet potato leaf curl virus* regulates brassinosteroid signaling pathway through interaction with AtBIN2 and affects male fertility in *Arabidopsis*. Front. Plant Sci. 8, 1689.

Bi, J., Yang, Y., Chen, B., Zhao, J., Chen, Z., Song, B., Chen, J., Yan, F., 2019. Retardation of the calvin cycle contributes to the reduced CO_2 assimilation ability of rice stripe virus-infected *N. benthamiana* and suppresses viral infection. Front. Microbiol. 210, 568.

Bird, S.W., Maynard, N.D., Covert, M.W., Kirkegaard, K., 2014. Nonlytic viral spread enhanced by autophagy components. Proc. Natl. Acad. Sci. U. S. A. 111 (36), 13081–13086.

Blasco-Moreno, B., de Campos-Mata, L., Böttcher, R., García-Martínez, J., Jungfleisch, J., Nedialkova, D.D., Chattopadhyay, S., Gas, M.E., Oliva, B., Pérez-Ortín, J.E., Leidel, S.A., Choder, M., Díez, J., 2019. The exonuclease Xrn1 activates transcription and translation of mRNAs encoding membrane proteins. Nat. Commun. 10 (1), 1298.

Blázquez, M.A., Nelson, D.C., Weijers, D., 2020. Evolution of plant hormone response pathways. Annu. Rev. Plant Biol. 71. https://doi.org/10.1146/annurev-arplant-050718-100309, in press.

Bortolamiol, D., Pazhouhandeh, M., Marrocco, K., Genschik, P., Ziegler-Graff, V., 2007. The Polerovirus F box protein P0 targets ARGONAUTE1 to suppress RNA silencing. Curr. Biol. 17 (18), 1615–1621.

Camborde, L., Planchais, S., Tournier, V., Jakubiec, A., Drugeon, G., Lacassagne, E., Pflieger, S., Chenon, M., Jupin, I., 2010. The ubiquitin-proteasome system regulates the accumulation of Turnip yellow mosaic virus RNA-dependent RNA polymerase during viral infection. Plant Cell 22 (9), 3142–3152.

Carbonell, A., Carrington, J.C., 2015. Antiviral roles of plant ARGONAUTES. Curr. Opin. Plant Biol. 27, 111–117.

Carr, J.P., Donnelly, R., Tungadi, T., Murphy, A.M., Jiang, S., Bravo-Cazar, A., Yoon, J.Y., Cunniffe, N.J., Glover, B.J., Gilligan, C.A., 2018. Viral manipulation of plant stress responses and host interactions with insects. Adv. Virus Res. 102, 177–197.

Casey, J.R., Grinstein, S., Orlowski, J., 2010. Sensors and regulators of intracellular pH. Nat. Rev. Mol. Cell Biol. 11 (1), 50–61.

Castillo, A.G., Kong, L.J., Hanley-Bowdoin, L., Bejarano, E.R., 2004. Interaction between a geminivirus replication protein and the plant sumoylation system. J. Virol. 78 (6), 2758–2769.

Champagne, J., Laliberté-Gagné, M.E., Leclerc, D., 2007. Phosphorylation of the termini of Cauliflower mosaic virus precapsid protein is important for productive infection. Mol. Plant Microbe Interact. 20 (6), 648–658.

Chapdelaine, Y., Kirk, D., Karsies, A., Hohn, T., Leclerc, D., 2002. Mutation of capsid protein phosphorylation sites abolishes cauliflower mosaic virus infectivity. J. Virol. 76 (22), 11748–11752.

Chen, I.H., Chiu, M.H., Cheng, S.F., Hsu, Y.H., Tsai, C.H., 2013. The glutathione transferase of Nicotiana benthamiana NbGSTU4 plays a role in regulating the early replication of Bamboo mosaic virus. New Phytol. 199 (3), 749–757.

Chen, Y., Chen, Q., Li, M., Mao, Q., Chen, H., Wu, W., Jia, D., Wei, T., 2017a. Autophagy pathway induced by a plant virus facilitates viral spread and transmission by its insect vector. PLoS Pathog. 13 (11), e1006727.

Chen, I.H., Tsai, A.Y., Huang, Y.P., Wu, I.F., Cheng, S.F., Hsu, Y.H., Tsai, C.H., 2017b. Nuclear-encoded plastidal carbonic anhydrase is involved in replication of *Bamboo mosaic virus* RNA in *Nicotiana benthamiana*. Front. Microbiol. 8, 2046.

Chen, H., Cao, Y., Li, Y., Xia, Z., Xie, J., Carr, J.P., Wu, B., Fan, Z., Zhou, T., 2017c. Identification of differentially regulated maize proteins conditioning Sugarcane mosaic virus systemic infection. New Phytol. 215 (3), 1156–1172.

Chen, S., Sun, X., Shi, Y., Wei, Y., Han, X., Li, H., Chen, L., Sun, B., Sun, H., Shi, Y., 2019. Cucurbit chlorotic yellows virus p22 protein interacts with cucumber SKP1LB1 and its F-box-like motif is crucial for silencing suppressor activity. Viruses 11 (9), 818.

Cheng, C.P., 2017. Host factors involved in the intracellular movement of *Bamboo mosaic virus*. Front. Microbiol. 8, 759.

Cheng, X., Wang, A., 2017. The Potyvirus silencing suppressor protein VPg mediates degradation of SGS3 via ubiquitination and autophagy pathways. J. Virol. 91 (1), e01478-16.

Cheng, C.P., Jaag, H.M., Jonczyk, M., Serviene, E., Nagy, P.D., 2007. Expression of the Arabidopsis Xrn4p 5′-3′ exoribonuclease facilitates degradation of tombusvirus RNA and promotes rapid emergence of viral variants in plants. Virology 368 (2), 238–248.

Cheng, S.F., Huang, Y.P., Chen, L.H., Hsu, Y.H., Tsai, C.H., 2013. Chloroplast phosphoglycerate kinase is involved in the targeting of Bamboo mosaic virus to chloroplasts in Nicotiana benthamiana plants. Plant Physiol. 163 (4), 1598–1608.

Cheng, X., Xiong, R., Li, Y., Li, F., Zhou, X., Wang, A., 2017. Sumoylation of turnip mosaic virus RNA polymerase promotes viral infection by counteracting the host NPR1-mediated immune response. Plant Cell 29 (3), 508–525.

Cheng, G., Yang, Z., Zhang, H., Zhang, J., Xu, J., 2020. Remorin interacting with PCaP1 impairs Turnip mosaic virus intercellular movement but is antagonised by VPg. New Phytol. 225 (5), 2122–2139.

Chenon, M., Camborde, L., Cheminant, S., Jupin, I., 2012. A viral deubiquitylating enzyme targets viral RNA-dependent RNA polymerase and affects viral infectivity. EMBO J. 31 (3), 741–753.

Choe, K.N., Moldovan, G.L., 2017. Forging ahead through darkness: PCNA, still the principal conductor at the replication fork. Mol. Cell 65, 380–392.

Choi, Y., Bowman, J.W., Jung, J.U., 2018. Autophagy during viral infection—a double-edged sword. Nat. Rev. Microbiol. 16 (6), 341–354.

Chuang, C., Prasanth, K.R., Nagy, P.D., 2015. Coordinated function of cellular DEAD-box helicases in suppression of viral RNA recombination and maintenance of viral genome integrity. PLoS Pathog. 11 (2), e1004680.

Chuang, C., Prasanth, K.R., Nagy, P.D., 2017. The glycolytic pyruvate kinase is recruited directly into the viral replicase complex to generate ATP for RNA synthesis. Cell Host Microbe 22 (5), 639–652.e7.

Chung, H.Y., Lacatus, G., Sunter, G., 2014. Geminivirus AL2 protein induces expression of, and interacts with, a calmodulin-like gene, an endogenous regulator of gene silencing. Virology 460–461, 108–118.

Clarke, B.D., Roby, J.A., Slonchak, A., Khromykh, A.A., 2015. Functional non-coding RNAs derived from the flavivirus 3′ untranslated region. Virus Res. 206, 53–61.

Csorba, T., Lózsa, R., Hutvágner, G., Burgyán, J., 2010. Polerovirus protein P0 prevents the assembly of small RNA-containing RISC complexes and leads to degradation of ARGONAUTE1. Plant J. 62 (3), 463–472.

Csorba, T., Kontra, L., Burgyán, J., 2015. Viral silencing suppressors: tools forged to fine-tune host-pathogen coexistence. Virology 479–480, 85–103.

De Ronde, D., Butterbach, P., Kormelink, R., 2014. Dominant resistance against plant viruses. Front. Plant Sci. 5, 307.

Deng, X.G., Zhu, T., Peng, X.J., Xi, D.H., Guo, H., Yin, Y., Zhang, D.W., Lin, H.H., 2016a. Role of brassinosteroid signaling in modulating Tobacco mosaic virus resistance in Nicotiana benthamiana. Sci. Rep. 6, 20579.

Deng, X.G., Zhu, T., Zou, L.J., Han, X.Y., Zhou, X., Xi, D.H., Zhang, D.W., Lin, H.H., 2016b. Orchestration of hydrogen peroxide and nitric oxide in brassinosteroid-mediated systemic virus resistance in Nicotiana benthamiana. Plant J. 85 (4), 478–493.

Derrien, B., Baumberger, N., Schepetilnikov, M., Viotti, C., De Cillia, J., Ziegler-Graff, V., Isono, E., Schumacher, K., Genschik, P., 2012. Degradation of the antiviral component ARGONAUTE1 by the autophagy pathway. Proc. Natl. Acad. Sci. U. S. A. 109 (39), 15942–15946.

Derrien, B., Clavel, M., Baumberger, N., Iki, T., Sarazin, A., Hacquard, T., Ponce, M.R., Ziegler-Graff, V., Vaucheret, H., Micol, J.L., Voinnet, O., Genschik, P., 2018. A suppressor screen for AGO1 degradation by the viral F-box P0 protein uncovers a role for AGO DUF1785 in sRNA duplex unwinding. Plant Cell 30 (6), 1353–1374.

Díez, J., Ishikawa, M., Kaido, M., Ahlquist, P., 2000. Identification and characterization of a host protein required for efficient template selection in viral RNA replication. Proc. Natl. Acad. Sci. U. S. A. 97 (8), 3913–3918.

Dilweg, I.W., Gultyaev, A.P., Olsthoorn, R.C., 2019. Structural features of an Xrn1-resistant plant virus RNA. RNA Biol. 16 (6), 838–845.

Elena, S.F., Fraile, A., García-Arenal, F., 2014. Evolution and emergence of plant viruses. Adv. Virus Res. 88, 161–191.

Endres, M.W., Gregory, B.D., Gao, Z., Foreman, A.W., Mlotshwa, S., Ge, X., Pruss, G.J., Ecker, J.R., Bowman, L.H., Vance, V., 2010. Two plant viral suppressors of silencing require the ethylene-inducible host transcription factor RAV2 to block RNA silencing. PLoS Pathog. 6 (1), e1000729.

Feng, Z., Xu, K., Kovalev, N., Nagy, P.D., 2019. Recruitment of Vps34 PI3K and enrichment of PI3P phosphoinositide in the viral replication compartment is crucial for replication of a positive-strand RNA virus. PLoS Pathog. 15 (1), e1007530.

Flobinus, A., Hleibieh, K., Klein, E., Ratti, C., Bouzoubaa, S., Gilmer, D., 2016. A viral noncoding RNA complements a weakened viral RNA silencing suppressor and promotes efficient systemic host infection. Viruses 8 (10), 272.

Flobinus, A., Chevigny, N., Charley, P.A., Seissler, T., Klein, E., Bleykasten-Grosshans, C., Ratti, C., Bouzoubaa, S., Wilusz, J., Gilmer, D., 2018. Beet necrotic yellow vein virus noncoding RNA production depends on a 5′→3′ Xrn exoribonuclease activity. Viruses 10 (3), 137.

Folimonova, S.Y., Tilsner, J., 2018. Hitchhikers, highway tolls and roadworks, the interactions of plant viruses with the phloem. Curr. Opin. Plant Biol. 43, 82–88.

Fu, S., Xu, Y., Li, C., Li, Y., Wu, J., Zhou, X., 2018. Rice stripe virus interferes with S-acylation of remorin and induces its autophagic degradation to facilitate virus infection. Mol. Plant 11 (2), 269–287.

Galão, R.P., Chari, A., Alves-Rodrigues, I., Lobão, D., Mas, A., Kambach, C., Fischer, U., Díez, J., 2010. LSm1-7 complexes bind to specific sites in viral RNA genomes and regulate their translation and replication. RNA 16 (4), 817–827.

Garcia-Ruiz, H., 2018. Susceptibility genes to plant viruses. Viruses 10 (9), 484.

Garcia-Ruiz, H., 2019. Host factors against plant viruses. Mol. Plant Pathol. 20 (11), 1588–1601.

Garcia-Ruiz, H., Carbonell, A., Hoyer, J.S., Fahlgren, N., Gilbert, K.B., Takeda, A., Giampetruzzi, A., Garcia Ruiz, M.T., McGinn, M.G., Lowery, N., Martinez Baladejo, M.T., Carrington, J.C., 2015. Roles and programming of Arabidopsis ARGONAUTE proteins during Turnip mosaic virus infection. PLoS Pathog. 11 (3), e1004755.

Gil-Monreal, M., Giuntoli, B., Zabalza, A., Licausi, F., Royuela, M., 2019. ERF-VII transcription factors induce ethanol fermentation in response to amino acid biosynthesis-inhibiting herbicides. J. Exp. Bot. 70 (20), 5839–5851.

Göertz, G.P., Abbo, S.R., Fros, J.J., Pijlman, G.P., 2018. Functional RNA during Zika virus infection. Virus Res. 254, 41–53.

Golisz, A., Sikorski, P.J., Kruszka, K., Kufel, J., 2013. Arabidopsis thaliana LSM proteins function in mRNA splicing and degradation. Nucleic Acids Res. 41 (12), 6232–6249.

Goodwin, C.M., Xu, S., Munger, J., 2015. Stealing the keys to the kitchen: viral manipulation of the host cell metabolic network. Trends Microbiol. 23 (12), 789–798.

Graus, D., Konrad, K.R., Bemm, F., Patir Nebioglu, M.G., Lorey, C., Duscha, K., Güthoff, T., Herrmann, J., Ferjani, A., Cuin, T.A., Roelfsema, M.R.G., Schumacher, K., Neuhaus, H.E., Marten, I., Hedrich, R., 2018. High V-PPase activity is beneficial under high salt loads, but detrimental without salinity. New Phytol. 219 (4), 1421–1432.

Gullberg, R.C., Jordan Steel, J., Moon, S.L., Soltani, E., Geiss, B.J., 2015. Oxidative stress influences positive strand RNA virus genome synthesis and capping. Virology 475, 219–229.

Gullner, G., Komives, T., Király, L., Schröder, P., 2018. Glutathione S-transferase enzymes in plant-pathogen interactions. Front. Plant Sci. 9, 1836.

Gunawardene, C.D., Newburn, L.R., White, K.A., 2019. A 212-nt long RNA structure in the Tobacco necrosis virus-D RNA genome is resistant to Xrn degradation. Nucleic Acids Res. 47 (17), 9329–9342.

Guo, H., Gu, L., Liu, F., Chen, F., Ge, F., Sun, Y., 2019a. Aphid-borne viral spread is enhanced by virus-induced accumulation of plant reactive oxygen species. Plant Physiol. 179 (1), 143–155.

Guo, Z., Li, Y., Ding, S.W., 2019b. Small RNA-based antimicrobial immunity. Nat. Rev. Immunol. 19 (1), 31–44.

Gy, I., Gasciolli, V., Lauressergues, D., Morel, J.B., Gombert, J., Proux, F., Proux, C., Vaucheret, H., Mallory, A.C., 2007. Arabidopsis FIERY1, XRN2, and XRN3 are endogenous RNA silencing suppressors. Plant Cell 19 (11), 3451–3461.

Hafrén, A., Lõhmus, A., Mäkinen, K., 2015. Formation of potato virus A-induced RNA granules and viral translation are interrelated processes required for optimal virus accumulation. PLoS Pathog. 11 (12), e1005314.

Hafrén, A., Macia, J.L., Love, A.J., Milner, J.J., Drucker, M., Hofius, D., 2017. Selective autophagy limits cauliflower mosaic virus infection by NBR1-mediated targeting of viral capsid protein and particles. Proc. Natl. Acad. Sci. U. S. A. 114 (10), E2026–E2035.

Hafrén, A., Üstün, S., Hochmuth, A., Svenning, S., Johansen, T., Hofius, D., 2018. Turnip mosaic virus counteracts selective autophagy of the viral silencing suppressor HCpro. Plant Physiol. 176 (1), 649–662.

Haimovich, G., Medina, D.A., Causse, S.Z., Garber, M., Millán-Zambrano, G., Barkai, O., Chávez, S., Pérez-Ortín, J.E., Darzacq, X., Choder, M., 2013. Gene expression is circular: factors for mRNA degradation also foster mRNA synthesis. Cell 153 (5), 1000–1011.

Hanley-Bowdoin, L., Bejarano, E.R., Robertson, D., Mansoor, S., 2013. Geminiviruses: masters at redirecting and reprogramming plant processes. Nat. Rev. Microbiol. 11, 777–788.

Hanssen, I.M., Lapidot, M., Thomma, B.P., 2010. Emerging viral diseases of tomato crops. Mol. Plant Microbe Interact. 23 (5), 539–548.

Hashimoto, M., Neriya, Y., Yamaji, Y., Namba, S., 2016. Recessive resistance to plant viruses: potential resistance genes beyond translation initiation factors. Front. Microbiol. 7, 1695.

Haxim, Y., Ismayil, A., Jia, Q., Wang, Y., Zheng, X., Chen, T., Qian, L., Liu, N., Wang, Y., Han, S., Cheng, J., Qi, Y., Hong, Y., Liu, Y., 2017. Autophagy functions as an antiviral mechanism against geminiviruses in plants. Elife 6, e23897.

He, L., Chen, X., Yang, J., Zhang, T., Li, J., Zhang, S., Zhong, K., Zhang, H., Chen, J., Yang, J., 2020. Rice black-streaked dwarf virus-encoded P5-1 regulates the ubiquitination activity of SCF E3 ligases and inhibits jasmonate signaling to benefit its infection in rice. New Phytol. 225 (2), 896–912.

Heinlein, M., 2015. Plant virus replication and movement. Virology 479–480, 657–671.

Hoover, H.S., Wang, J.C., Middleton, S., Ni, P., Zlotnick, A., Vaughan, R.C., Kao, C.C., 2016. Phosphorylation of the brome mosaic virus capsid regulates the timing of viral infection. J. Virol. 90 (17), 7748–7760.

Hu, Y., Li, Z., Yuan, C., Jin, X., Yan, L., Zhao, X., Zhang, Y., Jackson, A.O., Wang, X., Han, C., Yu, J., Li, D., 2015. Phosphorylation of TGB1 by protein kinase CK2 promotes barley stripe mosaic virus movement in monocots and dicots. J. Exp. Bot. 66 (15), 4733–4747.

Hu, T., Huang, C., He, Y., Castillo-González, C., Gui, X., Wang, Y., Zhang, X., Zhou, X., 2019. βC1 protein encoded in geminivirus satellite concertedly targets MKK2 and MPK4 to counter host defense. PLoS Pathog. 15 (4), e1007728.

Huang, Y.P., Chen, I.H., Tsai, C.H., 2017. Host factors in the infection cycle of *Bamboo mosaic virus*. Front. Microbiol. 8, 437.

Huang, Y.P., Huang, Y.W., Hsiao, Y.J., Li, S.C., Hsu, Y.H., Tsai, C.H., 2019a. Autophagy is involved in assisting the replication of Bamboo mosaic virus in Nicotiana benthamiana. J. Exp. Bot. 70 (18), 4657–4670.

Huang, D., Sun, Y., Ma, Z., Ke, M., Cui, Y., Chen, Z., Chen, C., Ji, C., Tran, T.M., Yang, L., Lam, S.M., Han, Y., Shu, G., Friml, J., Miao, Y., Jiang, L., Chen, X., 2019b. Salicylic acid-mediated plasmodesmal closure via Remorin-dependent lipid organization. Proc. Natl. Acad. Sci. U. S. A. 116 (42), 21274–21284.

Huang, Y.W., Hu, C.C., Tsai, C.H., Lin, N.S., Hsu, Y.H., 2019c. Nicotiana benthamiana Argonaute10 plays a pro-viral role in Bamboo mosaic virus infection. New Phytol. 224 (2), 804–817.

Hung, C.J., Huang, Y.W., Liou, M.R., Lee, Y.C., Lin, N.S., Meng, M., Tsai, C.H., Hu, C.C., Hsu, Y.H., 2014. Phosphorylation of coat protein by protein kinase CK2 regulates cell-to-cell movement of Bamboo mosaic virus through modulating RNA binding. Mol. Plant Microbe Interact. 27 (11), 1211–1225.

Huot, B., Yao, J., Montgomery, B.L., He, S.Y., 2014. Growth-defense tradeoffs in plants: a balancing act to optimize fitness. Mol. Plant 7 (8), 1267–1287.

Hyodo, K., Okuno, T., 2014. Host factors used by positive-strand RNA plant viruses for genome replication. J. Gen. Plant Pathol. 80, 123–135.

Hyodo, K., Okuno, T., 2016. Pathogenesis mediated by proviral host factors involved in translation and replication of plant positive-strand RNA viruses. Curr. Opin. Virol. 17, 11–18.

Hyodo, K., Hashimoto, K., Kuchitsu, K., Suzuki, N., Okuno, T., 2017a. Harnessing host ROS-generating machinery for the robust genome replication of a plant RNA virus. Proc. Natl. Acad. Sci. U. S. A. 114 (7), E1282–E1290.

Hyodo, K., Suzuki, N., Mise, K., Okuno, T., 2017b. Roles of superoxide anion and hydrogen peroxide during replication of two unrelated plant RNA viruses in Nicotiana benthamiana. Plant Signal. Behav. 12 (6), e1338223.

Hyodo, K., Suzuki, N., Okuno, T., 2019. Hijacking a host scaffold protein, RACK1, for replication of a plant RNA virus. New Phytol. 221 (2), 935–945.

Imura, Y., Molho, M., Chuang, C., Nagy, P.D., 2015. Cellular Ubc2/Rad6 E2 ubiquitin-conjugating enzyme facilitates tombusvirus replication in yeast and plants. Virology 484, 265–275.

Inaba, J.I., Nagy, P.D., 2018. Tombusvirus RNA replication depends on the TOR pathway in yeast and plants. Virology 519, 207–222.

Inaba, J.I., Xu, K., Kovalev, N., Ramanathan, H., Roy, C.R., Lindenbach, B.D., Nagy, P.D., 2019. Screening *Legionella* effectors for antiviral effects reveals Rab1 GTPase as a proviral factor coopted for tombusvirus replication. Proc. Natl. Acad. Sci. U. S. A. 116 (43), 21739–21747.

Ishibashi, K., Ishikawa, M., 2014. Mechanisms of tomato mosaic virus RNA replication and its inhibition by the host resistance factor Tm-1. Curr. Opin. Virol. 9, 8–13.

Ishibashi, K., Ishikawa, M., 2016. Replication of tobamovirus RNA. Annu. Rev. Phytopathol. 54, 55–78.

Ishibashi, K., Saruta, M., Shimizu, T., Shu, M., Anai, T., Komatsu, K., Yamada, N., Katayose, Y., Ishikawa, M., Ishimoto, M., Kaga, A., 2019. Soybean antiviral immunity conferred by dsRNase targets the viral replication complex. Nat. Commun. 10 (1), 4033.

Ivanov, K.I., Puustinen, P., Merits, A., Saarma, M., Mäkinen, K., 2001. Phosphorylation down-regulates the RNA binding function of the coat protein of potato virus A. J. Biol. Chem. 276 (17), 13530–13540.

Ivanov, K.I., Puustinen, P., Gabrenaite, R., Vihinen, H., Rönnstrand, L., Valmu, L., Kalkkinen, N., Mäkinen, K., 2003. Phosphorylation of the potyvirus capsid protein by protein kinase CK2 and its relevance for virus infection. Plant Cell 15 (9), 2124–2139.

Iwakawa, H.O., Mizumoto, H., Nagano, H., Imoto, Y., Takigawa, K., Sarawaneeyaruk, S., Kaido, M., Mise, K., Okuno, T., 2008. A viral noncoding RNA generated by cis-element-mediated protection against 5′->3′ RNA decay represses both cap-independent and cap-dependent translation. J. Virol. 82 (20), 10162–10174.

Iwakawa, H.O., Tajima, Y., Taniguchi, T., Kaido, M., Mise, K., Tomari, Y., Taniguchi, H., Okuno, T., 2012. Poly(A)-binding protein facilitates translation of an uncapped/non-polyadenylated viral RNA by binding to the 3′ untranslated region. J. Virol. 86 (15), 7836–7849.

Jaag, H.M., Nagy, P.D., 2009. Silencing of Nicotiana benthamiana Xrn4p exoribonuclease promotes tombusvirus RNA accumulation and recombination. Virology 386 (2), 344–352.

Jakubiec, A., Jupin, I., 2007. Regulation of positive-strand RNA virus replication: the emerging role of phosphorylation. Virus Res. 129 (2), 73–79.

Jia, Q., Liu, N., Xie, K., Dai, Y., Han, S., Zhao, X., Qian, L., Wang, Y., Zhao, J., Gorovits, R., Xie, D., Hong, Y., Liu, Y., 2016. CLCuMuB βC1 subverts ubiquitination by interacting with NbSKP1s to enhance geminivirus infection in Nicotiana benthamiana. PLoS Pathog. 12 (6), e1005668.

Jiang, S., Jiang, L., Yang, J., Peng, J., Lu, Y., Zheng, H., Lin, L., Chen, J., Yan, F., 2018. Over-expression of Oryza sativa Xrn4 confers plant resistance to virus infection. Gene 639, 44–51.

Jin, X., Cao, X., Wang, X., Jiang, J., Wan, J., Laliberté, J.F., Zhang, Y., 2018. Three-dimensional architecture and biogenesis of membrane structures associated with plant virus replication. Front. Plant Sci. 9, 57.

Jones, R.A.C., Naidu, R.A., 2019. Global dimensions of plant virus diseases: current status and future perspectives. Annu. Rev. Virol. 6 (1), 387–409.

Jungfleisch, J., Chowdhury, A., Alves-Rodrigues, I., Tharun, S., Díez, J., 2015. The Lsm1-7-Pat1 complex promotes viral RNA translation and replication by differential mechanisms. RNA 21 (8), 1469–1479.

Jungfleisch, J., Blasco-Moreno, B., Díez, J., 2016. Use of cellular decapping activators by positive-strand RNA viruses. Viruses 8 (12), 340.

Jungfleisch, J., Nedialkova, D.D., Dotu, I., Sloan, K.E., Martinez-Bosch, N., Brüning, L., Raineri, E., Navarro, P., Bohnsack, M.T., Leidel, S.A., Díez, J., 2017. A novel translational control mechanism involving RNA structures within coding sequences. Genome Res. 27 (1), 95–106.

Jupin, I., Ayach, M., Jomat, L., Fieulaine, S., Bressanelli, S., 2017. A mobile loop near the active site acts as a switch between the dual activities of a viral protease/deubiquitinase. PLoS Pathog. 13 (11), e1006714.

Kang, H.K., Yang, S.H., Lee, Y.P., Park, Y.I., Kim, S.H., 2012. A tobacco CBL-interacting protein kinase homolog is involved in phosphorylation of the N-terminal domain of the cucumber mosaic virus polymerase 2a protein. Biosci. Biotechnol. Biochem. 76 (11), 2101–2106.

Kellner, R., De la Concepcion, J.C., Maqbool, A., Kamoun, S., Dagdas, Y.F., 2017. ATG8 expansion: a driver of selective autophagy diversification? Trends Plant Sci. 222 (3), 204–214.

Kieft, J.S., Rabe, J.L., Chapman, E.G., 2015. New hypotheses derived from the structure of a flaviviral Xrn1-resistant RNA: conservation, folding, and host adaptation. RNA Biol. 12 (11), 1169–1177.

Kim, S.H., Palukaitis, P., Park, Y.I., 2002. Phosphorylation of cucumber mosaic virus RNA polymerase 2a protein inhibits formation of replicase complex. EMBO J. 21 (9), 2292–2300.

Kimura, S., Waszczak, C., Hunter, K., Wrzaczek, M., 2017. Bound by fate: the role of reactive oxygen species in receptor-like kinase signaling. Plant Cell 29 (4), 638–654.

Kloth, K.J., Kormelink, R., 2020. Defenses against virus and vector: a phloem-biological perspective on RTM- and SLI1-mediated resistance to potyviruses and aphids. Viruses 12 (2), 129.

Kobayashi, K., Sekine, K.T., Nishiguchi, M., 2014. Breakdown of plant virus resistance: can we predict and extend the durability of virus resistance? J. Gen. Plant Pathol. 80, 327–336.

Kovalev, N., Nagy, P.D., 2014. The expanding functions of cellular helicases: the tombusvirus RNA replication enhancer co-opts the plant eIF4AIII-like AtRH2 and the DDX5-like AtRH5 DEAD-box RNA helicases to promote viral asymmetric RNA replication. PLoS Pathog. 10 (4), e1004051.

Kovalev, N., Barajas, D., Nagy, P.D., 2012a. Similar roles for yeast Dbp2 and Arabidopsis RH20 DEAD-box RNA helicases to Ded1 helicase in tombusvirus plus-strand synthesis. Virology 432 (2), 470–484.

Kovalev, N., Pogany, J., Nagy, P.D., 2012b. A co-opted DEAD-box RNA helicase enhances tombusvirus plus-strand synthesis. PLoS Pathog. 8 (2), e1002537.

Kovalev, N., de Castro Martín, I.F., Pogany, J., Barajas, D., Pathak, K., Risco, C., Nagy, P.D., 2016. Role of viral RNA and co-opted cellular ESCRT-I and ESCRT-III factors in formation of tombusvirus spherules harboring the tombusvirus replicase. J. Virol. 90 (7), 3611–3626.

Kovalev, N., Inaba, J.I., Li, Z., Nagy, P.D., 2017. The role of co-opted ESCRT proteins and lipid factors in protection of tombusviral double-stranded RNA replication intermediate against reconstituted RNAi in yeast. PLoS Pathog. 13 (7), e1006520.

Kraft, J.J., Treder, K., Peterson, M.S., Miller, W.A., 2013. Cation-dependent folding of 3′ cap-independent translation elements facilitates interaction of a 17-nucleotide conserved sequence with eIF4G. Nucleic Acids Res. 41 (5), 3398–3413.

Kushner, D.B., Lindenbach, B.D., Grdzelishvili, V.Z., Noueiry, A.O., Paul, S.M., Ahlquist, P., 2003. Systematic, genome-wide identification of host genes affecting replication of a positive-strand RNA virus. Proc. Natl. Acad. Sci. U. S. A. 100 (26), 15764–15769.

Kushwaha, N.K., Mansi, B., Chakraborty, S., 2017. The replication initiator protein of a geminivirus interacts with host monoubiquitination machinery and stimulates transcription of the viral genome. PLoS Pathog. 13, e1006587.

Kushwaha, N.K., Hafrén, A., Hofius, D., 2019. Autophagy-virus interplay in plants: from antiviral recognition to proviral manipulation. Mol. Plant Pathol. 20 (9), 1211–1216.

Laliberté, J.F., Sanfaçon, H., 2010. Cellular remodeling during plant virus infection. Annu. Rev. Phytopathol. 48, 69–91.

Laliberté, J.F., Zheng, H., 2014. Viral manipulation of plant host membranes. Annu. Rev. Virol. 1 (1), 237–259.

Lee, C.C., Lin, T.L., Lin, J.W., Han, Y.T., Huang, Y.T., Hsu, Y.H., Meng, M., 2016. Promotion of bamboo mosaic virus accumulation in Nicotiana benthamiana by 5′→3′ exonuclease NbXRN4. Front. Microbiol. 6, 1508.

Lee, C.C., Wu, Y.J., Hsueh, C.H., Huang, Y.T., Hsu, Y.H., Meng, M., 2018. Mitogen-activated protein kinase phosphatase 1 reduces the replication efficiency of Bamboo mosaic virus in Nicotiana benthamiana. Mol. Plant Pathol. 19 (10), 2319–2332.

Li, F., Wang, A., 2018. RNA decay is an antiviral defense in plants that is counteracted by viral RNA silencing suppressors. PLoS Pathog. 14 (8), e1007228.

Li, F., Wang, A., 2019. RNA-targeted antiviral immunity: more than just RNA silencing. Trends Microbiol. 27 (9), 792–805.

Li, X.H., Valdez, P., Olvera, R.E., Carrington, J.C., 1997. Functions of the tobacco etch virus RNA polymerase (NIb): subcellular transport and protein-protein interaction with VPg/proteinase (NIa). J. Virol. 71 (2), 1598–1607.

Li, Z., Barajas, D., Panavas, T., Herbst, D.A., Nagy, P.D., 2008. Cdc34p ubiquitin-conjugating enzyme is a component of the tombusvirus replicase complex and ubiquitinates p33 replication protein. J. Virol. 82 (14), 6911–6926.

Li, F., Huang, C., Li, Z., Zhou, X., 2014. Suppression of RNA silencing by a plant DNA virus satellite requires a host calmodulin-like protein to repress RDR6 expression. PLoS Pathog. 10 (2), e1003921.

Li, F., Zhao, N., Li, Z., Xu, X., Wang, Y., Yang, X., Liu, S.S., Wang, A., Zhou, X., 2017. A calmodulin-like protein suppresses RNA silencing and promotes geminivirus infection by degrading SGS3 via the autophagy pathway in Nicotiana benthamiana. PLoS Pathog. 13 (2), e1006213.

Li, F., Zhang, C., Li, Y., Wu, G., Hou, X., Zhou, X., Wang, A., 2018. Beclin1 restricts RNA virus infection in plants through suppression and degradation of the viral polymerase. Nat. Commun. 9 (1), 1268.

Li, P., Liu, C., Deng, W.H., Yao, D.M., Pan, L.L., Li, Y.Q., Liu, Y.Q., Liang, Y., Zhou, X.P., Wang, X.W., 2019a. Plant begomoviruses subvert ubiquitination to suppress plant defenses against insect vectors. PLoS Pathog. 15 (2), e1007607.

Li, Y., Sun, Q., Zhao, T., Xiang, H., Zhang, X., Wu, Z., Zhou, C., Zhang, X., Wang, Y., Zhang, Y., Wang, X., Li, D., Yu, J., Dinesh-Kumar, S.P., Han, C., 2019b. Interaction between Brassica yellows virus silencing suppressor P0 and plant SKP1 facilitates stability of P0 in vivo against degradation by proteasome and autophagy pathways. New Phytol. 222 (3), 1458–1473.

Li, F., Zhang, M., Zhang, C., Zhou, X., 2020. Nuclear autophagy degrades a geminivirus nuclear protein to restrict viral infection in solanaceous plants. New Phytol. 225 (4), 1746–1761.
Lin, J.W., Ding, M.P., Hsu, Y.H., Tsai, C.H., 2007. Chloroplast phosphoglycerate kinase, a gluconeogenetic enzyme, is required for efficient accumulation of Bamboo mosaic virus. Nucleic Acids Res. 35 (2), 424–432.
Lin, W., Liu, Y., Molho, M., Zhang, S., Wang, L., Xie, L., Nagy, P.D., 2019. Co-opting the fermentation pathway for tombusvirus replication: compartmentalization of cellular metabolic pathways for rapid ATP generation. PLoS Pathog. 15 (10), e1008092.
Liu, Q., Liu, H., Gong, Y., Tao, Y., Jiang, L., Zuo, W., Yang, Q., Ye, J., Lai, J., Wu, J., Lübberstedt, T., Xu, M., 2017. An atypical thioredoxin imparts early resistance to sugarcane mosaic virus in maize. Mol. Plant 10 (3), 483–497.
Lõhmus, A., Hafrén, A., Mäkinen, K., 2017. Coat protein regulation by CK2, CPIP, HSP70, and CHIP is required for potato virus a replication and coat protein accumulation. J. Virol. 91 (3), e01316-16.
Lozano-Durán, R., Rosas-Díaz, T., Gusmaroli, G., Luna, A.P., Taconnat, L., Deng, X.W., Bejarano, E.R., 2011. Geminiviruses subvert ubiquitination by altering CSN-mediated derubylation of SCF E3 ligase complexes and inhibit jasmonate signaling in Arabidopsis thaliana. Plant Cell 23 (3), 1014–1032.
Lukhovitskaya, N., Ryabova, L.A., 2019. Cauliflower mosaic virus transactivator protein (TAV) can suppress nonsense-mediated decay by targeting VARICOSE, a scaffold protein of the decapping complex. Sci. Rep. 9 (1), 7042.
Ma, X., Zhou, Y., Moffett, P., 2019. Alterations in cellular RNA decapping dynamics affect tomato spotted wilt virus cap snatching and infection in Arabidopsis. New Phytol. 224 (2), 789–803.
Mair, A., Xu, S.L., Branon, T.C., Ting, A.Y., Bergmann, D.C., 2019. Proximity labeling of protein complexes and cell-type-specific organellar proteomes in *Arabidopsis* enabled by TurboID. Elife 8, e47864.
Mandadi, K.K., Scholthof, K.B., 2013. Plant immune responses against viruses: how does a virus cause disease? Plant Cell 25 (5), 1489–1505.
Martelli, G.P., Adams, M.J., Kreuze, J.F., Dolja, V.V., 2007. Family Flexiviridae: a case study in virion and genome plasticity. Annu. Rev. Phytopathol. 45, 73–100.
Martínez-Pérez, M., Aparicio, F., López-Gresa, M.P., Bellés, J.M., Sánchez-Navarro, J.A., Pallás, V., 2017. *Arabidopsis* m^6A demethylase activity modulates viral infection of a plant virus and the m^6A abundance in its genomic RNAs. Proc. Natl. Acad. Sci. U. S. A. 114 (40), 10755–10760.
Martinière, A., Bak, A., Macia, J.L., Lautredou, N., Gargani, D., Doumayrou, J., Garzo, E., Moreno, A., Fereres, A., Blanc, S., Drucker, M., 2013. A virus responds instantly to the presence of the vector on the host and forms transmission morphs. Elife 2, e00183.
Mas, A., Alves-Rodrigues, I., Noueiry, A., Ahlquist, P., Díez, J., 2006. Host deadenylation-dependent mRNA decapping factors are required for a key step in brome mosaic virus RNA replication. J. Virol. 80 (1), 246–251.
May, J.P., Yuan, X., Sawicki, E., Simon, A.E., 2018. RNA virus evasion of nonsense-mediated decay. PLoS Pathog. 14 (11), e1007459.
May, J.P., Johnson, P.Z., Ilyas, M., Gao, F., Simon, A.E., 2020. The multifunctional long-distance movement protein of *Pea Enation Mosaic Virus 2* protects viral and host transcripts from nonsense-mediated decay. mBio 11 (2), e00204-20.
Mei, Y., Wang, Y., Hu, T., Yang, X., Lozano-Duran, R., Sunter, G., Zhou, X., 2018a. Nucleocytoplasmic shuttling of geminivirus C4 protein mediated by phosphorylation and myristoylation is critical for viral pathogenicity. Mol. Plant 11 (12), 1466–1481.

Mei, Y., Yang, X., Huang, C., Zhang, X., Zhou, X., 2018b. Tomato leaf curl Yunnan virus-encoded C4 induces cell division through enhancing stability of cyclin D 1.1 via impairing NbSKη-mediated phosphorylation in Nicotiana benthamiana. PLoS Pathog. 14 (1), e1006789.

Meier, N., Hatch, C., Nagalakshmi, U., Dinesh-Kumar, S.P., 2019. Perspectives on intracellular perception of plant viruses. Mol. Plant Pathol. 20 (9), 1185–1190.

Meng, X., Zhang, S., 2013. MAPK cascades in plant disease resistance signaling. Annu. Rev. Phytopathol. 51, 245–266.

Michaeli, S., Clavel, M., Lechner, E., Viotti, C., Wu, J., Dubois, M., Hacquard, T., Derrien, B., Izquierdo, E., Lecorbeiller, M., Bouteiller, N., De Cilia, J., Ziegler-Graff, V., Vaucheret, H., Galili, G., Genschik, P., 2019. The viral F-box protein P0 induces an ER-derived autophagy degradation pathway for the clearance of membrane-bound AGO1. Proc. Natl. Acad. Sci. U. S. A. 116 (45), 22872–22883.

Miller, S., Krijnse-Locker, J., 2008. Modification of intracellular membrane structures for virus replication. Nat. Rev. Microbiol. 6 (5), 363–374.

Mills-Lujan, K., Andrews, D.L., Chou, C.W., Deom, C.M., 2015. The roles of phosphorylation and SHAGGY-like protein kinases in geminivirus C4 protein induced hyperplasia. PLoS One 10 (3), e0122356.

Miricescu, A., Goslin, K., Graciet, E., 2018. Ubiquitylation in plants: signaling hub for the integration of environmental signals. J. Exp. Bot. 69 (19), 4511–4527.

Nagy, P.D., 2016. Tombusvirus-host interactions: co-opted evolutionarily conserved host factors take center court. Annu. Rev. Virol. 3 (1), 491–515.

Nagy, P.D., Lin, W., 2020. Taking over cellular energy-metabolism for TBSV replication: the high ATP requirement of an RNA virus within the viral replication organelle. Viruses 12 (1), 56.

Nagy, P.D., Pogany, J., 2012. The dependence of viral RNA replication on co-opted host factors. Nat. Rev. Microbiol. 10 (2), 137–149.

Navarro, J.A., Sanchez-Navarro, J.A., Pallas, V., 2019. Key checkpoints in the movement of plant viruses through the host. Adv. Virus Res. 104, 1–64.

Nemes, K., Gellért, Á., Almási, A., Vági, P., Sáray, R., Kádár, K., Salánki, K., 2017. Phosphorylation regulates the subcellular localization of Cucumber Mosaic Virus 2b protein. Sci. Rep. 7 (1), 13444.

Nicaise, V., 2014. Crop immunity against viruses: outcomes and future challenges. Front. Plant Sci. 5, 660.

Niehl, A., Heinlein, M., 2019. Perception of double-stranded RNA in plant antiviral immunity. Mol. Plant Pathol. 20 (9), 1203–1210.

Nieto-Torres, J.L., Verdiá-Báguena, C., Castaño-Rodriguez, C., Aguilella, V.M., Enjuanes, L., 2015. Relevance of viroporin ion channel activity on viral replication and pathogenesis. Viruses 7 (7), 3552–3573.

Nishikiori, M., Ahlquist, P., 2018. Organelle luminal dependence of (+)strand RNA virus replication reveals a hidden druggable target. Sci. Adv. 4 (1), eaap8258.

Nishikiori, M., Meshi, T., Ishikawa, M., 2012. Guanylylation-competent replication proteins of Tomato mosaic virus are disulfide-linked. Virology 434 (1), 118–128.

Noack, L.C., Jaillais, Y., 2017. Precision targeting by phosphoinositides: how PIs direct endomembrane trafficking in plants. Curr. Opin. Plant Biol. 40, 22–33.

Noueiry, A.O., Ahlquist, P., 2003. Brome mosaic virus RNA replication: revealing the role of the host in RNA virus replication. Annu. Rev. Phytopathol. 41, 77–98.

Noueiry, A.O., Diez, J., Falk, S.P., Chen, J., Ahlquist, P., 2003. Yeast Lsm1p-7p/Pat1p deadenylation-dependent mRNA-decapping factors are required for brome mosaic virus genomic RNA translation. Mol. Cell. Biol. 23 (12), 4094–4106.

Olspert, A., Chung, B.Y., Atkins, J.F., Carr, J.P., Firth, A.E., 2015. Transcriptional slippage in the positive-sense RNA virus family Potyviridae. EMBO Rep. 16 (8), 995–1004.

Ouibrahim, L., Mazier, M., Estevan, J., Pagny, G., Decroocq, V., Desbiez, C., Moretti, A., Gallois, J.L., Caranta, C., 2014. Cloning of the Arabidopsis rwm1 gene for resistance to Watermelon mosaic virus points to a new function for natural virus resistance genes. Plant J. 79 (5), 705–716.

Ouibrahim, L., Rubio, A.G., Moretti, A., Montané, M.H., Menand, B., Meyer, C., Robaglia, C., Caranta, C., 2015. Potyviruses differ in their requirement for TOR signalling. J. Gen. Virol. 96, 2898–2903.

Pazhouhandeh, M., Dieterle, M., Marrocco, K., Lechner, E., Berry, B., Brault, V., Hemmer, O., Kretsch, T., Richards, K.E., Genschik, P., Ziegler-Graff, V., 2006. F-box-like domain in the polerovirus protein P0 is required for silencing suppressor function. Proc. Natl. Acad. Sci. U. S. A. 103 (6), 1994–1999.

Peltier, C., Klein, E., Hleibieh, K., D'Alonzo, M., Hammann, P., Bouzoubaa, S., Ratti, C., Gilmer, D., 2012. Beet necrotic yellow vein virus subgenomic RNA3 is a cleavage product leading to stable non-coding RNA required for long-distance movement. J. Gen. Virol. 93 (Pt. 5), 1093–1102.

Peng, J., Yang, J., Yan, F., Lu, Y., Jiang, S., Lin, L., Zheng, H., Chen, H., Chen, J., 2011. Silencing of NbXrn4 facilitates the systemic infection of Tobacco mosaic virus in Nicotiana benthamiana. Virus Res. 158 (1–2), 268–270.

Perea-Resa, C., Hernández-Verdeja, T., López-Cobollo, R., del Mar, C.M., Salinas, J., 2012. LSM proteins provide accurate splicing and decay of selected transcripts to ensure normal Arabidopsis development. Plant Cell 24 (12), 4930–4947.

Piroux, N., Saunders, K., Page, A., Stanley, J., 2007. Geminivirus pathogenicity protein C4 interacts with Arabidopsis thaliana shaggy-related protein kinase AtSKeta, a component of the brassinosteroid signalling pathway. Virology 362 (2), 428–440.

Pitzalis, N., Heinlein, M., 2017. The roles of membranes and associated cytoskeleton in plant virus replication and cell-to-cell movement. J. Exp. Bot. 69 (1), 117–132.

Pogany, J., Nagy, P.D., 2015. Activation of tomato bushy stunt virus RNA-dependent RNA polymerase by cellular heat shock protein 70 is enhanced by phospholipids in vitro. J. Virol. 89 (10), 5714–5723.

Pogany, J., Stork, J., Li, Z., Nagy, P.D., 2008. In vitro assembly of the Tomato bushy stunt virus replicase requires the host Heat shock protein 70. Proc. Natl. Acad. Sci. U. S. A. 105 (50), 19956–19961.

Poque, S., Pagny, G., Ouibrahim, L., Chague, A., Eyquard, J.P., Caballero, M., Candresse, T., Caranta, C., Mariette, S., Decroocq, V., 2015. Allelic variation at the rpv1 locus controls partial resistance to Plum pox virus infection in Arabidopsis thaliana. BMC Plant Biol. 15, 159.

Prasanth, K.R., Barajas, D., Nagy, P.D., 2015. The proteasomal Rpn11 metalloprotease suppresses tombusvirus RNA recombination and promotes viral replication via facilitating assembly of the viral replicase complex. J. Virol. 89 (5), 2750–2763.

Prasanth, K.R., Chuang, C., Nagy, P.D., 2017. Co-opting ATP-generating glycolytic enzyme PGK1 phosphoglycerate kinase facilitates the assembly of viral replicase complexes. PLoS Pathog. 13 (10), e1006689.

Prince, D.C., Drurey, C., Zipfel, C., Hogenhout, S.A., 2014. The leucine-rich repeat receptor-like kinase BRASSINOSTEROID INSENSITIVE1-ASSOCIATED KINASE1 and the cytochrome P450 PHYTOALEXIN DEFICIENT3 contribute to innate immunity to aphids in Arabidopsis. Plant Physiol. 164 (4), 2207–2219.

Restrepo, M.A., Freed, D.D., Carrington, J.C., 1990. Nuclear transport of plant potyviral proteins. Plant Cell 2 (10), 987–998.

Revers, F., García, J.A., 2015. Molecular biology of potyviruses. Adv. Virus Res. 92, 101–199.

Rodamilans, B., Valli, A., Mingot, A., San León, D., Baulcombe, D., López-Moya, J.J., García, J.A., 2015. RNA polymerase slippage as a mechanism for the production of frameshift gene products in plant viruses of the potyviridae family. J. Virol. 89 (13), 6965–6967.

Rohozinski, J., Hancock, J.M., 1996. Do light-induced pH changes within the chloroplast drive turnip yellow mosaic virus assembly? J. Gen. Virol. 77 (Pt. 2), 163–165.
Rojas, R.M., Gilbertson, R.L., 2008. Emerging plant viruses: a diversity of mechanisms and opportunities. In: Roossinck, M.J. (Ed.), Plant Virus Evolution. Springer, Berlin, Heidelberg, pp. 27–52.
Saleh, A., Withers, J., Mohan, R., Marqués, J., Gu, Y., Yan, S., Zavaliev, R., Nomoto, M., Tada, Y., Dong, X., 2015. Posttranslational modifications of the master transcriptional regulator NPR1 enable dynamic but tight control of plant immune responses. Cell Host Microbe 18 (2), 169–182.
Sanchez, E.L., Lagunoff, M., 2015. Viral activation of cellular metabolism. Virology 479–480, 609–618.
Sánchez-Durán, M.A., Dallas, M.B., Ascencio-Ibañez, J.T., Reyes, M.I., Arroyo-Mateos, M., Ruiz-Albert, J., Hanley-Bowdoin, L., Bejarano, E.R., 2011. Interaction between geminivirus replication protein and the SUMO-conjugating enzyme is required for viral infection. J. Virol. 85 (19), 9789–9800.
Sanfaçon, H., 2015. Plant translation factors and virus resistance. Viruses 7 (7), 3392–3419.
Sastry, S.K., Zitter, T.A., 2014. Management of virus and viroid diseases of crops in the tropics. In: Sastry, S.K. (Ed.), Plant Virus and Viroid Diseases in the Tropics. In: Epidemiology and Management, vol. 2. Springer, Dordrecht, pp. 149–480.
Sasvari, Z., Kovalev, N., Nagy, P.D., 2013. The GEF1 proton-chloride exchanger affects tombusvirus replication via regulation of copper metabolism in yeast. J. Virol. 87 (3), 1800–1810.
Sasvari, Z., Alatriste Gonzalez, P., Nagy, P.D., 2014. Tombusvirus-yeast interactions identify conserved cell-intrinsic viral restriction factors. Front. Plant Sci. 5, 383.
Schepetilnikov, M., Kobayashi, K., Geldreich, A., Caranta, C., Robaglia, C., Keller, M., Ryabova, L.A., 2011. Viral factor TAV recruits TOR/S6K1 signalling to activate reinitiation after long ORF translation. EMBO J. 30 (7), 1343–1356.
Schwechheimer, C., 2018. NEDD8-its role in the regulation of Cullin-RING ligases. Curr. Opin. Plant Biol. 45 (Pt. A), 112–119.
Serva, S., Nagy, P.D., 2006. Proteomics analysis of the tombusvirus replicase: Hsp70 molecular chaperone is associated with the replicase and enhances viral RNA replication. J. Virol. 80 (5), 2162–2169.
Sewelam, N., Kazan, K., Schenk, P.M., 2016. Global plant stress signaling: reactive oxygen species at the cross-road. Front. Plant Sci. 7, 187.
Signorelli, S., Tarkowski, Ł.P., Van den Ende, W., Bassham, D.C., 2019. Linking autophagy to abiotic and biotic stress responses. Trends Plant Sci. 24 (5), 413–430.
Simon, A.E., Miller, W.A., 2013. 3′ cap-independent translation enhancers of plant viruses. Annu. Rev. Microbiol. 67, 21–42.
Smakowska, E., Kong, J., Busch, W., Belkhadir, Y., 2016. Organ-specific regulation of growth-defense tradeoffs by plants. Curr. Opin. Plant Biol. 29, 129–137.
Steckelberg, A.L., Akiyama, B.M., Costantino, D.A., Sit, T.L., Nix, J.C., Kieft, J.S., 2018a. A folded viral noncoding RNA blocks host cell exoribonucleases through a conformationally dynamic RNA structure. Proc. Natl. Acad. Sci. U. S. A. 115 (25), 6404–6409.
Steckelberg, A.L., Vicens, Q., Kieft, J.S., 2018b. Exoribonuclease-resistant RNAs exist within both coding and noncoding subgenomic RNAs. mBio 9 (6), e02461-18.
Sun, Y.D., Folimonova, S.Y., 2019. The p33 protein of Citrus tristeza virus affects viral pathogenicity by modulating a host immune response. New Phytol. 221 (4), 2039–2053.
Sun, H., Shen, L., Qin, Y., Liu, X., Hao, K., Li, Y., Wang, J., Yang, J., Wang, F., 2018. CLC-Nt1 affects Potato Virus Y infection via regulation of endoplasmic reticulum luminal Ph. New Phytol. 220 (2), 539–552.

Tajima, Y., Iwakawa, H.O., Hyodo, K., Kaido, M., Mise, K., Okuno, T., 2017. Requirement for eukaryotic translation initiation factors in cap-independent translation differs between bipartite genomic RNAs of red clover necrotic mosaic virus. Virology 509, 152–158.

Tao, T., Zhou, C.J., Wang, Q., Chen, X.R., Sun, Q., Zhao, T.Y., Ye, J.C., Wang, Y., Zhang, Z.Y., Zhang, Y.L., Guo, Z.J., Wang, X.B., Li, D.W., Yu, J.L., Han, C.G., 2017. Rice black streaked dwarf virus P7-2 forms a SCF complex through binding to Oryza sativa SKP1-like proteins, and interacts with GID2 involved in the gibberellin pathway. PLoS One 12 (5), e0177518.

Thiel, H., Hleibieh, K., Gilmer, D., Varrelmann, M., 2012. The P25 pathogenicity factor of beet necrotic yellow vein virus targets the sugar beet 26S proteasome involved in the induction of a hypersensitive resistance response via interaction with an F-box protein. Mol. Plant Microbe Interact. 25 (8), 1058–1072.

To, E.E., Vlahos, R., Luong, R., Halls, M.L., Reading, P.C., King, P.T., Chan, C., Drummond, G.R., Sobey, C.G., Broughton, B.R.S., Starkey, M.R., van der Sluis, R., Lewin, S.R., Bozinovski, S., O'Neill, L.A.J., Quach, T., Porter, C.J.H., Brooks, D.A., O'Leary, J.J., Selemidis, S., 2017. Endosomal NOX2 oxidase exacerbates virus pathogenicity and is a target for antiviral therapy. Nat. Commun. 8 (1), 69.

Truniger, V., Miras, M., Aranda, M.A., 2017. Structural and functional diversity of plant virus 3'-cap-independent translation enhancers (3'-CITEs). Front. Plant Sci. 8, 2047.

Valli, A.A., Gallo, A., Rodamilans, B., López-Moya, J.J., García, J.A., 2018. The HCPro from the Potyviridae family: an enviable multitasking helper component that every virus would like to have. Mol. Plant Pathol. 19 (3), 744–763.

Verchot, J., 2016. Plant virus infection and the ubiquitin proteasome machinery: arms race along the endoplasmic reticulum. Viruses 8 (11), 314.

Wang, A., 2015. Dissecting the molecular network of virus-plant interactions: the complex roles of host factors. Annu. Rev. Phytopathol. 53, 45–66.

Wang, R.Y., Stork, J., Pogany, J., Nagy, P.D., 2009a. A temperature sensitive mutant of heat shock protein 70 reveals an essential role during the early steps of tombusvirus replication. Virology 394 (1), 28–38.

Wang, R.Y., Stork, J., Nagy, P.D., 2009b. A key role for heat shock protein 70 in the localization and insertion of tombusvirus replication proteins to intracellular membranes. J. Virol. 83 (7), 3276–3287.

Wang, Q., Tao, T., Han, Y., Chen, X., Fan, Z., Li, D., Yu, J., Han, C., 2013. Nonstructural protein P7-2 encoded by Rice black-streaked dwarf virus interacts with SKP1, a core subunit of SCF ubiquitin ligase. Virol. J. 10, 325.

Waszczak, C., Carmody, M., Kangasjärvi, J., 2018. Reactive oxygen species in plant signaling. Annu. Rev. Plant Biol. 69, 209–236.

Weber, P.H., Bujarski, J.J., 2015. Multiple functions of capsid proteins in (+) stranded RNA viruses during plant-virus interactions. Virus Res. 196, 140–149.

Whitfield, A.E., Falk, B.W., Rotenberg, D., 2015. Insect vector-mediated transmission of plant viruses. Virology 479–480, 278–289.

Withers, J., Dong, X., 2017. Post-translational regulation of plant immunity. Curr. Opin. Plant Biol. 38, 124–132.

Wu, X., Ye, J., 2020. Manipulation of jasmonate signaling by plant viruses and their insect vectors. Viruses 12 (2), 148.

Wu, J., Yang, R., Yang, Z., Yao, S., Zhao, S., Wang, Y., Li, P., Song, X., Jin, L., Zhou, T., Lan, Y., Xie, L., Zhou, X., Chu, C., Qi, Y., Cao, X., Li, Y., 2017. ROS accumulation and antiviral defence control by microRNA528 in rice. Nat. Plants 3, 16203.

Xiong, Y., Sheen, J., 2015. Novel links in the plant TOR kinase signaling network. Curr. Opin. Plant Biol. 28, 83–91.

Xiong, R., Wang, A., 2013. SCE1, the SUMO-conjugating enzyme in plants that interacts with NIb, the RNA-dependent RNA polymerase of Turnip mosaic virus, is required for viral infection. J. Virol. 87 (8), 4704–4715.

Xu, F.Q., Xue, H.W., 2019. The ubiquitin-proteasome system in plant responses to environments. Plant Cell Environ. 42 (10), 2931–2944.

Xu, J., Yang, J.Y., Niu, Q.W., Chua, N.H., 2006. Arabidopsis DCP2, DCP1, and VARICOSE form a decapping complex required for postembryonic development. Plant Cell 18, 3386–3398.

Yamaji, Y., Maejima, K., Ozeki, J., Komatsu, K., Shiraishi, T., Okano, Y., Himeno, M., Sugawara, K., Neriya, Y., Minato, N., Miura, C., Hashimoto, M., Namba, S., 2012. Lectin-mediated resistance impairs plant virus infection at the cellular level. Plant Cell 24 (2), 778–793.

Yang, Z., Li, Y., 2018. Dissection of RNAi-based antiviral immunity in plants. Curr. Opin. Virol. 32, 88–99.

Yang, M., Zhang, Y., Xie, X., Yue, N., Li, J., Wang, X.B., Han, C., Yu, J., Liu, Y., Li, D., 2018a. *Barley stripe mosaic virus* γb protein subverts autophagy to promote viral infection by disrupting the ATG7-ATG8 interaction. Plant Cell 30 (7), 1582–1595.

Yang, M., Li, Z., Zhang, K., Zhang, X., Zhang, Y., Wang, X., Han, C., Yu, J., Xu, K., Li, D., 2018b. Barley stripe mosaic virus γb interacts with glycolate oxidase and inhibits peroxisomal ROS production to facilitate virus infection. Mol. Plant 11 (2), 338–341.

Yang, J., Zhang, T., Li, J., Wu, N., Wu, G., Yang, J., Chen, X., He, L., Chen, J., 2020. Chinese wheat mosaic virus-derived vsiRNA-20 can regulate virus infection in wheat through inhibition of vacuolar- (H^+)-PPase induced cell death. New Phytol. 226 (1), 205–220.

Yoshida, T., Shiraishi, T., Hagiwara-Komoda, Y., Komatsu, K., Maejima, K., Okano, Y., Fujimoto, Y., Yusa, A., Yamaji, Y., Namba, S., 2019. The plant noncanonical antiviral resistance protein JAX1 inhibits potexviral replication by targeting the viral RNA-dependent RNA polymerase. J. Virol. 93 (3), e01506-18.

Zhang, X., Guo, H., 2017. mRNA decay in plants: both quantity and quality matter. Curr. Opin. Plant Biol. 35, 138–144.

Zhang, S., Klessig, D.F., 1998. Resistance gene N-mediated de novo synthesis and activation of a tobacco mitogen-activated protein kinase by tobacco mosaic virus infection. Proc. Natl. Acad. Sci. U. S. A. 95, 7433–7438.

Zhang, H., Xia, R., Meyers, B.C., Walbot, V., 2015. Evolution, functions, and mysteries of plant ARGONAUTE proteins. Curr. Opin. Plant Biol. 27, 84–90.

Zhang, X., Dong, K., Xu, K., Zhang, K., Jin, X., Yang, M., Zhang, Y., Wang, X., Han, C., Yu, J., Li, D., 2018. Barley stripe mosaic virus infection requires PKA-mediated phosphorylation of γb for suppression of both RNA silencing and the host cell death response. New Phytol. 218 (4), 1570–1585.

Zhang, Y., Song, G., Lal, N.K., Nagalakshmi, U., Li, Y., Zheng, W., Huang, P.J., Branon, T.C., Ting, A.Y., Walley, J.W., Dinesh-Kumar, S.P., 2019a. TurboID-based proximity labeling reveals that UBR7 is a regulator of N NLR immune receptor-mediated immunity. Nat. Commun. 10 (1), 3252.

Zhang, Z., He, G., Filipowicz, N.A., Randall, G., Belov, G.A., Kopek, B.G., Wang, X., 2019b. Host lipids in positive-strand RNA virus genome replication. Front. Microbiol. 10, 286.

Zhang, C., Wei, Y., Xu, L., Wu, K.C., Yang, L., Shi, C.N., Yang, G.Y., Chen, D., Yu, F.F., Xie, Q., Li, Y., Ding, S.W., Wu, J.G., 2020. A bunyavirus-inducible ubiquitin ligase targets RNA polymerase IV for degradation during viral pathogenesis in rice. Mol. Plant. https://doi.org/10.1016/j.molp.2020.02.010, in press.

Zhao, X., Wang, X., Dong, K., Zhang, Y., Hu, Y., Zhang, X., Chen, Y., Wang, X., Han, C., Yu, J., Li, D., 2015. Phosphorylation of Beet black scorch virus coat protein by PKA is required for assembly and stability of virus particles. Sci. Rep. 25, 11585.

Zhong, L., Shu, W., Dai, W., Gao, B., Xiong, S., 2017. Reactive oxygen species-mediated c-Jun NH2-terminal kinase activation contributes to hepatitis B virus X protein-induced autophagy via regulation of the Beclin-1/Bcl-2 interaction. J. Virol. 91 (15), e00001-17.

Zhou, T., Murphy, A.M., Lewsey, M.G., Westwood, J.H., Zhang, H.M., González, I., Canto, T., Carr, J.P., 2014. Domains of the cucumber mosaic virus 2b silencing suppressor protein affecting inhibition of salicylic acid-induced resistance and priming of salicylic acid accumulation during infection. J. Gen. Virol. 95 (Pt. 6), 1408–1413.

Zhu, H., Hu, F., Wang, R., Zhou, X., Sze, S.H., Liou, L.W., Barefoot, A., Dickman, M., Zhang, X., 2011. Arabidopsis Argonaute10 specifically sequesters miR166/165 to regulate shoot apical meristem development. Cell 145 (2), 242–256.

Ziegler-Graff, V., 2020. Molecular insights into host and vector manipulation by plant viruses. Viruses 12 (3), 263.

Zvereva, A.S., Golyaev, V., Turco, S., Gubaeva, E.G., Rajeswaran, R., Schepetilnikov, M.V., Srour, O., Ryabova, L.A., Boller, T., Pooggin, M.M., 2016. Viral protein suppresses oxidative burst and salicylic acid-dependent autophagy and facilitates bacterial growth on virus-infected plants. New Phytol. 211 (3), 1020–1034.

CHAPTER THREE

Modulation of disease severity by plant positive-strand RNA viruses: The complex interplay of multifunctional viral proteins, subviral RNAs and virus-associated RNAs with plant signaling pathways and defense responses

Hélène Sanfaçon*

Summerland Research and Development Centre, Agriculture and Agri-Food Canada, Summerland, BC, Canada

*Corresponding author: e-mail address: helene.sanfacon@canada.ca

Contents

Advances in Virus Research, Volume 107
ISSN 0065-3527
https://doi.org/10.1016/bs.aivir.2020.04.003

Abstract

Plant viruses induce a range of symptoms of varying intensity, ranging from severe systemic necrosis to mild or asymptomatic infection. Several evolutionary constraints drive virus virulence, including the dependence of viruses on host factors to complete their infection cycle, the requirement to counteract or evade plant antiviral defense responses and the mode of virus transmission. Viruses have developed an array of strategies to modulate disease severity. Accumulating evidence has highlighted not only the multifunctional role that viral proteins play in disrupting or highjacking plant factors, hormone signaling pathways and intracellular organelles, but also the interaction networks between viral proteins, subviral RNAs and/or other viral-associated RNAs that regulate disease severity. This review focusses on positive-strand RNA viruses, which constitute the majority of characterized plant viruses. Using well-characterized viruses with different genome types as examples, recent advances are discussed as well as knowledge gaps and opportunities for further research.

1. Introduction

Although the first discovered plant viruses were severe pathogens of economically important crops, the impact of virus infection on their host plants varies greatly ranging from damaging diseases to asymptomatic co-existence (Fraile and García-Arenal, 2016; Lefeuvre et al., 2019; Roossinck, 2015). When disease occurs, its severity is determined by the virulence of the specific viral strain, the possible presence of other viruses in mixed infections, the plant species/cultivar and environmental conditions (Mascia and Gallitelli, 2016; Moreno and López-Moya, 2020; Osterbaan and Fuchs, 2019). Tolerant interactions allow substantial accumulation of viruses without significant impact on the host fitness (Pagán and García-Arenal, 2018; Paudel and Sanfaçon, 2018). In some cases, the impact of symptomatic virus infections is mitigated by the emergence of asymptomatic leaves at later stages of infection in a phenomenon termed symptom recovery (Ghoshal and Sanfaçon, 2015).

Viruses are obligate parasites that depend on their host to complete their infection cycles (Mauck et al., 2019; Wang, 2015). They also often rely on vectors to achieve plant-to-plant transmission (Gallet et al., 2018). Viruses are highly adaptive due to a high mutation rate of their polymerases and their propensity for recombination or genome segment reassortment (Novella et al., 2014; Peck and Lauring, 2018; Simon-Loriere and Holmes, 2011; Sztuba-Solińska et al., 2011b). This allows them to expand their host range, by evolving to recruit plant factors for their replication cycle, to counteract or repurpose plant defense responses, to regulate disease severity and to facilitate vector transmission (McLeish et al., 2019).

Positive-strand RNA viruses constitute the majority of plant viruses, including many well-characterized model systems, and will be the focus of this review. Positive-strand RNA viruses have a distinct advantage over other types of viruses, in that the encapsidated genomic RNAs serve as messenger RNAs for the translation of viral proteins (Miras et al., 2017). Although positive-strand RNA viruses are generally described as cytoplasmic, their proteins have been shown to translocate into various sub-cellular compartments allowing them to remodel plant intracellular structures and manipulate plant signaling pathways (Laliberté and Sanfaçon, 2010; Medina-Puche and Lozano-Duran, 2019).

Recent reviews have discussed in detail the plant-virus interactions that facilitate virus multiplication and movement within the plant, and the variety of plant defense responses and viral counter-defense mechanisms (Carr et al., 2019a; Garcia-Ruiz, 2019; Medina-Puche and Lozano-Duran, 2019; Wang, 2015; Wu et al., 2019). Other recent reviews have examined the nature of symptom determinants in virus-infected plants (García and Pallas, 2015; Osterbaan and Fuchs, 2019), and the diversity of molecular mechanisms associated with symptom attenuation in tolerant interactions (Paudel and Sanfaçon, 2018). This review focusses on the molecular mechanisms by which (+)-strand RNA viruses enhance or mitigate symptom intensity in response to evolutionary constraints. A few well-characterized pathosystems are described in details to highlight the diversity of these mechanisms.

2. Evolutionary constraints driving disease severity

A priori, minimizing disease severity should be beneficial to plant viruses (Pagán and García-Arenal, 2018; Paudel and Sanfaçon, 2018). Indeed, viruses are obligate parasites and have evolved to highjack, modify or remodel plant cellular organelles, entire physiological pathways or specific proteins to facilitate the translation and replication of their genome and the systemic invasion of the plant (Laliberté and Sanfaçon, 2010; Nagy, 2016; Wang, 2015). Maintaining a relatively healthy status of the infected plant is imperative to ensure that the necessary plant resources remain available to viruses as the infection progresses. On the other hand, the extensive repurposing and/or alteration of host resources by viruses has been highlighted as one the major causes of symptom expression (Culver and

Padmanabhan, 2007; García and Pallas, 2015; Osterbaan and Fuchs, 2019). Thus, virus adaptation to a new host requires not only changes to the sequence or structure of their proteins (or nucleic acids) so that they can use cellular resources that may differ from those of a previous host, but also ensuring that the fitness cost to the host remains within acceptable limits. Often, this means reaching an optimized compromise by limiting the accumulation or toxicity of viral nucleic acids or proteins (Paudel and Sanfaçon, 2018). For viruses with a wide host range (termed generalist viruses), fine-tuning this balance in a variety of hosts may be extraordinarily challenging (Bedhomme et al., 2015; McLeish et al., 2019).

However, other evolutionary constraints are also at play. Many viruses depend on insect, mite, nematode or fungal vectors for plant-to-plant transmission (Gallet et al., 2018). This implies that they must accumulate to sufficient levels for vector acquisition. Recent advances have highlighted the extent by which viruses manipulate their hosts to optimize their transmission (Carr et al., 2018; Dietzgen et al., 2016; Gallet et al., 2018; Mauck et al., 2019). This includes interfering with plant defense responses targeted at insects, modifying the chemical composition of plant volatiles to make them more attractive to the insect vectors or inducing macroscopic plant modifications, such as leaf chlorotic lesions to provide visual cues for insect vectors or increased root surface to assist soil-borne vector transmission. On the other hand, viruses that are vertically transmitted must ensure that seed or pollen yield and viability is maintained. Consistently, these viruses generally cause mild diseases or persist in a symptomless manner, as exemplified by endornaviruses with a double-stranded RNA genome (Fukuhara, 2019).

Host adaptation also implies counteracting, evading or even repurposing antiviral defense responses. Plant defense responses are diverse and include ubiquitous basal and adaptive immune responses, such as salicylic acid (SA)-dependent basal immune responses (also termed pathogen-associated molecular pattern triggered immunity or PTI), adaptive antiviral RNA silencing, various RNA decay and protein degradation mechanisms, as well as specific responses determined by dominant or recessive resistance genes (Carr et al., 2019a; Hashimoto et al., 2016; Ismayil et al., 2019; Kushwaha et al., 2019; Li and Wang, 2019; Nicaise, 2017; Sanfaçon, 2015; Verchot, 2016b; Wu et al., 2019). Viruses encode specialized proteins that suppress RNA silencing (Csorba et al., 2015) and/or manipulate other defense responses (Carr et al., 2018; Li and Wang, 2019). They have also developed sophisticated mechanisms to counteract or repurpose plant RNA and protein degradation pathways (Ismayil et al., 2019; Kushwaha et al., 2019; Li and Wang, 2019). Because plant antiviral responses often overlap with universal gene regulation mechanisms that control plant metabolic processes, viral counter-defense responses can impede normal plant development. For example, many viral suppressors of RNA silencing (VSRs) also interfere with plant microRNAs (miRNAs) function causing massive transcriptome reprogramming (Csorba et al., 2015). Thus, the accumulation and/or activity of these VSRs is often tightly regulated.

Plant-virus interactions are impacted by the lifespan of the host plant and the biodiversity of the ecological system, both of which can be significantly reduced in agricultural settings (Roossinck and García-Arenal, 2015). Climatic conditions and the presence of other pathogens can also affect the outcome of virus infections (Jones, 2016; Paudel and Sanfaçon, 2018; Sanfaçon, 2017). Mixed plant virus infections are common and can result in synergistic or antagonistic interactions that impact disease severity (Mascia and Gallitelli, 2016; Moreno and López-Moya, 2020). Even in cases of single infection, viruses exist as variant clouds, termed quasi-species (Novella et al., 2014; Roossinck and Schneider, 2006).

The effect of these various and sometimes opposing evolutionary constraints on the evolution of virus virulence is exemplified in a pioneering study that reconstructed the phylogeny of symptom intensity following the emergence of maize streak virus (MSV), a single-strand DNA virus and member of the family *Geminiviridae* (Monjane et al., 2020). This study examined the symptom range of a large collection of isolates on various maize genotypes with different degrees of resistance to MSV and inferred the symptoms induced by presumed ancestral strains. The authors reconstituted infectious clones for several deduced ancestral strains and confirmed the predicted symptom impact. Chlorotic lesions are known to attract the MSV leafhopper vector. On the other hand, reduced chloroplast function can inflict a significant cost to the host. Remarkably, the study concluded that over the past 110 years, MSV populations have evolved to (1) increase chlorotic lesion sizes and maintain high virus titers in these lesions to facilitate virus transmission, and (2) decrease the intensity of the chlorotic symptoms and minimize the harm to the host (Monjane et al., 2020). While similar phylogenetic reconstructions of disease severity are not yet available for positive-strand RNA viruses, the availability of large collection of isolate sequences have allowed to infer the emergence of various viruses or virus families and the expansion of their host range (Gibbs et al., 2020; Lefeuvre et al., 2019; McLeish et al., 2019; Moury et al., 2017; Moury and Desbiez, 2020).

3. Regulating the impact of multifunctional viral proteins on plant intracellular structures and signaling pathways

In most characterized plant-virus interactions, symptom determinants have been mapped to viral proteins of different types, including but not limited to VSRs, coat proteins (CPs), movement proteins (MPs), proteases and replication proteins (Culver and Padmanabhan, 2007; García and Pallas, 2015; Osterbaan and Fuchs, 2019; Paudel and Sanfaçon, 2018). The causes of symptom induction can vary greatly. Virus replication and movement depends on the remodeling of the plant intracellular membrane network by viral proteins, creating stress in the associated organelles (e.g., endoplasmic reticulum, Golgi apparatus, chloroplasts) (Bhattacharyya and Chakraborty, 2017; Laliberté and Sanfaçon, 2010; Li et al., 2016; Verchot, 2016a; Zhao et al., 2016). Specific interactions between viral proteins and cellular proteins can also disrupt the normal function of these organelles, such as the chloroplast (Li et al., 2016; Qiu et al., 2018; Zhao et al., 2016). Signaling pathways controlled by phytohormones, such as SA, jasmonic acid (JA) and auxin, are also common targets of viral proteins (Collum and Culver, 2016). Most viral proteins are multifunctional and many have extensive networks of cellular interaction partners, as exemplified by the tomato bushy stunt virus p33 replication protein or the potyvirus HC-Pro protein (Nagy, 2016; Valli et al., 2018). Thus, the impact of viral proteins can be multipronged and affect multiple plant processes simultaneously or sequentially. In addition, viral proteins can act in concert or in opposition to fine-tune their effect on the plant metabolism, growth and development.

Adaptive mutations in viral symptom determinants can impact their activity (or activities) and mitigate the severity of the disease. Indeed, even single point mutations can have a profound impact on the disease outcome (Osterbaan and Fuchs, 2019). As will be highlighted in the selected examples below, another common way for viruses to reduce the global impact of their proteins is to self-attenuate (Ghoshal and Sanfaçon, 2015; Paudel and Sanfaçon, 2018; Wu et al., 2019). This can be achieved in several ways: (1) minimizing the accumulation of viral RNAs from which these proteins are translated, (2) reducing the translation rate of these viral RNAs, and (3) decreasing the stability of viral proteins. Viruses use plant RNA degradation pathways (RNA silencing or RNA decay) and protein degradation pathways (autophagy or the ubiquitin-proteasome pathways) to control the stability of their RNAs and proteins (Ismayil et al., 2019; Kushwaha et al., 2019; Li and Wang, 2019; Mäkinen et al., 2017; Verchot, 2016b; Xu et al., 2020). Translation rates of viral RNAs can be regulated by fine-tuning the use of the cellular translation machinery (Jaafar and Kieft, 2019; Miras et al., 2017) or by repurposing cellular defense responses that repress viral RNA translation (Ghoshal and Sanfaçon, 2015; Sanfaçon, 2015). In addition, the activities of viral symptom determinants can be modulated by counteracting activities of other viral proteins or RNAs, by regulated release from precursor polyproteins, or by post-translational modifications (Paudel and Sanfaçon, 2018).

4. Deploying subviral or virus-associated RNAs to manipulate symptom intensity

Plant viruses are known to give rise to or be associated with various RNA species that can impact their virulence (Shimura and Masuta, 2016). Sub-genomic RNAs (sgRNAs) are produced to translate internal open reading frames (ORFs), allowing to control the timing and efficiency of translation of distinct viral proteins (Sztuba-Solińska et al., 2011a). Defective RNAs (dRNAs) contain portions of the corresponding viral genomic RNAs and are produced by recombination during viral replication (Pathak and Nagy, 2009). Different types of virus-derived long non-coding RNAs (lncRNAs) have been identified that are produced by various mechanisms and often correspond to the 5′ or 3′ untranslated regions (UTRs) of the viral genome (Flobinus et al., 2018; Kang et al., 2019). Viral small-interfering RNAs (vsiRNAs) are produced by plant dicer-like (dcl) enzymes as a consequence of the activation of plant antiviral RNA silencing (Pooggin, 2018; Zhang et al., 2019). Finally, satellite RNAs do not share extensive homology to viral genomic RNAs, may or may not encode proteins and depend on the helper virus for their replication (Palukaitis, 2016). These subviral and virus-associated RNAs can impact the virus virulence in various ways (Shimura and Masuta, 2016). siRNAs derived from the viral genome or from satellite RNAs not only impact the stability or translation of viral RNAs, but they can also repress the expression of plant genes. DRNAs, lncRNAs and satellite RNAs can compete for replication with the viral genomic RNAs or they can stabilize the viral genomic RNAs accumulation by acting as sponges of RNA silencing enzymes. Recently, new roles for lncRNAs have been identified, for example in the repression of SA-dependent defense responses (Kang et al., 2019).

5. Selected examples of well-characterized plant-virus interactions

A few well-characterized plant-virus interactions are described in details below to illustrate the impact of viral proteins or viral-associated RNAs in reprogramming plant hormone signaling pathways, disrupting or repurposing plant defense responses or protein degradation pathways and repressing plant transcriptional or translational machineries to induce macroscopic symptoms and regulate their severity. These interactions also exemplify how the complex interplay between two or more viral proteins impacts the activity or accumulation of viral symptom determinants and the role played by subviral RNAs or viral-associated RNAs. The selected pathosystems were also chosen to represent different genome types, symptomatologies and host ranges, including crops.

5.1 Cucumber mosaic virus: Cross-talk between viral proteins, viral satellite RNAs and multiple plant signaling pathways

Cucumber mosaic virus (CMV, genus *Cucumovirus*, family *Bromoviridae*) is an economically important pathogen causing serious diseases on a wide range of hosts. CMV, first discovered 100 years ago, infects over 1000 plant species and is also one of the best studied plant viruses (Jacquemond, 2012; Mochizuki and Ohki, 2012; Salánki et al., 2018; Scholthof et al., 2011). Many strains of CMV have been characterized that are generally classified into two subgroups and differ in their severity and host range. CMV infection is also often associated with the presence of small linear non-coding satellite RNAs that can either exacerbate or attenuate symptoms. The symptoms induced by CMV strains on different hosts range from mild mosaic symptoms leading to symptom recovery, to severe developmental defects or systemic necrosis. CMV is transmitted by aphids and has become a model system to demonstrate how viruses manipulate plants to facilitate their transmission (Carr et al., 2019b). Study of natural strain evolution patterns under field conditions have highlighted the emergence and subsequent demise of necrotic strains due to the progressive selection of benign non-necrogenic satellite RNAs over severe necrogenic ones (Escriu et al., 2000). In experimental forced evolution of mixed infections, viral virulence evolution was shown to be influenced by a combination of factors, including

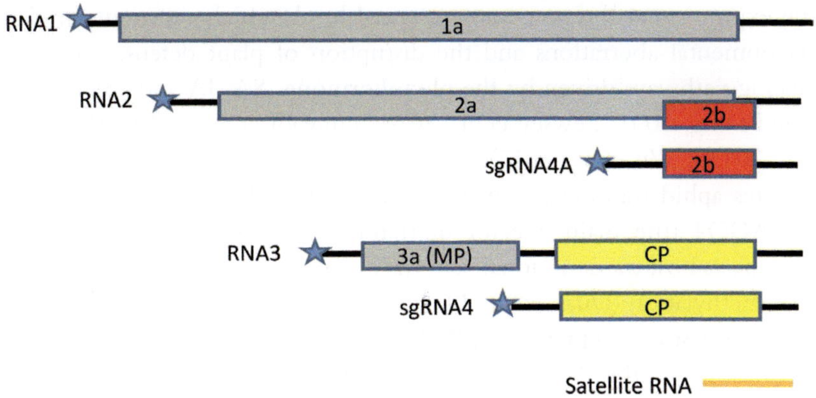

Fig. 1 Genomic organization of cucumber mosaic virus. The three genomic RNAs (RNA1 to 3) are depicted with the cap structure (star) at the 5′end. ORFs coding for viral proteins are shown with boxes. sgRNAs allowing the translation of the second ORF in dicistronic RNAs are also shown. Finally, a virus-associated satellite RNA is shown. Features discussed in depth in Section 5.1 are shown in color (ORFs coding for relevant proteins, and the associated satellite RNA).

the density of aphid and plant populations, the lifespan of the host and environmental conditions (Betancourt et al., 2013; Escriu et al., 2003; Hily et al., 2016). These experiments have also underlined evolutionary constraints that led to coadaptation of CMV proteins (Escriu et al., 2007; Fraile et al., 1997).

Like other members of the family *Bromoviridae*, CMV has a tripartite genome (Fig. 1) (Bujarski et al., 2019). RNA1 encodes the 1a replication protein, a putative methyltransferase and helicase. RNA2 contains two ORFs, the 2a RNA-dependent RNA polymerase and the small 2b protein, which is unique to cucumoviruses and some members of the genus *Ilarvirus*. RNA 3 is also bicistronic, encoding the MP and CP. Depending on the host and CMV strains, mutations in any of these proteins can impact symptom induction and/or severity (reviewed in Mochizuki and Ohki, 2012). However, 2b and CP are considered the main symptom determinants in most hosts. Both proteins are expressed from subgenomic RNAs: RNA4a for the translation of 2b, and RNA4 for the translation of CP.

The 2b protein is multifunctional. It is a VSR and interferes with several RNA silencing pathways. First, it binds to vsiRNAs and inhibits antiviral RNA silencing, an activity that requires 2b dimerization (González et al., 2012; Goto et al., 2007; Xu et al., 2013). Second, 2b interacts with ARGONAUTE 1 (AGO1, a key effector of RNA silencing) and interferes with the plant miRNA pathway, causing a major reprogramming of plant gene expression at the post-transcriptional level, which can result in plant developmental aberrations and the disruption of plant defense and stress signaling pathways driven by the phytohormones SA, JA and abscisic acid (Ji and Ding, 2001; Lewsey et al., 2010; Tungadi et al., 2019; Westwood et al., 2013b; Wu et al., 2017). The manipulation of JA defense responses facilitates aphid transmission (Carr et al., 2019a). Third, 2b also interacts with AGO4 (the main effector of transcriptional gene silencing, TGS) and impacts host transcriptional patterns by interfering with RNA-directed DNA methylation (Duan et al., 2012; González et al., 2010; Hamera et al., 2012). Interestingly, epigenetic DNA methylation changes, including hypomethylation of the AGO2 gene, were found to be associated with the induction of symptom recovery (Wang et al., 2018). These diverse silencing suppression activities of 2b vary with the specific CMV strain and with the host (reviewed in Mochizuki and Ohki, 2012; Salánki et al., 2018).

The 2b's function as a virulence factor is not only attributed to its silencing suppression activity. Indeed, the sub-cellular localization of 2b influences its activities as a VSR and as a virulence determinant in a differential manner (Du et al., 2014). 2b is partly shuttled into the nucleus and the rate of nuclear to cytoplasmic localization affects virulence (Du et al., 2014; González et al., 2010). The VSR activity of 2b is dependent on its cytoplasmic localization but other virulence functions take place in the nucleus. Phosphorylation of 2b regulates nuclear shuttling, revealing another layer of regulation of the protein activities (Nemes et al., 2017). Interestingly, 2b also plays a dual role in inducing the accumulation of SA and downstream defense responses in *Arabidopsis thaliana* and *Nicotiana benthamiana* and in mitigating the impact of these responses (Zhou et al., 2014). A direct interaction between the 2b of a necrotic CMV strain and a plant catalase (CAT3) was shown to contribute to necrosis severity associated with some CMV strains (Inaba et al., 2011; Murota et al., 2017). CAT3, which normally accumulates in the cytoplasm, minimizes the accumulation of toxic reactive oxygen species (ROS) during SA-dependent necrotic responses. Relocalization of CAT3 to the nucleus by 2b inhibited this function and intensified necrotic symptoms (Inaba et al., 2011). In addition, 2b induced proteasome-mediated destabilization of CAT3 in a strain-specific manner (Murota et al., 2017).

The accumulation and activity of 2b is tightly regulated using various self-attenuation mechanisms to minimize its impact on the host. Different plant transcriptome changes were noted in CMV-infected plants or in transgenic lines ectopically expressing 2b (Zhao et al., 2018), suggesting a regulatory role for other viral proteins or for plant proteins induced by viral infection. A plant calmodulin-like protein (rgs-CaM, also acting as an endogenous RNA silencing suppressor) is activated by the CMV-induced SA-dependent defense response and interacts with 2b to promote its degradation via autophagy (Jeon et al., 2017; Nakahara et al., 2012). The CMV 2a protein induces the basal immune PTI response (Westwood et al., 2013a) and thus may indirectly contribute to the degradation of 2b. On the other hand, the CMV MP suppresses PTI and minimizes ROS accumulation (Kong et al., 2018). The CMV CP, which accumulates to high levels at late stages of infection, inhibits the translation of 2b and promotes symptom recovery (Zhang et al., 2017). Interestingly, the phosphorylation state of the CMV CP impacts its ability to down-regulate 2b (Nemes et al., 2019). Since phosphorylation is reversible, this could contribute to the cycling mosaic symptoms observed in tobacco, in which symptomatic and asymptomatic phases of infection alternate through the course of infection (Nemes et al., 2019). The CMV 1a protein was also shown to restrict the 2b-directed miRNA regulation in double transgenic lines, affecting defense responses to aphids (Westwood et al., 2013a). From these combined results, it can be concluded that every CMV protein impacts the activity of 2b in direct or indirect ways.

Benign satellite RNAs can also counteract the disruption of the miRNA pathway caused by 2b (Feng et al., 2012). A satellite RNA (SD-satRNA) was shown to reduce the accumulation of 2b and of its coding subgenomic RNA and to mitigate the induction of yellowing symptoms (Hou et al., 2011). In addition, the satellite RNA recruited the RNA silencing machinery in co-infection with 2b-deficient CMV mutants, stimulating the production of satRNA-specific siRNAs and protecting viral RNAs from degradation (Hou et al., 2011). It was later shown that satRNA-derived siRNAs are able to bind 2b and saturate its binding capacity for other siRNAs or miRNAs (Shen et al., 2015). In this manner, presence of the satellite RNA could reverse alterations of miRNA and siRNA pathways caused by 2b, minimizing the impact on the host (Shen et al., 2015).

In addition to their modulating effects on 2b, the CP and satellite RNAs can also directly contribute to symptom formation. Comparison of two strains of CMV, CMV-M inducing chlorosis and CMV-Q inducing green mosaic, mapped the chlorosis symptom determinant to the CP (Qiu et al., 2018). The CP of CMV-M interacts with the precursor form of a chloroplast ferredoxin protein (Fd1) and hinders its transport to the chloroplast for maturation. In contrast, the CP of the mild isolate did not. It was concluded that the CP interference of Fd1 contributes to the induction of chlorosis by some CMV strains (Qiu et al., 2018). Finally, an siRNA derived from a CMV satellite RNA (Y-satRNA) was shown to induce chlorosis by targeting a chloroplast gene, tobacco magnesium protoporphyrin chelatase subunit I, which is involved in chlorophyll synthesis (Shimura et al., 2011; Smith et al., 2011).

Taken together, these studies reveal the complex regulatory schemes between CMV proteins and satellite RNAs, that co-exist to induce symptoms and mitigate disease severity. These include manipulation of multiple plant signaling pathways and repurposing of protein degradation and RNA silencing mechanisms to control the accumulation of the different players.

5.2 Plum pox virus: Concerted action of three multifunctional viral proteases and of other viral proteins

Members of the family *Potyviridae* represent the largest group of plant positive-strand RNA viruses and many have important economic impact (Wylie et al., 2017). It has been estimated that the first potyvirus emerged in Eurasia 15,000–30,000 years ago, and that potyviruses evolved concomitantly with agriculture intensification to expand their host range (Gibbs et al., 2020; Moury and Desbiez, 2020). Thus, potyviruses have a long-history of co-evolution with their hosts. Among notable potyviruses, plum pox virus (PPV, genus *Potyvirus*) infects wild and cultivated *Prunus* species and causes the devastating sharka disease (García et al., 2014; Rodamilans et al., 2020; Scholthof et al., 2011). PPV also infects experimental herbaceous hosts, including pea, *A. thaliana* and *N. benthamiana*. Our current knowledge of the molecular biology and pathogenesis of PPV is mostly derived from research conducted in these experimental hosts, although recent efforts have been directed at examining the impact of PPV infection in woody hosts (Rodamilans et al., 2020). There are more than 200 complete genome sequences available for PPV, which are regrouped into 10 strains based on phylogenetic relationships (Hajizadeh et al., 2019). Host range and pathogenicity varies with the PPV isolate and is not always correlated with the strain groupings. In some cases, host range expansion or differential symptom expression of PPV isolates has been attributed to single point mutations (Maliogka et al., 2012; Sáenz et al., 2001; Wallis et al., 2007). These mutations can be present in variants representing a subpopulation of PPV isolates and the complexity of variant populations probably contributes to the overall pathogenicity of the virus (Maliogka et al., 2012; Sáenz et al., 2001).

Although disease severity varies with the specific PPV isolate and with the host, the development of chlorotic or necrotic lesions has been associated with oxidative stress in both herbaceous and woody hosts (Díaz-Vivancos et al., 2006, 2008). The oxidative stress is concomitant with ultrastructural changes of the chloroplasts, increased accumulation of SA and ROS and disruption of several antioxidant enzymes. Consistently, treatment of peach plants with antioxidant compounds such as L-2-oxo-4-thiazolidine-carboxylic acid alleviates the oxidative stress caused by PPV and ameliorates symptoms (Clemente-Moreno et al., 2010, 2013). The oxidative stress is due to the activation of SA-dependent basal responses (Rodamilans et al., 2020). Indeed, transcriptomic and proteomic studies of PPV-infected plants reveal an induction of defense- and stress-related genes and a decrease in photosynthesis and carbohydrate-related genes (Nováková et al., 2018; Rodamilans et al., 2014; Rubio et al., 2015). The intensity of these expression changes is correlated with the PPV strain virulence (Nováková et al., 2018). Identification of translated mRNAs in phloem and non-phloem cells from PPV-infected leaves highlighted the phloem as a key player in the plant response to virus infection (Collum et al., 2020). Defense gene mRNAs were shown to be actively translated in phloem cells in response to rapid virus accumulation changes and during leaf maturation. In addition to SA-dependent signaling, other phytohormone signaling pathways are also disrupted in PPV-infected cells and the levels of JA, gibberellin and abscisic acid are altered (Dehkordi et al., 2018).

Like other members of the genus *Potyvirus*, the PPV genome consists of one positive-strand RNA (approximately 9770 nts) which encodes one large polyprotein (Fig. 2) (Rodamilans et al., 2020; Wylie et al., 2017). The polyprotein is cut by three viral proteases of different types (reviewed in Mann and Sanfaçon, 2019) The N-terminal protease, P1, is a serine protease that cleaves at a single site to release itself from the polyprotein. Immediately adjacent to P1 is HC-Pro, a multifunctional cysteine protease that also cleaves the polyprotein at a single site for auto-processing (Valli et al., 2018). The third protease, NIa-Pro, is a chymotrypsin-like cysteine protease and processes the polyprotein at the remaining cleavage sites. As will be described below, the processing efficiency of the three proteases is highly regulated allowing control of not only the timing of release of mature proteins and intermediate polyproteins, but also of the activity of these proteins. The C-terminal region of the polyprotein contains domains for replication proteins (CI, 6K2, NIa-VPg, NIa-Pro, NIb polymerase) and the CP. In addition, polymerase slippage in the P3 coding region allows

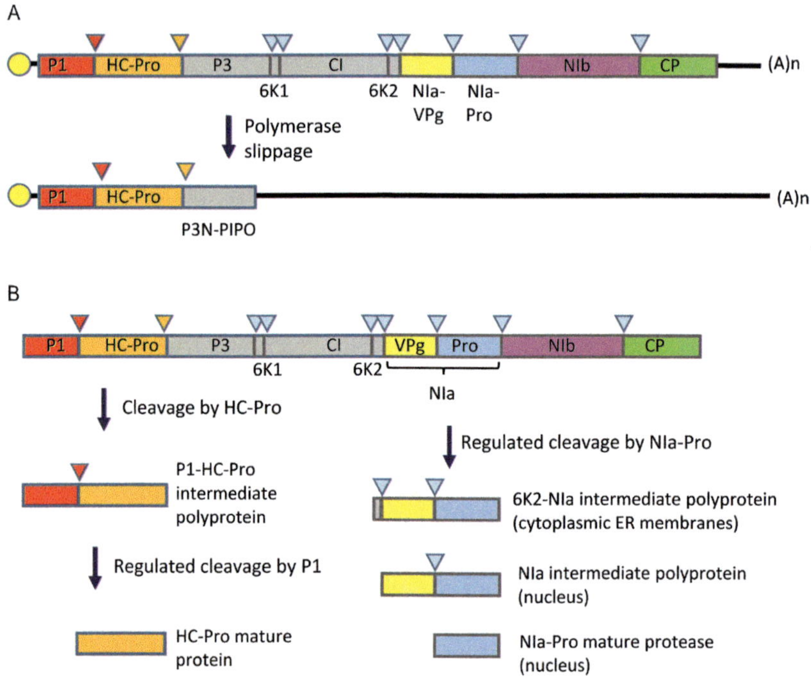

Fig. 2 Genomic organization of plum pox virus. (A) The genomic RNA is shown at the top with the box representing the ORF coding for the single large polyprotein. The yellow circle represents the viral VPg protein covalently attached to the 5′end of the RNA and (A)n represents the 3′ terminal poly A tail. The alternative RNA produced by polymerase slippage is shown at the bottom, along with the resulting polyprotein. Vertical lines in the ORF for the polyprotein represent the location of cleavage sites processed by the P1 protease (red arrowheads), HC-Pro (orange arrowheads) and NIa-Pro (blue arrowheads). Regions of the polyprotein coding for proteins discussed in depth in Section 5.2 are highlighted in color. (B) Regulated proteolytic cleavage of the main polyprotein. As discussed in the text, regulated cleavage by the P1 and NIa-Pro proteases allows the accumulation of intermediate polyproteins that contain two or more protein domains. These intermediate polyproteins are then slowly processed into the corresponding mature proteins. Please note that not all intermediate polyproteins and mature proteins derived from the main polyprotein are shown. Rather, the figure only depicts intermediate polyproteins and corresponding mature proteins that are discussed in detail in the text. The stable NIa intermediate polyprotein contains both the VPg and protease domains, while the mature protease is referred to as NIa-Pro.

the production of an alternative polyprotein, which is cleaved to produce another mature protein, P3N-PIPO (Olspert et al., 2015, 2016; Rodamilans et al., 2015). Symptom and host range determinants have been mapped to various regions of the polyproteins, including the P1 and HC-Pro proteases,

P3/P3N-PIPO, VPg, NIb and CP, as well as the cleavage site between 6K1 and CI (Calvo et al., 2014; Carbonell et al., 2013; Maliogka et al., 2012; Nagyová et al., 2012; Sáenz et al., 2000, 2001; Shan et al., 2015).

Like the CMV 2b protein, the HC-Pro protein of potyviruses is multifunctional. It is associated with the tip of the filamentous virus particles and coordinates various steps of the infection cycle, including translation, virus particle accumulation, cell-to-cell movement, antiviral RNA silencing suppression and aphid transmission (Valli et al., 2018). The HC-Pro of potato virus Y was the first characterized plant virus VSR, as evidenced by its ability to enhance virus accumulation and spread and to facilitate synergistic interactions with other viruses (Anandalakshmi et al., 1998; Pruss et al., 1997). Depending on the specific potyvirus, HC-Pro has been found to interact with various plant proteins including AGO1, proteasome subunit proteins, chloroplast proteins, translation factors and endogenous suppressors of silencing (such as rgs-CaM) (reviewed in Valli et al., 2018). Recently, the HC-Pro protein of potato virus A was also shown to interact with the plant *S*-adenosyl-L-methionine synthetase and *S*-adenosyl-L-homocysteine hydrolase, causing major disruption of the methionine cycle and of the downstream transcriptional and post-transcriptional RNA silencing pathways and reduced accumulation of the antioxidant glutathione (De et al., 2018; Ivanov et al., 2016; Mäkinen and De, 2019). HC-Pro also sequesters vsiRNAs and miRNAs. Finally, a recent study showed that the HC-Pro of turnip mosaic virus (TuMV) also counteracts the RNA decay mechanism by interacting with and inhibiting the plant exoribonuclease 4 (Xrn4) (Li and Wang, 2018). While these protein-protein and protein-RNA interactions may differ from virus to virus and have not all been confirmed for PPV, they highlight the diverse functions of potyvirus HC-Pros and their global impact on the plant metabolism and virus infection cycle. In the case of PPV, mutations in HC-Pro have been shown to impact virus virulence, hinder viral cell-to-cell movement in a host specific manner and prevent synergistic interactions with other viruses (González-Jara et al., 2005; Sáenz et al., 2001; Varrelmann et al., 2007). The PPV HC-Pro was also shown to stabilize the CP and enhance virus particle yield (Valli et al., 2014).

The VSR activity of the PPV HC-Pro is controlled by the processing activity of the adjacent P1 protease, providing a highly regulated viral self-attenuation mechanism. Indeed, the P1-HC-Pro intermediate polyprotein shows impaired VSR activity compared to the mature HC-Pro (Pasin et al., 2014). The variable N-terminal region of P1 was identified as a symptom and host range determinant (Maliogka et al., 2012). This region of P1 was shown to inhibit cleavage at the P1-HC-Pro junction and host factors, yet to be identified, are involved in this function (Pasin et al., 2014). Deletion of the N-terminal region of P1 accelerated the release of mature HC-Pro, enhanced virus accumulation in infected plants and exacerbated symptom expression (Pasin et al., 2014; Shan et al., 2018). Thus, although the N-terminal region of P1 is dispensable for infection, it was suggested that it is retained to modulate disease severity (Pasin et al., 2014). The results also highlight the coordinated action of two viral proteases.

The identification of a host range determinant at the 6K1-CI junction, which is cleaved by NIa-Pro, highlights that the NIa-Pro proteolytic activity also contributes to virus virulence through regulated cleavage of the polyprotein (Calvo et al., 2014). The host specific impact of mutations at the 6K1-CI cleavage site also implies a role for plant factors in regulating the activity of NIa-Pro. The sub-cellular localization of NIa-Pro is controlled by its regulated autoproteolytic release from two intermediate polyproteins, NIa (containing the NIa-VPg and NIa-Pro domains) and 6K2-NIa (containing the 6K2, NIa-VPg and NIa-Pro domains) (Fig. 2B). (Restrepo-Hartwig and Carrington, 1992). The 6K2-NIa polyprotein accumulates in cytoplasmic ER membranes, while the NIa stable intermediate polyprotein and the mature NIa-Pro protein are predominantly found in the nucleus. In addition, the NIa-Pro of TuMV was shown to promote aphid transmission and is relocalized to vacuoles in the presence of aphids (Bak et al., 2017; Casteel et al., 2014, 2015). The shuttling of NIa-Pro to various sub-cellular localizations provides opportunities for interactions with diverse plant proteins. Indeed, the NIa-Pro of potyviruses has a large network of interacting partners, as demonstrated by affinity purification of protein complexes associated with the tobacco etch virus NIa-Pro mature protein or with the NIa intermediate polyprotein (Martinez et al., 2016). Perhaps not surprisingly, many of the identified interacting partners were shown to be associated with stress responses and photosynthesis functions. Recently, the NIa-Pro of pepper mottle virus was also shown to interfere with TGS, by manipulating the methylation status of the plant DNA (Gong et al., 2020). Because NIa-Pro is the only potyvirus protease with trans-cleavage activity, a question of future interest will be to determine whether it also cleaves plant proteins to facilitate virus infection, counteract defense responses or influence symptom expression. This question is particularly relevant since NIa-Pro is closely related to the 3C protease of animal and human picornaviruses, a protease known to cleave host proteins, e.g., translation and transcription factors and components of the innate immune response (Jensen et al., 2015; Laitinen et al., 2016; Sun et al., 2016).

In addition to the three viral proteases, other potyviral proteins also contribute to the regulation of symptom severity. As mentioned above, this includes P3/P3N-PIPO, VPg, NIb and CP. The PPV CP was shown to suppress the oxidative stress response and the induction of marker genes associated with the basal PTI response (Nicaise and Candresse, 2017). The stability of the CP is impacted by the phosphorylation status of its N-terminal region, likely controlling the rate of CP degradation by the proteasome (Martínez-Turiño et al., 2018). The CP is also modified by *O*-GlcNAcylation (Martínez-Turiño et al., 2018). Both types of modifications are dispensable for virus viability, but impact the degree of virus accumulation in plant. Although phosphorylation of the core region of the CP was shown to influence virus particle assembly (Hervás et al., 2020), the impact of phosphorylation in the core or N-terminal region of the CP on other functions of the CP (e.g., inhibition of PTI) is not known. Recently published results on other potyviruses have also highlighted counter-defensive roles for the VPg protein (silencing suppression by enhanced degradation of SGS3, a plant protein associated with antiviral RNA silencing pathways) and for the NIb polymerase (suppression of host basal immune responses directed by sumoylation of NIb) (Cheng and Wang, 2017; Cheng et al., 2017; Rajamäki et al., 2014), although it is not known whether similar functions are conserved for the PPV VPg and NIb proteins. The potyvirus VPg proteins, including that of PPV, have many other interacting partners, including translation factor eIF4E isoforms, and facilitate the translation of viral RNAs and virus movement (Jiang and Laliberté, 2011; Sanfaçon, 2015). It was recently shown that the VPg of TuMV also interacts with a decapping enzyme (DCP2) to counteract RNA decay (Li and Wang, 2018). As demonstrated for TuMV, P3N-PIPO is a movement protein that interacts with plasmodesmata proteins and redirect replication complexes to the plasmodesmata (Chai et al., 2020; Wei et al., 2010). Although not a VSR, the P3N-PIPO of clover yellow vein virus was found to play a key role in synergistic interactions with other viruses (Hisa et al., 2014). The mechanism of action of the PPV P3 and/or P3N-PIPO proteins in regulating disease severity is not well understood, but may implicate some of the activities described above for other potyviruses.

The emerging picture on how PPV and other potyviruses regulate disease severity is complex. It includes regulated cleavage of the viral polyprotein by three proteases to control the release and activity of multifunctional effector viral proteins. It also implicates a role for post-translational modifications of viral proteins (as exemplified for CP and NIb) to regulate their stability and activities. Finally, a recent study has also suggested differential translation rates for various portions of the potato virus A polyprotein contributes to controlling the accumulation of different viral proteins during the course of infection, although how differential translation occurs is not known (Saha et al., 2019).

5.3 Citrus tristeza virus: Management of complex virus populations and plant immune responses in *Citrus* trees

As highlighted above, studying viral manipulation of disease severity in trees is difficult. This is especially true for RNA viruses with very large genomes, such as members of the family *Closteroviridae* (Fuchs et al., 2020). However, despite these problems, citrus tristeza virus (CTV, genus *Closterovirus*) has become a model system for the study of virus pathogenicity and of virus population evolution in citrus trees (Dawson et al., 2015; Ruiz-Ruiz et al., 2019; Scholthof et al., 2011). CTV exists in complex populations in the field and induces a variety of symptoms with different degrees of severity depending on the isolate and the host. The natural host range of CTV is restricted to cultivated citrus trees and related wild species. Symptoms induced by CTV include decline, stem pitting, seedling yellows and vein clearing but many CTV infections of citrus trees are also asymptomatic (Dawson et al., 2015). Perhaps because of the persistent nature of CTV infection in long-lived citrus trees, CTV populations are complex but stable and can include a mixture of 2, 3 or more strains in various ratio (Harper et al., 2015a). Using genome sequence data, CTV has been grouped into seven phylogenetically distinct strains. Strains present in mixed populations have been reported to cooperate to accomplish various steps of the infection cycle. For example, the T36 strain is restricted to the roots in some hosts but can successfully invade shoot tissues in the presence of the VT strain (but not in the presence of other strains) (Harper et al., 2015b). Aphids have been shown to transmit several strains at once and complementation between specific CTV strains have been reported to enhance aphid transmission rates (Harper et al., 2018). The rules regulating CTV populations are complex. Within an infected tree, related sequence variants (belonging to the same strain) are excluded, but sequence variants from different strains are not, a phenomenon referred to as superinfection exclusion (Folimonova et al., 2010). Sequence variants from different CTV strains can be found

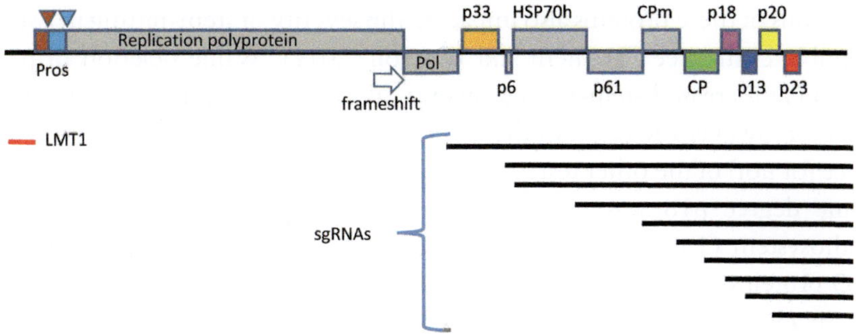

Fig. 3 Genomic organization of citrus tristeza virus. The genomic RNA is shown at the top with the boxes representing the different ORFs. The 5′ end of the RNA is likely capped, but this has not been confirmed experimentally. The first ORF is a polyprotein, which is cut at two sites in its N-terminal region by the two leader proteases. A ribosomal frameshift is required to produce a fusion protein that contains the RNA-dependent RNA polymerase domain (Pol). C-terminal ORFs are translated from a series of coterminal sgRNAs as shown. LMT1 is a lncRNA corresponding to the 5′UTR. Features discussed in depth in Section 5.3 are shown in color (ORFs coding for relevant proteins and LMT1 RNA).

co-infecting the same cells providing opportunities for virus evolution through recombination (Bergua et al., 2016). CTV is restricted to the phloem, shows limited or no cell-to-cell movement depending on the host, and relies on the plant vascular system for systemic invasion of the plant (Folimonova et al., 2008).

The CTV genome consists of a single large RNA (approximately 19,300 nts) and includes 12 ORFs (Fig. 3) (Dawson et al., 2015). ORF1 is translated as a polyprotein from the genomic RNA and is cleaved into mature proteins by two adjacent cysteine proteases, that arose through duplication of an ancestral protease (Kang et al., 2018). A series of 3′ coterminal subgenomic RNAs direct the translation of the 10 3′ terminal ORFs. Of the 19 predicted mature proteins encoded by CTV, 12 are core proteins conserved in other closteroviruses, while 5 (p33, p18, p13, p20 and p23) are unique to CTV (Dolja et al., 2006). All five non-core proteins have been implicated in host range expansion, viral counter defenses and disease severity. p20 and p23 have VSR activities along with the CP (Lu et al., 2004) and p23 was identified as a symptom determinant in susceptible hosts (Flores et al., 2013). The p33, p18 and p13 proteins are dispensable in two highly susceptible citrus trees but are required in different combinations for infection of other hosts (Tatineni et al., 2008, 2011). Single, double and triple deletions of these proteins also impacted the severity of stem-pitting in a susceptible citrus tree (Tatineni and Dawson, 2012). While deletion of p33 generally increased stem-pitting severity, deletion of p13 had the opposite effect. However the effect of these deletions was also tempered by the presence (or not) of the other tested proteins. A balance between p33, p13 and to some degree p18 was proposed to regulate the disease severity in this pathosystem (Tatineni and Dawson, 2012). The multifunctional p23 and p33 proteins are the best characterized CTV non-core proteins.

In addition to its VSR activity, p23 regulates the asymmetric accumulation of negative-strand and positive-strand RNAs to enhance translation of viral RNAs (Satyanarayana et al., 2002). It binds RNAs and also interacts with the plant glyceraldehyde 3-phosphate dehydrogenase to facilitate virus accumulation, and p23 possibly enhances virus movement (López et al., 2000; Ruiz-Ruiz et al., 2018). The p23 protein accumulates to the nucleus, nucleolus, Cajal bodies, plasmodesmata, and cytoplasm (Ruiz-Ruiz et al., 2013). Self-interaction occurs in the nucleolus, Cajal bodies and plasmodesmata, while interaction with glyceraldehyde 3-phosphate dehydrogenase is associated with cytoplasm aggregates (Ruiz-Ruiz et al., 2018). The VSR activity required most regions of the protein, but was only observed for p23 mutants that retained nucleolus localization (Ruiz-Ruiz et al., 2013). A zinc finger motif within a region of p23 enriched in basic amino acids is required for efficient RNA binding and silencing suppression (López et al., 2000; Ruiz-Ruiz et al., 2013), although the relationship between these two activities is not clear. vsiRNAs accumulate to high levels in CTV-infected Mexican lime and sweet orange, two susceptible hosts, suggesting that the CTV VSRs, including p23, act downstream of vsiRNA production, possibly by sequestering vsiRNAs, but more likely by inactivating one or several AGOs (Ruiz-Ruiz et al., 2011, 2019). Endogenous miRNA levels are disrupted in infected trees, possibly through the action of p23 or another VSR (Harper et al., 2019; Ruiz-Ruiz et al., 2011). One of these miRNA alterations was actually linked with disease severity; miR164 regulates the expression of CUC2 (cupped shaped cotyledon 2), a transcription factor associated with vascular differentiation (Yang et al., 2015). Interestingly, miR164 accumulation was reduced at the onset of CTV-induced stem pitting in *Citrus macrophylla* (Harper et al., 2019). However, the effect of CTV infection on miR164 accumulation is host specific. Indeed, miR164 is upregulated in infected Mexican lime and sweet orange but downregulated in CTV-infected sour orange (Ruiz-Ruiz et al., 2011).

The p23 protein was identified as a symptom determinant for yellows seedling syndrome in CTV-infected sour orange and grapefruit (Albiach-Marti et al., 2010). Constitutive transgenic expression of p23 resulted in various developmental defects that were dependent on the specific *Citrus* species, and included some of the phenotypes observed in natural infections plus additional defects (Fagoaga et al., 2005; Ghorbel et al., 2001). However, transgenic expression of p23 from a phloem-specific promoter in Mexican lime recapitulated more accurately the vein clearing and stem-pitting symptoms associated with the phloem limited CTV infection of the same host (Soler et al., 2015). In addition, the phenotype induced by the *p23* transgene was consistent with the CTV strain from which it was originated. A truncated p23 protein deficient in VSR activity induced symptoms similar to those induced by the full-length p23; uncoupling the pathogenic and VSR functions of p23 (Soler et al., 2015). It will be interesting to analyze miRNA expression patterns in transgenic lines expressing wild-type or VSR-deficient p23 variants and compare them with those observed in CTV-infected trees.

As mentioned above, p33 is dispensable to infect some citrus hosts, including the highly susceptible *Citrus macrophylla*, but it is required for systemic infection in other hosts, notably sour orange (Tatineni et al., 2011). A transmembrane domain in the p33 protein directs its association with the plasma membrane and with cytoplasmic vesicles and is required for sour orange infection (Kang et al., 2015). The p33 protein was also identified as a viral effector that induces ROS production and cell death in *C. macrophylla* (Sun and Folimonova, 2019). In this host, deletion of the p33 coding region allowed the mutant CTV to escape from the phloem and into immature xylem tracheid cells, and aberrant differentiation of the vascular system resulted (Sun and Folimonova, 2019). Thus, p33 contributes to restricting CTV to the phloem and limits the damage the virus causes to susceptible hosts. Another activity of p33 is its essential function in superinfection exclusion. Deletion of p33 eliminated superinfection exclusion at the plant level (Folimonova, 2012), although it did not prevent superinfection exclusion in single cells (Bergua et al., 2014). Other viral factors are also required for superinfection exclusion, notably the region of the genome encoding the two cysteine proteases (Atallah et al., 2016). The molecular mechanism(s) directing superinfection exclusion is not known, although it is likely not related to RNA silencing in this case as it does not depend on the induction of vsiRNAs (Folimonova et al., 2014). Given the multifunctional activities of p33 in viral movement and pathogenicity and given the known synergies between strains present in mixed populations, it is possible that the control of virus population complexity is one of the key elements regulating disease severity.

CTV infections are often associated with accumulation of various dRNAs that vary in size and differ in what portion of the CTV genome has been deleted (Bar-Joseph and Mawassi, 2013). The impact of these dRNAs on virus multiplication or symptomology is largely unknown and will be a fruitful area of research in the future. For example, it has been suggested that long dRNAs associated with most CTV isolates may act as siRNA sponges or decoys that counteract plant antiviral RNA silencing (Bar-Joseph and Mawassi, 2013). In addition to dRNAs, long non-coding RNAs derived from the 5′ region of the genome are also readily detected in CTV-infected trees (Gowda et al., 2003). One of the lncRNAs, LMT1, is indispensable for infection of citrus trees (Kang et al., 2019). In the model herbaceous host *N. benthamiana*, LMT1 suppresses the SA-dependent defense response and facilitates the movement of CTV. LMT1 also mitigates the accumulation of ROS by inducing the expression of an alternative oxidase, AOX-1a (Kang et al., 2019). Thus, the balance between the p33-dependent induction of ROS and the LMT1 counter-defense activity may play a key role in the establishment of a persistent infection.

Taken together, these studies highlight the complexity of the interactions between CTV strains and various citrus hosts. Results also indicate how unique CTV proteins and lncRNAs balance each other activities to overcome immune defense responses and regulate the complexity of viral population.

5.4 Beet necrotic yellow vein virus: Manipulation of plant auxin signaling leading to rhizomania disease in sugar beet

Beet necrotic yellow vein virus (BNYVV) is the type member of the genus *Benyvirus* (family *Benyviridae*), which also includes beet soil-borne mosaic virus (BSBMV) (Gilmer et al., 2017). Although BNYVV and BSBMV have similar host range, and are both transmitted by the soil-borne plasmodiophoromycete *Polymyxa betae*, they differ in the symptoms they induce (Laufer et al., 2018). BNYVV causes the devastating rhizomania disease in sugar beet, which is characterized by the deregulated proliferation of lateral roots and root hairs. In contrast, BSBMV does not cause symptoms in roots of sugar beet, but instead induces yellow vein banding or mosaic, as well as leaf distortion. It has been suggested that alteration of the root system by BNYV may facilitate its transmission by the soil-borne vector, which preferentially accumulates around expanding root hairs (Fernando Gil et al., 2018).

In plants, the formation of lateral roots is tightly regulated by auxin. Indeed, LATERAL ORGAN BOUNDARIES DOMAIN (LBD) transcription factors control the expression of expansins, which are proteins responsible for cell wall loosening/expansion (Lavenus et al., 2013). In BNYVV-infected sugar beet plants, LBD transcription factors, *EXPANSIN* genes, and genes associated with auxin biosynthesis are drastically upregulated, likely contributing to the excessive lateral root formation associated with rhizomania disease (Fan et al., 2015; Fernando Gil et al., 2018, 2020). Reprogramming of genes related to auxin signaling was also observed in BSBMV infected plants but to a lesser extent and some LBD transcription factor genes are only upregulated by BYNVV (Fernando Gil et al., 2020). Not surprisingly, expression of defense response genes (transcription factors and pathogenesis-related proteins) is also enhanced in response to either BYNVV or BSBMV infection (Fan et al., 2015; Fernando Gil et al., 2020).

The BNYVV genome consists of four or five RNAs, depending on the isolate (Fig. 4). BSBMV has four genomic RNAs. Both viruses share similar genomic organization with RNA1 and RNA2 encoding conserved core proteins implicated in replication, encapsidation and cell-to-cell movement (Gilmer et al., 2017). The other RNAs encode accessory proteins that share little sequence identity between the two viruses and have been implicated in vector transmission, cell-to-cell movement and symptom severity. The RNA3-encoded p25 protein of BYNVV has been identified as the main pathogenicity determinant and mutations of this protein can prevent lateral root formation (Tamada et al., 1999). Breaking of the sugar beet single dominant resistance gene *Rz1* was also mapped to mutations in a tetrad of amino acids (positions 67–70) in the BNYVV p25 protein, although resistance breakage was also attributed to the RNA5-encoded p26 protein in some isolates (Acosta-Leal et al., 2010; Bornemann and Varrelmann, 2013; Chiba et al., 2011; Decroës et al., 2015; Galein et al., 2018; Koenig et al., 2009). Resistance-breaking strains with mutations in the p25 tetrad motif displayed reduced fitness in susceptible hosts (Bornemann and Varrelmann, 2013; Galein et al., 2018), demonstrating opposing evolutionary constraints. In the field, BNYVV populations are complex with several virus types present, notably distinguished by the p25 tetrad motif, and the presence or absence of RNA5. These populations are dynamic and are influenced by the beet genotype (Bornemann and Varrelmann, 2013; Chiba et al., 2011; Galein et al., 2018; Weiland et al., 2019; Zhuo et al., 2015).

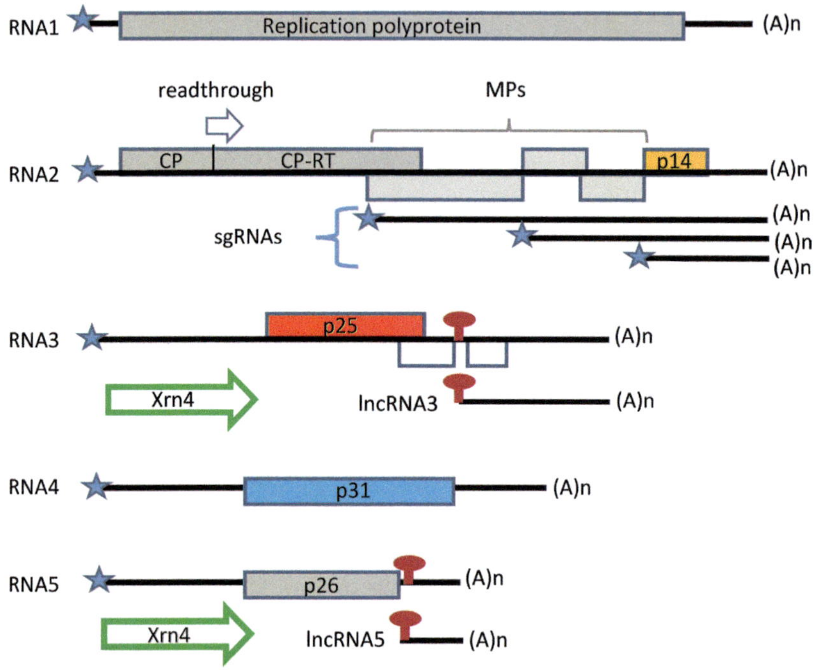

Fig. 4 Genomic organization of beet necrotic yellow vein virus. The five genomic RNAs are shown along with subviral RNAs (RNA2-derived sgRNAs) and lncRNAs derived from RNA3 and RNA5. The 5′ cap is shown with the blue star and (A)n represent the 3′ terminal polyA tail. ORFs are shown with the boxes. The coremin RNA structure that protects the lncRNAs from degradation by Xrn4 is shown with the red hairpin. Features discussed in depth in Section 5.4 are shown in color (ORFs coding for relevant proteins and coremin structure).

Transgenic expression of p25 in *A. thaliana* (a non-host) induced root branching, impacted hormonal regulation, in particular by auxin, and caused reprogramming of auxin-responsive genes. This recapitulated some of the symptoms and transcriptional changes observed in sugar beet plants infected with BNYVV (Peltier et al., 2011). BNYVV p25 was shown to interact with BvAUX28, a transcriptional repressor implicated in lateral root formation (Fernando Gil et al., 2018). BvAUX28 inhibits the activity of transcription factors ARF7 and ARF19, which regulate the expression of LBD transcription factors and expansins. In plants, auxin controls the activity of transcriptional repressors such as BvAUX28 by promoting their degradation (Lavenus et al., 2013). The p25 protein contains both a nuclear localization signal and a nuclear export signal and shuttles between the nucleus and the cytoplasm (Vetter et al., 2004). Interestingly, interaction with the BNYVV p25 protein was shown to prevent the nuclear localization of BvAUX28 and its interaction with ARF7 and ARF19 (Fernando Gil et al., 2018). Thus, it was suggested that p25 plays a functional role similar to that of auxin by releasing the transcriptional repression activity of BvAUX28 (Fernando Gil et al., 2018).

Other factors may also contribute to the regulation of lateral root formation in BNYVV-infected plants. Auxin accumulates to higher levels in BNYVV-infected sugar beets and in *A. thaliana* expressing p25, and a positive feedback loop that contributes to the excessive lateral root formation has been proposed (Fernando Gil et al., 2018; Peltier et al., 2011). Deep sequencing of miRNAs from BYNVV-infected sugar beet samples, along with analysis of the corresponding mRNA degradome, revealed enhanced accumulation of miRNAs involved in hormone signaling and plant defense responses, and down-regulation of their mRNA targets (Liu et al., 2017). Whether this is caused by viral RNA silencing suppression or by other factors is not known. The impact of these altered miRNA patterns on virus accumulation and/or symptom severity has not yet been elucidated. Virus accumulation in roots is also facilitated by root-specific RNA silencing suppression activities of the RNA2-encoded p14 protein (Andika et al., 2012). The RNA4-encoded p31 protein, which contributes to symptom severity, is not a VSR per se but enhances the ability of BNYVV to suppress RNA silencing in roots (Rahim et al., 2007). In addition to suppressing RNA silencing, BNYVV must also counteract necrotic defense responses activated by resistance genes. An interaction was confirmed between p25 and a sugar beet F-box containing kelch-repeat protein involved in the induction of necrotic hypersensitive responses, suggesting that p25 targets the 28S proteasome to counteract necrotic defense responses and assists virus accumulation (Thiel et al., 2012). This interaction is not affected by the nature of the tetrad motif in the p25 protein (Thiel et al., 2012), implying that other factors contribute to the resistance-breaking activity of the p25 protein. Several additional putative cellular interaction partners of p25 have been identified using yeast two-hybrid screens, although the role of these interactions in the virus infection cycle need to be clarified (Thiel and Varrelmann, 2009). Finally, the mechanism by which the RNA5-encoded p26 protein contributes to symptom severity remains unknown.

The p25 protein is emerging as a key orchestrator of the BNYVV-sugar beet interaction, leading to rhizomania disease by manipulating the plant auxin pathway, and by interacting with the proteasome to down-regulate the defense response. However, many questions remain. For example, it is not known how the accumulation and activity of the p25 protein is regulated. Given its interaction with components of the plant proteasome, it is possible that p25 is targeted to degradation by the proteasome, although no experimental evidence could be obtained that this is the case (Thiel et al., 2012). The activity of the p25 protein may also be controlled by its ability to self-aggregate, which is influenced by the nature of the tetrad motif (Klein et al., 2007). In addition, the translation rate of p25 is likely affected by the intrinsic instability of the BNYVV RNA3, which is targeted by the cellular RNA decay pathway (5′ to 3′ degradation by Xrn4) (Dilweg et al., 2019; Flobinus et al., 2018). Interestingly, a conserved hairpin structure (termed coremin) downstream of the p25 open reading frame stalls the Xrn4-directed degradation of RNA3 and allows the accumulation of a stable lncRNA corresponding to the RNA3 3′UTR (Dilweg et al., 2019; Flobinus et al., 2018). A similar coremin motif was identified in the RNA5 sequence. The molecular function of these lncRNAs is not completely clear, although the RNA3-derived lncRNA has been implicated in promoting the virus cell-to-cell movement (Peltier et al., 2012). It is possible that the lncRNAs interferes with one or several defense pathways, but this has not been studied in the BNYVV-sugar beet interaction.

5.5 Nepoviruses: Repurposing of diverse plant RNA silencing mechanisms to allow symptom recovery

Within the family *Secoviridae*, the genus *Nepovirus* is the largest (Fuchs et al., 2017; Thompson et al., 2017). Nepoviruses generally have a wide host range that can span deciduous trees and herbaceous plants and includes both cultivated crops and wild species. Plant-to-plant transmission is facilitated by vectors (soil nematodes for most, but not all, nepoviruses), and nepoviruses are also transmitted through seeds (Fuchs et al., 2017). Many nepoviruses invade and persist in the meristem, and in reproductive organs (Santovito et al., 2014). A typical outcome of nepovirus infection in herbaceous hosts is symptom recovery (Fuchs et al., 2017; Ghoshal and Sanfaçon, 2015). Although recovered leaves do not display visible symptoms, the virus persists at low levels as first described in *N. tabacum* plants infected with tobacco ringspot virus (Wingard, 1928).

The nepovirus genome consists of two RNAs each encoding a single polyprotein (Fig. 5) (Fuchs et al., 2017). RNA1 encodes replication proteins. RNA2 encodes the MP, CP and the homing protein, which assists the replication of RNA2. The RNA1-encoded cysteine protease is related to the PPV NIa protease and cleaves both polyproteins. The processing of

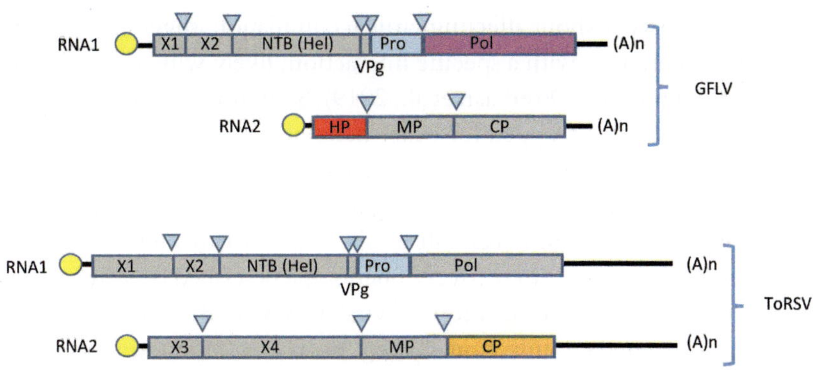

Fig. 5 Genomic organization of grapevine fanleaf virus (GFLV) and tomato ringspot virus (ToRSV). The two genomic RNAs are shown for each virus, along with the 5′ linked VPg protein (yellow circle) and the 3′ polyA tail (A)n. Each RNA includes one ORF (coding for a single polyprotein shown with the boxes). Cleavage sites recognized by the protease (Pro) are shown by the vertical lines and arrowheads. Regions of the polyproteins coding for proteins discussed in Section 5.5 are shown in color. HP: homing protein.

the RNA2 polyprotein is delayed because it depends on the enhanced activity of the mature protease, which must first be released from the RNA1 polyprotein (Chisholm et al., 2001). Individual cleavage sites on the two polyproteins are recognized at different efficiencies by the protease and this efficiency can also differ between virus species or even between isolates of the same species (Wetzel et al., 2013). Thus, as described for potyviruses, the tightly controlled proteolytic processing of the two polyproteins regulates the accumulation and activity of viral proteins at different stages of infection.

In the initial stages of the interactions between tomato ringspot virus (ToRSV) and either *N. benthamiana* or *N. tabacum* and between the F13 strain of grapevine fanleaf virus (GFLV) and *N. occidentalis*, the necrotic symptoms show the hallmarks of an SA-dependent response but only partially restrict virus movement (Jovel et al., 2007, 2011; Martin et al., 2018). Domain exchange between the mild GFLV strain GHu and the severe strain F13, identified the C-terminal region of the RNA2-encoded homing protein as an avirulence factor eliciting the necrotic response in this incompatible interaction (Martin et al., 2018). On the other hand, the symptom determinant for the non-necrotic vein clearing symptoms induced by the GFLV GHu strain in a compatible interaction with *N. benthamiana* was mapped to the C-terminal region of the RNA1-encoded polymerase (Osterbaan et al., 2019; Vigne et al., 2013). Interestingly, a single point mutation in this region of the polymerase was sufficient to modulate the symptom intensity without affecting virus accumulation, suggesting that the mutation is interfering with a specific interaction, likely with an as-yet unidentified plant protein (Osterbaan et al., 2019). Symptom determinants have not been conclusively mapped for other nepoviruses.

The association between symptom recovery and the induction of plant antiviral RNA silencing was first highlighted for the *N. clevelandii*-tomato black ring virus interaction (Ratcliff et al., 1997). In this interaction and in other nepovirus symptom recovery phenotype described later, recovered leaves show reduced accumulation of viral RNAs and sequence-specific resistance to secondary infection (Dong et al., 2010; Santovito et al., 2014; Siddiqui et al., 2008). A time-course study of *N. tabacum* infected with artichoke Italian latent virus showed that the virus titer is directly correlated with symptom severity and that a minimal threshold of virus accumulation is required before the plant RNA silencing mechanism is activated (Santovito et al., 2014). This suggests that nepoviruses evolved to use plant antiviral RNA silencing to moderate their accumulation and restore plant health, allowing them to persist at low levels and enabling meristem invasion and seed transmission.

In ToRSV-infected *N. benthamiana* plants, symptom recovery from the initial necrotic response is not strictly associated with reduced viral RNA accumulation (Ghoshal and Sanfaçon, 2014; Jovel et al., 2007; Paudel et al., 2018). At initial stages of the recovery, the viral RNA2 concentration in recovered leaves is similar to that detected in symptomatic leaves. However, RNA2-encoded viral proteins are detected at lower levels in recovered leaves than in symptomatic leaves due to reduced translation rates of viral RNA2 (Ghoshal and Sanfaçon, 2014). Symptom recovery and translation repression are prevented in plants silenced for AGO1, suggesting that AGO1 is involved in the translation repression mechanism (Ghoshal and Sanfaçon, 2014). Because plant miRNAs that down-regulate SA-dependent necrotic responses are loaded into AGO1-RISC complex (Li et al., 2017; Zhang et al., 2019), it is possible that AGO1 plays a dual role in the symptom recovery phenotype by preventing the over-accumulation of viral effectors of the necrotic response and by keeping this response at bay. The ToRSV CP has weak silencing suppression activity that was shown to counteract the translation repression of a reporter mRNA (Karran and Sanfaçon, 2014). The ToRSV CP was also shown to bind to and partially destabilize AGO1. Because that there are several closely related isoforms of AGO1 in *N. benthamiana*, further study would be required to determine which isoform(s) is implicated in symptom recovery from ToRSV infection and whether the CP targets one or more isoforms. The activity of the CP in the context of an actual virus infection would also need to be investigated to gain additional insights into the balance between the activation of plant antiviral RNA silencing and the partial viral counter-silencing that regulate viral protein accumulation and symptom expression. Finally, while AGO1 has been identified as a key orchestrator of the symptom recovery phenotype, other AGOs are also likely to be involved in the symptom recovery phenomenon. A recent study showed a spike of AGO2 accumulation at early stages of infection in ToRSV-infected plants (Paudel et al., 2018). While AGO2 mRNAs were upregulated in plants infected with mild or severe isolates, increased AGO2 protein accumulation was only observed in plants that later recovered from infection (i.e., plants infected with a mild isolate or plants grown at a higher temperature). Further examination revealed that the stability of the AGO2 protein is differentially regulated, raising the possibility that virus infection activates one or several cellular protein degradation pathways in an isolate- and temperature-dependent manner (Paudel et al., 2018). *N. benthamiana* plants deficient in AGO2 showed increased virus accumulation but still recovered from infection, likely because of functional overlap with other AGOs (Paudel et al., 2018).

Taken together, these studies highlight the diversity of plant RNA silencing mechanisms, the overlapping functions of plant AGOs and the regulating roles of viral proteins associated with symptom recovery in herbaceous hosts. Nepoviruses that infect deciduous trees (e.g., fruit trees) are often unevenly distributed and symptom intensity can vary greatly depending on the host, the viral strain or the environmental conditions. Unfortunately, the molecular mechanisms associated with the modulation of disease severity in these more complex hosts have not yet been studied.

6. Concluding remarks

The selected pathosystems described above exemplify how there is cross-talk between various viral components (RNAs, proteins) and associated molecules (satellite RNAs) to fine-tune the expression of disease-associated symptoms. Indeed, although viral virulence proteins can cause severe disease symptoms when constitutively expressed from a transgene, it is becoming clear that these proteins have different effects when they are expressed in the context of virus infection, which includes the presence of other mitigating virus components or of host components induced by virus infection. For example, the CMV 2b protein (Section 5.1) and the tombusvirus p19 protein (Kontra et al., 2016) have different effects on host transcriptomic changes or on miRNA-directed RNA silencing when expressed ectopically or in the context of virus infection. The strategies used by viruses to mitigate the impact of their virulence proteins are diverse and include controlling the stability of viral RNAs coding for virulence proteins (e.g., BNYVV RNAs, Section 5.4), the translation rate of virulence proteins (e.g., ToRSV, potyviruses, Sections 5.2 and 5.5), the timely release of virulence proteins from polyproteins (e.g., potyviruses, nepoviruses, Sections 5.2 and 5.5) or the stability of virulence proteins (all viruses described above, Sections 5.1–5.5). In other cases, viral proteins have opposing functions. For example, tobacco mosaic virus encodes both VSR activity (conferred by an amino acid sequence common to the 126 and 183 kDa proteins) and an enhancer of systemic RNA silencing (MP) (Conti et al., 2017; Vogler et al., 2008). As more viruses are characterized, new regulatory mechanisms will most likely be uncovered that allow viruses to control disease severity in response to the multiple evolutionary constraints they face.

The impact of virus infection on plant signaling pathways is diverse and complex. As described in the sections above, common pathways disrupted by viruses included RNA silencing and phytohormone signaling (SA, JA, auxin, etc.). On the other hand, plant-virus interactions are also host and virus specific, highlighting the necessity to examine a range of hosts and virus strains and avoid broad generalizations. For example, the plant RNA silencing mechanisms deployed against nepoviruses are very different depending on the virus species (Section 5.5). Studies with benyviruses also highlighted striking differences in the symptomatology and degree of manipulation of auxin pathways, depending on the virus species and on the host (Section 5.4). In addition, examples were provided in which virus variants (even when present in minor percentage) contribute to the pathogenesis of complex virus populations (CTV and PPV, Sections 5.3 and 5.2). While most studies have been conducted in model herbaceous hosts using infectious cDNA viral clones that correspond to a single sequence usually representing the main component of virus populations, they do not always parallel the more complex plant-virus interactions occurring in mixed virus populations. As described in Section 5.3, CTV has emerged as a model to study the regulation of viral population composition, and synergies between variants present in these populations. It will become imperative that future studies aim at reconstituting quasi-species for other plant viruses and examining mechanisms regulating plant disease severity in this context.

Although the impact of positive-strand RNA viruses on post-transcriptional RNA silencing regulatory pathways (including antiviral defenses), and on SA or JA-dependent defense responses is now well-established, recent studies have also highlighted that positive-strand RNA viruses affect other layers of plant defenses, in particular TGS. Thus, studying the impact of virus infection on epigenetic changes in the host, and their downstream effects on disease expression has attracted much recent attention (Wang et al., 2019). It is well established that plant DNA viruses, specifically the geminiviruses, are targeted by and counteract TGS (Hanley-Bowdoin et al., 2013; Raja et al., 2010; Ramesh et al., 2017). However, and as highlighted above, plant positive-strand RNA virus VSRs also interact with and inhibit enzymes of the plant TGS pathway, such as AGO4 or various methylation enzymes (e.g., CMV 2b and potyvirus HC-Pro proteins, Sections 5.1–5.2). In addition, symptom recovery from a tobamovirus infection requires components of both transcriptional and post-transcriptional RNA silencing (Kørner et al., 2018). DNA methylation also plays an important role in controlling the severity of symptoms induced by tobacco rattle virus in *A. thaliana* (Diezma-Navas et al., 2019). Thus, plant TGS mechanisms and viral counter-defenses against TGS are likely to play key roles in many other interactions between positive-strand RNA viruses and their hosts.

Another emerging area of interest is the influence of plant RNA decay mechanisms on virus infection (Li and Wang, 2019). As discussed in Section 5.2, potyviruses are targeted by RNA decay, and two TuMV VSRs (HC-Pro and VPg) were shown to counteract RNA decay by disrupting key enzymes of this pathway. An example of repurposing of RNA decay was provided by the BNYVV lncRNAs, which are produced by halting Xrn4-mediated decay using a stable RNA structure (termed coremin, Section 5.4). Interestingly, signature coremin structures were found in the genome of several other plant positive-strand RNA viruses suggesting that other viruses use similar mechanisms to produce functional subgenomic RNAs (Dilweg et al., 2019). Finally, other types of Xrn-resistant RNA structures also lead to the production of subgenomic RNAs from the genomes of tombusviruses (Gunawardene et al., 2019). Thus, plant RNA decay mechanisms regulate the stability and production of viral genomic or subgenomic RNAs from different types of positive-strand RNA viruses.

While the last decades of plant positive-strand RNA virus research have highlighted overlapping, complementary or opposing regulatory mechanisms that allows viruses (and their hosts) to control disease severity, it is likely that many more cross-talk points between viruses and their hosts are awaiting to be discovered, especially as we move toward more holistic approaches to the study of plant-virus interactions.

Acknowledgments

Research in my lab has been supported by Agriculture and Agri-Food Canada funding. The research was also supported in part by a discovery grant from the Natural Sciences and Engineering Research Council of Canada (stipends for Ph.D. students from the University of British Columbia working in my lab). Due to space limitations, only a few pathosystems could be discussed in details. I apologize to the authors of many excellent studies that could not be included in this review.

References

Acosta-Leal, R., Bryan, B.K., Smith, J.T., Rush, C.M., 2010. Breakdown of host resistance by independent evolutionary lineages of Beet necrotic yellow vein virus involves a parallel C/U mutation in its p25 gene. Phytopathology 100, 127–133.

Albiach-Marti, M.R., Robertson, C., Gowda, S., Tatineni, S., Belliure, B., Garnsey, S.M., Folimonova, S.Y., Moreno, P., Dawson, W.O., 2010. The pathogenicity determinant of *Citrus tristeza virus* causing the seedling yellows syndrome maps at the 3'-terminal region of the viral genome. Mol. Plant Pathol. 11, 55–67.

Anandalakshmi, R., Pruss, G.J., Ge, X., Marathe, R., Mallory, A.C., Smith, T.H., Vance, V.B., 1998. A viral suppressor of gene silencing in plants. Proc. Natl. Acad. Sci. U. S. A. 95, 13079–13084.

Andika, I.B., Kondo, H., Nishiguchi, M., Tamada, T., 2012. The cysteine-rich proteins of beet necrotic yellow vein virus and tobacco rattle virus contribute to efficient suppression of silencing in roots. J. Gen. Virol. 93, 1841–1850.

Atallah, O.O., Kang, S.H., El-Mohtar, C.A., Shilts, T., Bergua, M., Folimonova, S.Y., 2016. A 5'-proximal region of the *Citrus tristeza virus* genome encoding two leader proteases is involved in virus superinfection exclusion. Virology 489, 108–115.

Bak, A., Cheung, A.L., Yang, C., Whitham, S.A., Casteel, C.L., 2017. A viral protease relocalizes in the presence of the vector to promote vector performance. Nat. Commun. 8, 14493.

Bar-Joseph, M., Mawassi, M., 2013. The defective RNAs of Closteroviridae. Front. Microbiol. 4, 132.

Bedhomme, S., Hillung, J., Elena, S.F., 2015. Emerging viruses: why they are not jacks of all trades? Curr. Opin. Virol. 10, 1–6.

Bergua, M., Zwart, M.P., El-Mohtar, C., Shilts, T., Elena, S.F., Folimonova, S.Y., 2014. A viral protein mediates superinfection exclusion at the whole-organism level but is not required for exclusion at the cellular level. J. Virol. 88, 11327–11338.

Bergua, M., Kang, S.H., Folimonova, S.Y., 2016. Understanding superinfection exclusion by complex populations of *Citrus tristeza virus*. Virology 499, 331–339.

Betancourt, M., Escriu, F., Fraile, A., García-Arenal, F., 2013. Virulence evolution of a generalist plant virus in a heterogeneous host system. Evol. Appl. 6, 875–890.

Bhattacharyya, D., Chakraborty, S., 2017. Chloroplast: the Trojan horse in plant-virus interaction. Mol. Plant Pathol. 19, 504–518.

Bornemann, K., Varrelmann, M., 2013. Effect of sugar beet genotype on the *Beet necrotic yellow vein virus* P25 pathogenicity factor and evidence for a fitness penalty in resistance-breaking strains. Mol. Plant Pathol. 14, 356–364.

Bujarski, J., Gallitelli, D., García-Arenal, F., Pallás, V., Palukaitis, P., Reddy, M.K., Wang, A., ICTV Report Consortium, 2019. ICTV virus taxonomy profile: *Bromoviridae*. J. Gen. Virol. 100, 1206–1207.

Calvo, M., Malinowski, T., García, J.A., 2014. Single amino acid changes in the 6K1-CI region can promote the alternative adaptation of *Prunus*- and *Nicotiana*-propagated *Plum pox virus* C isolates to either host. Mol. Plant Microbe Interact. 27, 136–149.

Carbonell, A., Maliogka, V.I., de Jesús Pérez, J., Salvador, B., San León, D., García, J.A., Simón-Mateo, C., 2013. Diverse amino acid changes at specific positions in the N-terminal region of the coat protein allow *Plum pox virus* to adapt to new hosts. Mol. Plant Microbe Interact. 26, 1211–1224.

Carr, J.P., Donnelly, R., Tungadi, T., Murphy, A.M., Jiang, S., Bravo-Cazar, A., Yoon, J.Y., Cunniffe, N.J., Glover, B.J., Gilligan, C.A., 2018. Viral manipulation of plant stress responses and host interactions with insects. Adv. Virus Res. 102, 177–197.

Carr, J.P., Murphy, A.M., Tungadi, T., Yoon, J.Y., 2019a. Plant defense signals: players and pawns in plant-virus-vector interactions. Plant Sci. 279, 87–95.

Carr, J.P., Tungadi, T., Donnelly, R., Bravo-Cazar, A., Rhee, S.J., Watt, L.G., Mutuku, J.M., Wamonje, F.O., Murphy, A.M., Arinaitwe, W., Pate, A.E., Cunniffe, N.J., Gilligan, C.A., 2019b. Modelling and manipulation of aphid-mediated spread of non-persistently transmitted viruses. Virus Res. 277, 197845.

Casteel, C.L., Yang, C., Nanduri, A.C., De Jong, H.N., Whitham, S.A., Jander, G., 2014. The NIa-Pro protein of *Turnip mosaic virus* improves growth and reproduction of the aphid vector, *Myzus persicae* (green peach aphid). Plant J. 77, 653–663.

Casteel, C.L., De Alwis, M., Bak, A., Dong, H., Whitham, S.A., Jander, G., 2015. Disruption of ethylene responses by turnip mosaic virus mediates suppression of plant defense against the green peach aphid vector. Plant Physiol. 169, 209–218.

Chai, M., Wu, X., Liu, J., Fang, Y., Luan, Y., Cui, X., Zhou, X., Wang, A., Cheng, X., 2020. P3N-PIPO interacts with P3 via the shared N-terminal domain to recruit viral replication vesicles for cell-to-cell movement. J. Virol.

Cheng, X., Wang, A., 2017. The potyvirus silencing suppressor protein VPg mediates degradation of SGS3 via ubiquitination and autophagy pathways. J. Virol. 91, e01478-16.

Cheng, X., Xiong, R., Li, Y., Li, F., Zhou, X., Wang, A., 2017. Sumoylation of *Turnip mosaic virus* RNA polymerase promotes viral infection by counteracting the host NPR1-mediated immune response. Plant Cell 29, 508–525.

Chiba, S., Kondo, H., Miyanishi, M., Andika, I.B., Han, C., Tamada, T., 2011. The evolutionary history of *Beet necrotic yellow vein virus* deduced from genetic variation, geographical origin and spread, and the breaking of host resistance. Mol. Plant Microbe Interact. 24, 207–218.

Chisholm, J., Wieczorek, A., Sanfaçon, H., 2001. Expression and partial purification of recombinant tomato ringspot nepovirus 3C-like proteinase: comparison of the activity of the mature proteinase and the VPg-proteinase precursor. Virus Res. 79, 153–164.

Clemente-Moreno, M.J., Díaz-Vivancos, P., Barba-Espín, G., Hernández, J.A., 2010. Benzothiadiazole and l-2-oxothiazolidine-4-carboxylic acid reduce the severity of Sharka symptoms in pea leaves: effect on antioxidative metabolism at the subcellular level. Plant Biol. 12, 88–97.

Clemente-Moreno, M.J., Díaz-Vivancos, P., Rubio, M., Fernández-García, N., Hernández, J.A., 2013. Chloroplast protection in plum pox virus-infected peach plants by L-2-oxo-4-thiazolidine-carboxylic acid treatments: effect in the proteome. Plant Cell Environ. 36, 640–654.

Collum, T.D., Culver, J.N., 2016. The impact of phytohormones on virus infection and disease. Curr. Opin. Virol. 17, 25–31.

Collum, T.D., Stone, A.L., Sherman, D.J., Rogers, E.E., Dardick, C., Culver, J.N., 2020. Translatome profiling of plum pox virus-infected leaves in European plum reveals temporal and spatial coordination of defense responses in phloem tissues. Mol. Plant Microbe Interact. 33, 66–77.

Conti, G., Rodriguez, M.C., Venturuzzi, A.L., Asurmendi, S., 2017. Modulation of host plant immunity by Tobamovirus proteins. Ann. Bot. 119, 737–747.

Csorba, T., Kontra, L., Burgyán, J., 2015. Viral silencing suppressors: tools forged to fine-tune host-pathogen coexistence. Virology 479–480, 85–103.

Culver, J.N., Padmanabhan, M.S., 2007. Virus-induced disease: altering host physiology one interaction at a time. Annu. Rev. Phytopathol. 45, 221–243.

Dawson, W.O., Bar-Joseph, M., Garnsey, S.M., Moreno, P., 2015. *Citrus tristeza virus*: making an ally from an enemy. Annu. Rev. Phytopathol. 53, 137–155.

De, S., Chavez-Calvillo, G., Wahlsten, M., Mäkinen, K., 2018. Disruption of the methionine cycle and reduced cellular gluthathione levels underlie potex-potyvirus synergism in *Nicotiana benthamiana*. Mol. Plant Pathol. 19, 1820–1835.

Decroës, A., Clausse, M., Galein, V., Legreve, A., Wauters, A., Bragard, C., 2015. New BNYVV P25 Variants in Belgium. Commun. Agric. Appl. Biol. Sci. 80, 407–409.

Dehkordi, A.N., Rubio, M., Babaeian, N., Albacete, A., Martínez-Gómez, P., 2018. Phytohormone signaling of the resistance to *Plum pox virus* (PPV, sharka disease) induced by almond (*Prunus dulcis* (Miller) Webb) grafting to peach (*P. persica* L. Batsch). Viruses 10, 238.

Díaz-Vivancos, P., Rubio, M., Mesonero, V., Periago, P.M., Barceló, A.R., Martínez-Gómez, P., Hernández, J.A., 2006. The apoplastic antioxidant system in *Prunus*: response to long-term plum pox virus infection. J. Exp. Bot. 57, 3813–3824.

Díaz-Vivancos, P., Clemente-Moreno, M.J., Rubio, M., Olmos, E., García, J.A., Martínez-Gómez, P., Hernández, J.A., 2008. Alteration in the chloroplastic metabolism leads to ROS accumulation in pea plants in response to plum pox virus. J. Exp. Bot. 59, 2147–2160.

Dietzgen, R.G., Mann, K.S., Johnson, K.N., 2016. Plant virus-insect vector interactions: current and potential future research directions. Viruses 8, 303.

Diezma-Navas, L., Pérez-González, A., Artaza, H., Alonso, L., Caro, E., Llave, C., Ruiz-Ferrer, V., 2019. Crosstalk between epigenetic silencing and infection by tobacco rattle virus in *Arabidopsis*. Mol. Plant Pathol. 20, 1439–1452.

Dilweg, I.W., Gultyaev, A.P., Olsthoorn, R.C., 2019. Structural features of an Xrn1-resistant plant virus RNA. RNA Biol. 16, 838–845.

Dolja, V.V., Kreuze, J.F., Valkonen, J.P., 2006. Comparative and functional genomics of closteroviruses. Virus Res. 117, 38–51.

Dong, F., Mochizuki, T., Ohki, S.T., 2010. Tobacco ringspot virus persists in the shoot apical meristem but not in the root apical meristem of infected tobacco. Eur. J. Plant Pathol. 126, 117–122.

Du, Z., Chen, A., Chen, W., Liao, Q., Zhang, H., Bao, Y., Roossinck, M.J., Carr, J.P., 2014. Nuclear-cytoplasmic partitioning of cucumber mosaic virus protein 2b determines the balance between its roles as a virulence determinant and an RNA-silencing suppressor. J. Virol. 88, 5228–5241.

Duan, C.G., Fang, Y.Y., Zhou, B.J., Zhao, J.H., Hou, W.N., Zhu, H., Ding, S.W., Guo, H.S., 2012. Suppression of *Arabidopsis* ARGONAUTE1-mediated slicing, transgene-induced RNA silencing, and DNA methylation by distinct domains of the *Cucumber mosaic virus* 2b protein. Plant Cell 24, 259–274.

Escriu, F., Fraile, A., García-Arenal, F., 2000. Evolution of virulence in natural populations of the satellite RNA of *Cucumber mosaic virus*. Phytopathology 90, 480–485.

Escriu, F., Fraile, A., García-Arenal, F., 2003. The evolution of virulence in a plant virus. Evolution 57, 755–765.

Escriu, F., Fraile, A., García-Arenal, F., 2007. Constraints to genetic exchange support gene coadaptation in a tripartite RNA virus. PLoS Pathog. 3, e8.

Fagoaga, C., López, C., Moreno, P., Navarro, L., Flores, R., Peña, L., 2005. Viral-like symptoms induced by the ectopic expression of the p23 gene of *Citrus tristeza virus* are citrus specific and do not correlate with the pathogenicity of the virus strain. Mol. Plant Microbe Interact. 18, 435–445.

Fan, H., Zhang, Y., Sun, H., Liu, J., Wang, Y., Wang, X., Li, D., Yu, J., Han, C., 2015. Transcriptome analysis of *Beta macrocarpa* and Identification of differentially expressed transcripts in response to *Beet Necrotic Yellow Vein Virus* Infection. PLoS One 10, e0132277.

Feng, J., Lai, L., Lin, R., Jin, C., Chen, J., 2012. Differential effects of *Cucumber mosaic virus* satellite RNAs in the perturbation of microRNA-regulated gene expression in tomato. Mol. Biol. Rep. 39, 775–784.

Fernando Gil, J., Liebe, S., Thiel, H., Lennefors, B.L., Kraft, T., Gilmer, D., Maiss, E., Varrelmann, M., Savenkov, E.I., 2018. Massive up-regulation of LBD transcription factors and EXPANSINs highlights the regulatory programs of rhizomania disease. Mol. Plant Pathol. 19, 2333–2348.

Fernando Gil, J., Wibberg, D., Eini, O., Savenkov, E.I., Varrelmann, M., Liebe, S., 2020. Comparative transcriptome analysis provides molecular insights into the interaction of *Beet necrotic yellow vein virus* and *Beet soil-borne mosaic virus* with their host sugar beet. Viruses 12, 76.

Flobinus, A., Chevigny, N., Charley, P.A., Seissler, T., Klein, E., Bleykasten-Grosshans, C., Ratti, C., Bouzoubaa, S., Wilusz, J., Gilmer, D., 2018. Beet necrotic yellow vein virus noncoding RNA production depends on a 5'->3' Xrn exoribonuclease activity. Viruses 10, 137.

Flores, R., Ruiz-Ruiz, S., Soler, N., Sánchez-Navarro, J., Fagoaga, C., López, C., Navarro, L., Moreno, P., Peña, L., 2013. Citrus tristeza virus p23: a unique protein mediating key virus-host interactions. Front. Microbiol. 4, 98.

Folimonova, S.Y., 2012. Superinfection exclusion is an active virus-controlled function that requires a specific viral protein. J. Virol. 86, 5554–5561.

Folimonova, S.Y., Folimonov, A.S., Tatineni, S., Dawson, W.O., 2008. *Citrus tristeza virus*: survival at the edge of the movement continuum. J. Virol. 82, 6546–6556.

Folimonova, S.Y., Robertson, C.J., Shilts, T., Folimonov, A.S., Hilf, M.E., Garnsey, S.M., Dawson, W.O., 2010. Infection with strains of *Citrus tristeza virus* does not exclude superinfection by other strains of the virus. J. Virol. 84, 1314–1325.

Folimonova, S.Y., Harper, S.J., Leonard, M.T., Triplett, E.W., Shilts, T., 2014. Superinfection exclusion by *Citrus tristeza virus* does not correlate with the production of viral small RNAs. Virology 468-470C, 462–471.

Fraile, A., García-Arenal, F., 2016. Environment and evolution modulate plant virus pathogenesis. Curr. Opin. Virol. 17, 50–56.

Fraile, A., Alonso-Prados, J.L., Aranda, M.A., Bernal, J.J., Malpica, J.M., García-Arenal, F., 1997. Genetic exchange by recombination or reassortment is infrequent in natural populations of a tripartite RNA plant virus. J. Virol. 71, 934–940.

Fuchs, M., Schmitt-Keichinger, C., Sanfaçon, H., 2017. A renaissance in nepovirus research provides new insights into their molecular interface with hosts and vectors. Adv. Virus Res. 97, 61–105.

Fuchs, M., Bar-Joseph, M., Candresse, T., Maree, H.J., Martelli, G.P., Melzer, M.J., Menzel, W., Minafra, A., Sabanadzovic, S., ICTV Report Consortium, 2020. ICTV virus taxonomy profile: *Closteroviridae*. J. Gen. Virol. 101, 364–365.

Fukuhara, T., 2019. Endornaviruses: persistent dsRNA viruses with symbiotic properties in diverse eukaryotes. Virus Genes 55, 165–173.

Galein, Y., Legrève, A., Bragard, C., 2018. Long term management of rhizomania disease-insight into the changes of the *Beet necrotic yellow vein virus* RNA-3 observed under resistant and non-resistant sugar beet fields. Front. Plant Sci. 9, 795.

Gallet, R., Michalakis, Y., Blanc, S., 2018. Vector-transmission of plant viruses and constraints imposed by virus-vector interactions. Curr. Opin. Virol. 33, 144–150.
García, J.A., Pallas, V., 2015. Viral factors involved in plant pathogenesis. Curr. Opin. Virol. 11, 21–30.
García, J.A., Glasa, M., Cambra, M., Candresse, T., 2014. Plum pox virus and sharka: a model potyvirus and a major disease. Mol. Plant Pathol. 15, 226–241.
Garcia-Ruiz, H., 2019. Host factors against plant viruses. Mol. Plant Pathol. 20, 1588–1601.
Ghorbel, R., Lopez, C., Fagoaga, C., Moreno, P., Navarro, L., Flores, R., Pena, L., 2001. Transgenic citrus plants expressing the citrus tristeza virus p23 protein exhibit viral-like symptoms. Mol. Plant Pathol. 2, 27–36.
Ghoshal, B., Sanfaçon, H., 2014. Temperature-dependent symptom recovery in *Nicotiana benthamiana* plants infected with tomato ringspot virus is associated with reduced translation of viral RNA2 and requires ARGONAUTE 1. Virology 456–457, 188–197.
Ghoshal, B., Sanfaçon, H., 2015. Symptom recovery in virus-infected plants: revisiting the role of RNA silencing mechanisms. Virology 479–480, 167–179.
Gibbs, A.J., Hajizadeh, M., Ohshima, K., Jones, R.A.C., 2020. The potyviruses: an evolutionary synthesis is emerging. Viruses 12, 132.
Gilmer, D., Ratti, C., ICTV Report Consortium, 2017. ICTV virus taxonomy profile: *Benyviridae*. J. Gen. Virol. 98, 1571–1572.
Gong, Y.N., Tang, R.-Q., Zhang, Y., Peng, J., Xian, O.Y., Zhang, Z.-H., Zhang, S.-B., Zhang, D.-Y., Liu, H., Luo, X.-W., Liu, Y., 2020. The NIa-protease protein encoded by the *Pepper mottle virus* is a pathogenicity determinant and releases DNA methylation of *Nicotiana benthamiana*. Front. Microbiol. 11, 102.
González, I., Martínez, L., Rakitina, D.V., Lewsey, M.G., Atencio, F.A., Llave, C., Kalinina, N.O., Carr, J.P., Palukaitis, P., Canto, T., 2010. Cucumber mosaic virus 2b protein subcellular targets and interactions: their significance to RNA silencing suppressor activity. Mol. Plant Microbe Interact. 23, 294–303.
González, I., Rakitina, D., Semashko, M., Taliansky, M., Praveen, S., Palukaitis, P., Carr, J.P., Kalinina, N., Canto, T., 2012. RNA binding is more critical to the suppression of silencing function of *Cucumber mosaic virus* 2b protein than nuclear localization. RNA 18, 771–782.
González-Jara, P., Atencio, F.A., Martínez-García, B., Barajas, D., Tenllado, F., Díaz-Ruíz, J.R., 2005. A single amino acid mutation in the *Plum pox virus* helper component-proteinase gene abolishes both synergistic and RNA silencing suppression activities. Phytopathology 95, 894–901.
Goto, K., Kobori, T., Kosaka, Y., Natsuaki, T., Masuta, C., 2007. Characterization of silencing suppressor 2b of cucumber mosaic virus based on examination of its small RNA-binding abilities. Plant Cell Physiol. 48, 1050–1060.
Gowda, S., Ayllón, M.A., Satyanarayana, T., Bar-Joseph, M., Dawson, W.O., 2003. Transcription strategy in a *Closterovirus*: a novel 5'-proximal controller element of *Citrus tristeza virus* produces 5'- and 3'-terminal subgenomic RNAs and differs from 3' open reading frame controller elements. J. Virol. 77, 340–352.
Gunawardene, C.D., Newburn, L.R., White, K.A., 2019. A 212-nt long RNA structure in the Tobacco necrosis virus-D RNA genome is resistant to Xrn degradation. Nucleic Acids Res. 47, 9329–9342.
Hajizadeh, M., Gibbs, A.J., Amirnia, F., Glasa, M., 2019. The global phylogeny of *Plum pox virus* is emerging. J. Gen. Virol. 100, 1457–1468.
Hamera, S., Song, X., Su, L., Chen, X., Fang, R., 2012. Cucumber mosaic virus suppressor 2b binds to AGO4-related small RNAs and impairs AGO4 activities. Plant J. 69, 104–115.
Hanley-Bowdoin, L., Bejarano, E.R., Robertson, D., Mansoor, S., 2013. Geminiviruses: masters at redirecting and reprogramming plant processes. Nat. Rev. Microbiol. 11, 777–788.

Harper, S.J., Cowell, S.J., Dawson, W.O., 2015a. Finding balance: virus populations reach equilibrium during the infection process. Virology 485, 205–212.

Harper, S.J., Cowell, S.J., Dawson, W.O., 2015b. With a little help from my friends: complementation as a survival strategy for viruses in a long-lived host system. Virology 478, 123–128.

Harper, S.J., Cowell, S.J., Dawson, W.O., 2018. Bottlenecks and complementation in the aphid transmission of citrus tristeza virus populations. Arch. Virol. 163, 3373–3376.

Harper, S.J., Cowell, S.J., Dawson, W.O., 2019. Changes in host microRNA expression during citrus tristeza induced disease. J. Cit. Pathol. 6, 41164.

Hashimoto, M., Neriya, Y., Yamaji, Y., Namba, S., 2016. Recessive resistance to plant viruses: potential resistance genes beyond translation initiation factors. Front. Microbiol. 7, 1695.

Hervás, M., Navajas, R., Chagoyen, M., García, J.A., Martínez-Turiño, S., 2020. Phosphorylation-related cross-talk between distant regions of the core region of the coat protein contributes to virion assembly of plum pox virus. Mol. Plant Microbe Interact. 33, 653–667 MPMI10190305R.

Hily, J.M., Poulicard, N., Mora, M.A., Pagán, I., García-Arenal, F., 2016. Environment and host genotype determine the outcome of a plant-virus interaction: from antagonism to mutualism. New Phytol. 209, 812–822.

Hisa, Y., Suzuki, H., Atsumi, G., Choi, S.H., Nakahara, K.S., Uyeda, I., 2014. P3N-PIPO of clover yellow vein virus exacerbates symptoms in pea infected with white clover mosaic virus and is implicated in viral synergism. Virology 449, 200–206.

Hou, W.N., Duan, C.G., Fang, R.X., Zhou, X.Y., Guo, H.S., 2011. Satellite RNA reduces expression of the 2b suppressor protein resulting in the attenuation of symptoms caused by *Cucumber mosaic virus* infection. Mol. Plant Pathol. 12, 595–605.

Inaba, J., Kim, B.M., Shimura, H., Masuta, C., 2011. Virus-induced necrosis is a consequence of direct protein-protein interaction between a viral RNA-silencing suppressor and a host catalase. Plant Physiol. 156, 2026–2036.

Ismayil, A., Yang, M., Liu, Y., 2019. Role of autophagy during plant-virus interactions. Semin. Cell Dev. Biol. S1084–9521 (18), 30225–30228.

Ivanov, K.I., Eskelin, K., Bašić, M., De, S., Lõhmus, A., Varjosalo, M., Mäkinen, K., 2016. Molecular insights into the function of the viral RNA silencing suppressor HCPro. Plant J. 85, 30–45.

Jaafar, Z.A., Kieft, J.S., 2019. Viral RNA structure-based strategies to manipulate translation. Nat. Rev. Microbiol. 17, 110–123.

Jacquemond, M., 2012. Cucumber mosaic virus. Adv. Virus Res. 84, 439–504.

Jensen, L.M., Walker, E.J., Jans, D.A., Ghildyal, R., 2015. Proteases of human rhinovirus: role in infection. Methods Mol. Biol. 1221, 129–141.

Jeon, E.J., Tadamura, K., Murakami, T., Inaba, J.I., Kim, B.M., Sato, M., Atsumi, G., Kuchitsu, K., Masuta, C., Nakahara, K.S., 2017. rgs-CaM detects and counteracts viral RNA silencing suppressors in plant immune priming. J. Virol. 91, e00761-17.

Ji, L.H., Ding, S.W., 2001. The suppressor of transgene RNA silencing encoded by *Cucumber mosaic virus* interferes with salicylic acid-mediated virus resistance. Mol. Plant Microbe Interact. 14, 715–724.

Jiang, J., Laliberté, J.F., 2011. The genome-linked protein VPg of plant viruses-a protein with many partners. Curr. Opin. Virol. 1, 347–354.

Jones, R.A., 2016. Future scenarios for plant virus pathogens as climate change progresses. Adv. Virus Res. 95, 87–147.

Jovel, J., Walker, M., Sanfaçon, H., 2007. Recovery of *Nicotiana benthamiana* plants from a necrotic response induced by a nepovirus is associated with RNA silencing but not with reduced virus titer. J. Virol. 81, 12285–12297.

Jovel, J., Walker, M., Sanfaçon, H., 2011. Salicylic acid-dependent restriction of *Tomato ringspot virus* spread in tobacco is accompanied by a hypersensitive response, local RNA silencing, and moderate systemic resistance. Mol. Plant Microbe Interact. 24, 706–718.

Kang, S.H., Bak, A., Kim, O.K., Folimonova, S.Y., 2015. Membrane association of a non-conserved viral protein confers virus ability to extend its host range. Virology 482, 208–217.

Kang, S.H., Atallah, O.O., Sun, Y.D., Folimonova, S.Y., 2018. Functional diversification upon leader protease domain duplication in the *Citrus tristeza virus* genome: role of RNA sequences and the encoded proteins. Virology 514, 192–202.

Kang, S.H., Sun, Y.D., Atallah, O.O., Huguet-Tapia, J.C., Noble, J.D., Folimonova, S.Y., 2019. A long non-coding RNA of *Citrus tristeza virus*: role in the virus interplay with the host immunity. Viruses 11, 436.

Karran, R.A., Sanfaçon, H., 2014. *Tomato ringspot virus* coat protein binds to ARGONAUTE 1 and suppresses the translation repression of a reporter gene. Mol. Plant Microbe Interact. 27, 933–943.

Klein, E., Link, D., Schirmer, A., Erhardt, M., Gilmer, D., 2007. Sequence variation within Beet necrotic yellow vein virus p25 protein influences its oligomerization and isolate pathogenicity on *Tetragonia expansa*. Virus Res. 126, 53–61.

Koenig, R., Loss, S., Specht, J., Varrelmann, M., Lüddecke, P., Deml, G., 2009. A single U/C nucleotide substitution changing alanine to valine in the beet necrotic yellow vein virus P25 protein promotes increased virus accumulation in roots of mechanically inoculated, partially resistant sugar beet seedlings. J. Gen. Virol. 90, 759–763.

Kong, J., Wei, M., Li, G., Lei, R., Qiu, Y., Wang, C., Li, Z.H., Zhu, S., 2018. The *cucumber mosaic virus* movement protein suppresses PAMP-triggered immune responses in Arabidopsis and tobacco. Biochem. Biophys. Res. Commun. 498, 395–401.

Kontra, L., Csorba, T., Tavazza, M., Lucioli, A., Tavazza, R., Moxon, S., Tisza, V., Medzihradszky, A., Turina, M., Burgyán, J., 2016. Distinct effects of p19 RNA silencing suppressor on small RNA mediated pathways in plants. PLoS Pathog. 12, e1005935.

Kørner, C.J., Pitzalis, N., Peña, E.J., Erhardt, M., Vazquez, F., Heinlein, M., 2018. Crosstalk between PTGS and TGS pathways in natural antiviral immunity and disease recovery. Nat. Plants 4, 157–164.

Kushwaha, N.K., Hafrén, A., Hofius, D., 2019. Autophagy-virus interplay in plants: from antiviral recognition to proviral manipulation. Mol. Plant Pathol. 20, 1211–1216.

Laitinen, O.H., Svedin, E., Kapell, S., Nurminen, A., Hytönen, V.P., Flodström-Tullberg, M., 2016. Enteroviral proteases: structure, host interactions and pathogenicity. Rev. Med. Virol. 26, 251–267.

Laliberté, J.F., Sanfaçon, H., 2010. Cellular remodeling during plant virus infection. Annu. Rev. Phytopathol. 48, 69–91.

Laufer, M., Mohammad, H., Maiss, E., Richert-Pöggeler, K., Dall'Ara, M., Ratti, C., Gilmer, D., Liebe, S., Varrelmann, M., 2018. Biological properties of *Beet soil-borne mosaic virus* and *Beet necrotic yellow vein virus* cDNA clones produced by isothermal *in vitro* recombination: insights for reassortant appearance. Virology 518, 25–33.

Lavenus, J., Goh, T., Roberts, I., Guyomarc'h, S., Lucas, M., De Smet, I., Fukaki, H., Beeckman, T., Bennett, M., Laplaze, L., 2013. Lateral root development in *Arabidopsis*: fifty shades of auxin. Trends Plant Sci. 18, 450–458.

Lefeuvre, P., Martin, D.P., Elena, S.F., Shepherd, D.N., Roumagnac, P., Varsani, A., 2019. Evolution and ecology of plant viruses. Nat. Rev. Microbiol. 17, 632–641.

Lewsey, M.G., Murphy, A.M., Maclean, D., Dalchau, N., Westwood, J.H., Macaulay, K., Bennett, M.H., Moulin, M., Hanke, D.E., Powell, G., Smith, A.G., Carr, J.P., 2010. Disruption of two defensive signaling pathways by a viral RNA silencing suppressor. Mol. Plant Microbe Interact. 23, 835–845.

Li, F., Wang, A., 2018. RNA decay is an antiviral defense in plants that is counteracted by viral RNA silencing suppressors. PLoS Pathog. 14, e1007228.
Li, F., Wang, A., 2019. RNA-targeted antiviral immunity: more than just RNA silencing. Trends Microbiol. 27, 792–805.
Li, Y., Cui, H., Cui, X., Wang, A., 2016. The altered photosynthetic machinery during compatible virus infection. Curr. Opin. Virol. 17, 19–24.
Li, S., Castillo-González, C., Yu, B., Zhang, X., 2017. The functions of plant small RNAs in development and in stress responses. Plant J. 90, 654–670.
Liu, J.Y., Fan, H.Y., Wang, Y., Zhang, Y.L., Li, D.W., Yu, J.L., Han, C.G., 2017. Characterization of microRNAs of *Beta macrocarpa* and their responses to *Beet necrotic yellow vein virus* infection. PLoS One 12, e0186500.
López, C., Navas-Castillo, J., Gowda, S., Moreno, P., Flores, R., 2000. The 23-kDa protein coded by the 3'-terminal gene of citrus tristeza virus is an RNA-binding protein. Virology 269, 462–470.
Lu, R., Folimonov, A., Shintaku, M., Li, W.X., Falk, B.W., Dawson, W.O., Ding, S.W., 2004. Three distinct suppressors of RNA silencing encoded by a 20-kb viral RNA genome. Proc. Natl. Acad. Sci. U. S. A. 101, 15742–15747.
Mäkinen, K., De, S., 2019. The significance of methionine cycle enzymes in plant virus infections. Curr. Opin. Plant Biol. 50, 67–75.
Mäkinen, K., Lõhmus, A., Pollari, M., 2017. Plant RNA regulatory network and RNA granules in virus infection. Front. Plant Sci. 8, 2093.
Maliogka, V.I., Salvador, B., Carbonell, A., Sáenz, P., San León, D., Oliveros, J.C., Delgadillo, M.O., García, J.A., Simón-Mateo, C., 2012. Virus variants with differences in the P1 protein coexist in a *Plum pox virus* population and display particular host-dependent pathogenicity features. Mol. Plant Pathol. 13, 877–886.
Mann, K.S., Sanfaçon, H., 2019. Expanding repertoire of plant positive-strand RNA virus proteases. Viruses 11, 66.
Martin, I.R., Vigne, E., Berthold, F., Komar, V., Lemaire, O., Fuchs, M., Schmitt-Keichinger, C., 2018. The 50 distal amino acids of the 2A(HP) homing protein of *Grapevine fanleaf virus* elicit a hypersensitive reaction on *Nicotiana occidentalis*. Mol. Plant Pathol. 19, 731–743.
Martinez, F., Rodrigo, G., Aragones, V., Ruiz, M., Lodewijk, I., Fernandez, U., Elena, S.F., Daros, J.A., 2016. Interaction network of tobacco etch potyvirus NIa protein with the host proteome during infection. BMC Genomics 17, 87.
Martínez-Turiño, S., De Jesús Pérez, J., Hervás, M., Navajas, R., Ciordia, S., Udeshi, N.D., Shabanowitz, J., Hunt, D.F., García, J.A., 2018. Phosphorylation coexists with *O*-GlcNAcylation in a plant virus protein and influences viral infection. Mol. Plant Pathol. 19, 1427–1443.
Mascia, T., Gallitelli, D., 2016. Synergies and antagonisms in virus interactions. Plant Sci. 252, 176–192.
Mauck, K.E., Kenney, J., Chesnais, Q., 2019. Progress and challenges in identifying molecular mechanisms underlying host and vector manipulation by plant viruses. Curr. Opin. Insect Sci. 33, 7–18.
McLeish, M.J., Fraile, A., García-Arenal, F., 2019. Evolution of plant-virus interactions: host range and virus emergence. Curr. Opin. Virol. 34, 50–55.
Medina-Puche, L., Lozano-Duran, R., 2019. Tailoring the cell: a glimpse of how plant viruses manipulate their hosts. Curr. Opin. Plant Biol. 52, 164–173.
Miras, M., Miller, W.A., Truniger, V., Aranda, M.A., 2017. Non-canonical translation in plant RNA viruses. Front. Plant Sci. 8, 494.
Mochizuki, T., Ohki, S.T., 2012. *Cucumber mosaic virus*: viral genes as virulence determinants. Mol. Plant Pathol. 13, 217–225.

Monjane, A.L., Dellicour, S., Hartnady, P., Oyeniran, K.A., Owor, B.E., Bezeidenhout, M., Linderme, D., Syed, R.A., Donaldson, L., Murray, S., Rybicki, E.P., Kvarnheden, A., Yazdkhasti, E., Lefeuvre, P., Froissart, R., Roumagnac, P., Shepherd, D.N., Harkins, G.W., Suchard, M.A., Lemey, P., Varsani, A., Martin, D.P., 2020. Symptom evolution following the emergence of maize streak virus. Elife 9, e51984.

Moreno, A.B., López-Moya, J.J., 2020. When viruses play team sports: mixed infections in plants. Phytopathology 110, 29–48.

Moury, B., Desbiez, C., 2020. Host range evolution of potyviruses: a global phylogenetic analysis. Viruses 12, 111.

Moury, B., Fabre, F., Hébrard, E., Froissart, R., 2017. Determinants of host species range in plant viruses. J. Gen. Virol. 98, 862–873.

Murota, K., Shimura, H., Takeshita, M., Masuta, C., 2017. Interaction between *Cucumber mosaic virus* 2b protein and plant catalase induces a specific necrosis in association with proteasome activity. Plant Cell Rep. 36, 37–47.

Nagy, P.D., 2016. Tombusvirus-host interactions: co-opted evolutionarily conserved host factors take center court. Annu. Rev. Virol. 3, 491–515.

Nagyová, A., Kamencayová, M., Glasa, M., Subr, Z.W., 2012. The 3'-proximal part of the Plum pox virus P1 gene determinates the symptom expression in two herbaceous host plants. Virus Genes 44, 505–512.

Nakahara, K.S., Masuta, C., Yamada, S., Shimura, H., Kashihara, Y., Wada, T.S., Meguro, A., Goto, K., Tadamura, K., Sueda, K., Sekiguchi, T., Shao, J., Itchoda, N., Matsumura, T., Igarashi, M., Ito, K., Carthew, R.W., Uyeda, I., 2012. Tobacco calmodulin-like protein provides secondary defense by binding to and directing degradation of virus RNA silencing suppressors. Proc. Natl. Acad. Sci. U. S. A. 109, 10113–10118.

Nemes, K., Gellért, A., Almási, A., Vági, P., Sáray, R., Kádár, K., Salánki, K., 2017. Phosphorylation regulates the subcellular localization of Cucumber Mosaic Virus 2b protein. Sci. Rep. 7, 13444.

Nemes, K., Gellért, A., Bóka, K., Vági, P., Salánki, K., 2019. Symptom recovery is affected by Cucumber mosaic virus coat protein phosphorylation. Virology 536, 68–77.

Nicaise, V., 2017. Boosting innate immunity to sustainably control diseases in crops. Curr. Opin. Virol. 26, 112–119.

Nicaise, V., Candresse, T., 2017. *Plum pox virus* capsid protein suppresses plant pathogen-associated molecular pattern (PAMP)-triggered immunity. Mol. Plant Pathol. 18, 878–886.

Nováková, S., Danchenko, M., Skultety, L., Fialová, I., Lešková, A., Beke, G., Flores-Ramírez, G., Glasa, M., 2018. Photosynthetic and stress responsive proteins are altered more effectively in *Nicotiana benthamiana* infected with *Plum pox virus* aggressive PPV-CR versus mild PPV-C cherry-adapted isolates. J. Proteome Res. 17, 3114–3127.

Novella, I.S., Presloid, J.B., Taylor, R.T., 2014. RNA replication errors and the evolution of virus pathogenicity and virulence. Curr. Opin. Virol. 9, 143–147.

Olspert, A., Chung, B.Y., Atkins, J.F., Carr, J.P., Firth, A.E., 2015. Transcriptional slippage in the positive-sense RNA virus family *Potyviridae*. EMBO Rep. 16, 995–1004.

Olspert, A., Carr, J.P., Firth, A.E., 2016. Mutational analysis of the *Potyviridae* transcriptional slippage site utilized for expression of the P3N-PIPO and P1N-PISPO proteins. Nucleic Acids Res. 44, 7618–7629.

Osterbaan, L.J., Fuchs, M., 2019. Dynamic interactions between plant viruses and their hosts for symptom development. J. Plant Pathol. 101, 885–895.

Osterbaan, L.J., Choi, J., Kenney, J., Flasco, M., Vigne, E., Schmitt-Keichinger, C., Rebelo, A.R., Heck, M., Fuchs, M., 2019. The identity of a single residue of the RNA-dependent RNA polymerase of grapevine fanleaf virus modulates vein clearing in *Nicotiana benthamiana*. Mol. Plant Microbe Interact. 32, 790–801.

Pagán, I., García-Arenal, F., 2018. Tolerance to plant pathogens: theory and experimental evidence. Int. J. Mol. Sci. 19, e810.

Palukaitis, P., 2016. Satellite RNAs and satellite viruses. Mol. Plant Microbe Interact. 29, 181–186.

Pasin, F., Simón-Mateo, C., García, J.A., 2014. The hypervariable amino-terminus of P1 protease modulates potyviral replication and host defense responses. PLoS Pathog. 10, e1003985.

Pathak, K.B., Nagy, P.D., 2009. Defective interfering RNAs: foes of viruses and friends of virologists. Viruses 1, 895–919.

Paudel, D.B., Sanfaçon, H., 2018. Exploring the diversity of mechanisms associated with plant tolerance to virus infection. Front. Plant Sci. 9, 1575.

Paudel, D.B., Ghoshal, B., Jossey, S., Ludman, M., Fatyol, K., Sanfaçon, H., 2018. Expression and antiviral function of ARGONAUTE 2 in *Nicotiana benthamiana* plants infected with two isolates of tomato ringspot virus with varying degrees of virulence. Virology 524, 127–139.

Peck, K.M., Lauring, A.S., 2018. Complexities of viral mutation rates. J. Virol. 92, e01031-17.

Peltier, C., Schmidlin, L., Klein, E., Taconnat, L., Prinsen, E., Erhardt, M., Heintz, D., Weyens, G., Lefebvre, M., Renou, J.P., Gilmer, D., 2011. Expression of the *Beet necrotic yellow vein virus* p25 protein induces hormonal changes and a root branching phenotype in *Arabidopsis thaliana*. Transgenic Res. 20, 443–466.

Peltier, C., Klein, E., Hleibieh, K., D'Alonzo, M., Hammann, P., Bouzoubaa, S., Ratti, C., Gilmer, D., 2012. Beet necrotic yellow vein virus subgenomic RNA3 is a cleavage product leading to stable non-coding RNA required for long-distance movement. J. Gen. Virol. 93, 1093–1102.

Pooggin, M.M., 2018. Small RNA-omics for plant virus identification, virome reconstruction, and antiviral defense characterization. Front. Microbiol. 9, 2779.

Pruss, G., Ge, X., Shi, X.M., Carrington, J.C., Bowman Vance, V., 1997. Plant viral synergism: the potyviral genome encodes a broad-range pathogenicity enhancer that transactivates replication of heterologous viruses. Plant Cell 9, 859–868.

Qiu, Y., Zhang, Y., Wang, C., Lei, R., Wu, Y., Li, X., Zhu, S., 2018. *Cucumber mosaic virus* coat protein induces the development of chlorotic symptoms through interacting with the chloroplast ferredoxin I protein. Sci. Rep. 8, 1205.

Rahim, M.D., Andika, I.B., Han, C., Kondo, H., Tamada, T., 2007. RNA4-encoded p31 of beet necrotic yellow vein virus is involved in efficient vector transmission, symptom severity and silencing suppression in roots. J. Gen. Virol. 88, 1611–1619.

Raja, P., Wolf, J.N., Bisaro, D.M., 2010. RNA silencing directed against geminiviruses: post-transcriptional and epigenetic components. Biochim. Biophys. Acta 1799, 337–351.

Rajamäki, M.L., Streng, J., Valkonen, J.P., 2014. Silencing suppressor protein VPg of a potyvirus interacts with the plant silencing-related protein SGS3. Mol. Plant Microbe Interact. 27, 1199–1210.

Ramesh, S.V., Sahu, P.P., Prasad, M., Praveen, S., Pappu, H.R., 2017. Geminiviruses and plant hosts: a closer examination of the molecular arms race. Viruses 9, E256.

Ratcliff, F., Harrison, B.D., Baulcombe, D.C., 1997. A similarity between viral defense and gene silencing in plants. Science 276, 1558–1560.

Restrepo-Hartwig, M.A., Carrington, J.C., 1992. Regulation of nuclear transport of a plant potyvirus protein by autoproteolysis. J. Virol. 66, 5662–5666.

Rodamilans, B., San Léon, D., Mühlberger, L., Candresse, T., Neumüller, M., Oliveros, J.C., García, J.A., 2014. Transcriptomic analysis of *Prunus domestica* undergoing hypersensitive response to *Plum pox virus* infection. PLoS One 9, e100477.

Rodamilans, B., Valli, A., Mingot, A., San León, D., Baulcombe, D., López-Moya, J.J., García, J.A., 2015. RNA polymerase slippage as a mechanism for the production of frameshift gene products in plant viruses of the *Potyviridae* family. J. Virol. 89, 6965–6967.

Rodamilans, B., Valli, A., García, J.A., 2020. Molecular plant-plum pox virus interactions. Mol. Plant Microbe Interact. 33, 6–17.

Roossinck, M.J., 2015. Plants, viruses and the environment: ecology and mutualism. Virology 479–480C, 271–277.

Roossinck, M.J., García-Arenal, F., 2015. Ecosystem simplification, biodiversity loss and plant virus emergence. Curr. Opin. Virol. 10, 56–62.

Roossinck, M.J., Schneider, W.L., 2006. Mutant clouds and occupation of sequence space in plant RNA viruses. Curr. Top. Microbiol. Immunol. 299, 337–348.

Rubio, M., Rodríguez-Moreno, L., Ballester, A.R., de Moura, M.C., Bonghi, C., Candresse, T., Martínez-Gómez, P., 2015. Analysis of gene expression changes in peach leaves in response to *Plum pox virus* infection using RNA-Seq. Mol. Plant Pathol. 16, 164–176.

Ruiz-Ruiz, S., Navarro, B., Gisel, A., Peña, L., Navarro, L., Moreno, P., Di Serio, F., Flores, R., 2011. Citrus tristeza virus infection induces the accumulation of viral small RNAs (21-24-nt) mapping preferentially at the 3'-terminal region of the genomic RNA and affects the host small RNA profile. Plant Mol. Biol. 75, 607–619.

Ruiz-Ruiz, S., Soler, N., Sánchez-Navarro, J., Fagoaga, C., López, C., Navarro, L., Moreno, P., Peña, L., Flores, R., 2013. *Citrus tristeza virus* p23: determinants for nucleolar localization and their influence on suppression of RNA silencing and pathogenesis. Mol. Plant Microbe Interact. 26, 306–318.

Ruiz-Ruiz, S., Spanò, R., Navarro, L., Moreno, P., Peña, L., Flores, R., 2018. Citrus tristeza virus co-opts glyceraldehyde 3-phosphate dehydrogenase for its infectious cycle by interacting with the viral-encoded protein p23. Plant Mol. Biol. 98, 363–373.

Ruiz-Ruiz, S., Navarro, B., Peña, L., Navarro, L., Moreno, P., Di Serio, F., Flores, R., 2019. *Citrus tristeza virus:* host RNA silencing and virus counteraction. Methods Mol. Biol. 2015, 195–207.

Sáenz, P., Cervera, M.T., Dallot, S., Quiot, L., Quiot, J.B., Riechmann, J.L., García, J.A., 2000. Identification of a pathogenicity determinant of *Plum pox virus* in the sequence encoding the C-terminal region of protein P3+6K(1). J. Gen. Virol. 81, 557–566.

Sáenz, P., Quiot, L., Quiot, J.B., Candresse, T., García, J.A., 2001. Pathogenicity determinants in the complex virus population of a *Plum pox virus* isolate. Mol. Plant Microbe Interact. 14, 278–287.

Saha, S., Hafren, A., Mäkinen, K., 2019. Dynamics of protein accumulation from the 3' end of viral RNA are different from those in the rest of the genome in potato virus A infection. J. Virol. 93, e00721-19.

Salánki, K., Gellért, A., Nemes, K., Divéki, Z., Balázs, E., 2018. Molecular modeling for better understanding of cucumovirus pathology. Adv. Virus Res. 102, 59–88.

Sanfaçon, H., 2015. Plant translation factors and virus resistance. Viruses 7, 3392–3419.

Sanfaçon, H., 2017. Grand challenge in plant virology: understanding the impact of plant viruses in model plants, in agricultural crops, and in complex ecosystems. Front. Microbiol. 8, 860.

Santovito, E., Mascia, T., Siddiqui, S.A., Minutillo, S.A., Valkonen, J.P., Gallitelli, D., 2014. Infection cycle of *Artichoke Italian latent virus* in tobacco plants: meristem invasion and recovery from disease symptoms. PLoS One 9, e99446.

Satyanarayana, T., Gowda, S., Ayllón, M.A., Albiach-Martí, M.R., Rabindran, S., Dawson, W.O., 2002. The p23 protein of *Citrus tristeza virus* controls asymmetrical RNA accumulation. J. Virol. 76, 473–483.

Scholthof, K.B., Adkins, S., Czosnek, H., Palukaitis, P., Jacquot, E., Hohn, T., Hohn, B., Saunders, K., Candresse, T., Ahlquist, P., Hemenway, C., Foster, G.D., 2011. Top 10 plant viruses in molecular plant pathology. Mol. Plant Pathol. 12, 938–954.

Shan, H., Pasin, F., Valli, A., Castillo, C., Rajulu, C., Carbonell, A., Simón-Mateo, C., García, J.A., Rodamilans, B., 2015. The Potyviridae P1a leader protease contributes to host range specificity. Virology 476, 264–270.

Shan, H., Pasin, F., Tzanetakis, I.E., Simón-Mateo, C., García, J.A., Rodamilans, B., 2018. Truncation of a P1 leader proteinase facilitates potyvirus replication in a non-permissive host. Mol. Plant Pathol. 19, 1504–1510.

Shen, W.X., Au, P.C., Shi, B.J., Smith, N.A., Dennis, E.S., Guo, H.S., Zhou, C.Y., Wang, M.B., 2015. Satellite RNAs interfere with the function of viral RNA silencing suppressors. Front. Plant Sci. 6, 281.

Shimura, H., Masuta, C., 2016. Plant subviral RNAs as a long noncoding RNA (lncRNA): analogy with animal lncRNAs in host-virus interactions. Virus Res. 212, 25–29.

Shimura, H., Pantaleo, V., Ishihara, T., Myojo, N., Inaba, J., Sueda, K., Burgyán, J., Masuta, C., 2011. A viral satellite RNA induces yellow symptoms on tobacco by targeting a gene involved in chlorophyll biosynthesis using the RNA silencing machinery. PLoS Pathog. 7, e1002021.

Siddiqui, S.A., Sarmiento, C., Kiisma, M., Koivumäki, S., Lemmetty, A., Truve, E., Lehto, K., 2008. Effects of viral silencing suppressors on tobacco ringspot virus infection in two *Nicotiana* species. J. Gen. Virol. 89, 1502–1508.

Simon-Loriere, E., Holmes, E.C., 2011. Why do RNA viruses recombine? Nat. Rev. Microbiol. 9, 617–626.

Smith, N.A., Eamens, A.L., Wang, M.B., 2011. Viral small interfering RNAs target host genes to mediate disease symptoms in plants. PLoS Pathog. 7, e1002022.

Soler, N., Fagoaga, C., López, C., Moreno, P., Navarro, L., Flores, R., Peña, L., 2015. Symptoms induced by transgenic expression of p23 from *Citrus tristeza virus* in phloem-associated cells of Mexican lime mimic virus infection without the aberrations accompanying constitutive expression. Mol. Plant Pathol. 16, 388–399.

Sun, Y.D., Folimonova, S.Y., 2019. The p33 protein of *Citrus tristeza virus* affects viral pathogenicity by modulating a host immune response. New Phytol. 221, 2039–2053.

Sun, D., Chen, S., Cheng, A., Wang, M., 2016. Roles of the picornaviral 3C proteinase in the viral life cycle and host cells. Viruses 8, 82.

Sztuba-Solińska, J., Stollar, V., Bujarski, J.J., 2011a. Subgenomic messenger RNAs: mastering regulation of (+)-strand RNA virus life cycle. Virology 412, 245–255.

Sztuba-Solińska, J., Urbanowicz, A., Figlerowicz, M., Bujarski, J.J., 2011b. RNA-RNA recombination in plant virus replication and evolution. Annu. Rev. Phytopathol. 49, 415–443.

Tamada, T., Uchino, H., Kusume, T., Saito, M., 1999. RNA 3 deletion mutants of beet necrotic yellow vein virus do not cause rhizomania disease in sugar beets. Phytopathology 89, 1000–1006.

Tatineni, S., Dawson, W.O., 2012. Enhancement or attenuation of disease by deletion of genes from *Citrus tristeza virus*. J. Virol. 86, 7850–7857.

Tatineni, S., Robertson, C.J., Garnsey, S.M., Bar-Joseph, M., Gowda, S., Dawson, W.O., 2008. Three genes of *Citrus tristeza virus* are dispensable for infection and movement throughout some varieties of citrus trees. Virology 376, 297–307.

Tatineni, S., Robertson, C.J., Garnsey, S.M., Dawson, W.O., 2011. A plant virus evolved by acquiring multiple nonconserved genes to extend its host range. Proc. Natl. Acad. Sci. U. S. A. 108, 17366–17371.

Thiel, H., Varrelmann, M., 2009. Identification of *Beet necrotic yellow vein virus* P25 pathogenicity factor-interacting sugar beet proteins that represent putative virus targets or components of plant resistance. Mol. Plant Microbe Interact. 22, 999–1010.

Thiel, H., Hleibieh, K., Gilmer, D., Varrelmann, M., 2012. The P25 pathogenicity factor of *Beet necrotic yellow vein virus* targets the sugar beet 26S proteasome involved in the induction of a hypersensitive resistance response via interaction with an F-box protein. Mol. Plant Microbe Interact. 25, 1058–1072.

Thompson, J.R., Dasgupta, I., Fuchs, M., Iwanami, T., Karasev, A.V., Petrzik, K., Sanfaçon, H., Tzanetakis, I., van der Vlugt, R., Wetzel, T., Yoshikawa, N., ICTV Report Consortium, 2017. ICTV virus taxonomy profile: *Secoviridae*. J. Gen. Virol. 98, 529–531.

Tungadi, T., Donnelly, R., Qing, L., Iqbal, J., Murphy, A.M., Pate, A.E., Cunniffe, N.J., Carr, J.P., 2019. Cucumber mosaic virus 2b proteins inhibit virus-induced aphid resistance in tobacco. Mol. Plant Pathol. 21, 250–257.

Valli, A., Gallo, A., Calvo, M., de Jesús Pérez, J., García, J.A., 2014. A novel role of the potyviral helper component proteinase contributes to enhance the yield of viral particles. J. Virol. 88, 9808–9818.

Valli, A.A., Gallo, A., Rodamilans, B., López-Moya, J.J., García, J.A., 2018. The HCPro from the *Potyviridae* family: an enviable multitasking Helper component that every virus would like to have. Mol. Plant Pathol. 19, 744–763.

Varrelmann, M., Maiss, E., Pilot, R., Palkovics, L., 2007. Use of pentapeptide-insertion scanning mutagenesis for functional mapping of the plum pox virus helper component proteinase suppressor of gene silencing. J. Gen. Virol. 88, 1005–1015.

Verchot, J., 2016a. How does the stressed out ER find relief during virus infection? Curr. Opin. Virol. 17, 74–79.

Verchot, J., 2016b. Plant virus infection and the ubiquitin proteasome machinery: arms race along the endoplasmic reticulum. Viruses 8, E314.

Vetter, G., Hily, J.M., Klein, E., Schmidlin, L., Haas, M., Merkle, T., Gilmer, D., 2004. Nucleo-cytoplasmic shuttling of the beet necrotic yellow vein virus RNA-3-encoded p25 protein. J. Gen. Virol. 85, 2459–2469.

Vigne, E., Gottula, J., Schmitt-Keichinger, C., Komar, V., Ackerer, L., Belval, L., Rakotomalala, L., Lemaire, O., Ritzenthaler, C., Fuchs, M., 2013. A strain-specific segment of the RNA-dependent RNA polymerase of grapevine fanleaf virus determines symptoms in *Nicotiana* species. J. Gen. Virol. 94, 2803–2813.

Vogler, H., Kwon, M.O., Dang, V., Sambade, A., Fasler, M., Ashby, J., Heinlein, M., 2008. *Tobacco mosaic virus* movement protein enhances the spread of RNA silencing. PLoS Pathog. 4, e1000038.

Wallis, C.M., Stone, A.L., Sherman, D.J., Damsteegt, V.D., Gildow, F.E., Schneider, W.L., 2007. Adaptation of plum pox virus to a herbaceous host (*Pisum sativum*) following serial passages. J. Gen. Virol. 88, 2839–2845.

Wang, A., 2015. Dissecting the molecular network of virus-plant interactions: the complex roles of host factors. Annu. Rev. Phytopathol. 53, 45–66.

Wang, C., Wang, C., Xu, W., Zou, J., Qiu, Y., Kong, J., Yang, Y., Zhang, B., Zhu, S., 2018. Epigenetic changes in the regulation of *Nicotiana tabacum* response to *Cucumber mosaic virus* infection and symptom recovery through single-base resolution methylomes. Viruses 10, 402.

Wang, C., Wang, C., Zou, J., Yang, Y., Li, Z., Zhu, S., 2019. Epigenetics in the plant-virus interaction. Plant Cell Rep. 38, 1031–1038.

Wei, T., Zhang, C., Hong, J., Xiong, R., Kasschau, K.D., Zhou, X., Carrington, J.C., Wang, A., 2010. Formation of complexes at plasmodesmata for potyvirus intercellular movement is mediated by the viral protein P3N-PIPO. PLoS Pathog. 6, e1000962.

Weiland, J.J., Bornemann, K., Neubauer, J.D., Khan, M.F.R., Bolton, M.D., 2019. Prevalence and distribution of beet necrotic yellow vein virus strains in North Dakota and Minnesota. Plant Dis. 103, 2083–2089.

Westwood, J.H., Groen, S.C., Du, Z., Murphy, A.M., Anggoro, D.T., Tungadi, T., Luang-In, V., Lewsey, M.G., Rossiter, J.T., Powell, G., Smith, A.G., Carr, J.P., 2013a. A trio of viral proteins tunes aphid-plant interactions in *Arabidopsis thaliana*. PLoS One 8, e83066.

Westwood, J.H., McCann, L., Naish, M., Dixon, H., Murphy, A.M., Stancombe, M.A., Bennett, M.H., Powell, G., Webb, A.A., Carr, J.P., 2013b. A viral RNA silencing suppressor interferes with abscisic acid-mediated signalling and induces drought tolerance in *Arabidopsis thaliana*. Mol. Plant Pathol. 14, 158–170.

Wetzel, T., Chisholm, J., Dupuis-Maguiraga, L., Bassler, A., Sanfaçon, H., 2013. *In vitro* and *in vivo* evidence for differences in the protease activity of two arabis mosaic nepovirus isolates and their impact on the infectivity of chimeric cDNA clones. Virology 446, 102–111.

Wingard, S.A., 1928. Hosts and symptoms of ring spot, a virus disease of plants. J. Agric. Res. 37, 127–153.

Wu, D., Qi, T., Li, W.X., Tian, H., Gao, H., Wang, J., Ge, J., Yao, R., Ren, C., Wang, X.B., Liu, Y., Kang, L., Ding, S.W., Xie, D., 2017. Viral effector protein manipulates host hormone signaling to attract insect vectors. Cell Res. 27, 402–415.

Wu, X., Valli, A., Garcia, J.A., Zhou, X., Cheng, X., 2019. The tug-of-war between plants and viruses: great progress and many remaining questions. Viruses 11, 203.

Wylie, S.J., Adams, M., Chalam, C., Kreuze, J., López-Moya, J.J., Ohshima, K., Praveen, S., Rabenstein, F., Stenger, D., Wang, A., Zerbini, F.M., ICTV Report Consortium, 2017. ICTV virus taxonomy profile: *Potyviridae*. J. Gen. Virol. 98, 352–354.

Xu, A., Zhao, Z., Chen, W., Zhang, H., Liao, Q., Chen, J., Carr, J.P., Du, Z., 2013. Self-interaction of the cucumber mosaic virus 2b protein plays a vital role in the suppression of RNA silencing and the induction of viral symptoms. Mol. Plant Pathol. 14, 803–812.

Xu, M., Mazur, M.J., Tao, X., Kormelink, R., 2020. Cellular RNA hubs: friends and foes of plant viruses. Mol. Plant Microbe Interact. 33, 40–54.

Yang, C., Liu, T., Bai, F., Wang, N., Pan, Z., Yan, X., Peng, S., 2015. miRNAome analysis associated with anatomic and transcriptomic investigations reveal the polar exhibition of corky split vein in boron deficient *Citrus sinensis*. Mol. Genet. Genomics 290, 1639–1657.

Zhang, X.P., Liu, D.S., Yan, T., Fang, X.D., Dong, K., Xu, J., Wang, Y., Yu, J.L., Wang, X.B., 2017. Cucumber mosaic virus coat protein modulates the accumulation of 2b protein and antiviral silencing that causes symptom recovery *in planta*. PLoS Pathog. 13, e1006522.

Zhang, B., Li, W., Zhang, J., Wang, L., Wu, J., 2019. Roles of small RNAs in virus-plant interactions. Viruses 11, 827.

Zhao, J., Zhang, X., Hong, Y., Liu, Y., 2016. Chloroplast in plant-virus interaction. Front. Microbiol. 7, 1565.

Zhao, J.H., Liu, X.L., Fang, Y.Y., Fang, R.X., Guo, H.S., 2018. CMV2b-dependent regulation of host defense pathways in the context of viral infection. Viruses 10, 618.

Zhou, T., Murphy, A.M., Lewsey, M.G., Westwood, J.H., Zhang, H.M., González, I., Canto, T., Carr, J.P., 2014. Domains of the cucumber mosaic virus 2b silencing suppressor protein affecting inhibition of salicylic acid-induced resistance and priming of salicylic acid accumulation during infection. J. Gen. Virol. 95, 1408–1413.

Zhuo, N., Jiang, N., Zhang, C., Zhang, Z.Y., Zhang, G.Z., Han, C.G., Wang, Y., 2015. Genetic diversity and population structure of beet necrotic yellow vein virus in China. Virus Res. 205, 54–62.

CHAPTER FOUR

Host protein chaperones, RNA helicases and the ubiquitin network highlight the arms race for resources between tombusviruses and their hosts

Peter D. Nagy*
Department of Plant Pathology, University of Kentucky, Lexington, KY, United States
*Corresponding author: e-mail address: pdnagy2@uky.edu

Contents

Abstract

Positive-strand RNA viruses need to arrogate many cellular resources to support their replication and infection cycles. These viruses co-opt host factors, lipids and subcellular membranes and exploit cellular metabolites to built viral replication organelles in infected cells. However, the host cells have their defensive arsenal of factors to protect themselves from easy exploitation by viruses. In this review, the author discusses an emerging arms race for cellular resources between viruses and hosts, which occur during the early events of virus-host interactions. Recent findings with tomato bushy stunt virus and its hosts revealed that the need of the virus to exploit and co-opt given

Advances in Virus Research, Volume 107
ISSN 0065-3527
https://doi.org/10.1016/bs.aivir.2020.06.006

members of protein families provides an opportunity for the host to deploy additional members of the same or associated protein family to interfere with virus replication. Three examples with well-established heat shock protein 70 and RNA helicase protein families and the ubiquitin network will be described to illustrate this model on the early arms race for cellular resources between tombusviruses and their hosts. We predict that arms race for resources with additional cellular protein families will be discovered with tombusviruses. These advances will fortify research on interactions among other plant and animal viruses and their hosts.

1. Introduction

Positive-strand (+)RNA viruses of eukaryotes replicate in the cytosol of infected cells by assembling numerous membrane-bound viral replicase complexes (VRCs). The VRCs are usually harbored inside elaborate virus-induced membranous structures, called viral replication organelles (VROs) (Fernandez de Castro et al., 2016; Paul and Bartenschlager, 2015; Shulla and Randall, 2016; Wang, 2015). The VROs consist of single-membrane or double membrane vesicles, convoluted membranes and other structures for a group of (+)RNA viruses that include coronaviruses, enteroviruses and hepatitis C virus (Altan-Bonnet, 2017; Belov and Sztul, 2014; Nagy and Pogany, 2012). Another group of (+)RNA viruses, such as alphaviruses, most flaviviruses, the insect-infecting nodaviruses and the plant-infecting tombusviruses and bromoviruses, deform membranes, creating numerous vesicle-like structures, also called spherules, with small opening toward the cytosol (Brinton, 2013; den Boon and Ahlquist, 2010; Ertel et al., 2017; Ishibashi and Ishikawa, 2016; Neufeldt et al., 2018; Zhang et al., 2019). The spherules could be formed by using membranes derived from the ER, mitochondria, plasma membrane or peroxisomes (Laliberte and Sanfacon, 2010; Shulla and Randall, 2016; Wang, 2015). Overall, the VROs sequester viral and co-opted host components to a confined place to achieve robust viral RNA replication. The VROs organize temporally and spatially the viral replication process and other distinct steps during the infection cycle. The VROs also protect the viral RNA and proteins from degradation and facilitate the avoidance of recognition by the host immune system.

Our knowledge on the biogenesis of VROs and the individual VRCs and overall on the interactions between viruses and their hosts are quickly progressing based on a few model (+)RNA viruses and other important human pathogenic viruses (Altan-Bonnet, 2017; Nagy, 2016; Nagy and Pogany, 2012; Neufeldt et al., 2018; Wang, 2015; Zhang et al., 2019).

In this article, we will use the example of the small plant tombusviruses to visualize how the arms race for control of resources between viruses and their hosts include conserved host protein families.

Tomato bushy stunt virus (TBSV) and other tombusviruses have less than 5 kb (+)RNA genomes, which code for five proteins, two of which are the viral replication proteins, termed p33 and $p92^{pol}$ (Nagy, 2008, 2011; Nagy and Pogany, 2010). The replication proteins are essential for TBSV replication, the biogenesis of VROs and the formation of spherule structures utilizing peroxisomal membranes in plants and yeast, a model host (Barajas et al., 2009a; McCartney et al., 2005). Several genome-wide and proteome-wide screens in combination with lipidomics and bioinformatics approaches led to the identification of over 500 yeast genes, which affected TBSV replication (Nagy, 2016, 2017; Nagy et al., 2014). More detailed, mechanistic studies with cell-free replicase reconstitution assays, yeast knockout and knockdown mutants or temperature-sensitive mutants, and validation experiments in a plant host revealed the intricate interaction of TBSV with the hosts through co-opting numerous mostly-conserved host proteins (Li and Nagy, 2011; Nagy, 2016; Nagy et al., 2012). As discussed below, the hosts also respond by using proteins frequently from the same or associated protein families to greatly interfere with TBSV replication, thus creating a never-ending arms race with the virus.

The article will follow a scheme that presents the idea of how the need of the virus to hijack host factors creates opportunities for the host to use similar host proteins that interfere with virus replication. In other words, the arms race for control of resources between the virus and host could involve different members of the same protein families. Three examples with well-established protein families will be described to illustrate this model on the arms race between tombusviruses and their hosts.

Description of the arms race between viruses and hosts frequently starts with explaining what pathogen associated molecular patterns (PAMPs) are recognized by the host, which is mostly based on specific features present in viral proteins and the viral RNAs. The detection of PAMPs then triggers RNAi and other immune responses in plants and animals (Caplan et al., 2008; Chow et al., 2018; Ding, 2010; Rehwinkel and Gack, 2020; Soosaar et al., 2005). However, viruses well-adapted to their hosts are able to suppress the effective immune responses through expressing suppressor or effector proteins to stave off the antiviral defenses (Chow and Kagan, 2018; Goertz et al., 2018; Kuka and Iannacone, 2018; Li and Ding, 2006). Often the hosts will have another level of antiviral responses in the arms race. For example, plants might select for resistance (R) genes against the most pathogenic viruses or animals produce specialized immune cells guarding against future infections by the same virus (Schwerk et al., 2020; Soosaar et al., 2005; Tassetto et al., 2017). Note that the arms race described in this article is placed prior to the above powerful virus–host interactions, which are rather costly to the hosts and not all individuals are protected by such expensive defensive measures in a population due to genetic diversity. The arms race to exploit cellular resources by the invading virus and the counteractive defensive strategy deployed by the host/host cell to protect those resources from the virus takes place rather early during infections. However, this defensive strategy by the host might be effective against most, poorly or not-well adapted viruses. If the virus succeeds in overtaking this early defensive line of the cells, then the host cell will activate the PAMP-based defensive responses. Taken together, based on the level of adaptation between the virus and hosts, arms races occur at several levels and at various costs for the hosts.

2. The heat shock protein 70 chaperone network

The first example on the arms race between tombusviruses and their hosts will involve the Hsp70 chaperone network. The Hsp70 family of protein chaperones is highly conserved, present in all domains of life forms and required for protein quality control and maintenance of cellular homeostasis (Kim et al., 2013; Moran Luengo et al., 2019; Rosenzweig et al., 2019). Hsp70s are highly abundant, even in the absence of stress. They are ATP-dependent molecular chaperone machines to fold unfolded proteins, refold misfolded protein substrates, protein aggregates or large protein assemblies. Cytosolic Hsp70s are also involved in disassembly of large protein complexes or protein-nucleic acid complexes (Mayer, 2013). These include virions, clathrin coats, or pre-replication complex of bacteriophage λ (Rosenzweig et al., 2019). Hsp70s often cooperate with other accessory co-chaperones or Hsp90 in multiple protein refolding and maturation processes. The co-chaperones include the family of J-domain proteins and tetratricopeptide repeat (TPR) domain-containing proteins, which help determine the fate of client proteins (Genest et al., 2019).

Importantly, Hsp70 and co-chaperones play critical roles in RNA virus replication with ever-increasing numbers of documented cases, such for several flaviviruses (Bozzacco et al., 2016; Taguwa et al., 2015, 2019), influenza virus (Cao et al., 2014; Manzoor et al., 2014), and a couple of plant viruses (Chenon et al., 2012; Lamm et al., 2017; Lohmus et al., 2017;

Nagy et al., 2011; Verchot, 2012; Wang et al., 2018). Based on these discoveries, the findings described below in case of TBSV and its hosts could be relevant for a wide range of viruses.

3. The pro-viral roles of co-opted protein chaperones in TBSV replication

Although viruses are relatively simple parasitic entities, they still require amazing level of coordination and regulation of biological processes during their infection cycles. Therefore, they functionally repurpose co-opted cellular proteins to facilitate viral replication. For example, the p92^{pol} RdRp of TBSV is originally inactive as an RdRp after translation by the host ribosomes (Pogany and Nagy, 2012). The p92^{pol} requires binding to membranes and the viral (+)RNA to become activated as an RdRp. The co-opted Hsp70 chaperone is absolutely essential for this activation process, albeit the p33-domain in p92^{pol} also plays a yet undefined role in the activation process (Pogany and Nagy, 2012, 2015). Moreover, Hsp70 facilitates the intracellular localization and the proper insertion of p33 and p92^{pol} replication proteins into subcellular membranes, which is needed for the stability of both replication proteins (Wang et al., 2009a,b). In vitro replicase reconstitution experiments unambiguously revealed the absolute essential contribution of Hsp70 members to TBSV replication (Pogany et al., 2008). Hsp70 is also required for the virion assembly process within or close to the VROs (Alam and Rochon, 2015, 2017). Additional members of the Hsp70 chaperone network have also been shown to facilitate TBSV replication (Table 1), but the detailed mechanisms are not yet dissected.

These co-opted Hsp70s require lots of ATP molecules for their functions within VROs. Interestingly, the ATP is generated locally by the co-opted glycolytic enzymes within the VROs (Chuang et al., 2017; Nagy and Lin, 2020; Prasanth et al., 2017). TBSV also recruits the fermentation enzymes, possibly to replenish the NAD+ pool, which is required for continuous operation of the glycolytic enzymes within VROs (Lin et al., 2019). Through recruiting the aerobic glycolytic pathway into VROs for rapid ATP generation, TBSV might be able to fully exploit the co-opted Hsp70s for efficient VRO formation and robust viral replication to overwhelm the host cells (Nagy and Lin, 2020). Co-opting the Hsp70 chaperone network with the glycolytic enzymes into VROs elegantly illustrates how TBSV could utilize the cell own resources against cellular antiviral responses via the efficient biogenesis of VROs.

Table 1 Chaperone network genes likely involved in the arms race between TBSV and its hosts.

Pro-viral effect

Gene	Function	Pro-viral function	Homologs
SSA1/2	Hsp70, constitutive	Activation of p92 RdRp	Hsp70-1/2
SSA3/4	Hsp70, heat-shock induced	Partially redundant with Ssa1/2	
SSB1	Hsp70, ribosome associated	Binds to p92	
HSC82	Hsp90, constitutive	Binds to p92	Hsp90
HSP104	Hsp104, disaggregase	Binds to p92	Hsp104

Antiviral effect

Gene	Function	Antiviral function	Homologs
CPR7	TPR domain, cyclophilin	Inhibition of VRC assembly	Cyp40
CNS1	TPR domain, co-chaperone	Inhibition of VRC assembly	TTC4 oncogene
STI1	TPR1 domain, co-chaperone	Inhibition of CIRV VRC assembly	Hop
CPR6	TPR domain, cyclophilin	Weak inhibitor of VRC assembly	Cyp40

Undefined effect/function

Gene	Function	Viral target	Homologs
SGT2	TPR domain, co-chaperone	Interacts with p33	hSGT-like
JJJ1	Co-chaperone	Interacts with p33/p92/(+) RNA over-expression stimulatory	Hsp40
JJJ3	Co-chaperone	Interacts with p33	Hsp40
YDJ1	Co-chaperone	Over-expression inhibitory	Hsp40
CDC37	Co-chaperone	ts-mutant, reduced viral replication	
SEC17	TPR domain	Inhibition of TBSV replication in vitro	

Genes shown in bold face are characterized in details in TBSV replication.

4. The antiviral roles of co-opted protein co-chaperones in TBSV replication

Like other powerful molecular machines, Hsp70s and other protein chaperones are also associated with accessory proteins, which help tune their activities (Rosenzweig et al., 2019). Therefore, the host cells could utilize these accessory proteins as antiviral factors. Indeed, genome-wide screens in yeast unearthed protein co-chaperones associated with Hsp70 or Hsp90 chaperones as inhibitors of TBSV replication (Table 1). For example, the Cyp40-like Cpr7 yeast cyclophilin with a TPR domain turned out to be a strong inhibitor of TBSV replication. Cpr7 or its TPR-domain alone bind to the RPR RNA-binding region in p33 and $p92^{pol}$ replication proteins, which likely prevent the replication proteins to recruit the viral (+)RNA into replication and other functions within VRCs (Fig. 1) (Lin et al., 2012). Interestingly, a second yeast Cyp40-like cyclophilin Cpr6 with a TPR domain had only weak restriction function on TBSV replication in a cell-free assay or in yeast. However, Cpr6 binds to a different region in the viral replication proteins, not the RPR domain bound by Cpr7 (Lin et al., 2012). Moreover, Cpr6 has a strong PPIase activity, whereas Cpr7 acts as a co-chaperone. These differences between the two yeast Cyp40-like cyclophilins might explain that only Cpr7 is a strong tombusvirus restriction factor. On the contrary, three Cyp40-like cyclophilins from *Arabidopsis* plants were found to inhibit tombusvirus replication in vitro (Lin et al., 2012). These findings suggest that plants have more Cyp40-like cyclophilins to combat tombusvirus replication than yeast.

The human Ttc4 oncogene-like yeast Cns1 co-chaperone also contains a TPR domain in its N-terminal region (Schopf et al., 2019). Cns1 can activate the ATPase function of Hsp70 (Hainzl et al., 2004). Cns1 strongly interacts with the TBSV replication proteins via its extended TPR-domain, which results in blocking VRC assembly (Fig. 1) (Lin and Nagy, 2013). This inhibitory effect of Cns1 on TBSV replication is likely due to binding of Cns1 to the p33-p33/p92 interaction domain in the viral replication proteins. The interaction between the viral replication proteins via the p33-p33/p92 interaction domain is absolutely critical for VRC formation (Panavas et al., 2005a; Rajendran and Nagy, 2006). Unlike Cpr7, which inhibited the replication of both tombusviruses and the unrelated insect-infecting nodaviruses, Cns1 seems to be TBSV-specific restriction factor (Lin et al., 2012; Lin and Nagy, 2013).

Another interesting tombusvirus restriction factor is the Hop-like Sti1 co-chaperone of Hsp70 and Hsp90 (Xu et al., 2014). Sti1 and the plant

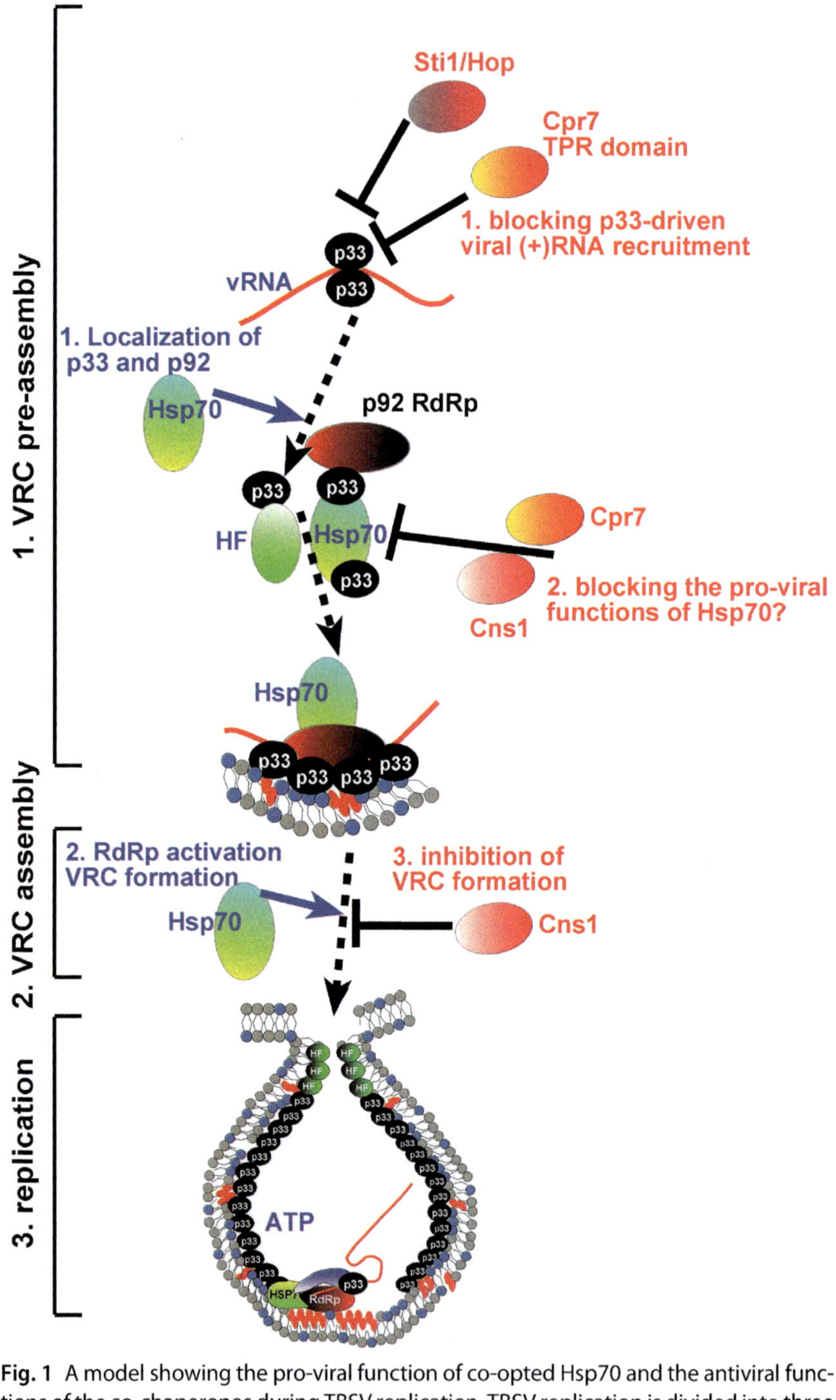

Fig. 1 A model showing the pro-viral function of co-opted Hsp70 and the antiviral functions of the co-chaperones during TBSV replication. TBSV replication is divided into three sequential steps, as shown on the left. Based on published data, we propose that TBSV co-opts the cytosolic Hsp70 chaperones to help perform several pro-viral functions indicated in blue letters on the left. The hosts use several co-chaperones to block the pro-viral function of Hsp70 and the viral replication proteins at several steps (as indicated with red letters), thus reducing viral replication. See further details in the main text.

Hop-1 have three TPR domains and TPR1 is the most important in providing the antiviral effect by binding to the RPR RNA-binding domain in the p36 and $p95^{pol}$ replication proteins of carnation Italian ringspot virus (CIRV), a tombusvirus with mitochondrial membrane-based VROs. Surprisingly, Sti1 and Hop1 only acted as weak restriction factors against TBSV (peroxisomal membrane-based VROs) in yeast or in vitro (Xu et al., 2014). Thus, unlike Cpr7 and Cns1 co-chaperones, which restrict the replication of both TBSV and CIRV, Sti1/Hop seems only to be a strong CIRV restriction factor. Subcellular localization data indicated that the mitochondrial CIRV is targeted more efficiently than the peroxisomal TBSV by Sti1, suggesting that accessibility of Sti1 in various subcellular locations might differ (Lin et al., 2012; Lin and Nagy, 2013; Xu et al., 2014).

Based on our discoveries, we propose that the need of TBSV to hijack Hsp70 functions from the cells during viral replication opens up opportunities for the hosts to fight back by utilizing protein chaperone-associated proteins and co-chaperones. All the three above described TPR-domain containing proteins are co-chaperones of Hsp70 and Hsp90 chaperones. This observation suggests that the co-chaperones might inhibit tombusvirus replication not only by directly binding to the viral replication proteins, but via inhibiting the pro-viral functions of the co-opted Hsp70 chaperone. Indeed, for example, TPR-domain containing proteins are predicted to be present in all eukaryotes in large numbers, including ~200 in humans and ~250 in *Arabidopsis* (Haslbeck et al., 2013; Prasad et al., 2010). An emerging picture is that the antiviral co-chaperones, and cyclophilins have to be targeted into action at the very early stage of TBSV infection, before the viral infection overwhelms the host cells defensive potential.

Altogether, this creates an arms race between the virus and its host for the access to the Hsp70 chaperone network. Through utilizing members of the cellular co-chaperones, the host interferes with the "easy" exploitation of Hsp70 chaperone machine for tombusvirus replication.

5. Cellular RNA helicases and RNA chaperones

The second example on the arms race between tombusviruses and their hosts will present the protein families involved in RNA structure modulation. Similar to their protein counterparts, the numerous cellular RNAs also require the guidance of specialized army of proteins that perform their folding, maturation, transport and ultimately degradation. The largest families of proteins involved with cellular RNAs are RNA helicases and RNA chaperones, which often function as ribonucleoprotein complexes

(Guenther and Jankowsky, 2009; Jankowsky, 2011; Sloan and Bohnsack, 2018). Whereas RNA helicases use ATP to modulate RNA structures/ functions, RNA chaperones perform those functions through binding to RNA targets without using ATP (Gilman et al., 2017; Linder and Jankowsky, 2011; Sharma and Jankowsky, 2014). Similar to cellular processes that depend on RNAs, (+)RNA viruses also require the involvement of RNA helicases and/or RNA chaperones to promote their replication and the infection process (Huang et al., 2010; Li and Nagy, 2011; Ranji and Boris-Lawrie, 2010; Taschuk and Cherry, 2020). The hosts also deploy cellular DEAD-box helicases, such as DICER, RIG-I or MDA5, which are involved in sensing the viral RNAs and initiating antiviral immune signaling or the RNAi pathway to destroy viruses (Ding, 2010; Garcia-Ruiz et al., 2010; Schroder, 2011; Taschuk and Cherry, 2020).

6. The pro-viral roles of co-opted DEAD-box helicases and RNA chaperones in TBSV replication

Similar to the viral p92^{pol} replication protein, the TBSV (+)RNA needs to be recruited to the site of replication (Monkewich et al., 2005; Pogany et al., 2005). The (+)RNA template selection for replication is performed by the p33 replication protein via specifically recognizing an internal extended stem-loop structure with a C·C mismatch within an internal loop (Fig. 2) (Monkewich et al., 2005; Pogany et al., 2005). The TBSV (+)RNA needs also "be activated" prior to functioning as a template at the sites of replication (Pathak et al., 2011, 2012). This is because the 3′ end of TBSV (+)RNA is embedded in an RNA secondary structure not readily recognized by the RdRp. An upstream RNA silencer element binds to the 3'terminal promoter, forming a partially dsRNA structure (Pogany et al., 2003). After recruitment into the sites of replication, co-opted RNA chaperones, namely eEF1A and eEF1Bγ, open up the above secondary structures and bring in the p92 RdRp to initiate minus-strand synthesis (Fig. 2) (Li et al., 2010; Sasvari et al., 2011).

Dissociation of the p92 RdRp from the template at the end of minus-strand synthesis is greatly affected by a DEAD-box helicase, the DDX3-like Ded1 in yeast, and RH20 in plants (Chuang et al., 2015). The formation of dsRNA during minus-strand synthesis leads to an inactive template (Kovalev et al., 2014), which has to be partially unwound by a DEAD-box helicase, Ded1 and Dbp2 in yeast, and RH20 in plants, prior to initiation of plus-strand synthesis by the p92 RdRp (Kovalev et al., 2012a,b). Since, the initiation on dsRNA template has to be repeated by ~100 times,

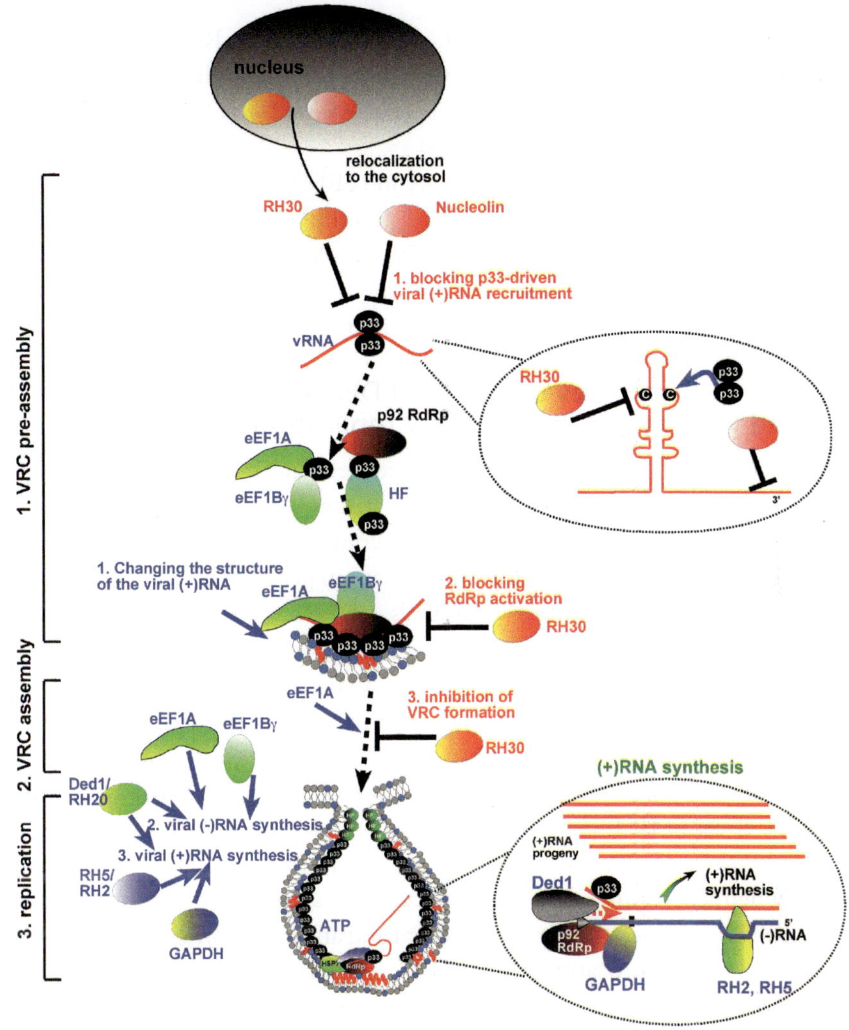

Fig. 2 A model of the arms race between TBSV and its hosts based on RNA helicases and RNA chaperones. The pro-viral functions of the co-opted RNA helicases and RNA chaperones versus the antiviral functions of additional RNA helicases and RNA chaperones during TBSV replication are shown schematically. The co-opted RNA chaperones (translation elongation factors) facilitate refolding of viral (+)RNA, followed by assisting TBSV (−)RNA synthesis as shown on the left. The co-opted RNA helicases facilitate the termination of (−)RNA synthesis, then the following initiation of (+)-strand synthesis and asymmetrical replication with RNA chaperone contribution of Glyceraldehyde 3-phosphate dehydrogenase (GAPDH). Some of the details are enlarged on the right. On the contrary, the cellular DDX17-like RH30 helicase interferes with several steps during TBSV replication as shown in red on the right side of the panel. The RNA chaperone Nucleolin also inhibits the recruitment of the TBSV (+)RNA into replication. The RII(+)-SL *cis*-acting RNA element with the critical internal C·C mismatch, which specifically binds to the p33 replication protein only when the stem-loop structure is formed, is shown in the circle. The cytosolic pools of RH30 and Nucleolin are essential for the antiviral activity. See further details in the main text.

a second set of DEAD-box helicases, the eIF4AIII-like RH2 and DDX5-like RH5 in plants, facilitate the "recycling" of the viral RdRp with the help of RNA replication enhancers present on the viral (−)RNA (Kovalev and Nagy, 2014; Panavas and Nagy, 2005).

In contrast with the co-opted eEF1A and eEF1Bγ RNA chaperones, the co-opted DEAD-box helicases require plentiful ATP for their pro-viral functions. As described above for Hsp70, the ATP is generated locally by the co-opted glycolytic enzymes within the membranous VROs (Chuang et al., 2017; Nagy and Lin, 2020; Prasanth et al., 2017).

6.1 The antiviral roles of co-opted DEAD-box helicases and RNA chaperones in TBSV replication

The need for TBSV to recruit conserved cellular RNA chaperones and DEAD-box helicases for viral replication makes the virus "vulnerable" to those cellular DEAD-box helicases, which could inhibit the replication process. This possibility might be really important in plants, which encode more than twice as many helicases as animals and yeast do. Accordingly, the DDX17-like RH30 helicase from *Arabidopsis* has been shown to be recruited into the VROs and strongly inhibits TBSV replication (Fig. 2) (Wu and Nagy, 2019). The inhibition is due to unwinding of the critical *cis*-acting secondary element in the viral (+)RNA, which is required for binding to p33 replication protein, followed by selection for replication. Since this is an early step, the effect of RH30 was found to be dominant over the co-opted pro-viral helicases, which act at latter steps in the replication process. By altering the biochemical features of DEAD-box helicases, either by domain analysis or by creating chimeras between pro-viral and antiviral helicases, it might be possible to obtain new antiviral tools by engineering or through evolution, called arms race between viruses and their hosts.

In addition to the well-characterized DEAD-box helicases (see above), plants also deploy RNA chaperones, such as Nucleolin, against TBSV (Jiang et al., 2010). Nucleolin (Nsr1 in yeast) has been shown to be recruited into the VROs and strongly inhibits TBSV replication in yeast and plants (Jiang et al., 2010). The inhibition is due to binding of Nucleolin to a *cis*-acting secondary element in the viral (+)RNA, which is inhibitory to recruitment of TBSV (+)RNA for replication (Fig. 2). Nucleolin, like other RNA chaperones, modify RNA structures, but without using ATP. Because TBSV has its own RNA chaperone, the p33 replication protein (Stork et al., 2011), and co-opt translation elongation factors, which also have RNA chaperone function, it is possible that the hosts will utilize an arsenal of RNA chaperones against invading viruses. Accordingly, genome-wide screens in yeast did identify numerous RNA binding host proteins that inhibited TBSV replication (Jiang et al., 2006; Li et al., 2009; Panavas et al., 2005b; Prasanth et al., 2016; Shah Nawaz-Ul-Rehman et al., 2012). The antiviral functions of the identified cellular RNA-binding proteins have not been defined (Table 2). Further research will highlight whether there is ongoing arms race between viruses and their hosts using RNA chaperones.

An emerging concept is that the antiviral DEAD-box helicases and RNA chaperones have to be targeted into action at the very early stage of TBSV infection to block viral replication (Jiang et al., 2010; Wu and Nagy, 2019). During late stages, the viral infection overwhelms the host cells defensive potential at this level.

Cellular RNA chaperones and DEAD-box helicases affect translation and replication of many RNA viruses (Garbelli et al., 2011; Huang et al., 2010; Ranji and Boris-Lawrie, 2010; Upadya et al., 2014). For example, host DEAD-box helicases have been shown to affect translation of viral proteins

Table 2 DEAD-box helicases and RNA chaperones likely involved in the arms race between TBSV and its hosts.

Pro-viral effect

Gene	Function	Pro-viral function	Homologs
DED1	DEAD-box helicase	(−) and (+)-strand synthesis	DDX3/***AtRH20***
DBP2	DEAD-box helicase	(+)-strand synthesis	p68/***AtRH20***
AtRH2	DEAD-box helicase	(+)-strand synthesis	eIF4AIII/Sc*FAL1*
AtRH5	DEAD-box helicase	(+)-strand synthesis	DDX5/Sc*DBP3*
TEF1/2	Translation elongation factor	(−)-strand synthesis, p92 binding	**eEF1A**
TEF4	Translation elongation factor	(−)-strand synthesis	**eEF1Bγ**
TDH2/3	GAPDH, glycolysis	(+)-strand synthesis	GAPDH/***At*GAPC**

Antiviral effect

Gene	Function	Antiviral function	Homologs
AtRH30	DEAD-box helicase	Inhibition of (+)RNA recruitment	DDX17
NSP1	Ribosomal RNA maturation	Inhibition of (+)RNA recruitment	**Nucleolin**

Continued

Table 2 DEAD-box helicases and RNA chaperones likely involved in the arms race between TBSV and its hosts.—cont'd

Undefined effect/function

Gene	Function	Viral target	Homologs
TIF1	Translation initiation factor eIF4A DEAD-box helicase	Interacts with p33	eIF4A
TIF11	Translation initiation factor eIF1A DEAD-box helicase	Interacts with p33/p92	eIF1A
PBP2	heterogeneous nuclear RNP K protein	Over-expression is stimulatory	
HAS1	ATP-dependent RNA helicase	Over-expression is inhibitory	

Genes shown in bold face are characterized in details in TBSV replication.

(Bolinger et al., 2010; Noueiry et al., 2000; Watanabe et al., 2009); viral RNA replication (Goh et al., 2004; Huang et al., 2010; Lawrence and Rieder, 2009; Morohashi et al., 2011; Upadhyay et al., 2013); subgenomic RNA synthesis (Gimenez-Barcons et al., 2013); and virus assembly (Xu and Hobman, 2012). Rift Valley fever virus is restricted by the DDX17 helicase (Moy et al., 2014). Moreover, DDX3 blocks Dengue virus infections, whereas DDX21 helicase inhibits influenza A virus (Chen et al., 2014; Li et al., 2015; Schroder, 2011; Zhang et al., 2011). Human immunodeficiency virus (HIV-1) infections are affected by several helicases providing either pro-viral or antiviral functions (Lorgeoux et al., 2012; Yasuda-Inoue et al., 2013). Thus, the emerging idea is that cellular DEAD-box helicases are important co-opted host factors for several viruses, whereas other cellular DEAD-box helicases play active roles in restricting RNA virus replication. Taken together, cellular DEAD-box helicases perform critical roles in virus-host interactions with many more discoveries expected in the future.

7. Comparison of protein chaperone and RNA helicase/chaperone functions

It seems that TBSV hijacks two highly-conserved protein family members from the host cells: the protein chaperoning Hsp70 proteins and co-chaperones and the RNA chaperoning DEAD-box helicases. The similarities and parallels between these proteins in function for the virus are interesting: (i) they utilize ATP for their function, which are generated by co-opted glycolytic enzymes within VROs; (ii) they are recruited to the sites of TBSV replication (VROs); (iii) they are directly involved with the viral replication proteins/RNA templates; (iv) and they are permanently part of the VRCs. The major difference between them is the substrate selection: Hsp70s work with the viral replication proteins, whereas the DEAD-box helicases target the viral RNAs as substrates. However, both protein family members contribute pro-viral as well as antiviral functions. This highlights that these proteins families are core members of the arms race between viruses and their hosts. They could be also potential central players in strategies to develop novel antiviral approaches.

8. The ubiquitin network

The last example on the arms race between tombusviruses and their hosts will involve the ubiquitin network. Many cellular proteins have multiple functions at different subcellular locations. Post-translational modifications are frequently being utilized by the host to expand the functionality and stability of host proteins. Ubiquitination, the addition of the conserved 76 amino acids long ubiquitin to the lysine residues in target proteins, are among the most common post-translational modifications (Kwon and Ciechanover, 2017; Nielsen and MacGurn, 2020; Varshavsky, 2017). The protein ubiquitination process consists of multiple enzymes acting sequentially. The process includes an E1 ubiquitin-activating enzyme, then the E2 ubiquitin-conjugating enzyme, in collaboration with an E3 ubiquitin-ligase enzyme. The covalently-linked ubiquitin can be removed by de-ubiquitinases. Interestingly, protein mono-ubiquitination (addition of a single ubiquitin) frequently affects the subcellular localization and other functions of the client proteins, whereas poly-ubiquitination mainly serves as a protein degradation signal for the proteasome machinery (Berner et al., 2018; Isaacson and Ploegh, 2009).

Viruses also utilize their proteins for multiple functions in different subcellular locations, therefore, the ubiquitin network is also central in the arms race between viruses and their hosts. Many examples are documented on the ubiquitin-directed proteasome-based degradation of viral proteins, or viral protein-driven degradation of host proteins in case of animal RNA viruses (Choi et al., 2013; Lee et al., 2019; Liu and Tan, 2020; Llamas-Gonzalez et al., 2019; Luo, 2016; Xu et al., 2017). The role of the ubiquitin network is also drawn major attention in studies of innate immunity against animal RNA viruses (Garcia-Sastre, 2017; Rajsbaum and Garcia-Sastre, 2013).

Plant RNA viruses are also capable of taking advantage of the ubiquitin pathways for regulating their replication and interactions with hosts (Camborde et al., 2010; Verchot, 2016). In summary, the ubiquitin network determines many facets of the arms race and it is one of the hot topics in virus-host interactions.

9. The roles of the host ubiquitin network in TBSV replication and in the arms race with hosts

An emerging area in virus–host interaction is the role of the host ubiquitin network. On one hand, TBSV hijacks the yeast Rad6 (Ubc2 in plants) and Cdc34 E2 ubiquitin-conjugating enzymes in order to mono- and bi-ubiquitinate the p33 replication protein (Fig. 3) (Imura et al., 2015; Li et al., 2008). In combination with the presence of the so-called late domain sequence in p33 (Fig. 3), the ubiquitination of p33 replication protein promotes the recruitment of cellular Vps23 ESCRT-I protein and Bro1 accessory protein into the VROs (Barajas and Nagy, 2010). This in turn, is needed to further recruit additional ESCRT-I and ESCRT-III proteins and Vps4 AAA ATPase into the VROs (Barajas et al., 2009a, 2014). The coordinated actions of these ESCRT proteins with p33 and p92^{pol} replication proteins and the viral (+)RNA are needed for membrane deformation and the biogenesis of the spherule-like VRCs (Barajas et al., 2014; Kovalev et al., 2016, 2017). This is important to provide a semi-closed membranous structure of VRCs, which is beneficial to protect the viral RNAs from the host ribonucleases (Fig. 3).

Another characterized member of the ubiquitin network co-opted by TBSV is Rpn11 (Regulatory Particle Non-ATPase) metalloprotease subunit of 19S regulatory particle in the proteasome. Rpn11 is a critical component of the proteasome due to its function in coupling deubiquitination and degradation of proteasome substrates. Multiple pro-viral functions have been documented for the co-opted cellular Rpn11 deubiquitinase, including promotion of the recruitment of the DDX3-like yeast Ded1 and the homologous plant RH20 DEAD-box helicases into the VROs (Prasanth et al., 2016). In addition, Rpn11 contributes to the stability of the viral replication proteins. Overall, Rpn11 is a key co-opted host factor for the correct VRC assembly that greatly affects both viral replication and RNA recombination (Prasanth et al., 2016).

On the other hand, the need to hijack ubiquitination network components, however, renders TBSV vulnerable for the antiviral activities of additional components of this network. Indeed, the NEDD4-like Rsp5 E3

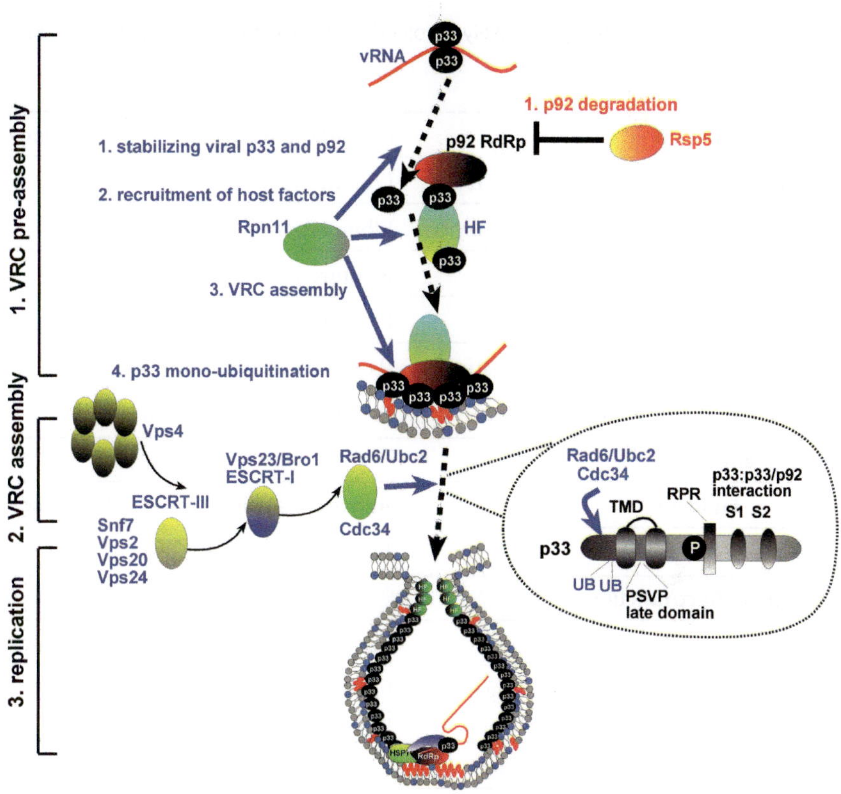

Fig. 3 A model showing the pro-viral functions versus the antiviral functions of the ubiquitin network during TBSV replication. The cellular Rpn11 deubiquitinase is a key co-opted host factor, which helps with the stabilization of the viral replication proteins, and facilitates the subversion of Ded1/RH20 RNA helicases and needed for the correct VRC assembly. In addition, the co-opted Rad6, Ubc2 and Cdc34 E2 ubiquitin conjugating enzymes (shown in blue) bi-ubiquitinate the p33 replication protein as shown in a circle on the right side of the panel. These events together with the late domain in p33 facilitate the hijacking of Vps23 and Bro1 ESCRT proteins, and ultimately the ESCRT machinery to deform intracellular membranes into spherule-like structures, as shown schematically. However, the host also utilize the ubiquitin network for antiviral functions as shown by the action of Rsp5 E3 ubiquitin ligase, resulting in degradation of the viral p92 replication protein, as shown on the right side of the panel. See further details in the main text.

ubiquitin-ligase could enter the VROs based on direct binding to the viral replication proteins, which then leads to the degradation of $p92^{pol}$ RdRp protein (Barajas et al., 2009b). This event greatly inhibits TBSV replication. The previous high throughput screens have identified several additional components of the ubiquitination network, which seem to inhibit TBSV replication in yeast (Table 3). Therefore, we propose that the host ubiquitination

Table 3 Ubiquitin network genes likely involved in the arms race between TBSV and its hosts.

Pro-viral effect

Gene	Function	Pro-viral function	Homologs
RAD6	Ubiquitin-conjugating enzyme	Ubiquitination of p33	*UBC2*
CDC34	Ubiquitin-conjugating enzyme	Ubiquitination of p33	
BRE1	Ubiquitin-protein ligase	Required for replication	
DOA4	Protein deubiquitination	Required for replication	
LGE1	Protein mono-ubiquitination	Required for replication	
RPN11	Protein deubiquitination	Required for replication	*RPN11*

Antiviral effect

Gene	Function	Antiviral function	Homologs
RSP5	E3 Ubiquitin-protein ligase	Degradation of p92	*NEDD4*

Undefined effect/function

Gene	Function	Viral target	Homologs
ASI1	E3 Ubiquitin-protein ligase	Over-expression is inhibitory	
UBA1	Ubiquitin activating E1 enzyme	Binds to p33	
UBP3	Ubiquitin-specific protease	Suppresses recombination	
UBP8	Ubiquitin-specific protease	Over-expression is inhibitory	
UBP10	Ubiquitin-specific protease	Binds to p33/p92/(+) RNA	
UBP15	Ubiquitin-specific protease	Binds to p33	
ATG7	Ubiquitin activating E1 enzyme Autophagy-related protein	Over-expression is inhibitory	
BUL2	Rsp5p E3-ubiquitin ligase complex	Over-expression is stimulatory	
HUL4	E3 Ubiquitin-protein ligase-like	Over-expression is stimulatory	
PIB1	E3 Ubiquitin-protein ligase	Over-expression is stimulatory	

Genes shown in bold face are characterized in details in TBSV replication.

network could emerge as a major frontier among viruses and their hosts. Further studies are needed to dissect the mechanisms of their anti-TBSV functions in yeast and plants (Table 3). Taken together, the exploitation of the host ubiquitination network seems to be central in the on going arms race between TBSV and its hosts.

10. Summary

Viruses require resources from the host cells in order to replicate efficiently and complete their infection cycles. Viruses also need to hide from the host antivirus-surveillance network and degradative apparatus. To accomplish all the requirements, viruses co-opt numerous host proteins belonging to several protein families. However, the hosts could use similar members of the same or associated protein families to interfere with the viral replication processes. The rapid changes in the viral genomes in evolutionary terms create a never-ending arms race between the virus and hosts for the exploitation of members of several host protein families. These protein families could be potential central players in strategies to develop novel antiviral approaches that favor the hosts against pathogenic viruses.

Acknowledgments

The authors thank Dr. Z. Feng, Ms. M. Molho and Ms. P. Gonzalez for valuable comments. This work was supported by the National Science Foundation (IOS-1922895), and the USDA hatch grant (KY012042) to P.D.N.

References

Alam, S.B., Rochon, D., 2015. Cucumber necrosis virus recruits cellular heat shock protein 70 homologs at several stages of infection. J. Virol. 90 (7), 3302–33017.

Alam, S.B., Rochon, D., 2017. Evidence that Hsc70 is associated with cucumber necrosis virus particles and plays a role in particle disassembly. J. Virol. 91 (2), JVI.01555-16.

Altan-Bonnet, N., 2017. Lipid tales of viral replication and transmission. Trends Cell Biol. 27 (3), 201–213.

Barajas, D., Nagy, P.D., 2010. Ubiquitination of tombusvirus p33 replication protein plays a role in virus replication and binding to the host Vps23p ESCRT protein. Virology 397 (2), 358–368.

Barajas, D., Jiang, Y., Nagy, P.D., 2009a. A unique role for the host ESCRT proteins in replication of Tomato bushy stunt virus. PLoS Pathog. 5 (12), e1000705.

Barajas, D., Li, Z., Nagy, P.D., 2009b. The Nedd4-type Rsp5p ubiquitin ligase inhibits tombusvirus replication by regulating degradation of the p92 replication protein and decreasing the activity of the tombusvirus replicase. J. Virol. 83 (22), 11751–11764.

Barajas, D., Martin, I.F., Pogany, J., Risco, C., Nagy, P.D., 2014. Noncanonical role for the host Vps4 AAA+ ATPase ESCRT protein in the formation of tomato bushy stunt virus replicase. PLoS Pathog. 10 (4), e1004087.

Belov, G.A., Sztul, E., 2014. Rewiring of cellular membrane homeostasis by picornaviruses. J. Virol. 88 (17), 9478–9489.

Berner, N., Reutter, K.R., Wolf, D.H., 2018. Protein quality control of the endoplasmic reticulum and ubiquitin-proteasome-triggered degradation of aberrant proteins: yeast pioneers the path. Annu. Rev. Biochem. 87, 751–782.

Bolinger, C., Sharma, A., Singh, D., Yu, L., Boris-Lawrie, K., 2010. RNA helicase A modulates translation of HIV-1 and infectivity of progeny virions. Nucleic Acids Res. 38 (5), 1686–1696.

Bozzacco, L., Yi, Z., Andreo, U., Conklin, C.R., Li, M.M., Rice, C.M., MacDonald, M.R., 2016. Chaperone-assisted protein folding is critical for yellow fever virus NS3/4A cleavage and replication. J. Virol. 90 (6), 3212–3228.

Brinton, M.A., 2013. Replication cycle and molecular biology of the West Nile virus. Viruses 6 (1), 13–53.

Camborde, L., Planchais, S., Tournier, V., Jakubiec, A., Drugeon, G., Lacassagne, E., Pflieger, S., Chenon, M., Jupin, I., 2010. The ubiquitin-proteasome system regulates the accumulation of Turnip yellow mosaic virus RNA-dependent RNA polymerase during viral infection. Plant Cell 22 (9), 3142–3152.

Cao, M., Wei, C., Zhao, L., Wang, J., Jia, Q., Wang, X., Jin, Q., Deng, T., 2014. DnaJA1/Hsp40 is co-opted by influenza A virus to enhance its viral RNA polymerase activity. J. Virol. 88 (24), 14078–14089.

Caplan, J.L., Mamillapalli, P., Burch-Smith, T.M., Czymmek, K., Dinesh-Kumar, S.P., 2008. Chloroplastic protein NRIP1 mediates innate immune receptor recognition of a viral effector. Cell 132 (3), 449–462.

Chen, G., Liu, C.H., Zhou, L., Krug, R.M., 2014. Cellular DDX21 RNA helicase inhibits influenza A virus replication but is counteracted by the viral NS1 protein. Cell Host Microbe 15 (4), 484–493.

Chenon, M., Camborde, L., Cheminant, S., Jupin, I., 2012. A viral deubiquitylating enzyme targets viral RNA-dependent RNA polymerase and affects viral infectivity. EMBO J. 31 (3), 741–753.

Choi, A.G., Wong, J., Marchant, D., Luo, H., 2013. The ubiquitin-proteasome system in positive-strand RNA virus infection. Rev. Med. Virol. 23 (2), 85–96.

Chow, J., Kagan, J.C., 2018. The fly way of antiviral resistance and disease tolerance. Adv. Immunol. 140, 59–93.

Chow, K.T., Gale Jr., M., Loo, Y.M., 2018. RIG-I and other RNA sensors in antiviral immunity. Annu. Rev. Immunol. 36, 667–694.

Chuang, C., Prasanth, K.R., Nagy, P.D., 2015. Coordinated function of cellular DEAD-box helicases in suppression of viral RNA recombination and maintenance of viral genome integrity. PLoS Pathog. 11 (2), e1004680.

Chuang, C., Prasanth, K.R., Nagy, P.D., 2017. The glycolytic pyruvate kinase is recruited directly into the viral replicase complex to generate ATP for RNA synthesis. Cell Host Microbe 22 (5), 639–652, e7.

den Boon, J.A., Ahlquist, P., 2010. Organelle-like membrane compartmentalization of positive-strand RNA virus replication factories. Annu. Rev. Microbiol. 64, 241–256.

Ding, S.W., 2010. RNA-based antiviral immunity. Nat. Rev. Immunol. 10 (9), 632–644.

Ertel, K.J., Benefield, D., Castano-Diez, D., Pennington, J.G., Horswill, M., den Boon, J.A., Otegui, M.S., Ahlquist, P., 2017. Cryo-electron tomography reveals novel features of a viral RNA replication compartment. Elife 6, 25940.

Fernandez de Castro, I., Tenorio, R., Risco, C., 2016. Virus assembly factories in a lipid world. Curr. Opin. Virol. 18, 20–26.

Garbelli, A., Radi, M., Falchi, F., Beermann, S., Zanoli, S., Manetti, F., Dietrich, U., Botta, M., Maga, G., 2011. Targeting the human DEAD-box polypeptide 3 (DDX3) RNA helicase as a novel strategy to inhibit viral replication. Curr. Med. Chem. 18 (20), 3015–3027.

Garcia-Ruiz, H., Takeda, A., Chapman, E.J., Sullivan, C.M., Fahlgren, N., Brempelis, K.J., Carrington, J.C., 2010. Arabidopsis RNA-dependent RNA polymerases and dicer-like proteins in antiviral defense and small interfering RNA biogenesis during Turnip Mosaic Virus infection. Plant Cell 22 (2), 481–496.

Garcia-Sastre, A., 2017. Ten strategies of interferon evasion by viruses. Cell Host Microbe 22 (2), 176–184.

Genest, O., Wickner, S., Doyle, S.M., 2019. Hsp90 and Hsp70 chaperones: collaborators in protein remodeling. J. Biol. Chem. 294 (6), 2109–2120.

Gilman, B., Tijerina, P., Russell, R., 2017. Distinct RNA-unwinding mechanisms of DEAD-box and DEAH-box RNA helicase proteins in remodeling structured RNAs and RNPs. Biochem. Soc. Trans. 45 (6), 1313–1321.

Gimenez-Barcons, M., Alves-Rodrigues, I., Jungfleisch, J., Van Wynsberghe, P.M., Ahlquist, P., Diez, J., 2013. The cellular decapping activators LSm1, Pat1, and Dhh1 control the ratio of subgenomic to genomic Flock House virus RNAs. J. Virol. 87 (11), 6192–6200.

Goertz, G.P., McNally, K.L., Robertson, S.J., Best, S.M., Pijlman, G.P., Fros, J.J., 2018. The methyltransferase-like domain of chikungunya virus nsP2 inhibits the interferon response by promoting the nuclear export of STAT1. J. Virol. 92 (17).

Goh, P.Y., Tan, Y.J., Lim, S.P., Tan, Y.H., Lim, S.G., Fuller-Pace, F., Hong, W., 2004. Cellular RNA helicase p68 relocalization and interaction with the hepatitis C virus (HCV) NS5B protein and the potential role of p68 in HCV RNA replication. J. Virol. 78 (10), 5288–5298.

Guenther, U.P., Jankowsky, E., 2009. Helicase multitasking in ribosome assembly. Mol. Cell 36 (4), 537–538.

Hainzl, O., Wegele, H., Richter, K., Buchner, J., 2004. Cns1 is an activator of the Ssa1 ATPase activity. J. Biol. Chem. 279 (22), 23267–23273.

Haslbeck, V., Eckl, J.M., Kaiser, C.J., Papsdorf, K., Hessling, M., Richter, K., 2013. Chaperone-interacting TPR proteins in Caenorhabditis elegans. J. Mol. Biol. 425 (16), 2922–2939.

Huang, T.S., Wei, T., Laliberte, J.F., Wang, A., 2010. A host RNA helicase-like protein, AtRH8, interacts with the potyviral genome-linked protein, VPg, associates with the virus accumulation complex, and is essential for infection. Plant Physiol. 152 (1), 255–266.

Imura, Y., Molho, M., Chuang, C., Nagy, P.D., 2015. Cellular Ubc2/Rad6 E2 ubiquitin-conjugating enzyme facilitates tombusvirus replication in yeast and plants. Virology 484, 265–275.

Isaacson, M.K., Ploegh, H.L., 2009. Ubiquitination, ubiquitin-like modifiers, and deubiquitination in viral infection. Cell Host Microbe 5 (6), 559–570.

Ishibashi, K., Ishikawa, M., 2016. Replication of Tobamovirus RNA. Annu. Rev. Phytopathol. 54, 55–78.

Jankowsky, E., 2011. RNA helicases at work: binding and rearranging. Trends Biochem. Sci. 36 (1), 19–29.

Jiang, Y., Serviene, E., Gal, J., Panavas, T., Nagy, P.D., 2006. Identification of essential host factors affecting tombusvirus RNA replication based on the yeast Tet promoters Hughes collection. J. Virol. 80 (15), 7394–7404.

Jiang, Y., Li, Z., Nagy, P.D., 2010. Nucleolin/Nsr1p binds to the 3' noncoding region of the tombusvirus RNA and inhibits replication. Virology 396 (1), 10–20.

Kim, Y.E., Hipp, M.S., Bracher, A., Hayer-Hartl, M., Hartl, F.U., 2013. Molecular chaperone functions in protein folding and proteostasis. Annu. Rev. Biochem. 82, 323–355.

Kovalev, N., Nagy, P.D., 2014. The expanding functions of cellular helicases: the tombusvirus RNA replication enhancer co-opts the plant eIF4AIII-like AtRH2 and the DDX5-like AtRH5 DEAD-box RNA helicases to promote viral asymmetric RNA replication. PLoS Pathog. 10 (4), e1004051.

Kovalev, N., Barajas, D., Nagy, P.D., 2012a. Similar roles for yeast Dbp2 and Arabidopsis RH20 DEAD-box RNA helicases to Ded1 helicase in tombusvirus plus-strand synthesis. Virology 432 (2), 470–484.
Kovalev, N., Pogany, J., Nagy, P.D., 2012b. A co-opted DEAD-box RNA helicase enhances tombusvirus plus-strand synthesis. PLoS Pathog. 8 (2), e1002537.
Kovalev, N., Pogany, J., Nagy, P.D., 2014. Template role of double-stranded RNA in tombusvirus replication. J. Virol. 88 (10), 5638–5651.
Kovalev, N., de Castro Martin, I.F., Pogany, J., Barajas, D., Pathak, K., Risco, C., Nagy, P.D., 2016. Role of viral RNA and co-opted cellular ESCRT-I and ESCRT-III factors in formation of tombusvirus spherules harboring the tombusvirus replicase. J. Virol. 90 (7), 3611–3626.
Kovalev, N., Inaba, J.I., Li, Z., Nagy, P.D., 2017. The role of co-opted ESCRT proteins and lipid factors in protection of tombusviral double-stranded RNA replication intermediate against reconstituted RNAi in yeast. PLoS Pathog. 13 (7), e1006520.
Kuka, M., Iannacone, M., 2018. Viral subversion of B cell responses within secondary lymphoid organs. Nat. Rev. Immunol. 18 (4), 255–265.
Kwon, Y.T., Ciechanover, A., 2017. The ubiquitin code in the ubiquitin-proteasome system and autophagy. Trends Biochem. Sci. 42 (11), 873–886.
Laliberte, J.F., Sanfacon, H., 2010. Cellular remodeling during plant virus infection. Annu. Rev. Phytopathol. 48, 69–91.
Lamm, C.E., Kraner, M.E., Hofmann, J., Bornke, F., Mock, H.P., Sonnewald, U., 2017. Hop/Sti1—a two-faced cochaperone involved in pattern recognition receptor maturation and viral infection. Front. Plant Sci. 8, 1754.
Lawrence, P., Rieder, E., 2009. Identification of RNA helicase A as a new host factor in the replication cycle of foot-and-mouth disease virus. J. Virol. 83 (21), 11356–11366.
Lee, J., Chan, S.T., Kim, J.Y., Ou, J.J., 2019. Hepatitis C virus induces the ubiquitin-editing enzyme A20 via depletion of the transcription factor upstream stimulatory factor 1 to support its replication. MBio 10 (4).
Li, F., Ding, S.W., 2006. Virus counterdefense: diverse strategies for evading the RNA-silencing immunity. Annu. Rev. Microbiol. 60, 503–531.
Li, Z., Nagy, P.D., 2011. Diverse roles of host RNA binding proteins in RNA virus replication. RNA Biol. 8 (2), 305–315.
Li, Z., Barajas, D., Panavas, T., Herbst, D.A., Nagy, P.D., 2008. Cdc34p ubiquitin-conjugating enzyme is a component of the tombusvirus replicase complex and ubiquitinates p33 replication protein. J. Virol. 82 (14), 6911–6926.
Li, Z., Pogany, J., Panavas, T., Xu, K., Esposito, A.M., Kinzy, T.G., Nagy, P.D., 2009. Translation elongation factor 1A is a component of the tombusvirus replicase complex and affects the stability of the p33 replication co-factor. Virology 385 (1), 245–260.
Li, Z., Pogany, J., Tupman, S., Esposito, A.M., Kinzy, T.G., Nagy, P.D., 2010. Translation elongation factor 1A facilitates the assembly of the tombusvirus replicase and stimulates minus-strand synthesis. PLoS Pathog. 6 (11), e1001175.
Li, G., Feng, T., Pan, W., Shi, X., Dai, J., 2015. DEAD-box RNA helicase DDX3X inhibits DENV replication via regulating type one interferon pathway. Biochem. Biophys. Res. Commun. 456 (1), 327–332.
Lin, J.Y., Nagy, P.D., 2013. Identification of novel host factors via conserved domain search: Cns1 cochaperone is a novel restriction factor of tombusvirus replication in yeast. J. Virol. 87 (23), 12600–12610.
Lin, J.Y., Mendu, V., Pogany, J., Qin, J., Nagy, P.D., 2012. The TPR domain in the host Cyp40-like cyclophilin binds to the viral replication protein and inhibits the assembly of the tombusviral replicase. PLoS Pathog. 8 (2), e1002491.
Lin, W., Liu, Y., Molho, M., Zhang, S., Wang, L., Xie, L., Nagy, P.D., 2019. Co-opting the fermentation pathway for tombusvirus replication: compartmentalization of cellular metabolic pathways for rapid ATP generation. PLoS Pathog. 15 (10), e1008092.

Linder, P., Jankowsky, E., 2011. From unwinding to clamping—the DEAD box RNA helicase family. Nat. Rev. Mol. Cell Biol. 12 (8), 505–516.

Liu, Y., Tan, X., 2020. Viral manipulations of the cullin-RING ubiquitin ligases. Adv. Exp. Med. Biol. 1217, 99–110.

Llamas-Gonzalez, Y.Y., Campos, D., Pascale, J.M., Arbiza, J., Gonzalez-Santamaria, J., 2019. A functional ubiquitin-proteasome system is required for efficient replication of new World Mayaro and Una alphaviruses. Viruses 11 (4), 370.

Lohmus, A., Hafren, A., Makinen, K., 2017. Coat protein regulation by CK2, CPIP, HSP70, and CHIP is required for potato virus A replication and coat protein accumulation. J. Virol. 91 (3), JVI.01316-16.

Lorgeoux, R.P., Guo, F., Liang, C., 2012. From promoting to inhibiting: diverse roles of helicases in HIV-1 Replication. Retrovirology 9, 79.

Luo, H., 2016. Interplay between the virus and the ubiquitin-proteasome system: molecular mechanism of viral pathogenesis. Curr. Opin. Virol. 17, 1–10.

Manzoor, R., Kuroda, K., Yoshida, R., Tsuda, Y., Fujikura, D., Miyamoto, H., Kajihara, M., Kida, H., Takada, A., 2014. Heat shock protein 70 modulates influenza A virus polymerase activity. J. Biol. Chem. 289 (11), 7599–7614.

Mayer, M.P., 2013. Hsp70 chaperone dynamics and molecular mechanism. Trends Biochem. Sci. 38 (10), 507–514.

McCartney, A.W., Greenwood, J.S., Fabian, M.R., White, K.A., Mullen, R.T., 2005. Localization of the tomato bushy stunt virus replication protein p33 reveals a peroxisome-to-endoplasmic reticulum sorting pathway. Plant Cell 17 (12), 3513–3531.

Monkewich, S., Lin, H.X., Fabian, M.R., Xu, W., Na, H., Ray, D., Chernysheva, O.A., Nagy, P.D., White, K.A., 2005. The p92 polymerase coding region contains an internal RNA element required at an early step in tombusvirus genome replication. J. Virol. 79 (8), 4848–4858.

Moran Luengo, T., Mayer, M.P., Rudiger, S.G.D., 2019. The Hsp70-Hsp90 chaperone cascade in protein folding. Trends Cell Biol. 29 (2), 164–177.

Morohashi, K., Sahara, H., Watashi, K., Iwabata, K., Sunoki, T., Kuramochi, K., Takakusagi, K., Miyashita, H., Sato, N., Tanabe, A., Shimotohno, K., Kobayashi, S., Sakaguchi, K., Sugawara, F., 2011. Cyclosporin A associated helicase-like protein facilitates the association of hepatitis C virus RNA polymerase with its cellular cyclophilin B. PLoS One 6 (4), e18285.

Moy, R.H., Cole, B.S., Yasunaga, A., Gold, B., Shankarling, G., Varble, A., Molleston, J.M., tenOever, B.R., Lynch, K.W., Cherry, S., 2014. Stem-loop recognition by DDX17 facilitates miRNA processing and antiviral defense. Cell 158 (4), 764–777.

Nagy, P.D., 2008. Yeast as a model host to explore plant virus-host interactions. Annu. Rev. Phytopathol. 46, 217–242.

Nagy, P.D., 2011. The roles of host factors in tombusvirus RNA recombination. Adv. Virus Res. 81, 63–84.

Nagy, P.D., 2016. Tombusvirus-host interactions: co-opted evolutionarily conserved host factors take center court. Annu. Rev. Virol. 3 (1), 491–515.

Nagy, P.D., 2017. Exploitation of a surrogate host, Saccharomyces cerevisiae, to identify cellular targets and develop novel antiviral approaches. Curr. Opin. Virol. 26, 132–140.

Nagy, P.D., Lin, W., 2020. Taking over cellular energy-metabolism for TBSV replication: the high ATP requirement of an RNA virus within the viral replication organelle. Viruses 12 (1), 56.

Nagy, P.D., Pogany, J., 2010. Global genomics and proteomics approaches to identify host factors as targets to induce resistance against Tomato bushy stunt virus. Adv. Virus Res. 76, 123–177.

Nagy, P.D., Pogany, J., 2012. The dependence of viral RNA replication on co-opted host factors. Nat. Rev. Microbiol. 10 (2), 137–149.

Nagy, P.D., Wang, R.Y., Pogany, J., Hafren, A., Makinen, K., 2011. Emerging picture of host chaperone and cyclophilin roles in RNA virus replication. Virology 411 (2), 374–382.

Nagy, P.D., Barajas, D., Pogany, J., 2012. Host factors with regulatory roles in tombusvirus replication. Curr. Opin. Virol. 2 (6), 685–692.

Nagy, P.D., Pogany, J., Lin, J.Y., 2014. How yeast can be used as a genetic platform to explore virus-host interactions: from 'omics' to functional studies. Trends Microbiol. 22 (6), 309–316.

Neufeldt, C.J., Cortese, M., Acosta, E.G., Bartenschlager, R., 2018. Rewiring cellular networks by members of the Flaviviridae family. Nat. Rev. Microbiol. 16 (3), 125–142.

Nielsen, C.P., MacGurn, J.A., 2020. Coupling conjugation and deconjugation activities to achieve cellular ubiquitin dynamics. Trends Biochem. Sci. 45 (5), 427–439.

Noueiry, A.O., Chen, J., Ahlquist, P., 2000. A mutant allele of essential, general translation initiation factor DED1 selectively inhibits translation of a viral mRNA. Proc. Natl. Acad. Sci. U. S. A 97 (24), 12985–12990.

Panavas, T., Nagy, P.D., 2005. Mechanism of stimulation of plus-strand synthesis by an RNA replication enhancer in a tombusvirus. J. Virol. 79 (15), 9777–9785.

Panavas, T., Hawkins, C.M., Panaviene, Z., Nagy, P.D., 2005a. The role of the p33:p33/p92 interaction domain in RNA replication and intracellular localization of p33 and p92 proteins of Cucumber necrosis tombusvirus. Virology 338 (1), 81–95.

Panavas, T., Serviene, E., Brasher, J., Nagy, P.D., 2005b. Yeast genome-wide screen reveals dissimilar sets of host genes affecting replication of RNA viruses. Proc. Natl. Acad. Sci. U. S. A 102 (20), 7326–7331.

Pathak, K.B., Pogany, J., Nagy, P.D., 2011. Non-template functions of the viral RNA in plant RNA virus replication. Curr. Opin. Virol. 1 (5), 332–338.

Pathak, K.B., Pogany, J., Xu, K., White, K.A., Nagy, P.D., 2012. Defining the roles of cis-acting RNA elements in tombusvirus replicase assembly in vitro. J. Virol. 86 (1), 156–171.

Paul, D., Bartenschlager, R., 2015. Flaviviridae replication organelles: oh, what a tangled web we weave. Annu. Rev. Virol. 2 (1), 289–310.

Pogany, J., Nagy, P.D., 2012. p33-independent activation of a truncated p92 RNA-dependent RNA polymerase of Tomato bushy stunt virus in yeast cell-free extract. J. Virol. 86 (22), 12025–12038.

Pogany, J., Nagy, P.D., 2015. Activation of tomato bushy stunt virus RNA-dependent RNA polymerase by cellular heat shock protein 70 is enhanced by phospholipids in vitro. J. Virol. 89 (10), 5714–5723.

Pogany, J., Fabian, M.R., White, K.A., Nagy, P.D., 2003. A replication silencer element in a plus-strand RNA virus. EMBO J. 22 (20), 5602–5611.

Pogany, J., White, K.A., Nagy, P.D., 2005. Specific binding of tombusvirus replication protein p33 to an internal replication element in the viral RNA is essential for replication. J. Virol. 79 (8), 4859–4869.

Pogany, J., Stork, J., Li, Z., Nagy, P.D., 2008. In vitro assembly of the Tomato bushy stunt virus replicase requires the host heat shock protein 70. Proc. Natl. Acad. Sci. U. S. A 105 (50), 19956–19961.

Prasad, B.D., Goel, S., Krishna, P., 2010. In silico identification of carboxylate clamp type tetratricopeptide repeat proteins in Arabidopsis and rice as putative co-chaperones of Hsp90/Hsp70. PLoS One 5 (9), e12761.

Prasanth, K.R., Kovalev, N., de Castro Martin, I.F., Baker, J., Nagy, P.D., 2016. Screening a yeast library of temperature-sensitive mutants reveals a role for actin in tombusvirus RNA recombination. Virology 489, 233–242.

Prasanth, K.R., Chuang, C., Nagy, P.D., 2017. Co-opting ATP-generating glycolytic enzyme PGK1 phosphoglycerate kinase facilitates the assembly of viral replicase complexes. PLoS Pathog. 13 (10), e1006689.

Rajendran, K.S., Nagy, P.D., 2006. Kinetics and functional studies on interaction between the replicase proteins of Tomato bushy stunt virus: requirement of p33:p92 interaction for replicase assembly. Virology 345 (1), 270–279.

Rajsbaum, R., Garcia-Sastre, A., 2013. Viral evasion mechanisms of early antiviral responses involving regulation of ubiquitin pathways. Trends Microbiol. 21 (8), 421–429.

Ranji, A., Boris-Lawrie, K., 2010. RNA helicases: emerging roles in viral replication and the host innate response. RNA Biol. 7 (6), 775–787.

Rehwinkel, J., Gack, M.U., 2020. RIG-I-like receptors: their regulation and roles in RNA sensing. Nat. Rev. Immunol.

Rosenzweig, R., Nillegoda, N.B., Mayer, M.P., Bukau, B., 2019. The Hsp70 chaperone network. Nat. Rev. Mol. Cell Biol. 20 (11), 665–680.

Sasvari, Z., Izotova, L., Kinzy, T.G., Nagy, P.D., 2011. Synergistic roles of eukaryotic translation elongation factors 1Bgamma and 1A in stimulation of tombusvirus minus-strand synthesis. PLoS Pathog. 7 (12), e1002438.

Schopf, F.H., Huber, E.M., Dodt, C., Lopez, A., Biebl, M.M., Rutz, D.A., Muhlhofer, M., Richter, G., Madl, T., Sattler, M., Groll, M., Buchner, J., 2019. The co-chaperone Cns1 and the recruiter protein Hgh1 link Hsp90 to translation elongation via chaperoning elongation factor 2. Mol. Cell 74 (1), 73–87e8.

Schroder, M., 2011. Viruses and the human DEAD-box helicase DDX3: inhibition or exploitation? Biochem. Soc. Trans. 39 (2), 679–683.

Schwerk, J., Negash, A., Savan, R., Gale Jr., M., 2020. Innate immunity in hepatitis C virus infection. Cold Spring Harb. Perspect. Med.

Shah Nawaz-Ul-Rehman, M., Martinez-Ochoa, N., Pascal, H., Sasvari, Z., Herbst, C., Xu, K., Baker, J., Sharma, M., Herbst, A., Nagy, P.D., 2012. Proteome-wide overexpression of host proteins for identification of factors affecting tombusvirus RNA replication: an inhibitory role of protein kinase C. J. Virol. 86 (17), 9384–9395.

Sharma, D., Jankowsky, E., 2014. The Ded1/DDX3 subfamily of DEAD-box RNA helicases. Crit. Rev. Biochem. Mol. Biol. 49 (4), 343–360.

Shulla, A., Randall, G., 2016. (+) RNA virus replication compartments: a safe home for (most) viral replication. Curr. Opin. Microbiol. 32, 82–88.

Sloan, K.E., Bohnsack, M.T., 2018. Unravelling the mechanisms of RNA helicase regulation. Trends Biochem. Sci. 43 (4), 237–250.

Soosaar, J.L., Burch-Smith, T.M., Dinesh-Kumar, S.P., 2005. Mechanisms of plant resistance to viruses. Nat. Rev. Microbiol. 3 (10), 789–798.

Stork, J., Kovalev, N., Sasvari, Z., Nagy, P.D., 2011. RNA chaperone activity of the tombusviral p33 replication protein facilitates initiation of RNA synthesis by the viral RdRp in vitro. Virology 409 (2), 338–347.

Taguwa, S., Maringer, K., Li, X., Bernal-Rubio, D., Rauch, J.N., Gestwicki, J.E., Andino, R., Fernandez-Sesma, A., Frydman, J., 2015. Defining Hsp70 subnetworks in dengue virus replication reveals key vulnerability in flavivirus infection. Cell 163 (5), 1108–1123.

Taguwa, S., Yeh, M.T., Rainbolt, T.K., Nayak, A., Shao, H., Gestwicki, J.E., Andino, R., Frydman, J., 2019. Zika virus dependence on host Hsp70 provides a protective strategy against infection and disease. Cell Rep. 26 (4), 906–920e3.

Taschuk, F., Cherry, S., 2020. DEAD-box helicases: sensors, regulators, and effectors for antiviral defense. Viruses 12 (2), 181.

Tassetto, M., Kunitomi, M., Andino, R., 2017. Circulating immune cells mediate a systemic RNAi-based adaptive antiviral response in drosophila. Cell 169 (2), 314–325e13.

Upadhyay, A., Dixit, U., Manvar, D., Chaturvedi, N., Pandey, V.N., 2013. Affinity capture and identification of host cell factors associated with hepatitis C virus (+) strand subgenomic RNA. Mol. Cell. Proteomics 12 (6), 1539–1552.

Upadya, M.H., Aweya, J.J., Tan, Y.J., 2014. Understanding the interaction of hepatitis C virus with host DEAD-box RNA helicases. World J. Gastroenterol. 20 (11), 2913–2926.

Varshavsky, A., 2017. The ubiquitin system, autophagy, and regulated protein degradation. Annu. Rev. Biochem. 86, 123–128.

Verchot, J., 2012. Cellular chaperones and folding enzymes are vital contributors to membrane bound replication and movement complexes during plant RNA virus infection. Front. Plant Sci. 3, 275.

Verchot, J., 2016. Plant virus infection and the ubiquitin proteasome machinery: arms race along the endoplasmic reticulum. Viruses 8 (11), 314.

Wang, A., 2015. Dissecting the molecular network of virus-plant interactions: the complex roles of host factors. Annu. Rev. Phytopathol. 53, 45–66.

Wang, R.Y., Stork, J., Nagy, P.D., 2009a. A key role for heat shock protein 70 in the localization and insertion of tombusvirus replication proteins to intracellular membranes. J. Virol. 83 (7), 3276–3287.

Wang, R.Y., Stork, J., Pogany, J., Nagy, P.D., 2009b. A temperature sensitive mutant of heat shock protein 70 reveals an essential role during the early steps of tombusvirus replication. Virology 394 (1), 28–38.

Wang, X., Cao, X., Liu, M., Zhang, R., Zhang, X., Gao, Z., Zhao, X., Xu, K., Li, D., Zhang, Y., 2018. Hsc70-2 is required for Beet black scorch virus infection through interaction with replication and capsid proteins. Sci. Rep. 8 (1), 4526.

Watanabe, Y., Ohtaki, N., Hayashi, Y., Ikuta, K., Tomonaga, K., 2009. Autogenous translational regulation of the Borna disease virus negative control factor X from polycistronic mRNA using host RNA helicases. PLoS Pathog. 5 (11), e1000654.

Wu, C.Y., Nagy, P.D., 2019. Blocking tombusvirus replication through the antiviral functions of DDX17-like RH30 DEAD-box helicase. PLoS Pathog. 15 (5), e1007771.

Xu, Z., Hobman, T.C., 2012. The helicase activity of DDX56 is required for its role in assembly of infectious West Nile virus particles. Virology 433 (1), 226–235.

Xu, K., Lin, J.Y., Nagy, P.D., 2014. The hop-like stress-induced protein 1 cochaperone is a novel cell-intrinsic restriction factor for mitochondrial tombusvirus replication. J. Virol. 88 (16), 9361–9378.

Xu, Q., Zhu, N., Chen, S., Zhao, P., Ren, H., Zhu, S., Tang, H., Zhu, Y., Qi, Z., 2017. E3 ubiquitin ligase Nedd4 promotes Japanese encephalitis virus replication by suppressing autophagy in human neuroblastoma cells. Sci. Rep. 7, 45375.

Yasuda-Inoue, M., Kuroki, M., Ariumi, Y., 2013. Distinct DDX DEAD-box RNA helicases cooperate to modulate the HIV-1 rev function. Biochem. Biophys. Res. Commun. 434 (4), 803–808.

Zhang, Z., Kim, T., Bao, M., Facchinetti, V., Jung, S.Y., Ghaffari, A.A., Qin, J., Cheng, G., Liu, Y.J., 2011. DDX1, DDX21, and DHX36 helicases form a complex with the adaptor molecule TRIF to sense dsRNA in dendritic cells. Immunity 34 (6), 866–878.

Zhang, Z., He, G., Filipowicz, N.A., Randall, G., Belov, G.A., Kopek, B.G., Wang, X., 2019. Host lipids in positive-strand RNA virus genome replication. Front. Microbiol. 10, 286.

CHAPTER FIVE

Update on immunopathology of bornavirus infections in humans and animals

Daniel Nobach[a], Jana Müller[a], Dennis Tappe[b], Christiane Herden[a,c,*,†]
[a]Institute of Veterinary Pathology, Justus-Liebig-University Giessen, Giessen, Germany
[b]Bernhard Nocht Institute for Tropical Medicine, Hamburg, Germany
[c]Center for Brain, Mind and Behavior, Justus-Liebig-University Giessen, Giessen, Germany
*Corresponding author: e-mail address: christiane.herden@vetmed.uni-giessen.de

Contents

† Senior author.

Advances in Virus Research, Volume 107
ISSN 0065-3527
https://doi.org/10.1016/bs.aivir.2020.06.004

Abstract

Knowledge on bornaviruses has expanded tremendously during the last decade through detection of novel bornaviruses and endogenous bornavirus-like elements in many eukaryote genomes, as well as by confirmation of insectivores as reservoir species for classical Borna disease virus 1 (BoDV-1). The most intriguing finding was the demonstration of the zoonotic potential of lethal human bornavirus infections caused by a novel bornavirus of different squirrel species (variegated squirrel 1 bornavirus, VSBV-1) and by BoDV-1 known as the causative agent for the classical Borna disease in horses and sheep. Whereas a T cell-mediated immunopathology has already been confirmed as key disease mechanism for infection with BoDV-1 by experimental studies in rodents, the underlying pathomechanisms remain less clear for human bornavirus infections, infection with other bornaviruses or infection of reservoir species. Thus, an overview of current knowledge on the pathogenesis of bornavirus infections focusing on immunopathology is given.

1. Introduction and history

The family *Bornaviridae* within the order *Mononegavirales* has been growing tremendously during recent years due to the discovery of several genetically distant new species and currently comprises (as of 2019) three genera: *Carbovirus*, *Cultervirus* and *Orthobornavirus* (Amarasinghe et al., 2018, 2019). Of the genus *Orthobornavirus*, two species, *Mammalian 1 orthobornavirus* and *Mammalian 2 orthobornavirus*, affect and cause disease in mammalian hosts. Belonging to *Mammalian 1 orthobornavirus*, Borna disease virus 1 (BoDV-1) represents the prototype and most prevalent bornavirus in mammals. It has already been known for a long time as causative agent of fatal neurological Borna disease (BD) mainly in horses and sheep (reviewed in Herden et al., 2013). Other orthobornaviruses, e.g., of the species *Psittaciform 1 orthobornavirus* and *Waterbird 1 orthobornavirus*, infect avian species including psittacine birds and water fowl (reviewed in Boatright-Horowitz, 2020; Herden et al., 2013; Hoppes and Shivaprasad, 2020; Payne et al., 2012; Tizard et al., 2016a). Bornaviruses in reptiles and fish have recently been found by means of next-generation sequencing (Hyndman et al., 2018; Shi et al., 2018).

Recently, the zoonotic potential of mammalian orthobornaviruses was unequivocally confirmed. Thus, a novel bornavirus initially discovered in variegated squirrels (variegated squirrel bornavirus-1; VSBV-1) was confirmed in four human cases of fatal encephalitis (Hoffmann et al., 2015; Tappe et al., 2018) and BoDV-1 was identified as causative agent in at least

18 naturally acquired sporadic or transplant-associated fatal human encephalitis cases (Coras et al., 2019; Korn et al., 2018; Liesche et al., 2019; Niller et al., 2020; Rubbenstroth et al., 2019; Schlottau et al., 2018). It should be emphasized that these human encephalitis cases are quite different from previous studies dealing with potential human BoDV-1 infections (Bode et al., 1996; Dürrwald et al., 2007; Hornig et al., 2012; Lieb and Staeheli, 2001; Richt et al., 1997; Richt and Rott, 2001; Rott et al., 1985; Rubbenstroth et al., 2019; Schwemmle et al., 1999; Wolff et al., 2006).

Bornaviruses are typical single-stranded non-segmented negative-sensed and enveloped RNA viruses, but with unique features regarding genome organization, viral life cycle and the ability to establish a non-cytolytic persistent infection (reviewed by Herden et al., 2013). Surprisingly, endogenous genome-integrated bornavirus-like sequences have been detected in a wide range of mainly vertebrate species (Belyi et al., 2010; Horie et al., 2010; Katzourakis and Gifford, 2010).

Historically, BD in horses was already described in the 1700s (Sind, 1767; Zwick, 1939) and was initially named, e.g., as "Hitzige Kopfkrankheit (German; "Hot-headed disease"), but later designated Borna disease due to major outbreaks in the years 1894–1896 in the district around the town of Borna in Saxony, Germany (Zwick, 1939). In animals, horses and sheep are most frequently affected by this fatal and severe neurological disease, but classical BD was also occasionally described in many other animal species (reviewed by Herden et al., 2013). Experimental studies, most of them carried out in the last century, confirmed the wide host range of BoDV-1 infections (reviewed by Danner, 1982; Heinig, 1969; Herden et al., 2013; Ludwig et al., 1985).

To date, serological studies provide evidence for a widespread, possibly worldwide distribution of orthobornaviruses. However, clinical BD in animals is bound to geographically restricted, endemic areas of Germany, Switzerland, Liechtenstein and Austria, suggesting the existence of a natural reservoir for a long time. Until now, only the bicolored white-toothed shrew (*Crocidura leucodon*) has been confirmed as a natural reservoir in these areas (Bourg et al., 2013; Dürrwald et al., 2014; Hilbe et al., 2006; Weissenböck et al., 2017). For VSBV-1, naturally infected squirrel species (Prevost's squirrel, *Callosciurus prevostii* and variegated squirrel, *Sciurus variegatoides*) are the presumed natural reservoir (Petzold et al., 2019; Schlottau et al., 2017a). Whereas shrews and squirrels exhibit a widespread viral organ distribution with virus shedding, but without clinical signs or severe neuroinflammation, in accidental dead-end host species, such as horses or humans, VSBV-1 and BoDV-1 exhibit a strict neurotropism and typically cause a severe non-purulent meningoencephalitis.

The first novel bornaviruses were detected in the first decade of the 21st century in parrots with proventricular dilatation disease (PDD) (Honkavuori et al., 2008; Kistler et al., 2008). Typical clinical signs of the worldwide occurring PDD are gastrointestinal dysfunctions and associated wasting with or without neurological symptoms. The etiological agents, avian orthobornaviruses, have been detected worldwide in many captive parrots, but also in other non-psittacine species (reviewed in Boatright-Horowitz, 2020; Herden et al., 2013; Hoppes et al., 2010; Payne et al., 2012). For avian orthobornavirus infections, non-purulent inflammation of the enteric autonomic and central nervous system accompanied by viral persistence, but also high variability of clinical outcome, is quite typical.

The key disease mechanism—as proven for BoDV-1 infections—is a T cell-mediated delayed-type hypersensitivity reaction with a key role of immunopathogenic $CD8^+$ and $CD4^+$ T cells. This has convincingly been shown in many experimental rodent studies (reviewed in Herden et al., 2013). Thus, infection does not cause a protective immune response with virus elimination, but typically, long-term virus persistence is established. For avian orthobornavirus infections an immunopathological mechanism is also assumed but needs to be further investigated. Recent initial data in human VSBV-1 infection point to potentially comparable immunopathology, but also potential differences. Thus, further characterization of virus- or host-specific disease mechanisms in accidental dead-end host or reservoir species is urgently needed, especially since no curative therapy is currently available.

The review focuses on current knowledge about the pathogenesis of bornavirus infections with special emphasis on available data concerning immunopathology, as well as gaps that have to be filled in order to open the avenue for successful treatment options.

2. Classification, genome organization, viral life cycle and virus-host interactions

Bornaviruses are typical single-stranded non-segmented negative-sensed and enveloped RNA viruses of the order *Mononegavirales*, but exhibit unique biological features (reviewed in Herden et al., 2013).

2.1 Taxonomy

Segregation into three genera (*Carbovirus*, *Cultervirus* and *Orthobornavirus*) within the family *Bornaviridae*, order *Mononegavirales* has recently been implemented (Afonso et al., 2016; Amarasinghe et al., 2018, 2019).

The genus *Orthobornavirus* now includes eight species. The virus BoDV-1 as well as Borna disease virus 2 (BoDV-2) belong to the species *Mammalian 1 orthobornavirus*, whereas VSBV-1 belongs to the species *Mammalian 2 orthobornavirus*. Avian orthobornaviruses within this genus include *Passeriform 1 orthobornavirus* (canary bornaviruses 1, 2 and 3), *Passeriform 2 orthobornavirus* (estrildid finch bornavirus 1), *Psittaciform 1 orthobornavirus* (parrot bornavirus 1–4, 7; PaBV-1–4, 7) and *Waterbird 1 orthobornavirus* (aquatic bird bornavirus 1 and 2). Formerly unclassified parrot bornavirus 5 is now assigned to the species *Psittaciform 2 orthobornavirus*. The *Elapid 1 orthobornavirus* includes a reptile bornavirus, Loveridge's garter snake virus 1. Unclassified members of the genus *Orthobornavirus* are parrot bornavirus 6 and 8, munia bornavirus 1, as well as Gaboon viper virus 1. The newly formed genus *Carbovirus* comprises two species: *Queensland carbovirus* (jungle carpet python virus, JCPV) and *Southwest carbovirus* (southwest carpet python virus, SWCPV). The third genus, *Cultervirus*, only includes the *Sharpbelly cultervirus* (Wuhan sharpbelly bornavirus, WhSBV).

2.2 Genome organization

Bornaviruses have a single-stranded non-segmented negative-sense RNA genome of approximately 8.9kb (Amarasinghe et al., 2019; Briese et al., 1992, 1994; Cubitt and de la Torre, 1994; Cubitt et al., 1994; Cubitt et al., 2001; Honkavuori et al., 2008; Hyndman et al., 2018; Kistler et al., 2008; Nowotny et al., 2000; Pleschka et al., 2001; Schneemann et al., 1994; Schneider et al., 1994, reviewed by Herden et al., 2013) At least six different proteins are encoded by six major open reading frames: the nucleoprotein (N, isoforms p40 and p38), the phosphoprotein (P, p24), a small non-structural polypeptide (X, p10), the matrix protein (M, p16), the glycoprotein (G, p57; gp94 when glycosylated) and the RNA-dependent RNA polymerase (L).

Orthobornaviruses have a typical genome organization (3′-N-X-P-M-G-L-5′) with mostly overlapping open reading frames, lack of specific intergenic regions and short noncoding complementary sequences at their termini (Schneemann et al., 1994; Schneider et al., 1994; Walker et al., 2000, reviewed by Herden et al., 2013) In contrast, the gene order of G and M genes in cultervirus and carbovirus genomes appear to be swapped (3'-N-X-P-G-M-L-5′) (Amarasinghe et al., 2019; Hyndman et al., 2018), but carbovirus genomes also contain predicted noncoding regions at their 3′ and 5′ ends (Hyndman et al., 2018).

Further research is warranted to confirm recent *in silico* findings and elucidate the relevance of similarities and differences found between viruses of these genera concerning their genome organization, viral life cycle and pathogenicity.

2.3 Genetic diversity

In general, isolates of BoDV-1 are highly conserved (>95% identity at the nucleotide level) with few sequence variations corresponding to five geographical clusters in endemic areas (Binz et al., 1994; Dürrwald et al., 2006; Herzog et al., 1997; Kolodziejek et al., 2005). In comparison to BoDV-1, the VSBV-1 genome only shares approx. 68% pairwise nucleotide sequence identity (Hoffmann et al., 2015). Avian orthobornaviruses display a much higher genetic variability compared with BoDV-1 (less than 70% identity at the nucleotide level) and comparing the gene sequences with each other, approx. 50–90% identity at the nucleotide level is present (He et al., 2014; Honkavuori et al., 2008; Kistler et al., 2008; Kuhn et al., 2015).

Carboviruses share 69% and less than 38% pairwise nucleotide identity with each other and with other bornaviruses, respectively (Amarasinghe et al., 2018). The WhSBV genome shares less than 37% pairwise nucleotide sequence identity with any of the other bornaviruses (Amarasinghe et al., 2019).

2.4 Viral life cycle

The viral cycle of infection, replication and transcription have been most extensively studied for BoDV-1, as the type species of the genus *Orthobornavirus* (reviewed in Herden et al., 2013; Honda and Tomonaga, 2013, see also Honda et al., 2017; Yanai et al., 2017).

In contrast to other RNA viruses, BoDV-1 employs unique intranuclear replication and transcription strategies and is able to establish non-cytolytic persistence *in vitro* and *in vivo* (reviewed in Herden et al., 2013). Gene expression is tightly regulated by transcriptional read-through leading to formation of several subgenomic RNA transcripts which are further differentially spliced, production of short leader RNAs from the noncoding 3′ terminus and limited nuclear export of viral mRNAs (reviewed in Herden et al., 2013, see also Honda et al., 2017b). Recently, splicing of N transcripts has been described, leading to different subcellular localization of the spliced mRNA, thereby contributing to the regulated replication efficiency of BoDV-1 (Kojima et al., 2019).

BoDV-1-N, -P and -L are part of the ribonucleoprotein (RNP) complex and essential for transcription and replication (reviewed in Herden et al., 2013 and Richt et al., 2007). For this, the N to P ratio is important for controlling BoDV-1 polymerase activity in the acute and chronic persistent phase (Perez et al., 2003; Schneider et al., 2004).

Virus infection cycle is initiated by binding to one or more yet unknown cellular receptors and internalization by clathrin-mediated, energy-dependent endocytosis followed by pH-dependent entry via intracellular vesicles and fusion with the host endosomal membrane (Clemente and de la Torre, 2009; Eickmann et al., 2005; Gonzalez-Dunia et al., 1998; Lennartz et al., 2016; Perez et al., 2001; Schneider et al., 1997). Receptor binding and fusion with the host endosomal membrane is mediated by cleavage products of the glycoprotein G into a C-terminal and N-terminal subunit that needs to be processed by cellular protease furin (Eickmann et al., 2005; Gonzalez-Dunia et al., 1998; Kiermayer et al., 2002; Richt et al., 1998). Cleavage of G appears to be needed for production of infectious virions (Eickmann et al., 2005; Richt et al., 1998). Additionally, receptor-independent cell-to-cell spread has been described (Cubitt and de la Torre, 1994; Gosztonyi et al., 1993; Ludwig et al., 1993), but still seems to be dependent upon G (Bajramovic et al., 2003; Lennartz et al., 2016). Assembly and release of BoDV-1 is accomplished by packaging of genomic RNA by N (Hock et al., 2010; Rudolph et al., 2003) and assembly with P and L to form the RNP complex. The viral matrixprotein M interacts with the RNP complex (Chase et al., 2007; Honda and Tomonaga, 2013; Jamali et al., 2011; Mayer et al., 2005) and infectious particles are successfully formed with N, P, M and G, whereas X protein is not essential for this (Perez and de la Torre, 2005).

The nonstructural protein X acts as a regulator for the RNA-dependent RNA polymerase activity by binding to P (Mayer et al., 2005; Perez et al., 2003; Poenisch et al., 2004, 2008; Schneider et al., 2003; Schwardt et al., 2005; Schwemmle et al., 1998). The viral M is essential for assembly and budding (Kraus et al., 2001, 2005).

2.5 Endogenous bornavirus-like elements

Endogenous bornavirus-like elements (EBL) similar to bornaviral genes were first described in eukaryote genomes about a decade ago, leading to the strong suggestion of a long-standing co-evolution of bornaviruses with vertebrates (Belyi et al., 2010; Gilbert et al., 2014; Horie et al., 2010, 2016; Horie and Tomonaga, 2018; Katzourakis and Gifford, 2010; Kobayashi et al., 2016). Further investigations into potential relevance and function will unravel their potential impact on pathogenesis and immunopathology of bornavirus infections.

Endogenous viral elements by retroviruses have already long been described, but non-retrovirus derived sequences similar to bornaviral genes N, M, G and L have been found just recently in vertebrate genomes and termed EBLN, EBLM, EBLG and EBLL, respectively (Belyi et al., 2010; Horie et al., 2010; Katzourakis and Gifford, 2010). EBL have been described in a wide range of species, mainly vertebrates, such as primates, rodents, chiroptera, afrotheria, insectivora, marsupials, cetaceans, birds, reptiles, fish, and one carnivore species, but also in an insect and a spider (Belyi et al., 2010; Cui and Wang, 2015; Cui et al., 2014; Gilbert et al., 2014; Horie et al., 2010, 2013; Hyndman et al., 2018; Katzourakis and Gifford, 2010; Mukai et al., 2018). EBL are the only integrated non-retroviral endogenous viral elements in the human genome identified so far. Integration is most likely mediated by cellular retrotransposons and EBL originating from viral nucleoprotein mRNA are most frequently detected (Belyi et al., 2010; Cui et al., 2014; Gilbert et al., 2014; Horie et al., 2010, 2013, 2016; Kobayashi et al., 2016). Human EBLN are best studied (He et al., 2016; Myers et al., 2016; Sofuku et al., 2015) and further links to potential functions have been made (reviewed in Honda, 2017; Horie, 2017). EBLL are less frequently found (Belyi et al., 2010; Katzourakis and Gifford, 2010; Mukai et al., 2018), but have been further studied in bats (reviewed by Skirmuntt et al., 2020, see also Cui and Wang, 2015; Horie et al., 2016; Mukai et al., 2018). Elements similar to viral M and G are even less common and elements derived from the X protein and phosphoprotein P have not been described so far. Distribution tendencies of different EBL among species and contributing factors are discussed by Horie and Tomonaga, 2018.

The range of species harboring EBL only partly overlaps with current knowledge about natural, accidental and experimental hosts of bornaviruses. Absence of EBL in accidental dead-end hosts such as horse, sheep and most bird species, absence of known exogenous bornaviruses in some species with EBL such as bats (Nobach and Herden, 2020) and afrotherians, as well as experimental data (Fujino et al., 2014; Parrish et al., 2015) point to possible antiviral functions of EBL. These antiviral functions might be exerted by the EBL DNA element itself, derived RNA or potentially encoded protein (reviewed in Horie, 2017). Whether some interaction of EBL contributes to the pathogenesis, especially immunopathology, was hypothesized as a putative immune-tolerance by expression of EBL-derived proteins (reviewed by Horie, 2017), but awaits experimental evidence. Other proposed antiviral mechanisms might include dominant-negative protein effects on exogenous bornaviruses as shown *in vitro* for a squirrel EBL in human cells (Fujino et al., 2014), RNA interference by small noncoding RNAs, such as piwi-interacting RNAs (piRNAs) with evidence gathered in rodents and primates (Girard et al., 2006; Lau et al., 2006; Parrish et al., 2015) or function as long noncoding RNAs (lncRNAs) (Sofuku et al., 2015). Recently, differential expression of lncRNAs in BoDV-1 infected cell cultures was reported (Sun et al., 2019). Thus, hypothetically, the host immune system might be modulated by EBL or their products (Horie et al., 2016; Sofuku et al., 2015, reviewed by Honda and Tomonaga, 2016).

2.6 Virus-host interactions

Versatile further bornavirus-host interactions have been described so far, ranging from interference with basic cellular signaling pathways, modulation of cellular functions, e.g., cellular enzymes, modulation of neuronal functions and effects on the host cell metabolism (Suberbielle et al., 2008, reviewed in Gonzalez-Dunia et al., 2005; Herden et al., 2013; Planz et al., 2009; Pleschka, 2008; Tizard et al., 2016a). Many of these effects are mediated by viral proteins, e.g., N, P or X and could contribute to various disease aspects, e.g., viral life cycle, modulation and evasion from the host immune response, impairment of neuronal functions or interaction with apoptotic processes (see below). Many of these data have been generated *in vitro*, also employing cell culture adapted viruses. Thus, translation into the natural human or animal infection of dead-end host or into the reservoir host requires caution and proof of concept. Only few interactions potentially related to immunopathology are briefly mentioned in the upcoming section or in the section on the immunopathology of mammalian orthobornaviruses.

2.6.1 Type I interferon system

Similar as other viruses, interference of BoDV-1 with the induction of the antiviral interferon (IFN) system is a major feature to avoid innate immune responses and to enable infection and persistence, which also contributes to immunopathology and vice versa (Randall and Goodbourn, 2008). Recently, this has also been shown for avian orthobornaviruses (Reuter et al., 2010, 2016). In intranasally BoDV-1-infected rats no upregulation of IFN-α was seen and in mice constitutively expressing IFN-α BoDV-1 replication was diminished, which was confirmed *in vitro* employing various cell cultures (Hallensleben and Staeheli, 1999; Shankar et al., 1992; Staeheli et al., 2001). Genetic ablation of the type I IFN receptor in mice also did not affect viral replication and spread (Staeheli et al., 2001). Recently, induction of the IFN system was found in rat astrocytes and microglia, as well as in mouse neurons, but not in rat neurons after BoDV-1 infection, indicating cell and host species specificities (Lin et al., 2013). Trimming of 5′ terminus of the viral genome hinders recognition through retinoic acid-inducible gene I (RIG-I), which subsequently avoids type I IFN induction (Habjan et al., 2008; Rosario et al., 2005, reviewed in Schneider et al., 2007). This also contributes to an adjusted viral spread. In this scenario, BoDV-1-P functions as decoy substrate for TANK-binding kinase 1 (TBK1), competes with interferon regulatory factor (IFR)-3/-7 and thereby suppresses IFN-ß induction, inducible nitric oxidase synthase (iNOS) synthesis and potentially also nuclear factor kappa-light-chain-enhancer of activated B cells (NF-κB) activity (Bourteele et al., 2005; Peng et al., 2007; Unterstab et al., 2005, reviewed in Planz et al., 2009). Furthermore, interaction with the microRNAs miR-122 and miR-155 was shown for the viral phosphoprotein, which also contributes to the inhibition of type I IFN expression in persistently infected cells (Qian et al., 2010; Zhai et al., 2013). An impact of the viral X protein on type I IFN signaling has also been reported (Wensman et al., 2013). Additionally, it was shown that BoDV-1-N interferes with the IRF-7 pathway and consequently inhibits IFN-α/β expression (Song et al., 2013). Within the viral nucleoprotein N, a NF-κB-inhibitory sequence has been found that blocks the antiviral NF-κB activation with transcriptional activity for interferons or other cytokines (Makino et al., 2015, reviewed in Planz et al., 2009). An antiviral effect of NF-κB has been shown, since a constitutive activation of NF-κB suppressed viral replication (Bourteele et al., 2005). Moreover, in neonatally BoDV-1-infected rats an upregulation of the miR-146 was found and interpreted as an inhibitor for NF-κB activation as well (Zhang et al., 2019). In BoDV-1-infected tumor necrosis factor (TNF)-transgenic mice and cultures thereof, there was no nuclear translocation of NF-κB in hippocampal neurons, which furthermore substantiates the viral strategy to inhibit its activation (Brachthäuser et al., 2013; Schaudien, 2007).

2.6.2 Viral persistence

The well-known factors that contribute to establishment of viral persistence and evasion from the immune system such as conservation of the genome among virus species, tightly regulated replication and transcription strategies, masking of antigenic epitopes, neurotropism of bornaviruses, and intraaxonal transport contribute to the immunopathology. New data provided evidence for a stable interaction of BoDV-1 with nuclear histones in the host cell as well as the modulation of epigenetic signaling as shown by the effect of BoDV-1-P on cellular histone acetyltransferases (Bonnaud et al., 2015; Matsumoto et al., 2012). Proteomic studies suggest that BoDV-1 is involved in pathways of histone acetylation (Liu et al., 2014). The RNP complex association with the chromosomes of the host cell can serve as an elegant way of spread during mitosis and to circumvent recognition of the immune system, again facilitating persistence (Bonnaud et al., 2015; Charlier et al., 2013; Matsumoto et al., 2012). This also employs high mobility group box 1 protein (HMGB1) for stabilization again via interaction with the viral P. Recently, it was shown that BoDV-1 employs the cellular A to I editing by double-stranded RNA-specific adenosine deaminase (ADAR2) for efficient viral replication, to appear as 'self' RNA in the nucleus and to avoid antiviral innate immune responses (Yanai et al., 2020).

2.6.3 Modulation of neuronal functions and apoptosis

Clinical signs and neurodevelopmental alterations after BoDV-1 infection were proposed to be based also on interference with neuronal functions such as synaptic plasticity, excitability and communication, neurotransmitter/neuropeptide release, but also with the neurotrophin system (reviewed in Gonzalez-Dunia et al., 2005; Planz et al., 2009; Tizard et al., 2016a). For this, viral interactions with PKC and the Raf/MEK/ERK cascade as prototype of the MAP kinase (MAPK) are important (reviewed in Planz et al., 2009). The viral phosphoprotein serves as decoy substrate for the ε-isoform of PKC, similar to TBK1, thereby competing for phosphorylation and subsequently impairing the recycling of synaptic vesicles (Prat et al., 2009; Schwemmle et al., 1997; Volmer et al., 2006, reviewed in Planz et al., 2009). The Raf/MEK/ERK cascade, among others, is stimulated by neurotrophins such as nerve growth factor and brain-derived neurotrophic factor (BDNF, reviewed in Gonzalez-Dunia et al., 2005). Interaction of BoDV-1 with neutrophins led to an increase in viral replication (Carbone et al., 1993; Ibrahim et al., 2002) and in persistently BoDV-1-infected cell cultures, the MAPK cascade was constitutively activated (Hans et al., 2001; Planz et al., 2001, reviewed in Planz et al., 2009). However, effects of nerve growth factor and BDNF as exogenous stimuli via MAPK signaling were inhibited in neuronal cultures leading to disturbances in synaptic plasticity (Hans et al., 2001, 2004; Peng et al., 2007; Volmer et al., 2006, 2007, reviewed in Planz et al., 2009). Proteomic studies further support a putative involvement in pathways of energy metabolism and MAPK signal transduction (Liu et al., 2014). An interference with neurogenesis, but not astrogliogenesis might further contribute to impairment of neuronal plasticity (Brnic et al., 2012) and it was recently shown that the viral phosphoprotein specifically impairs γ-aminobutyric acid (GABA)-ergic neurogenesis (Scordel et al., 2015).

Moreover, the viral phosphoprotein interferes with HMGB-1 and thereby suppresses p53 transcriptional activity and apoptosis (Kamitani et al., 2001; Zhang et al., 2003, reviewed in Gonzalez-Dunia et al., 2005; Planz et al., 2009). Most likely, this contributes to viral persistence and immune evasion as well. BoDV-1-X also serves as inhibitor of apoptotic processes by association with mitochondrial antiviral signaling protein and blocking the intrinsic caspase activation (Honda and Tomonaga, 2013; Li et al., 2013; Poenisch et al., 2004, 2009; Schwardt et al., 2005). In neonatally BoDV-1-infected rats, however, poly (ADP-ribose) polymerase 1 was upregulated, which could regulate microglia activation and caspase 3 and therefore could contribute to the apoptotic processes in the brain (Williams et al., 2008). On the other hand, BoDV-1-X shows mitochondrial-driven neuroprotective effects in several studies (Ferré et al., 2016; Szelechowski et al., 2014).

3. Mammalian orthobornaviruses

The most intriguing finding in bornavirus research was the recent evidence of the zoonotic capacity of the mammalian orthobornaviruses VSBV-1 and BoDV-1 (Hoffmann et al., 2015; Korn et al., 2018; Liesche et al., 2019; Niller et al., 2020; Schlottau et al., 2018). In addition, many other novel bornaviruses have been detected in the last decade, but knowledge on the underlying pathogenesis and immunopathology is still based predominantly on experimental studies carried out at the end of the last century, employing BoDV-1 infection in different rodent models. Some comparative data characterizing the natural infection in accidental dead-end hosts such as horses and very recently in humans are now available. It should be emphasized that the human orthobornavirus infections causing lethal encephalitis are quite different from previous studies dealing with potential human BoDV-1 infections and there is general consensus that BoDV-1 is not associated with human psychiatric illness (Bode et al., 1996;

Dürrwald et al., 2007; Hoffmann et al., 2015; Hornig et al., 2012; Lieb and Staeheli, 2001; Lipkin et al., 2011; Richt and Rott, 2001; Rubbenstroth et al., 2019; Schlottau et al., 2018).

The outcome of infection with VBSV-1 and BoDV-1 in accidental dead-end hosts such as horses or humans differs strikingly from the course of infection in reservoir species (BoDV-1: *C. leucodon*, VSBV-1: potentially variegated squirrels (*S. variegatoides*) and Prevost's squirrels (*C. prevostii*)). Clinical signs, tissue tropism, viral distribution and most likely underlying immunopathology is completely different. In accidental dead-end hosts, VSBV-1 and BoDV-1 exhibit a strict neurotropism without viral transmission but persistence, and infection typically causes a severe non-purulent meningoencephalitis. For BoDV-1 it is known that a T cell-mediated delayed type hypersensitivity reaction with a key role of immunopathogenic $CD8^+$ and $CD4^+$ T cells represents the key disease mechanism (reviewed in Herden et al., 2013; Staeheli, 2002; Stitz et al., 2002). Recent initial data in humans infected with VSBV-1 point to potentially comparable immunopathology, but also potential differences (Tappe et al., 2019b). In contrast, in reservoir species, VSBV-1 and BoDV-1 spreads and persists in neural and extra-neural tissues and is continuously shed via secretions, excretions and skin scaling (Bourg et al., 2013; Hoffmann et al., 2015; Nobach et al., 2015; Petzold et al., 2019; Schlottau et al., 2017a). Clinical signs, significant inflammatory or degenerative lesions in the central nervous system (CNS) or any other organ are typically lacking. It has already been reported that natural reservoir hosts seem to tolerate the infection, do not show any or only mild pathologic lesions and clinical signs, thus infection is asymptomatic (Calisher et al., 2006; Mackenzie and Jeggo, 2013; Mandl et al., 2015).

Immune competence, age, genetic and conditional makeup of the respective host and further specific features of the particular virus, e.g., pathogenicity, composition of the virus isolate, dose of infection and way of entry might contribute to the different scenarios in accidental dead-end and reservoir hosts. Sound bases to explain the variable course of the infection can be conducted from the rat model where neonatal, immune incompetent Lewis rats develop a comparable disease pattern as present in reservoir species indicating immune tolerance. In contrast, BoDV-1 infection in adult immune competent Lewis rats closely resembles the outcome found in accidental dead-end host (Gosztonyi and Ludwig, 1984, reviewed in Herden et al., 2013; Rott and Becht, 1995; Staeheli et al., 2000) based on immunopathology. However, in both scenarios, the virus establishes long-term persistence, either in many organs or restricted to the CNS.

Many open questions regarding immunopathology-based disease mechanisms, addressing the described differences in disease outcome in reservoir species versus accidental dead-end hosts, remain to be solved. Since no curative therapy or suitable vaccination strategy is currently available, due to the zoonotic potential of VSBV-1 and BoDV-1 and the growing evidence of fatal human cases, there is an urgent need to unravel underlying disease mechanisms.

3.1 Variegated squirrel bornavirus 1 (VSBV-1)

3.1.1 Epidemiology and host spectrum

In 2015, the novel zoonotic VSBV-1 was detected in three squirrel breeders with fatal meningoencephalitis and the source of infection could be traced back to a contact squirrel (*S. variegatoides)* in one of the holdings (Hoffmann et al., 2015). So far, one additional case of VSBV-1 infection in a zoo worker with a fatal limbic encephalitis has been confirmed (Tappe et al., 2018). Retrospectively, two further fatal human encephalitis cases among squirrel breeders are highly suspicious to be caused by VSBV-1 as well (Tappe et al., 2019a). Investigations in several squirrel holdings demonstrate the occurrence of VSBV-1 in Germany, the Netherlands and Croatia (Schlottau et al., 2017a, 2017b). Variegated squirrels (*S. variegatoides*, prevalence of 3–3.3%) or Prevost's squirrels (*C. prevostii*, prevalence of 8.8–13.6%) are most commonly affected, but other species like *Callosciurus finlaysonii* and *Tamiops swinhoei* (within the subfamily *Callosciurinae*), and *Sciurus granatensis* (within the subfamily *Sciurinae*) can also be infected (Schlottau et al., 2017a, 2017b).

All VSBV-1-infected squirrels found so far were non-indigenous and housed in European husbandries, without any evidence of infection in indigenous and invasive squirrel species (Schlottau et al., 2017a, 2017b; Schulze et al., 2020). Animal exchange and trade between owners and breeders of these exotic squirrels happens frequently, resulting in difficulties with infection tracing (Tappe et al., 2019a). So far, it seems possible that all infected animals could be linked to only one common trading source. The endemic origin of VSBV-1 is still unknown, because positive species from the subfamilies *Sciurinae* and *Callosciurinae* originate from Central America and Southeast Asia, respectively (Schlottau et al., 2017a). Two hypotheses can be considered: firstly, VSBV-1 was imported with squirrels from Central America or Southeast Asia, and secondly, a still unknown reservoir exists in Central Europe (Schulze et al., 2020).

3.1.2 Clinical signs

Human VSBV-1-infected patients, one 45 year-old female, and three 62–72 year-old males, suffered from shivers, fever, confusion, mental and movement slowing, gait disturbances (ataxia), muscle twitching and finally coma (Hoffmann et al., 2015; Tappe et al., 2018). The male patients had underlying medical conditions such as hypertension, diabetes or obesity. All four patients died within 2–4 months after disease onset. The respective clinical signs fit the anatomical localization of the virus, e.g., patients with cerebral cortex infection had muscle twitching, patients with cerebellar and basal ganglia infection showed ataxia. Virus-specific antibodies were found in serum and cerebrospinal fluid (CSF) (Hoffmann et al., 2015; Tappe et al., 2018). A putative epitope to these antibodies was detected on the viral nucleoprotein (Tappe et al., 2018). CSF analyses revealed lymphocytic pleocytosis, protein elevation, blood-brain-barrier dysfunction, and local antibody production typical for a viral infection. The extent of lactate elevations in all patients is normally not seen with viral encephalitis, but might be due to the severe inflammation and malacia in the CNS (Tappe et al., 2019b).

Interestingly, most severe CSF and histological aberrations were seen in the patient with the highest viral load, shortest duration of disease, and most medical preconditions. The incubation period of VSBV-1 is unknown, similar as for BoDV-1 in humans.

In contrast to human infections, no clinical signs were noted in the VSBV-1-infected squirrel species. However, virus-specific serum antibodies can be found in positive squirrels (Schlottau et al., 2017b).

3.1.3 Ways of entry, transmission, spread and shedding

There is limited knowledge on potential VSBV-1 shedding from squirrels, transmission pathways and infection routes to humans or for maintenance in the potential reservoir population. Since bites and scratches have been reported in at least one VSBV-1-infected squirrel breeder, a potential virus entry via skin wounds has already been discussed (Hoffmann et al., 2015). Whether other ways of entry, as already described for BoDV-1 (e.g., intranasal), also account for VSBV-1 infection, has to be investigated. For avian orthobornaviruses, wounds or skin defects with subsequent infection of adjacent nerve fibers and viral spread to the CNS have already been documented (Heckmann et al., 2017; Leal de Araujo et al., 2017).

Oropharyngeal swabs of VSBV-1-infected squirrels have high viral loads hinting to shedding by saliva, but viral RNA can also be detected in nasal and eye swabs (Schlottau et al., 2017b). This fits to the widespread tissue distribution of VSBV-1 (see below) in these animals that can enable potential virus shedding via secretions and excretions such as urine, feces, and skin (Schlottau et al., 2017b). In principle, shedding of infectious virus depends on presence in secretory or excretory cells, like epithelial cells, which has been shown for most of the organ systems in squirrels, including skin, intestine and urinary bladder (Petzold et al., 2019). Transmission in squirrel husbandries might also be mediated by scratches or bites or indirectly by contaminated litter in cages (Schlottau et al., 2017a).

3.1.4 Histopathology

In general, VBSV-1 infected patients displayed a severe panencephalitis (Tappe et al., 2019b), accompanied by edema, malacia, strong astroglial and microglial activation, and with typical mononuclear perivascular cuffing and parenchymal infiltrates (Fig. 1A). Most commonly, the temporal and insular lobes, basal ganglia and midbrain were randomly affected and showed inflammatory infiltrates in the white and gray matter. Areas of malacia contained neutrophils, foamy activated microglia and macrophages. Foamy cells, which either represent amoeboid microglia or macrophages, as well as few bushy microglia were strongly HLA-DR- and CD68-positive. Astrocytes were strongly activated as well, with variably enlarged protoplasmic appearance, partly with bizarre appearance and gemistocytes, all characterized by GFAP-positivity. Satellitosis, neuronophagia, single cell necrosis of Purkinje cells, loss of hippocampal pyramidal cell layer or focally extensive hemorrhage and lipofuscin deposition was additionally present. Other available organs did not show any inflammatory or degenerative lesions.

Opposed to lesions seen in human patients, VSBV-1 caused only mild mononuclear infiltration in infected squirrel brains (Fig. 2A, Petzold et al., 2019; Schlottau et al., 2017b.

3.1.5 Virus distribution in human and animal hosts

In general, infection was detected by an array of different methods, including whole-genome sequencing, different RT-qPCR protocols, antigen and RNA staining by immunohistochemistry and in situ hybridization, complemented by serological methods for presence of virus-specific antibodies in serum and CSF (Hoffmann et al., 2015; Schlottau et al., 2017a, 2017b). In VSBV-1-infected human patients viral RNA and antigen was present in all available brain regions: cerebral cortex, basal ganglia, hippocampus, midbrain, cerebellum, and olfactory bulb (Hoffmann et al., 2015; Tappe et al., 2018, 2019b), with lowest viral loads in the cerebellum, similar as seen

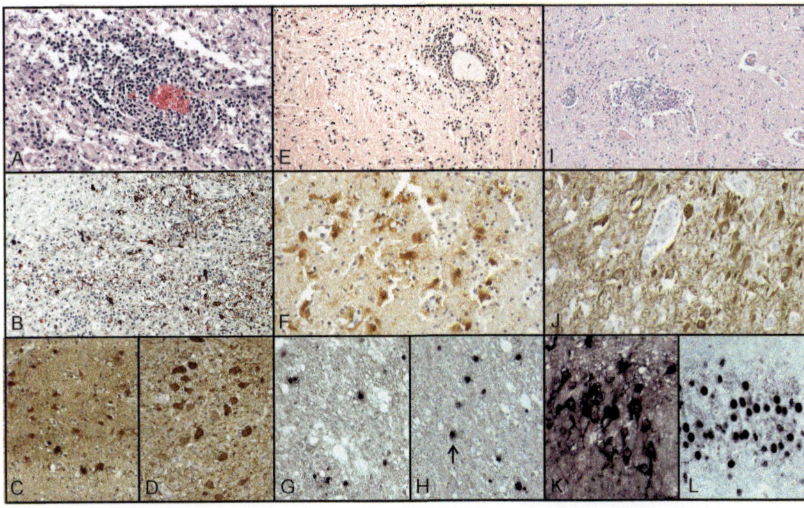

Fig. 1 Comparison of orthobornavirus infection in human and equine accidental dead-end hosts. Overview on histological lesions and virus distribution in the brain. (A–D) Human VSBV-1 infection is characterized by severe encephalitis with mononuclear perivascular infiltrates in the brain (A) and detection of VSBV-1-P (B and C) and VSBV-1-X (D) in the brain by immunohistochemistry. Viral antigen is present in the nuclei, cytoplasm and processes of brain cells. (E–H) Human BoDV-1 infection is characterized by severe encephalitis with mononuclear perivascular infiltrates in the brain (E) and detection of BoDV-1-N (F) and BoDV-1 mRNA encoding BoDV-1-N (G) and respective genomic RNA (H) in the brain by immunohistochemistry and in situ hybridization. Viral antigen is present in the nuclei, cytoplasm and processes of brain cells. Viral mRNA and genomic RNA are located in nuclei of brain cells and only few cells display weak cytoplasmic signals for the viral mRNA (arrow). (I–K) Equine BoDV-1 infection is characterized by severe encephalitis with mononuclear perivascular infiltrates in the brain (I) and detection of BoDV-1-N (J) and BoDV-1 mRNA encoding BoDV-1-N (K) and respective genomic RNA (L) in the brain by immunohistochemistry and in situ hybridization. Viral antigen is present in the nuclei, cytoplasm and processes of brain cells. Viral mRNA is present predominantly in the cytoplasm and genomic RNA is restricted to the nuclei of brain cells.

for BoDV-1 infections of animals (Carbone et al., 1987; Herden et al., 1999, 2000). In all investigated patients, virus antigen was detected in neurons, astrocytes, some gemistocytes and ependymal cells, and was located in the nucleus, cytoplasm and in the neuropil (Fig. 1B–D). No viral antigen was found in peripheral organs available, indicating a comparable neurotropism as seen in accidental dead-end hosts for BoDV-1.

In contrast, in infected squirrels, VSBV-1 RNA and antigen were found in a wide range of tissues with highest loads in the CNS, but also in peripheral organs like heart, lungs, kidneys, oropharyngeal swabs, skin, and the urinary bladder (Fig. 2B–F, Hoffmann et al., 2015; Petzold et al., 2019;

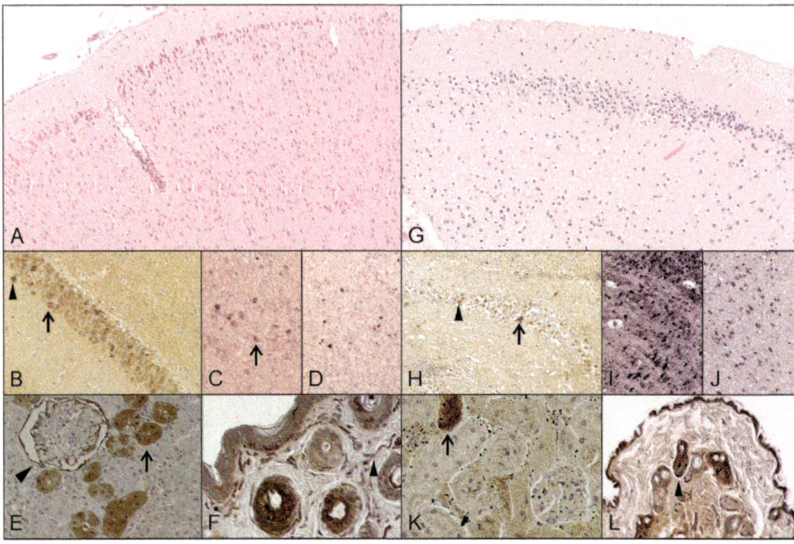

Fig. 2 Comparison of orthobornavirus infection in reservoir hosts for VSBV-1 infection (Prevost's squirrel) and BoDV-1 infection (bicolored white-toothed shrew). Overview on histological lesions and virus distribution. A-F VSBV-1 infection in a Prevost's squirrel is characterized by a mild non-purulent encephalitis with few mononuclear perivascular infiltrates in the brain (A) and detection of VSBV-1-P (B), VSBV-1 mRNA encoding VSBV-1-N (C) and VSBV-1 genomic RNA (D) in the brain by immunohistochemistry and in situ hybridization. Viral antigen is present in the nuclei (arrowheads) and cytoplasm of brain cells (arrow, neurons). Viral mRNA and genomic RNA are located in nuclei of brain cells and only few cells display weak cytoplasmic signals for the viral mRNA (arrow). Virus, demonstrated by detection of VSBV-1-P, is also present in peripheral organs like kidney in tubular epithelium (E, arrow) and glomerula (E, arrowhead) and skin (F) in epithelial cells, sebaceous glands, hair follicles and nerve fibers (F, arrowhead). G-L BoDV-1 infection in bicolored white-toothed shrews. In the brain, there are no histological lesions (G), despite detection of BoDV-1-P (H), viral mRNA encoding BoDV-1-N (I) and respective genomic RNA (J) by immunohistochemistry and in situ hybridization. Viral antigen is present in the nuclei (arrowhead) and cytoplasm of brain cells (arrow, neurons). Viral mRNA is present predominantly in the cytoplasm and genomic RNA is restricted to the nuclei of brain cells. Virus, demonstrated by detection of BoDV-1-P, is also present in peripheral organs like kidney in tubular epithelium (K, arrow) and skin (L) in epithelial cells, sebaceous glands (L, arrowhead), hair follicles and nerve fibers.

Schlottau et al., 2017a, 2017b). VSBV-1 in infected squirrels demonstrates a wide cell tropism, including several cell types of the central and peripheral nervous system, but also epithelial and mesenchymal cells of several secretory organs (Fig. 2B–F, for details see Petzold et al., 2019). This widespread distribution pattern closely resembles the virus dissemination in BoDV-1-infected *C. leucodon* and points to a role of the *Callosciurinae* and *Sciurinae* as reservoir hosts capable of shedding infectious virus via several routes.

3.1.6 Immunopathology in humans

To date, the brain tissue of four VSBV-1-infected patients was further analyzed to elucidate the type of immune cell infiltration and brain reaction patterns. Inflammatory infiltrates were found in the areas positive for VSBV-1 RNA and antigen and consisted mainly of $CD4^+$ and $CD8^+$ T cells, accompanied by macrophages, neutrophils and a few perivascular B cells (Fig. 3, Tappe et al., 2019b). In general, this fits the immune cell composition seen in BoDV-1-infected accidental dead-end hosts. The occurrence of neutrophils might be due to the areas of malacia in which they were found. $CD20^+$ B cells were mainly located in perivascular cuffs and only few of them were found in the tissue otherwise (Fig. 3A). In contrast, $CD3^+$ T cells were also present abundantly in the brain parenchyma and consisted of $CD4^+$ helper T cells and $CD8^+$ cytotoxic T cells in rather equal amounts (Fig. 3B and C). A predominance of $CD3^+$ T cells in the parenchyma, accompanied by fewer macrophages, B cells and possibly neutrophils, as well as $CD4^+$ and $CD8^+$ T cells has been detected in naturally BoDV-1-infected horses (Bilzer et al., 1995; Caplazi and Ehrensperger, 1998) and experimentally BoDV-1-infected mice and rats (Deschl et al., 1990; Hausmann et al., 1999; Kramer et al., 2012; Stitz et al., 2002). In contrast to the findings in the VSBV-1-infected patients, horses and rats depicted $CD4^+$ T cells predominantly in the perivascular infiltrates, while $CD8^+$ T cells also spread into the parenchyma. These discrepancies might be attributed to the time point of infection, as horses have to be euthanized early in the course of disease.

Adjacent to neurons and glial cells, granzyme B-positive immune cells were detected in the VSBV-1-infected brains (Fig. 3D). Widespread caspase 3-positive cells were additionally present and shrinkage of brain cells adjacent to $CD8^+$ T lymphocytes and granzyme B-positive cells indicated signs of a potential cytotoxic response and apoptosis (Fig. 3D). In addition, perivascular mononuclear cells, some neurons, astrocytes and rod cells throughout the affected areas expressed p53. In BoDV-1-infected mice it was shown that $CD8^+$ T cells employ IFN-γ and not perforin and that cytolysis does not occur (Hausmann et al., 2001, 2005b, 2001, 2005b). In neonatally BoDV-1-infected rats, hippocampal caspase 3 activation and neuronal apoptosis has also been observed (Weissenböck et al., 2000; Williams et al., 2008), however infection took place at an immune incompetent stage.

Thus, the type of the severe inflammatory response points to a comparable T cell-mediated immunopathology of human VSBV-1 infection as described in natural and experimental BoDV-1 infection of accidental dead-end hosts. The rather rapid clinical course in one patient could be linked to a higher viral load and more pronounced inflammation.

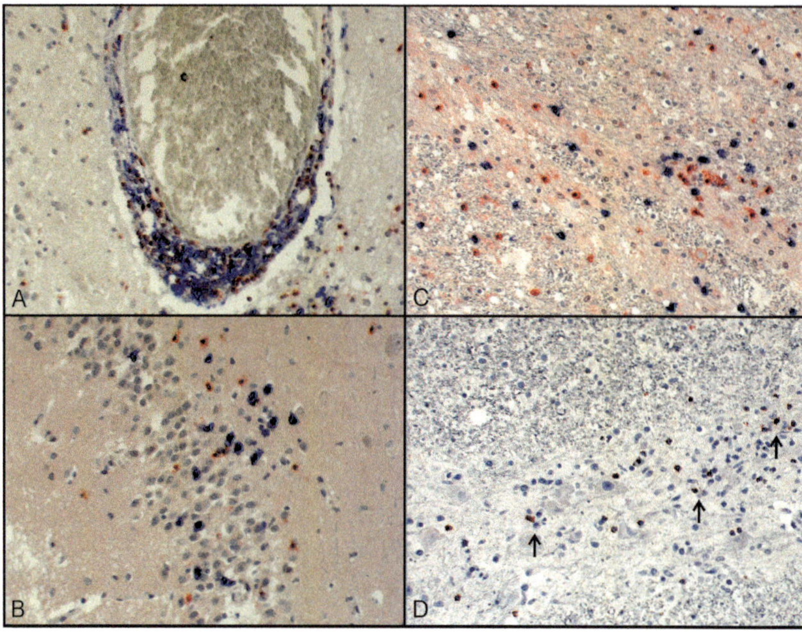

Fig. 3 Immunohistological differentiation of immune cells in the brain of VSBV-1 infected human. (A) Demonstration of T cells (CD3, red) and B cells (CD20, blue) in a perivascular infiltrate in the brain. B cells are prominent in the perivascular infiltration and T cells are also present in high numbers in the parenchyma. (B and C) Distribution of $CD4^+$ T cells (red) and $CD8^+$ T cells (blue) in hippocampus (B) and cerebral cortex (C). $CD4^+$ T cells and $CD8^+$ T cells are present in nearly equal amounts in the parenchyma. (D) T cells (granzyme B-positive) adjacent to neurons (arrow) and fewer smaller cells.

Whether observed differences could be explained by a potential pro-inflammatory state of the CNS due to the underlying medical conditions known to be associated with low level inflammation (Aguilar-Valles et al., 2015; D'Mello et al., 2009), or simply reflect a longer duration of inflammation and clinical disease has to be further investigated. In BoDV-1-infected, TNF overexpressing mice, the pro-inflammatory milieu of the CNS triggers inflammatory processes and comparable unusual glia activation (Kramer et al., 2012).

The wide-spread virus distribution in VSBV-1 infected squirrels (Petzold et al., 2019; Schlottau et al., 2017a, 2017b) could indicate an immune tolerance similar to the BoDV-1 reservoir or neonatally infected rats. However, the mild mononuclear meningoencephalitis (Petzold et al., 2019; Schlottau et al., 2017b) and the presence of serum antibodies might indicate incomplete immune tolerance and/or virus-host adaption. More detailed studies, e.g., on the immune status of the infected squirrels, as well as reactivity and specificity of the invading brain cells, are definitely needed but are hampered due to the lack of respective squirrel-specific tools. The recent successful isolation of a squirrel VSBV-1 isolate (Schlottau et al., 2017b) could contribute to fill these gaps, but respective animal models for the analysis of potential immunopathology are also definitely needed.

3.2 Borna disease virus-1 (BoDV-1)

3.2.1 Epidemiology and host spectrum

Shortly after the detection of the novel VSBV-1 in 2015, human lethal encephalitis cases caused by BoDV-1 were published in 2018 by independent research groups (Korn et al., 2018; Schlottau et al., 2018). The first five described cases were either transplant-associated or naturally acquired cases in previously healthy young persons. To date, at least 18 human fatal encephalitis cases have been found, including cases found by retrospective analyses back to 1996 (Coras et al., 2019; Korn et al., 2018; Liesche et al., 2019; Niller et al., 2020; Schlottau et al., 2018). Interestingly, all human cases known so far originated from endemic areas from Bavaria, Southern Germany, and virus genome sequencing revealed a close relationship to viruses that have occurred in same areas in horse and sheep indicating transmission from the local reservoir (Niller et al., 2020). Phylogenetic analyses point to multiple independent spill-over infections, most likely from the BoDV-1 reservoir *C. leucodon* (Niller et al., 2020; Rubbenstroth et al., 2019).

Most commonly horses and sheep are affected by natural BD. Natural and experimental infections provided evidence that other Equidae, Artiodactyla (like cattle, goats and alpacas), and less commonly a wide range of pet and zoo animals can be infected (reviewed in Herden et al., 2013). However, susceptibility can depend on host species and even species breeds or laboratory strains (Danner, 1982; Rott and Becht, 1995; Wu et al., 2013).

BD-affected horses are typical accidental dead-end hosts (Schmidt et al., 1999). Epidemiology of BD has pointed towards a natural reservoir for a long time (reviewed in Dürrwald et al., 2006) that was confirmed for several endemic areas, where *C. leucodon* have been found as indigenous reservoir hosts (Bourg et al., 2013; Dürrwald et al., 2014; Hilbe et al., 2006; Puorger et al., 2010; Weissenböck et al., 2017).

3.2.2 Clinical signs

The clinical signs after human BoDV-1 infection are very similar to human VSBV-1 infection, and also comparable to those in animal accidental dead-end hosts such as horses (Grabner and Fischer, 1991; Hoffmann et al., 2015; Niller et al., 2020; Schlottau et al., 2018; Tappe et al., 2018, reviewed in Herden et al., 2013). In recent retrospective studies, 13 additional BoDV-1 infected human cases were detected between 1999 and 2019 (Liesche et al., 2019; Niller et al., 2020). No sex or age predilection was found. Living in rural areas, outdoor or agricultural work and animal contacts, e.g., cats have been considered as potential risk factors for infection (Niller et al., 2020).

In case of the three transplant-associated BoDV-1 cases reported by Schlottau et al., 2018, organs were received from the same donor, aged 70 years, who also originated from a virus-endemic area of Bavaria; he died without any signs of neurologic diseases or of an active infection. The kidneys were allocated to a 66-year-old man and a 74-year-old woman suffering from end-stage renal disease. The liver was allocated to a 65-year-old male suffering from hepatocellular carcinoma and cirrhosis (Schlottau et al., 2018). Both kidney recipients received common post transplantation immunosuppressive medication containing also antithymocyte globulin (ATG). They developed progressive, ascending, flaccid, severe weakness of all limbs, starting 3–4 months post-transplantation. Signs worsened to irreversible coma and death occurred on day 208 and 179, respectively. One of the patients and two other patients with lethal BoDV-1 infection initially presented with symptoms consistent with a Guillain-Barré syndrome (Coras et al., 2019; Liesche et al., 2019; Schlottau et al., 2018). Three months post transplantation, the liver-graft-recipient showed facial palsy and cognitive deficits, and he was diagnosed with progressive leukoencephalopathy with small vessel involvement, affection of the retina, progressive optic nerve atrophy and a strong BoDV-1 specific seroconversion in serum and CSF (Schlottau et al., 2018). He has survived so far with sequelae from the CNS lesions.

Beside these transplant-associated cases, signs in previously healthy young persons were comparable and consisted of fever, headache, confusion, psychomotor slowing, myoclonus, unsteady gait, deep coma and deteriorating brain stem reflexes leading to death (Korn et al., 2018; Liesche et al., 2019; Niller et al., 2020). Involvement of the spinal cord could be present, which was not noted in the VBSV-1 infected patients. These acute cases in supposedly immune competent patients had a rapid course with death within 14–57 days after hospitalization. This resembles the acute infection in other dead-end hosts, such as horses, whereas the course of disease lasted longer in the immunocompromised transplant recipients, most likely due to the immunosuppressive treatment (53–99 days, Coras et al., 2019; Korn et al., 2018; Liesche et al., 2019; Niller et al., 2020; Schlottau et al., 2018). High serum antibody titers developed in almost all infected patients with variable time courses, whereas CSF samples usually showed lower titers which increased at later time points (Korn et al., 2018; Niller et al., 2020; Schlottau et al., 2018). No viral RNA was amplified from serum samples (Niller et al., 2020). In all BoDV-1 infected patients, CSF presented with typical changes as already described for the human VBSV-1 patients.

Animal accidental dead-end hosts, like horses and sheep, typically develop a severe neurological disorder upon BoDV-1 infection, resulting in death within 1–4 weeks after onset of symptoms and a mortality of about 90% (reviewed in Herden et al., 2013). However, serological evidence suggests frequent inapparent infections in horses and sheep (reviewed in Herden et al., 2013). Clinical signs of BD can include a variety of neurologic symptoms affecting psyche, sensorium, sensibility, motility and autonomous nervous system. Presentation depends on the course of disease and the brain area affected by inflammation. Clinical signs can also vary among host species, but are comparable to the symptoms in horses (Bode et al., 1994; Caplazi et al., 1994, 1999; Ludwig et al., 1985; Metzler et al., 1978; Otta and Jentzsch, 1960; Rott and Becht, 1995). No inflammatory infiltrate or degenerative lesions occur in the CNS of the reservoir host despite high viral loads (Bourg et al., 2013; Dürrwald et al., 2014; Puorger et al., 2010).

3.2.3 Ways of entry, transmission, spread and shedding

Potential routes of BoDV-1 transmission from its reservoir host to humans and other accidental dead-end hosts as well as maintenance in the reservoir population remains elusive. Parallels to natural dead-end host infections in animals could point to an intranasal infection also in humans, from shrew excretions or secretions and subsequent intra-axonal retrograde spread along the olfactory route to the limbic system. Moreover, iatrogenic BoDV-1 transmission by solid organ transplantation was shown (Niller et al., 2020; Schlottau et al., 2018). In most of the patients a similar viral neurotropism was noted and also indicated no natural transmission between humans. However, in the transplant-associated human cases, the explanted kidney harbored viral RNA and antigen, whereas the pretransplantation tissue did not (Schlottau et al., 2018). Together with the weak seropositivity of the donor this might argue for a subclinical infection state or an early phase of infection.

So far, the incubation period of BoDV-1-induced disease in humans is unknown. In naturally infected horses it is assumed to last several weeks to months, whereas it can be as short as two weeks in experimentally infected animals depending on the route of infection.

3.2.4 Histopathology

In general, in human BoDV-1-infected patients, a non-purulent (meningo-) panencephalomyelitis with detection of BoDV-1 eosinophilic, spherical intranuclear Joest-Degen inclusion bodies is quite typical, similarly as in naturally and experimentally BoDV-1-infected animals and in the VSBV-1-infected humans (Fig. 1E, Herden et al., 1999, 2000; Hoffmann et al., 2015; Liesche et al., 2019; Schlottau et al., 2018; Tappe et al., 2019b; Werner-Keiss et al., 2008). However, a variable distribution pattern of the inflammatory infiltrates has been noted (Liesche et al., 2019). Thus, either the basal nuclei, hippocampal region or upper brain could mainly be affected, but a more uniform distribution was also present. The cerebellum was less affected, but loss of Purkinje cells could occur. Perivascular lymphocytic infiltrates were accompanied by parenchymal lymphoplasmahistiocytic infiltrates (Fig. 1E), microglial nodules, gliosis with severely enlarged astrocytes and randomly distributed malacia (Coras et al., 2019; Korn et al., 2018; Liesche et al., 2019; Schlottau et al., 2018). Peripheral nerves can exhibit severe lymphocytic neuritis with necrosis as seen in the kidney recipient with the Guillain-Barré symptoms (Liesche et al., 2019). However, this was not found in the VSBV-1-infected patients, as no peripheral nerves were available for examination.

In animals, the pattern and type of inflammatory lesions in the CNS are well described (Gosztonyi and Ludwig, 1984; Joest and Degen, 1909, reviewed in Herden et al., 2013). The main affected brain areas are the olfactory bulb, basal cortex, caudate nucleus, thalamus, hippocampus, and periventricular areas of the medulla oblongata. A severe non-purulent poliomeningoencephalomyelitis with perivascular and parenchymal immune cell infiltrates accompanied by activation of astrocytes and microglia is quite characteristic (Fig. 1I). Pathognomonic eosinophilic intranuclear 'Joest-Degen' inclusion bodies in neurons are diagnostic (Joest and Degen, 1909) but occur infrequently to date. Additionally, loss of pyramidal cells of the hippocampus and very rarely degeneration of retinal neurons have been described (Dietzel et al., 2007). In the reservoir *C. leucodon*, no inflammatory or degenerative lesions in the CNS were detected (Fig. 2G).

3.2.5 Virus distribution

Most of the human BoDV-1 cases depict a similar neurotropism as known for animals serving as accidental dead-end host such as horses or after experimental infection of immune competent rats (Herden et al., 1999; Rott and Becht, 1995; Staeheli et al., 2000; Werner-Keiss et al., 2008). As already described for VSBV-1, infection was consistently detected by an array of different methods. In few cases, peripheral nerves harbored viral RNA and viral antigen, as was also demonstrated in the explanted kidney (Liesche et al., 2019; Niller et al., 2020; Schlottau et al., 2018). In general, brain cells such as neurons, glial cells, some ependymal cells, and Schwann cells harbored viral antigens in the nucleus, cytoplasm and processes, and viral antigen was also present in the neuropil and also in the neurohypophysis (Fig. 1F, Liesche et al., 2019; Schlottau et al., 2018). In line with the viral replication cycle, genomic RNA was restricted to the nucleus and the respective viral mRNA was also predominantly present in the nucleus, typical for a chronic persistent stage of the infection as known for naturally and experimentally BoDV-1 infected animals or even avian orthobornavirus infections (Fig. 1G and H, Algermissen, 2010; Al-Ibadi, 2015; Lebelt and Hagenau, 1996; Schlottau et al., 2018; Werner-Keiss et al., 2008). In the explanted kidney, viral antigen and RNA were found focally in tubular epithelial cells. Viral antigen was found in nerve fibers as well. Liesche et al. (2019) found interindividual differences regarding the viral distribution pattern and the intensity of the inflammatory changes with a strong accentuation of the basal nuclei in the brain of two patients compared to a more global distribution in other brains. The CSF was only inconsistently RT-PCR positive with low Ct values. Similar as in BoDV-1 infected, immune competent animals, no virus was found in non-neuronal cells in peripheral organs in the naturally acquired cases (Herden et al., 1999; Lebelt and Hagenau, 1996; Liesche et al., 2019; Niller et al., 2020; Werner-Keiss et al., 2008).

In animal infections, it was convincingly shown that BoDV-1 migrates from olfactory nerve endings to the olfactory bulbs to the brain with preference of certain regions like the limbic system (Kupke et al., 2019; Morales et al., 1988; Shankar et al., 1992). Primary target cells in natural and experimental animal infections are similar to the human VSBV-1 and BoDV-1 cases with preference for neurons and astrocytes (Fig. 1J–L).

In contrast, BoDV-1 spreads and persists in the reservoir *C. leucodon* in neural and extraneural tissues (Fig. 2H–L). BoDV-1 antigen can be found in several cell types of the central and peripheral nervous system, but also in mesenchymal cells like smooth muscle cells and a wide range of epithelial cell including epithelial cells of skin and sebaceous glands, tubulus cells of kidney, bronchial epithelium and epithelial cells of salivary and lacrimal glands. The virus distribution in secretory and excretory organs leads to continuous shedding by secretions, excretions and skin scaling (Bourg et al., 2013; Dürrwald et al., 2014; Nobach et al., 2015; Puorger et al., 2010).

3.2.6 Immunopathology in humans

Whether a similar T cell-mediated immunopathology is responsible for clinical disease and neuroinflammation in BoDV-1 infected humans as in animal models awaits further investigations. With the recent availability of a human BoDV-1 isolate (Niller et al., 2020) further *in vitro* and *in vivo* studies can now be conducted in order to address many open questions in human bornavirus immunopathology, including the comparable clinical and histological picture of a severe non-purulent meningoencephalitis with virus persistence in immunocompromised and apparently healthy patients (Coras et al., 2019; Korn et al., 2018; Liesche et al., 2019; Niller et al., 2020; Schlottau et al., 2018). Moreover, since many animal studies have been carried out using species-adapted or even cell culture virus for infection, the human BoDV-1 isolate opens the avenue to reliably study its pathogenicity, and could circumvent possible viral changes during the adaptation processes *in vivo* or *in vitro* (Ackermann et al., 2007; Nishino et al., 2002; Okayama et al., 2009; Siemetzki et al., 2009).

In a recent study by Liesche et al. (2019), most infiltrating lymphocytes were $CD3^+$ T cells with dominance of $CD4^+$ and lesser $CD8^+$ T cells. This is different from the human VSBV-1 infections where $CD4^+$ and $CD8^+$ T cells occurred in equal amounts (Tappe et al., 2019b). Whether this points to a more $CD4^+$ T cell-based immunopathology, as found in studies with immune competent Lewis rats (Richt et al., 1994; Stitz et al., 2002), remains elusive. However, an important role for $CD8^+$ T cells has already been assumed for human VBSV-1 immunopathology and in other rat and mouse studies (Tappe et al., 2019b; Chevalier et al., 2011; Hausmann et al., 1999, reviewed in Stitz et al., 2002). It was shown in animal experiments that $CD4^+$ and $CD8^+$ T cells can exert protective effects but are also the main drivers of the immunopathological process (Hallensleben et al., 1998; Hausmann et al., 1999; Richt et al., 1994; Stitz et al., 2002). Similar to the human VSBV-1 cases, the lymphocytic infiltrate was accompanied by strong macrophage/microglial activation and astrogliosis (Coras et al., 2019; Korn et al., 2018; Liesche et al., 2019) that contribute to maintenance of the inflammatory process as known for animal BoDV-1 infections (Herden et al., 2005; Ovanesov et al., 2006, 2008a).

In the case of the lymphocytic neuritis of the kidney recipient with Guillain-Barré symptoms, infiltrates consisted predominantly of $CD8^+$ T cells accompanied by fewer $CD4^+$ T cells (Liesche et al., 2019; Schlottau et al., 2018), probably indicating a cytotoxic effect.

Although parallels in clinical signs, histological lesions, virus distribution and persistence with natural and experimental BoDV-1 infections exist, the definite underlying pathogenesis and immunopathology in human BoDV-1 infections has to be further detailed. Whether the medical preconditions in case of the organ recipients might have contributed to the outcome as discussed for human VSBV-1 infections remains elusive. The kidney transplant recipients received the T cell depleting agent ATG which suggests that uncompromised T cell immunity might be required for the control of BoDV-1 in the early phase of infection. In contrast, immunosuppression may have facilitated viral replication and spread as seen in experimentally infected immunocompromised rats (Stitz et al., 1991). However, whether all human BoDV-1 and VSBV-1 infections end up in severe or fatal encephalitis, or whether different outcomes, e.g., mild or asymptomatic courses of disease even with virus elimination, might occur is unknown and requires further investigations, especially regarding the role of the underlying immunopathology.

To date there are no curative treatment or promising vaccination strategies for mammalian orthobornavirus infection available. Ribavirin and favipiravir (T-705) showed inhibitory effects against a broad spectrum of bornaviruses *in vitro* and in experimental animal infections whereas a postulated effect of amantadine could not be confirmed (Cubitt and de la Torre, 1997; Hallensleben et al., 1997; Jordan et al., 1999; Mizutani et al., 1998; Reuter et al., 2010; Tokunaga et al., 2017). Only the human liver recipient had been treated with ribavirin during the late phase of disease, but whether this contributed to his survival remains speculative (Schlottau et al., 2018). Recently, an additive antiviral effect of favipiravir and a siRNA cocktail was shown (Teng et al., 2019).

3.3 Immunopathology in animals

Mammalian orthobornavirus infections can run quite different courses of disease, either in reservoir species or in accidental dead-end hosts, regarding viral spread and shedding, clinical disease, initiation of inflammation and brain cell reaction with immunopathology as a key disease mechanism in dead-end hosts. To date, many factors remain elusive, but host factors including genetic makeup, status of the immune system at the time of infection, virus-host cell interaction, as well as virus factors such as route and dose of infection and composition of the respective virus species may also play essential roles. Many of these factors cannot be determined in natural infections, so the basic knowledge on immunopathology of orthobornavirus infection was gained using animal models. In the best studied rodent models for BoDV-1 infection, remarkable differences in clinical course, virus distribution and immunopathology have been described (reviewed in Herden et al., 2013; Staeheli, 2002; Stitz et al., 2002; Tizard et al., 2016a). In rats, infection of newborn immune incompetent animals results in a persistent CNS infection with virus spread to peripheral organs without obvious neurological signs, but with disturbances in learning and memory. In adult rats, virus infection also runs a persistent course, but it is restricted to the CNS and associated with a severe non-purulent meningoencephalitis and biphasic neurologic signs comparable to the disease outcome observed in humans and equines. In mouse models, course of infection also depends on the age, however, it is vice versa to rats as neonatal mice develop neurological signs with non-purulent meningoencephalitis, viral persistence and immunopathology, but adult mice of most strains do not show clinical signs or inflammation in the CNS despite high virus titers (Hallensleben et al., 1998; Kao et al., 1984; Rubin et al., 1993). In the mouse model, the availability of genetically altered mice can provide elegant opportunities to get deeper understanding of the underlying immunopathology. Due to the remarkable differences between rat and mouse models, the immunopathogenesis of these models is hence described separately. The detailed explanation of other animal models such as gerbils, rabbits, tupaias, non-human primates and bank voles can be found in the respective literature (reviewed in Herden et al., 2013; Tizard et al., 2016a).

Beside the unequivocally important role of host conditions, intrinsic viral factors should also be considered for differences in clinical outcome of infection. Species-specific virus adaption by serial passages in both, rats and mice, have shown to enhance neurovirulence of BoDV-1, which was attributed to a higher fidelity of the viral RNA-dependent RNA polymerase and few nucleotide exchanges in the viral glycoprotein gene (Ackermann et al., 2007; Nishino et al., 2002; Nitzschke, 1963; Okayama et al., 2009). Interestingly, distinct virus preparations might lead to other outcomes, e.g., a Guillain-Barré-like neuropathy in rats (Biesenbach et al., 1990) or an obesity syndrome in rats without obvious neurologic disease with inflammation and viral antigen and RNA restricted to septum, hippocampus, amygdala and ventromedian hypothalamus (Herden et al., 2000; Ludwig et al., 1988).

In natural virus field isolates there is high identity on nucleotide level between endemic clusters (Dürrwald et al., 2006), nevertheless, they are capable of inducing completely different courses of disease depending on infection of reservoir species or accidental dead-end hosts. This has been recently confirmed for the analyses of human and equine BoDV-1 cases from endemic areas in Bavaria (Bourg et al., 2013; Niller et al., 2020). This again substantiates the essential role of the host, especially the status of the immune system.

3.3.1 BoDV-1 infection of adult rats

In general, Lewis rats have been proven to be a reliable and susceptible animal model to study the immunopathology of bornavirus infection due to their enhanced susceptibility to immune-mediated disorders based on deficiencies in the hypothalamic-pituitary adrenal axis (Cizza and Sternberg, 1994). In contrast, other rat strains, e.g., Wistar rats and black-hooded rats are less susceptible, and a genetic base for resistance to BD has been reported (Herzog et al., 1991; Hirano et al., 1983; Pletnikov et al., 2002; Wu et al., 2013).

Intracerebral BoDV-1 infection of susceptible adult Lewis rats typically leads to a biphasic neurological disorder with an incubation period of approximately 20–40 days post infection (dpi) (Narayan et al., 1983a, 1983b). In the first stage hyperactivity, aggressiveness and exaggerated startle responses occur and in later stages, starting around 60 dpi, somnolence, apathy, neurologic symptoms affecting motility (paralysis, dystonia, dyskinesia, stereotyped and self-mutilation behavior) and blindness have been noted (reviewed in Herden et al., 2013). Histologically, a non-purulent poliomeningoencephalitis with astrogliosis and activated microglia is regularly present with most pronounced inflammation in the hippocampus, thalamus, olfactory bulb, cerebral cortex, caudate nucleus and periventricular areas (Deschl et al., 1990; Gosztonyi and Ludwig, 1995; Herden et al., 2000; Narayan et al., 1983a, 1983b). This does not fit with the varying distribution pattern in human BoDV-1 infection but is comparable to the inflammatory pattern in human VSBV-1 infection and equine BoDV-1 infection (Herden et al., 1999; Liesche et al., 2019; Tappe et al., 2019b). With increase of inflammation, dependent upon the virus used for infection, neuronal necrosis might occur in the cerebral cortex or hippocampus (Deschl et al., 1990; Herden et al., 2000). Intranuclear Joest-Degen inclusion bodies, as described before, can also be found. The observed blindness in later stages is caused by a non-purulent chorioretinitis with degeneration of rods and cones (Krey et al., 1979; Narayan et al., 1983a, 1983b). Interestingly, demyelination does not occur. The onset of clinical signs coincides with inflammation in the brain (Carbone et al., 1987; Herden et al., 2000; Narayan et al., 1983b). Maximum inflammation is typically reached around 40 dpi (Deschl et al., 1990; Herden et al., 2000; Narayan et al., 1983a; Stitz et al., 1991). The change in clinical signs is associated with a decrease of inflammation despite ongoing astrogliosis, microglia activation and viral persistence (Carbone et al., 1987; Gosztonyi and Ludwig, 1995; Hatalski et al., 1998; Herden et al., 2000, 2005; Narayan et al., 1983a; Werner-Keiss et al., 2008, reviewed in Stitz et al., 2002). In addition, high titer of virus-specific antibodies and infectious virus are continuously present. Thus, induction of a robust immune response does not result in viral clearance but in clinical neurological disease. Due to the latter fact, several studies in the last decades of the last century concentrated on the immunopathology of BoDV-1 infection. For this purpose, the composition of immune cell infiltrates was initially detailed. In brains of experimentally infected adult Lewis rats, infiltrates were composed of $CD4^+$ and $CD8^+$ T cells and macrophages with B-cells and plasma cells occurring at later time points (Deschl et al., 1990; Narayan et al., 1983b). During the entire investigation period macrophages and $CD4^+$ T cells were most prominent while $CD8^+$ T cells occurred in lesser number (Deschl et al., 1990). Typically, $CD8^+$ T cell distribution in the parenchyma and $CD4^+$ T cell accumulation primarily in perivascular cuffs was found as in naturally infected animals and humans (Bilzer et al., 1995; Bilzer and Stitz, 1994; Caplazi and Ehrensperger, 1998; Deschl et al., 1990; Liesche et al., 2019; Planz et al., 1993). This is different from the initial studies in the human VSBV-1 cases where $CD4^+$ and $CD8^+$ T cells were found in equal amounts (Tappe et al., 2019b). In BoDV-1-infected horses, the cellular infiltrates consist of macrophages, $CD4^+$ and $CD8^+$ T cells, and plasma cells with increased expression of MHC class I- and class II-antigens.

In further functional studies administration of the T cell inhibitor cyclosporine or dexamethasone before BoDV-1 infection was able to prevent clinical disease, but not viral persistence with high titers of infectious virus and virus-specific antibodies (reviewed in Herden et al., 2013; Stitz et al., 2002). This was also recently described for avian orthobornavirus infections (Hameed et al., 2018). On the other hand, BoDV-1 infection of immune incompetent, immunosuppressed or athymic rats did not result in clinical neurological disease, but virus persistence. Adoptive transfer of spleen cells from diseased rats, or of BoDV-1 specific $CD4^+$ T cell lines triggered classical disease and inflammation in the immune compromised infected animals (reviewed in Staeheli, 2002; Stitz et al., 2002). However, when $CD4^+$ T cells were applied before infection, only transient encephalitis without overt clinical disease and virus elimination might be induced with induction of virus-clearing by $CD8^+$ T cells (Nöske et al., 1998; Richt et al., 1994, reviewed in Stitz et al., 2002). Moreover, antibody treatment against $CD4^+$ and $CD8^+$ T cells led both to neuroprotection and abrogation of severe immunopathology (Bilzer and Stitz, 1994; Planz et al., 1993; Stitz et al., 1992) which fits with fact that virus-specific $CD4^+$ and $CD8^+$ T cells against the viral nucleoprotein N occur during infection (Richt et al., 1994, reviewed in Stitz et al., 2002).

$CD8^+$ T cells isolated from BoDV-1-infected brains of acutely diseased rats exhibit MHC I-restricted cytotoxic activity *in vitro* and could induce immunopathology *in vivo* which was not possible anymore when they were isolated in the chronic phase of the infection (Sobbe et al., 1997). This might be due to a downregulation of their avidity (Engelhardt et al., 2005). In addition to that, BoDV-1 infection does not cause cytopathic effects *in vitro* which furthermore led to the assumption of a primarily T cell-mediated pathology in immunocompetent adult rats.

B cells and virus-specific antibodies do not seem to play an important role in the immunopathogenesis of BoDV-1 infection since antibodies against N or P lack protective capacity as also found in human and avian species (Briese et al., 1995; Furrer et al., 2001; Heffels-Redmann et al., 2011; Niller et al., 2020). The role of neutralizing virus-specific antibodies has been discussed controversially. Adoptive transfer of immunoglobulins from BoDV-1 infected adult rats to immunocompromised recipients did not induce immunopathology but was able to restrict virus infection to the nervous tissue (Richt and Rott, 2001; Stitz et al., 1998).

Cytokines and other mediators either released by brain cells or by the inflammatory cells definitely play an essential role in the immunopathological process in adult Lewis rats. Thus, in dexamethasone treated rats, upregulation of TNF, interleukin (IL) 6, MIP1ß and the chemokine mob-1 mRNA was found and interpreted as cytokine response of the brain cells (Morimoto et al., 1996). After BoDV-1 infection of adult Lewis rats, a typical pro-inflammatory cytokine panel (TNF, IL-1α, IL-2, IL-6, IFN-ɣ) as well as $CD4^+$, $CD8^+$ and iNOS mRNA was detected (Hatalski et al., 1998; de la Torre, 2002; Shankar et al., 1992; Stitz et al., 1995; Zheng et al., 1993). In the later phases of the infection, correlating with the decrease of inflammation, an upregulation of IL-4, lower level of TNF accompanied by NK cells, B cells, activated microglia but continuous $CD8^+$ T cells and IFN-ɣ mRNA was noted and was thus interpreted as switch from a Th1- to a more Th2- based immune response (Hatalski et al., 1998). This fits with the increase of intrathecally produced IgG antibodies during the chronic disease phase and could indicate a role for $CD8^+$ T cells and IFN-ɣ for amelioration of the inflammation. The regulatory cytokine transforming growth factor (TGF) β was also upregulated and expression of the TGF-ß family signaling components were modulated during infection (Hatalski et al., 1998; Nishino et al., 2009; Plata-Salamán et al., 1999). Whether the anti-inflammatory cytokines IL-4 and TGF-β-marker originated from T cells, microglia or astrocytes has to be further investigated. Activation of T cells can also be triggered by many other factors, e.g., neurotransmitter such as glutamine/glutamate and acetylcholine or activated Th17 cells. An interference of BoDV-1 infection with the glutamate, cholinergic, dopaminergic, GABAergic system or dynorphine system has been described (Billaud et al., 2000; Brachthäuser et al., 2013; Gies et al., 1998, 2001; Nouri et al., 2014; Pacheco et al., 2010; Peng et al., 2008; Pikor et al., 2015; Scordel et al., 2015; Solbrig et al., 1996; Zhang et al., 2014; reviewed in Tizard et al., 2016a).

The upregulation of brain inflammation coincides with a long lasting activation of astrocytes and microglia even in the chronic phase of the disease (Carbone et al., 1991a; Deschl et al., 1990; Gosztonyi and Ludwig, 1995; Herden et al., 2000, 2005; Ovanesov et al., 2008a). Even in neonatally BoDV-1 infected rats, an upregulation of microglia occurs, most likely due to the virally induced cytokine upregulation in brain cells (Dietzschold et al., 2001; Weissenböck et al., 2000). Microglial activation precedes astroglia reaction but seems to depend on infection of neurons and activation of astrocytes (Herden et al., 2005; Ovanesov et al., 2006, 2008a). This could also contribute to neuronal loss via pro-inflammatory cytokines and reactive oxygen species. Since astrocytes upregulate chemokines early after infection this might also contribute to the recruitment of lymphocytes into the CNS (Ovanesov et al., 2006, 2008b; Sauder et al., 2000). The cytokine pattern in the acute phase fits also with a classical pro-inflammatory M1 polarization pattern of macrophages, microglia and recently also astrocytes (reviewed in Jha et al., 2016) whereas in the chronic phase the cytokine profile is compatible with an anti-inflammatory M2 polarization. There are presumably also manifold cytokine-neurotransmitter/neuropeptide interaction, e.g., TNF is able to induce glutamate production in activated microglia but inhibits glutamate uptake by astrocytes as found in BoDV-1-infected astrocytes

(Billaud et al., 2000; Takeuchi et al., 2006). However, the role of glutamate excitotoxicity in BoDV-1 infection for neuronal damage has been discussed controversially (Ovanesov et al., 2007; Richter et al., 2009).

So far, control of BoDV-1 infection by the immune system has been achieved by early adoptive transfer of virus-specific T cells or antigen specific priming (Hausmann et al., 2005b; Henkel et al., 2005; Lewis et al., 1999; Richt et al., 1994), but other treatment options especially in the light of the current human lethal BoDV-1 infections are definitely needed. For the immunopathology of BoDV-1 infection in adult rats, the delayed-type hypersensitivity response with $CD4^+$ T cell dependency and operative $CD8^+$ T cells mechanisms together with upregulation of cytokines and several other inflammatory mediators represents the key mechanism. However, both T cell types can exert protective effects but are also a trigger of the immunopathology. Several other factors than described here and potentially so far incompletely analyzed, like dysregulations of cellular functions, interference with signaling pathways or cellular proteins most likely contribute additionally to the immunopathological processes.

3.3.2 BoDV-1 infection of the neonatal rat

BoDV-1 infection of neonatal rats does not lead to overt immunopathology. Despite lifelong viral persistence, no overt inflammatory infiltrates in the brain, but disturbances of learning and memory, have been described (Hirano et al., 1983; Lancaster et al., 2007; Narayan et al., 1983a; Pletnikov et al., 1999, 2001; Rubin et al., 1999). Compared to normal littermates, rats infected with BDV neonatally show an altered circadian rhythm, have a smaller size, exhibit increased motor activity, decreased anxiety response, stereotyped behaviors and reduced play interactions (Bautista et al., 1994, 1995; Carbone et al., 1991b; Dittrich et al., 1989; Hornig et al., 1999; Pletnikov et al., 1999; Rubin et al., 1999). Therefore, also in the BoDV-1 infection of neonatal rats subtle clinical and histological changes occur which might be mediated by the virus infection itself as the early cytokine upregulation described before (Morimoto et al., 1996; Plata-Salamán et al., 1999). Despite significant astrogliosis and microglia activation, the overall CNS architecture of neonatally infected rats appears intact. However, apoptotic losses of granule cells of the dentate gyrus, Purkinje cell subsets, and cortical pyramidal neurons were detected that could be prevented by soluble factors (reviewed in Herden et al., 2013, see also Wu et al., 2013). This can explain the aforementioned behavioral changes.

In the CNS a brief increase of T cell infiltration and of pro-inflammatory cytokine, chemokine, and chemokine receptor transcripts expression was observed between 28 and 42 dpi (reviewed in Herden et al., 2013), accompanied by microglial proliferation and expression of MHC class I and class II, ICAM, CD4 and CD8 molecules (Weissenböck et al., 2000). B cells were also activated, modest serum antibody titers persist for more than 10 months (Carbone et al., 1991b). The relatively low level of T and B cell response without relationship to the histological lesions do not sufficiently explain the symptoms, indicating other virus host interactions for the apoptotic processes and their regulation (see virus host interactions, Hornig et al., 1999; Weissenböck et al., 2000; Williams et al., 2008). Further studies revealed an increase of pro-apoptotic factors (FAS, ICE-caspase-1) and a decrease of anti-apoptotic factors such as bcl-x on mRNA levels (Hornig et al., 1999). Moreover, excitotoxic stimulation, including activation of glutamatergic circuitry similar to the processes in adult rats was also assumed to trigger neuronal apoptosis.

In summary, neonatal BoDV-1-infected rats do not develop overt immunopathology most likely due to immune tolerance and CNS dysfunction may be linked to a variety of direct viral effects on morphogenesis, neurogenesis and induction of apoptotic processes (see virus host interactions).

3.3.3 BoDV-1 infection of mice

In contrast to the rat model, BoDV-1 infection of certain neonatal mice strains leads to neurological signs and typical non-purulent meningoencephalitis (Hallensleben et al., 1998) whereas adult mice of most strains do not develop clinical signs or inflammation in the CNS despite high virus titers (Kao et al., 1984; Rubin et al., 1993). This was explained by a strict neurotropism of the virus with antigen contact in the periphery at time points of immune competence in contrast to the immune tolerance in the neonatal rat (Hallensleben and Staeheli, 1999; Hausmann et al., 1999). Incidence and severity of clinical disease after BoDV-1 infections varies considerably between different mouse strains. In susceptible mouse strains, neonatal BoDV-1 infection results in clinical signs such as abnormal motility (abnormal hind limb postures up to paraparesis) in the later stages, associated with immune cell infiltration in the brain and glial activation, mainly in the cerebral cortex and hippocampus similar to the adult rat model (Hallensleben et al., 1998). The MHC H-2k allele in MRL strains is associated with severity of clinical disease, but the incidence of disease is most likely associated with other genetic factors (Hallensleben et al., 1998; Hausmann et al., 1999). In MRL mice, $CD4^+$ and $CD8^+$ T cells invade the brain and mainly $CD8^+$ T cells recognize the BoDV-1-N peptide $_{129}$TELEISSI$_{136}$ as a dominant epitope (Hausmann et al., 1999; Schamel et al., 2001). However, cytolysis does not occur and perforin did not affect virus distribution or development of clinical signs (Hausmann et al., 2001). This was substantiated by the observation that $_{129}$TELEISSI$_{136}$ specific $CD8^+$ T cells employ IFN-γ and not perforin for the control of the infection and elimination from neurons, and that $CD8^+$ T cells can recognize viral antigens on neurons and astrocytes via MHC I restriction (Baur et al., 2008; Hausmann et al., 2005a, b). Transgenic expression of BoDV-1-N in neurons or astrocytes in B10.BR mice did not result in clinical disease and the neuronal expression of N prevented induction of BoDV-1-N specific $CD8^+$ T cells (Rauer et al., 2004). Moreover, downregulation of the functional avidity of virus-specific $CD8^+$ T cells in experimentally infected mice were able to control the inflammatory reaction thereby facilitating viral persistence (Engelhardt et al., 2005). Similar as for the rat model, $CD8^+$ T cell were also able to exert protective effects if they were present before massive viral loads in the brain (Hausmann et al., 2005a). An upregulation of pro-inflammatory cytokines such TNF, IL-1, IL-6, IFN-γ and IL-10 but no elevation of Il-5 or TGF-β has been described (Kramer et al., 2012), which is comparable, except for TGF-β, with the rat model. Thus, an immunopathology mediated by MHC class I-restricted $CD8^+$ T cells became evident employing inflammatory mediators and cytokines (Hallensleben et al., 1998; Hausmann et al., 1999, 2001; Kramer et al., 2012; Richter et al., 2007; Schamel et al., 2001). Overexpression of cytokines such as IL-12 or TNF in less susceptible mouse strains could serve as trigger for the immune-mediated disorder and clinical disease such as epileptic seizures in case of TNF overexpression (Freude et al., 2002; Hofer et al., 2004; Kramer et al., 2012). In IL-12 transgenic mice, the IL-12 effect was mediated via IFN-γ. For IFN-γ neuroprotective effects were shown, it protected cells from BoDV-1 infection but did not eliminate the infection and prevented neuronal loss mediated by $CD8^+$ T cells when they recognized viral antigen-expressing astrocytes (Friedl et al., 2004; Hausmann et al., 2005b; Hofer et al., 2004; Richt and Stitz, 1992; Richter et al., 2009).

A behavioral syndrome similar to that of neonatally infected rats has been described after expression of high transgenic glial expression of BoDV-1-P in mice, which caused anxiety and impaired long-term memory. In these mice decreased synaptic density, serotonin receptors, and levels of BDNF result in behavioral disturbances (Kamitani et al., 2003). Recent studies indicate a major role for the phosphorylation of the viral P by PKC (see also virus host interaction, Bétourné et al., 2018).

In summary, in the BoDV-1 infected mouse model, other age dependencies as in the rat model are evident but an immunopathology mediated by antiviral MHC class I-restricted $CD8^+$ T cells complemented by upregulation of cytokines and other inflammatory mediators orchestrate as major players in clinical disease and inflammation in the CNS.

4. Avian orthobornaviruses

4.1 Epidemiology and host spectrum

Avian orthobornaviruses were first detected in the first decade of the 21th century as etiological agents of devastating proventricular dilatation disease (PDD) in birds (Honkavuori et al., 2008; Kistler et al., 2008). Avian orthobornaviruses represent a heterogenous group (see classification) that have been detected in more than 80 captive and free-ranging psittacine and non-psittacine species worldwide, including pet and wild birds, but also critically endangered species such as the Spix's macaw (*Cyanopsitta spixii*) (Kessler et al., 2020; Marton et al., 2015; Nielsen et al., 2018; Philadelpho et al., 2014; Sa-Ardta et al., 2019 and reviewed in Herden et al., 2013). Infections in non-psittacine species occur in songbirds (canaries, finches) and free-ranging waterfowl such as swans, geese or wild ducks (Delnatte et al., 2011, 2013; Guo et al., 2013; Hoppes et al., 2010; Payne et al., 2011a, 2011b; Rubbenstroth et al., 2013, 2014; Weissenböck et al., 2009b). However, parrot bornavirus 2 and 4 (PaBV-2 and 4; *Psittacine 1 orthobornaviruses*) have been detected most frequently (Rubbenstroth et al., 2016; Weissenböck et al., 2009a). Whether a virus reservoir as known for the mammalian bornaviruses exists and whether wild birds play a role in this scenario needs to be further investigated.

4.2 Clinical signs

Affected birds typically present with signs of progressive gastrointestinal tract dysfunction and/or neurological symptoms and virus detection was possible in 60–100% of PDD-affected birds investigated (reviewed in Boatright-Horowitz, 2020; Herden et al., 2013; Hoppes et al., 2013; Payne et al., 2012; Rossi et al., 2018; Tizard et al., 2016b). Clinical signs depend on host species and age and are variable, but often include weight loss up to cachexia, anorexia, lethargy, dysphagia, regurgitation and neurologic signs affecting sensorium and motility (depression, ataxia, tremor, seizures, and motor or proprioceptive deficits). Blindness has been described in a few cases (Steinmetz et al., 2008). The disease usually ends in death, comparable to accidental dead-end hosts of mammalian orthobornaviruses. Besides the classical clinical course, clinically inconspicuous infected animals, continuous or intermittent RNA shedding or sudden death might occur (reviewed in Herden et al., 2013). The latter might be associated with cardiac disease (Leal de Araujo et al., 2020). In some studies 15–40% healthy animals were found to be positive for PaBV (Heffels-Redmann et al., 2011 and reviewed in Herden et al., 2013; Hoppes and Shivaprasad, 2020; Rossi et al., 2018). Recently, a potential association with feather-damaging behavior in psittacines based on the presence of virus-specific antibodies was found (Fluck et al., 2019; Horie et al., 2012).

Surveillance studies in aviaries with PDD cases have shown that birds displaying high viral RNA loads and high PaBV-specific antibody titers are more prone to develop clinical manifest PDD (Heffels-Redmann et al., 2012). However, virus-specific antibodies do not have any protective effect as already confirmed for mammalian orthobornavirus infections (Heffels-Redmann et al., 2012). Moreover, the mechanisms for disease onset are still poorly understood.

4.3 Ways of entry, transmission, spread, shedding

To date, the natural route of infection and incubation period is unclear, but epidemiological data supports horizontal transmission (Rubbenstroth et al., 2016). Cockatiels and canaries have been used as experimental models of avian orthobornavirus infections. Canaries normally do not develop clinical signs, but they can be induced reliably in cockatiels. Experimentally, the incubation period ranged from 20 to 60 dpi up to 200 dpi which fits to the observation in some natural exposures (Gancz et al., 2009; Gray et al., 2010; Kistler et al., 2010; Leal de Araujo et al., 2017, 2019; Payne et al., 2011b; Piepenbring et al., 2012, 2016). Virus-specific serum antibodies were detected between seven and approximately 60 dpi in experimental studies. Viral RNA was found in swabs between 20 and 72 dpi, depending on the route of inoculation and the bird species used (reviewed in Herden et al., 2013). Virus was detected earliest after intracerebral infection in cockatiels. In naturally infected animals, viral RNA can be present in feces, cloacal and crop swabs which argue for bird-to-bird transmission and has also been observed in free ranging waterfowl (Delnatte et al., 2014b). Recent studies indicate skin lesions as one portal of entry since an oronasal route with or without additional mucosal trauma was not successful (Heckmann et al., 2017; Leal de Araujo et al., 2017; Runge et al., 2017). Experimentally, the intramuscular and subcutaneous route was effective, even with low virus doses (Runge et al., 2017). However, some animals did not get infected despite close contact with infected birds (Heffels-Redmann et al., 2012; Kistler et al., 2010; Raghav et al., 2010; Rubbenstroth et al., 2014).

Vertical transmission has been discussed due to the detection of avian orthobornaviral RNA and specific antibodies in embryos in one study, with virus restriction to the brain after egg infection, but productive infection has not been shown (Delnatte et al., 2014a; Kerski et al., 2012; Lierz et al., 2011; Monaco et al., 2012; Wüst, 2017).

Recently, spread of PaBV-2 after inoculation in the pectoral muscle was analyzed in cockatiels and revealed spread to the brain via the brachial plexus and spinal cord, accompanied by non-purulent inflammation (Leal de Araujo et al., 2017, 2019). Subsequent rostral to caudal distribution within the CNS followed as well as spread to the enteric nervous system with late occurrence also in mesenchymal and epithelial cells (Leal de Araujo et al., 2017, 2019).

Avian orthobornavirus infections can lead to a wide variation in the clinical outcome as correlation of clinical signs, gross and histological lesions, detection of viral RNA and antigen is quite variable. Thus, healthy animals with typical histological lesions at necropsy, clinically inconspicuous birds with widespread virus distribution or restriction to the nervous tissue and any variation thereof has been described (Delnatte et al., 2014b; Kloet and Dorrestein, 2009; Lierz et al., 2009; Piepenbring et al., 2012; Raghav et al., 2010; Villanueva et al., 2010; Wünschmann et al., 2011). Virus-specific antibodies also can be found variably in clinically healthy or sick animals (Fluck et al., 2019; Guo et al., 2014; Heffels-Redmann et al., 2011, 2012; Herzog et al., 2010; Kloet and Dorrestein, 2009; Kloet et al., 2011, reviewed in Hoppes et al., 2013). Asymptomatic carriers may contribute to virus dissemination. Recently, it was shown that chicks infected with PaBV-4 on the first day with of life did not develop clinical PDD despite virus persistence (Gartner et al., 2019).

4.4 Histopathology

Histologically, PDD is characterized by non-purulent inflammation and gliosis (Berhane et al., 2001; Gregory et al., 1996; Mannl et al., 1987;

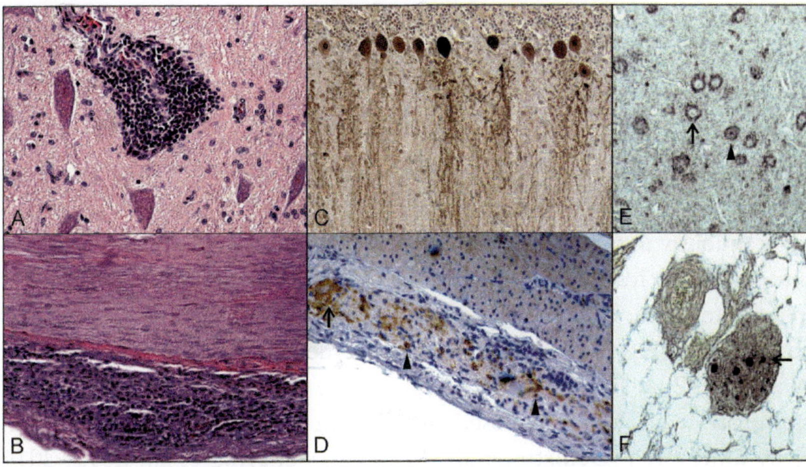

Fig. 4 Avian orthobornavirus infection. Overview on histological lesions and virus distribution in PaBV-4-infection of a cockatiel. (A and B) Infection with PaBV-4 is characterized by encephalitis with mononuclear perivascular infiltrates in the brain (A) and ganglionitis with mononuclear infiltrates in an enteric ganglion (B). (C and D) Detection of PaBV-4-P by immunohistochemistry in the cerebellum (C) and enteric ganglion (D) demonstrates the widespread virus distribution. Viral antigen is present in the nuclei, cytoplasm and processes of brain cells (C) as well as in nerve fibers (D, arrowheads) and enteric neurons (D, arrow). (E and F) Detection of PaBV-4 mRNA in the brain (E) and PaBV-4 genomic RNA in an epicardial ganglion (F) by in situ hybridization. Viral mRNA is present in the cytoplasm (E, arrow) and in some nuclei of brain cells (E, arrowhead) while genomic RNA is restricted to the neuronal nuclei (F, arrow).

Shivaprasad et al., 1995). Characteristic mononuclear infiltrates in ganglia and nerves of the enteric autonomous nervous system of proventriculus and ventriculus, but also of esophagus, crop and duodenum as well as non-purulent encephalitis, myelitis and/or ganglioradiculoneuritis of spinal nerves have been reported (Fig. 4A and B) (Berhane et al., 2001; Gregory et al., 1994; Keller et al., 2010; Kistler et al., 2010; Ouyang et al., 2009; Payne et al., 2011b; Piepenbring et al., 2012; Raghav et al., 2010; Wünschmann et al., 2011, reviewed in Herden et al., 2013; Hoppes et al., 2010). Involvement of the enteric nervous system causes gastrointestinal dysfunction, including paralytic dilatation of the esophagus and proventriculus. Similar inflammation in peripheral nerves, myocardium, cardiac conductive tissue, adrenal glands or eyes can be also present (Berhane et al., 2001; Shivaprasad et al., 1995). For proventricular dilatation, a role of the gut microbiota, dysfunction of the vagus nerve, other digestive processes or virally targeted interstitial cells of Cajal have been proposed (Tizard et al., 2016b).

4.5 Virus distribution

Avian orthobornaviruses depict a comparable neurotropism as the mammalian counterparts, but also exhibit affinity for the gastrointestinal tract and peripheral organs. The tissue tropism might depend at least in part on the respective virus species or even particular virus. Typically, a widespread tissue distribution is found, with viral RNA and/or antigen being present most frequently in the brain, spinal cord and gastrointestinal tract (crop, proventriculus, gizzard, small intestine), as well as in peripheral organs such as adrenal gland, eye, heart, liver, kidney, spleen, pancreas, lungs, gonads, thyroid and skin (Fig. 4C–F; reviewed in Herden et al., 2013; Hoppes and Shivaprasad, 2020; Payne et al., 2012). A role for the vagus nerve for viral dissemination has been proposed since major infected visceral organs are vagally innervated (Delnatte et al., 2014b; Piepenbring et al., 2012, 2016; Raghav et al., 2010 and reviewed in Rossi et al., 2018; Tizard et al., 2016b). Besides neuronal cells, virus can also be present in mesenchymal and epithelial cells (Al-Ibadi, 2015; Piepenbring et al., 2012, 2016; Raghav et al., 2010; Weissenböck et al., 2009a; Wünschmann et al., 2011) as seen in neonatally BoDV-1-infected rats and bornaviral reservoir species. Experimentally, differences in disease pattern were noted after experimental infection with either PaBV-2 or PaBV-4 in cockatiels, with slower infection kinetics, less viral RNA shedding and more variable tissue distribution in PaBV-2-infected animals (Al-Ibadi, 2015; Piepenbring et al., 2012, 2016; Runge et al., 2017). Differences in infectious dose, age of the animals or sensitivity of methods could have contributed to this observation as well (Runge et al., 2017).

4.6 Immunopathology

Whether and which immunopathological mechanisms play an important role as known for mammalian orthobornaviruses is not completely understood. Parallels in viral persistence and inflammatory lesion pattern might point to a comparable pathogenesis, but the variability of clinical course and virus dissemination is strikingly different. This might argue for involvement of further, so far unknown, factors that could be either host-related or virally based or a combination. Potential differences in the pathogenicity of the respective avian orthobornavirus as seen for PaBV-2 and PaBV-4, extent of contact to infected carriers as well as age, immune competence and bird species could have an impact on the course of disease.

Similar to mammalian orthobornaviruses, onset of clinical neurological symptoms seems to be correlated with the occurrence of $CD3^+$ T cells in the brain (Payne et al., 2012). Furthermore, lack of correlation with virus load and severity of clinical symptoms argued for an immunopathological process (reviewed in Rossi et al., 2018; Tizard et al., 2016b). Typically, inflammatory infiltrates consist of lymphocytes (further characterized as $CD8^+$ T cells), macrophages and variable number of plasma cells (Al-Ibadi, 2015; Piepenbring et al., 2012, 2016; Rossi et al., 2018) indicating a major role of the adaptive immune system. In contrast, asymptomatic birds depict lower numbers of $CD3^+$ lymphocytes with variable numbers of plasma cells and macrophages, as well as few $CD8^+$ cells.

Recent studies using cyclosporine A treatment did not show significant differences in the development of lesions, but considerably higher viral RNA loads in more organs were seen in treated animals after challenge with PaBV-2 (Hameed et al., 2018). However, when cyclosporine A was given previous to or simultaneously to the infection with PaBV-2, virus persistence but absence of clinical PDD was noted (Hameed et al., 2018).

Similar to mammalian orthobornaviruses, an interference with the type I IFN signaling was assumed since treatment with IFN-α was able to inhibit PaBV-2 and PaBV-4 *in vitro* (Reuter et al., 2010). Also similar to the mammalian counterparts, avian orthobornaviruses seem to have evolved strategies to circumvent the innate immune system recognition by RIG-1 due to trimming of viral genomic 5′-phosphates (see virus-host interaction, Habjan et al., 2008; Reuter et al., 2010; Schneider et al., 2005). Ribavirin efficiently reduced the load of PaBV-2 and PaBV-4 *in vitro*, induced type I IFN signaling and a synergistic effect of ribavirin and IFN-α against avian cells also was found (Musser et al., 2015; Reuter et al., 2016). Favipiravir (T-705), an antiviral comparable to ribavirin, has recently been applied in an *in vitro* study and high doses were able to reduce PaBV-4 RNA levels more effectively than ribavirin (Tokunaga et al., 2017).

Due to the assumption of a T cell-mediated immunopathogenesis comparable to mammalian orthobornaviruses, any vaccine strategy should aim to elicit a functional and protective T cell response without exacerbating the disease. For this purpose, viral vector vaccines have recently been designed and tested in various prime boost strategies (Olbert et al., 2016; Rall et al., 2019; Runge et al., 2017). Recombinant Newcastle disease virus (NDV) and modified vaccinia virus (MVA) for expressing viral nucleoprotein and phosphoprotein of either PaBV-4 or canary bornavirus (CnBv-2) (Olbert et al., 2016; Rall et al., 2019; Runge et al., 2017) have been used in cockatiels and canaries and revealed reduced viral shedding and viral tissue loads after homologous challenge (Olbert et al., 2016). However, clinical PDD and viral persistence was not prevented, but high viral doses were used. This changed by using the vector vaccines in combination or alone in a heterologous challenge regime. Interestingly, the combination of both vectors and the MVA/PaBV-4 vector alone provided protection against heterologous challenge with PaBV-2 in a lower dose in cockatiels (Rall et al., 2019; Runge et al., 2017). Whether this might be due to the boost regime, slower kinetic of PaBV-2, lower dose of infection or age of the animals needs to be further investigated. In addition, neither vector elicited obvious immunopathogenetic responses or disease severity in persistently infected birds, but did not reduce the viral load (Rall et al., 2019). This is different from similar studies in persistently BoDV-1-infected rats or mice where vaccination with viral N-expressing vectors caused immunopathology (Hausmann et al., 2005a). In another study, recombinant PaBV-1-N was given solely or in addition to inactivated virus to cockatiels and did not lead to reduced viral titers or protection against infection, but immunized animals developed reduced PDD typical lesions (Hameed et al., 2018). Therefore, it was speculated that their vaccine regime rather induced a Th2-dependent immune response, which was further supported by the effect of the cyclosporine A treatment (see above).

In summary, to date, further research detailing the impact of underlying immunopathology is definitely needed, since parallels, but also differences to mammalian orthobornavirus T cell-mediated immunopathology have been noted. To date, curative therapy is not yet available for avian orthobornavirus infections. However, recent vaccine approaches are promising to open the avenue for successful protection strategies in the future.

5. Novel bornaviruses

Apart from well-studied members of the orthobornaviruses, much less is known about other members of the family *Bornaviridae*, especially regarding recently discovered species.

Loveridge's garter snake virus 1 (LGSV-1) of the genus Elapid 1 orthobornavirus was discovered by a deep sequencing survey in a wild caught animal without any further description of potential pathology and pathogenesis. Additionally, fragments of another putative exogenous reptile bornavirus (Gaboon viper virus 1, GaVV-1) were detected in the transcriptome of a viper venome gland (Francischetti et al., 2004; Horie et al., 2010). Analysis of the available sequences and initial *in vitro* studies suggest similar cellular distribution of X and P proteins, but differences in their function compared with BoDV-1 and avian orthobornaviruses (Fujino et al., 2012).

Jungle carpet python virus (JCPV) and Southwest carpet python virus (SWCPV) of the genus Carbovirus were discovered by metagenomic and PCR screening in Australian pythons (Hyndman et al., 2018). Virus detection was variably associated with neurological signs such as progressive caudal paresis, flaccid paralysis, delayed righting reflex and incoordination. Histological lesions consisted of non-purulent encephalitis and ganglioneuritis. Detection of virus in animals without neurological signs or histopathological lesions is in accordance with the description of avian orthobornavirus infection in other species (reviewed in Hoppes et al., 2013; Payne et al., 2012; Tizard et al., 2016b). Analysis of JCPV and SWCPV genomes, their putative mRNAs and proteins, as well as comparison with Orthobornaviruses revealed remarkable differences and phylogenetically places carboviruses closer to EBL found in mammalian genomes. Further studies are needed to elucidate pathogenesis, host spectrum and putative immunopathology of carboviruses.

The Wuhan sharpbelly bornavirus (WhSBV) of the genus Cultervirus was discovered by metagenomic analysis only (Amarasinghe et al., 2019; Shi et al., 2018). The host species is assumed to be the ray-finned fish *Hemiculter leucisculus* (Shi et al., 2018). Genome organization of WhSBV suggests that WhSBV forms a distinct clade with the EBLs.

In the future, further new bornaviruses will likely be discovered as new methods and instruments are available, but whether they could indeed cause disease, and specifically immunopathology remains unknown.

6. Conclusions

Interestingly, the remarkable increase in novel bornaviruses and the zoonotic capacity strengthens the need for further detailed investigation on the underlying disease processes, especially potential immunopathology. In a rodent model the exclusive role of T cells has convincingly been confirmed. They exert their effects together with cytokines and other inflammatory mediators to target several cellular functions and signaling pathways to ensure host cell survival and viral persistence. In this respect, it seems noteworthy that T cells can exert protective effects but also represent the main players for the immunopathology and concurrent clinical disease and inflammation in the nervous system. Whether a similar T cell-mediated hypersensitivity reaction or induction of immune tolerance also account for other mammalian orthobornavirus infections remains to be solved. However, this serves as important groundwork to open the avenue for further antiviral strategies and potential therapeutic approaches that are currently not available.

Acknowledgments

This work was supported by grants from the German Federal Ministry of Education and research within the research consortium "ZooBoCo" (Grant No. 01KI1722E), and by "Graduiertenstipendium der Justus-Liebig-Universität Giessen". We thank Jana Petzold and Basim Al-Ibadi for support in the creation of the figures for squirrel and avian orthobornavirus infections, respectively. As a result of space limitations, some excellent studies could not be included in this review and main original work can be found in the respective cited reviews. We apologize to the colleagues of these works.

References

Ackermann, A., Staeheli, P., Schneider, U., 2007. Adaptation of Borna disease virus to new host species attributed to altered regulation of viral polymerase activity. J. Virol. 81 (15), 7933–7940.

Afonso, C.L., Amarasinghe, G.K., Banyai, K., Bao, Y., Basler, C.F., Bavari, S., Bejerman, N., Blasdell, K.R., Briand, F.-X., et al., 2016. Taxonomy of the order Mononegavirales: update 2016. Arch. Virol. 161 (8), 2351–2360.

Aguilar-Valles, A., Inoue, W., Rummel, C., Luheshi, G.N., 2015. Obesity, adipokines and neuroinflammation. Neuropharmacology 96 (Pt. A), 124–134.

Algermissen, D., 2010. Nachweis von Borna Disease Virus-spezifischen Proteinen und deren subgenomischer RNA bei natürlich infizierten Pferden. Diss. Tierärztliche Hochschule Hannover, Hannover.

Al-Ibadi, B.I.H., 2015. Avian Borna Virus in Psittacine Birds: Viral Distribution, Tropism and Immune Response. PhD thesis. Justus Liebig Universität Gießen, Gießen.

Amarasinghe, G.K., Aréchiga Ceballos, N.G., Banyard, A.C., Basler, C.F., Bavari, S., Bennett, A.J., Blasdell, K.R., Briese, T., Bukreyev, A., et al., 2018. Taxonomy of the order Mononegavirales: update 2018. Arch. Virol. 163 (8), 2283–2294.

Amarasinghe, G.K., Ayllón, M.A., Bào, Y., Basler, C.F., Bavari, S., Blasdell, K.R., Briese, T., Brown, P.A., Bukreyev, A., et al., 2019. Taxonomy of the order Mononegavirales: update 2019. Arch. Virol. 164 (7), 1967–1980.

Bajramovic, J.J., Münter, S., Syan, S., Nehrbass, U., Brahic, M., Gonzalez-Dunia, D., 2003. Borna disease virus glycoprotein is required for viral dissemination in neurons. J. Virol. 77 (22), 12222–12231.

Baur, K., Rauer, M., Richter, K., Pagenstecher, A., Götz, J., Hausmann, J., Staeheli, P., 2008. Antiviral CD8 T cells recognize Borna disease virus antigen transgenically expressed in either neurons or astrocytes. J. Virol. 82 (6), 3099–3108.

Bautista, J.R., Schwartz, G.J., La de Torre, J.C., Moran, T.H., Carbone, K.M., 1994. Early and persistent abnormalities in rats with neonatally acquired Borna disease virus infection. Brain Res. Bull. 34 (1), 31–40.

Bautista, J.R., Rubin, S.A., Moran, T.H., Schwartz, G.J., Carbone, K.M., 1995. Developmental injury to the cerebellum following perinatal Borna disease virus infection. Dev. Brain Res. 90 (1–2), 45–53.

Belyi, V.A., Levine, A.J., Skalka, A.M., 2010. Unexpected inheritance: multiple integrations of ancient bornavirus and ebolavirus/marburgvirus sequences in vertebrate genomes. PLoS Pathog. 6 (7), e1001030.

Berhane, Y., Smith, D.A., Newman, S., Taylor, M., Nagy, E., Binnington, B., Hunter, B., 2001. Peripheral neuritis in psittacine birds with proventricular dilatation disease. Avian Pathol. 30 (5), 563–570.

Bétourné, A., Szelechowski, M., Thouard, A., Abrial, E., Jean, A., Zaidi, F., Foret, C., Bonnaud, E.M., Charlier, C.M., et al., 2018. Hippocampal expression of a virus-derived protein impairs memory in mice. Proc. Natl. Acad. Sci. U. S. A. 115 (7), 1611–1616.

Biesenbach, W., Herzog, S., Frese, K., 1990. Peripheral neuropathy in experimental Borna disease virus infection in Lewis rats. Schweiz. Arch. Tierheilkd. 132 (8), 415.

Billaud, J.N., Ly, C., Phillips, T.R., La de Torre, J.C., 2000. Borna disease virus persistence causes inhibition of glutamate uptake by feline primary cortical astrocytes. J. Virol. 74 (22), 10438–10446.

Bilzer, T., Stitz, L., 1994. Immune-mediated brain atrophy. CD8+ T cells contribute to tissue destruction during Borna disease. J. Immunol. 153 (2), 818–823.

Bilzer, T., Planz, O., Lipkin, W.I., Stitz, L., 1995. Presence of CD4+ and CD8+ T cells and expression of MHC class I and MHC class II antigen in horses with Borna disease virus-induced encephalitis. Brain Pathol. 5 (3), 223–230.

Binz, T., Lebelt, J., Niemann, H., Hagenau, K., 1994. Sequence analyses of the p24 gene of Borna disease virus in naturally infected horse, donkey and sheep. Virus Res. 34 (3), 281–289.

Boatright-Horowitz, S.L., 2020. Avian Bornaviral Ganglioneuritis: current debates and unanswered questions. Vet. Med. Int. 2020, 6563723.

Bode, L., Dürrwald, R., Ludwig, H., 1994. Borna virus infections in cattle associated with fatal neurological disease. Vet. Rec. 135 (12), 283–284.

Bode, L., Dürrwald, R., Rantam, F.A., Ferszt, R., Ludwig, H., 1996. First isolates of infectious human Borna disease virus from patients with mood disorders. Mol. Psychiatry 1 (3), 200–212.

Bonnaud, E.M., Szelechowski, M., Bétourné, A., Foret, C., Thouard, A., Gonzalez-Dunia, D., Malnou, C.E., 2015. Borna disease virus phosphoprotein modulates epigenetic signaling in neurons to control viral replication. J. Virol. 89 (11), 5996–6008.

Bourg, M., Herzog, S., Encarnação, J.A., Nobach, D., Lange-Herbst, H., Eickmann, M., Herden, C., 2013. Bicolored white-toothed shrews as reservoir for borna disease virus, Bavaria, Germany. Emerg. Infect. Dis. 19 (12), 2064–2066.

Bourteele, S., Oesterle, K., Pleschka, S., Unterstab, G., Ehrhardt, C., Wolff, T., Ludwig, S., Planz, O., 2005. Constitutive activation of the transcription factor NF-kappaB results in impaired borna disease virus replication. J. Virol. 79 (10), 6043–6051.

Brachthäuser, L., Eisel, U., Herden, C., 2013. Borna disease virus infection of TNF-transgenic neurons: in vitro model for dysregulation of the anticonvulsive dynorphin system. Clin. Neuropathol. 32, 417–418.

Briese, T., de la Torre, J.C., Lewis, A., Ludwig, H., Lipkin, W.I., 1992. Borna disease virus, a negative-strand RNA virus, transcribes in the nucleus of infected cells. Proc. Natl. Acad. Sci. U. S. A. 89 (23), 11486–11489.

Briese, T., Schneemann, A., Lewis, A.J., Park, Y.S., Kim, S., Ludwig, H., Lipkin, W.I., 1994. Genomic organization of Borna disease virus. Proc. Natl. Acad. Sci. U. S. A. 91 (10), 4362–4366.

Briese, T., Hatalski, C.G., Kliche, S., Park, Y.S., Lipkin, W.I., 1995. Enzyme-linked immunosorbent assay for detecting antibodies to Borna disease virus-specific proteins. J. Clin. Microbiol. 33 (2), 348–351.

Brnic, D., Stevanovic, V., Cochet, M., Agier, C., Richardson, J., Montero-Menei, C.N., Milhavet, O., Eloit, M., Coulpier, M., 2012. Borna disease virus infects human neural progenitor cells and impairs neurogenesis. J. Virol. 86 (5), 2512–2522.

Calisher, C.H., Childs, J.E., Field, H.E., Holmes, K.V., Schountz, T., 2006. Bats: important reservoir hosts of emerging viruses. Clin. Microbiol. Rev. 19 (3), 531–545.

Caplazi, P., Ehrensperger, F., 1998. Spontaneous Borna disease in sheep and horses: immunophenotyping of inflammatory cells and detection of MHC-I and MHC-II antigen expression in Borna encephalitis lesions. Vet. Immunol. Immunopathol. 61 (2–4), 203–220.

Caplazi, P., Waldvogel, A., Stitz, L., Braun, U., Ehrensperger, F., 1994. Borna disease in naturally infected cattle. J. Comp. Pathol. 111 (1), 65–72.

Caplazi, P., Melzer, K., Goetzmann, R., Rohner-Cotti, A., Bracher, V., Zlinszky, K., Ehrensperger, F., 1999. Borna disease in Switzerland and in the principality of Liechtenstein. Schweiz. Arch. Tierheilkd. 141 (11), 521–527.

Carbone, K.M., Duchala, C.S., Griffin, J.W., Kincaid, A.L., Narayan, O., 1987. Pathogenesis of Borna disease in rats: evidence that intra-axonal spread is the major route for virus dissemination and the determinant for disease incubation. J. Virol. 61 (11), 3431–3440.

Carbone, K.M., Moench, T.R., Lipkin, W.I., 1991a. Borna disease virus replicates in astrocytes, Schwann cells and ependymal cells in persistently infected rats: location of viral genomic and messenger RNAs by in situ hybridization. J. Neuropathol. Exp. Neurol. 50 (3), 205–214.

Carbone, K.M., Park, S.W., Rubin, S.A., Waltrip, R.W., Vogelsang, G.B., 1991b. Borna disease: association with a maturation defect in the cellular immune response. J. Virol. 65 (11), 6154–6164.

Carbone, K.M., Rubin, S.A., Sierra-Honigmann, A.M., Lederman, H.M., 1993. Characterization of a glial cell line persistently infected with Borna disease virus (BDV): influence of neurotrophic factors on BDV protein and RNA expression. J. Virol. 67 (3), 1453–1460.

Charlier, C.M., Wu, Y.-J., Allart, S., Malnou, C.E., Schwemmle, M., Gonzalez-Dunia, D., 2013. Analysis of Borna disease virus trafficking in live infected cells by using a virus encoding a tetracysteine-tagged p protein. J. Virol. 87 (22), 12339–12348.

Chase, G., Mayer, D., Hildebrand, A., Frank, R., Hayashi, Y., Tomonaga, K., Schwemmle, M., 2007. Borna disease virus matrix protein is an integral component of the viral ribonucleoprotein complex that does not interfere with polymerase activity. J. Virol. 81 (2), 743–749.

Chevalier, G., Suberbielle, E., Monnet, C., Duplan, V., Martin-Blondel, G., Farrugia, F., Le Masson, G., Liblau, R., Gonzalez-Dunia, D., 2011. Neurons are MHC class I-dependent targets for CD8 T cells upon neurotropic viral infection. PLoS Pathog. 7 (11), e1002393.

Cizza, G., Sternberg, E.M., 1994. The role of the hypothalamic-pituitary-adrenal axis in susceptibility to autoimmune/inflammatory disease. Immunomethods 5 (1), 73–78.

Clemente, R., de la Torre, J.C., 2009. Cell entry of Borna disease virus follows a clathrin-mediated endocytosis pathway that requires Rab5 and microtubules. J. Virol. 83 (20), 10406–10416.

Coras, R., Korn, K., Kuerten, S., Huttner, H.B., Ensser, A., 2019. Severe bornavirus-encephalitis presenting as Guillain-Barré-syndrome. Acta Neuropathol. 137, 1017–1019.

Cubitt, B., de la Torre, J.C., 1997. Amantadine does not have antiviral activity against Borna disease virus. Arch. Virol. 142 (10), 2035–2042.

Cubitt, B., de la Torre, J.C., 1994. Borna disease virus (BDV), a nonsegmented RNA virus, replicates in the nuclei of infected cells where infectious BDV ribonucleoproteins are present. J. Virol. 68 (3), 1371–1381.

Cubitt, B., Oldstone, C., de la Torre, J.C., 1994. Sequence and genome organization of Borna disease virus. J. Virol. 68 (3), 1382–1396.

Cubitt, B., Ly, C., de la Torre, J.C., 2001. Identification and characterization of a new intron in Borna disease virus. J. Gen. Virol. 82 (Pt. 3), 641–646.

Cui, J., Wang, L.-F., 2015. Genomic mining reveals deep evolutionary relationships between Bornaviruses and bats. Viruses 7 (11), 5792–5800.

Cui, J., Zhao, W., Huang, Z., Jarvis, E.D., Gilbert, M.T.P., Walker, P.J., Holmes, E.C., Zhang, G., 2014. Low frequency of paleoviral infiltration across the avian phylogeny. Genome Biol. 15 (12), 539.

Danner, K., 1982. Borna-Virus und Borna-Infektionen: Vom Miasma zum Modell. Enke, Stuttgart.

De la Torre, J.C., 2002. Molecular biology of Borna disease virus and persistence. Front. Biosci. 7, d569–d579.

Delnatte, P., Berkvens, C., Kummrow, M., Smith, D.A., Campbell, D., Crawshaw, G., Ojkic, D., DeLay, J., 2011. New genotype of avian bornavirus in wild geese and trumpeter swans in Canada. Vet. Rec. 169 (4), 108.

Delnatte, P., Ojkic, D., DeLay, J., Campbell, D., Crawshaw, G., Smith, D.A., 2013. Pathology and diagnosis of avian bornavirus infection in wild Canada geese (Branta canadensis), trumpeter swans (Cygnus buccinator) and mute swans (Cygnus olor) in Canada: a retrospective study. Avian Pathol. 42 (2), 114–128.

Delnatte, P., Nagy, E., Ojkic, D., Crawshaw, G., Smith, D.A., 2014a. Investigation into the possibility of vertical transmission of avian bornavirus in free-ranging Canada geese (Branta canadensis). Avian Pathol. 43 (4), 301–304.

Delnatte, P., Nagy, É., Ojkic, D., Leishman, D., Crawshaw, G., Elias, K., Smith, D.A., 2014b. Avian bornavirus in free-ranging waterfowl: prevalence of antibodies and cloacal shedding of viral RNA. J. Wildl. Dis. 50 (3), 512–523.

Deschl, U., Stitz, L., Herzog, S., Frese, K., Rott, R., 1990. Determination of immune cells and expression of major histocompatibility complex class II antigen in encephalitic lesions of experimental Borna disease. Acta Neuropathol. 81 (1), 41–50.

Dietzel, J., Kuhrt, H., Stahl, T., Kacza, J., Seeger, J., Weber, M., Uhlig, A., Reichenbach, A., Grosche, A., et al., 2007. Morphometric analysis of the retina from horses infected with the Borna disease virus. Vet. Pathol. 44 (1), 57–63.

Dietzschold, B., Morimoto, K., Hooper, D.C., 2001. Mechanisms of virus-induced neuronal damage and the clearance of viruses from the CNS. Curr. Top. Microbiol. Immunol. 253, 145–155.

Dittrich, W., Bode, L., Ludwig, H., Kao, M., Schneider, K., 1989. Learning deficiencies in Borna disease virus-infected but clinically healthy rats. Biol. Psychiatry 26 (8), 818–828.

D'Mello, C., Le, T., Swain, M.G., 2009. Cerebral microglia recruit monocytes into the brain in response to tumor necrosis factor alpha signaling during peripheral organ inflammation. J. Neurosci. 29 (7), 2089–2102.

Dürrwald, R., Kolodziejek, J., Muluneh, A., Herzog, S., Nowotny, N., 2006. Epidemiological pattern of classical Borna disease and regional genetic clustering of Borna disease viruses point towards the existence of to-date unknown endemic reservoir host populations. Microbes Infect. 8 (3), 917–929.

Dürrwald, R., Kolodziejek, J., Herzog, S., Nowotny, N., 2007. Meta-analysis of putative human bornavirus sequences fails to provide evidence implicating Borna disease virus in mental illness. Rev. Med. Virol. 17 (3), 181–203.

Dürrwald, R., Kolodziejek, J., Weissenböck, H., Nowotny, N., 2014. The bicolored white-toothed shrew Crocidura leucodon (HERMANN 1780) is an indigenous host of mammalian Borna disease virus. PLoS One 9 (4), e93659.

Eickmann, M., Kiermayer, S., Kraus, I., Gössl, M., Richt, J.A., Garten, W., 2005. Maturation of Borna disease virus glycoprotein. FEBS Lett. 579 (21), 4751–4756.

Engelhardt, K.R., Richter, K., Baur, K., Staeheli, P., Hausmann, J., 2005. The functional avidity of virus-specific CD8+ T cells is down-modulated in Borna disease virus-induced immunopathology of the central nervous system. Eur. J. Immunol. 35 (2), 487–497.

Ferré, C.A., Davezac, N., Thouard, A., Peyrin, J.-M., Belenguer, P., Miquel, M.-C., Gonzalez-Dunia, D., Szelechowski, M., 2016. Manipulation of the N-terminal sequence of the Borna disease virus X protein improves its mitochondrial targeting and neuroprotective potential. FASEB J. 30 (4), 1523–1533.

Fluck, A., Enderlein, D., Piepenbring, A., Heffels-Redmann, U., Herzog, S., Pieper, K., Herden, C., Lierz, M., 2019. Correlation of avian bornavirus-specific antibodies and viral ribonucleic acid shedding with neurological signs and feather-damaging behaviour in psittacine birds. Vet. Rec. 184 (15), 476.

Francischetti, I.M.B., My-Pham, V., Harrison, J., Garfield, M.K., Ribeiro, J.M.C., 2004. Bitis gabonica (Gaboon viper) snake venom gland: toward a catalog for the full-length transcripts (cDNA) and proteins. Gene 337, 55–69.

Freude, S., Hausmann, J., Hofer, M., Pham-Mitchell, N., Campbell, I.L., Staeheli, P., Pagenstecher, A., 2002. Borna disease virus accelerates inflammation and disease associated with transgenic expression of interleukin-12 in the central nervous system. J. Virol. 76 (23), 12223–12232.

Friedl, G., Hofer, M., Auber, B., Sauder, C., Hausmann, J., Staeheli, P., Pagenstecher, A., 2004. Borna disease virus multiplication in mouse organotypic slice cultures is site-specifically inhibited by gamma interferon but not by interleukin-12. J. Virol. 78 (3), 1212–1218.

Fujino, K., Horie, M., Honda, T., Nakamura, S., Matsumoto, Y., Francischetti, I.M.B., Tomonaga, K., 2012. Evolutionarily conserved interaction between the phosphoproteins and X proteins of bornaviruses from different vertebrate species. PLoS One 7 (12), e51161.

Fujino, K., Horie, M., Honda, T., Merriman, D.K., Tomonaga, K., 2014. Inhibition of Borna disease virus replication by an endogenous bornavirus-like element in the ground squirrel genome. Proc. Natl. Acad. Sci. U. S. A. 111 (36), 13175–13180.

Furrer, E., Bilzer, T., Stitz, L., Planz, O., 2001. Neutralizing antibodies in persistent Borna disease virus infection: prophylactic effect of gp94-specific monoclonal antibodies in preventing encephalitis. J. Virol. 75 (2), 943–951.

Gancz, A.Y., Kistler, A.L., Greninger, A.L., Farnoushi, Y., Mechani, S., Perl, S., Berkowitz, A., Perez, N., Clubb, S., et al., 2009. Experimental induction of proventricular dilatation disease in cockatiels (Nymphicus hollandicus) inoculated with brain homogenates containing avian bornavirus 4. Virol. J. 6, 100.

Gartner, A.M., Enderlein, D., Link, J., Bücking, B., Herzog, S., Petzold, J., Malberg, S., Herden, C., Lierz, M., 2019. Age dependent development and clinical characteristics of parrot bornavirus-4 (PaBV-4) infection in cockatiels (Nymphicus hollandicus). In: Proceedings of the 4th International Conference on Avian, Herpetological and Exotic mammal medicine 2019, International Conference on Avian, Herpetological and Exotic mammal, London.

Gies, U., Bilzer, T., Stitz, L., Staiger, J.F., 1998. Disturbance of the cortical cholinergic innervation in Borna disease prior to encephalitis. Brain Pathol. 8 (1), 39–48.

Gies, U., Görcs, T.J., Mulder, J., Planz, O., Stitz, L., Bilzer, T., Luiten, P.G., Harkany, T., 2001. Cortical cholinergic decline parallels the progression of Borna virus encephalitis. Neuroreport 12 (17), 3767–3772.

Gilbert, C., Meik, J.M., Dashevsky, D., Card, D.C., Castoe, T.A., Schaack, S., 2014. Endogenous hepadnaviruses, bornaviruses and circoviruses in snakes. Proc. R. Soc. B Biol. Sci. 281, 20141122.

Girard, A., Sachidanandam, R., Hannon, G.J., Carmell, M.A., 2006. A germline-specific class of small RNAs binds mammalian Piwi proteins. Nature 442 (7099), 199–202.

Gonzalez-Dunia, D., Cubitt, B., de la Torre, J.C., 1998. Mechanism of Borna disease virus entry into cells. J. Virol. 72 (1), 783–788.

Gonzalez-Dunia, D., Volmer, R., Mayer, D., Schwemmle, M., 2005. Borna disease virus interference with neuronal plasticity. Virus Res. 111 (2), 224–234.

Gosztonyi, G., Ludwig, H., 1984. Borna disease of horses. An immunohistological and virological study of naturally infected animals. Acta Neuropathol. 64 (3), 213–221.

Gosztonyi, G., Ludwig, H., 1995. Borna disease—neuropathology and pathogenesis. Curr. Top. Microbiol. Immunol. 190, 39–73.

Gosztonyi, G., Dietzschold, B., Kao, M., Rupprecht, C.E., Ludwig, H., Koprowski, H., 1993. Rabies and Borna disease. A comparative pathogenetic study of two neurovirulent agents. Lab. Investig. 68 (3), 285–295.

Grabner, A., Fischer, A., 1991. Symptomatology and diagnosis of Borna encephalitis of horses. A case analysis of the last 13 years. Tierarztl. Prax. 19 (1), 68–73.

Gray, P., Hoppes, S., Suchodolski, P., Mirhosseini, N., Payne, S., Villanueva, I., Shivaprasad, H.L., Honkavuori, K.S., Lipkin, W.I., et al., 2010. Use of avian bornavirus isolates to induce proventricular dilatation disease in conures. Emerg. Infect. Dis. 16 (3), 473–479.

Gregory, C.R., Latimer, K.S., Niagro, F.D., Ritchie, B.W., Campagnoil, R.P., Norton, T.M., McManamon, R., Greenacre, C.B., 1994. A review of proventricular dilatation syndrome. J. Assoc. Avian Vet. 8, 69–75.

Gregory, C.R., Latimer, K.S., Campagnoli, R.P., Ritchie, B.W., 1996. Histologic evaluation of the crop for diagnosis of proventricular dilatation syndrome in psittacine birds. J. Vet. Diagn. Invest. 8 (1), 76–80.

Guo, J., Baroch, J., Randall, A., Tizard, I., 2013. Complete genome sequence of an avian Bornavirus isolated from a healthy Canadian Goose (Branta canadensis). Genome Announc. 1 (5), e00839-13.

Guo, J., Payne, S., Zhang, S., Turner, D., Tizard, I., Suchodolski, P., 2014. Avian bornaviruses: diagnosis, isolation, and genotyping. Curr. Protoc. Microbiol. 34, 15I.1.1–15I.1.33.

Habjan, M., Andersson, I., Klingström, J., Schümann, M., Martin, A., Zimmermann, P., Wagner, V., Pichlmair, A., Schneider, U., et al., 2008. Processing of genome 5' termini as a strategy of negative-strand RNA viruses to avoid RIG-I-dependent interferon induction. PLoS One 3 (4), e2032.

Hallensleben, W., Staeheli, P., 1999. Inhibition of Borna disease virus multiplication by interferon: cell line differences in susceptibility. Arch. Virol. 144 (6), 1209–1216.

Hallensleben, W., Zocher, M., Staeheli, P., 1997. Borna disease virus is not sensitive to amantadine. Arch. Virol. 142 (10), 2043–2048.

Hallensleben, W., Schwemmle, M., Hausmann, J., Stitz, L., Volk, B., Pagenstecher, A., Staeheli, P., 1998. Borna disease virus-induced neurological disorder in mice: infection of neonates results in immunopathology. J. Virol. 72 (5), 4379–4386.

Hameed, S.S., Guo, J., Tizard, I., Shivaprasad, H.L., Payne, S., 2018. Studies on immunity and immunopathogenesis of parrot bornaviral disease in cockatiels. Virology 515, 81–91.

Hans, A., Syan, S., Crosio, C., Sassone-Corsi, P., Brahic, M., Gonzalez-Dunia, D., 2001. Borna disease virus persistent infection activates mitogen-activated protein kinase and blocks neuronal differentiation of PC12 cells. J. Biol. Chem. 276 (10), 7258–7265.

Hans, A., Bajramovic, J.J., Syan, S., Perret, E., Dunia, I., Brahic, M., Gonzalez-Dunia, D., 2004. Persistent, noncytolytic infection of neurons by Borna disease virus interferes with ERK 1/2 signaling and abrogates BDNF-induced synaptogenesis. FASEB J. 18 (7), 863–865.

Hatalski, C.G., Hickey, W.F., Lipkin, W.I., 1998. Evolution of the immune response in the central nervous system following infection with Borna disease virus. J. Neuroimmunol. 90 (2), 137–142.

Hausmann, J., Hallensleben, W., de la Torre, J.C., Pagenstecher, A., Zimmermann, C., Pircher, H., Staeheli, P., 1999. T cell ignorance in mice to Borna disease virus can be overcome by peripheral expression of the viral nucleoprotein. Proc. Natl. Acad. Sci. U. S. A. 96 (17), 9769–9774.

Hausmann, J., Schamel, K., Staeheli, P., 2001. CD8(+) T lymphocytes mediate Borna disease virus-induced immunopathology independently of perforin. J. Virol. 75 (21), 10460–10466.

Hausmann, J., Baur, K., Engelhardt, K.R., Fischer, T., Rziha, H.-J., Staeheli, P., 2005a. Vaccine-induced protection against Borna disease in wild-type and perforin-deficient mice. J. Gen. Virol. 86 (Pt. 2), 399–403.

Hausmann, J., Pagenstecher, A., Baur, K., Richter, K., Rziha, H.-J., Staeheli, P., 2005b. CD8 T cells require gamma interferon to clear Borna disease virus from the brain and prevent immune system-mediated neuronal damage. J. Virol. 79 (21), 13509–13518.

He, M., An, T.-Z., Teng, C.-B., 2014. Evolution of mammalian and avian bornaviruses. Mol. Phylogenet. Evol. 79, 385–391.

He, P., Sun, L., Zhu, D., Zhang, H., Zhang, L., Guo, Y., Liu, S., Zhou, J., Xu, X., et al., 2016. Knock-down of endogenous bornavirus-like nucleoprotein 1 inhibits cell growth and induces apoptosis in human oligodendroglia cells. Int. J. Mol. Sci. 17 (4), 435.

Heckmann, J., Enderlein, D., Piepenbring, A.K., Herzog, S., Heffels-Redmann, U., Malberg, S., Herden, C., Lierz, M., 2017. Investigation of different infection routes of parrot bornavirus in cockatiels. Avian Dis. 61 (1), 90–95.

Heffels-Redmann, U., Enderlein, D., Herzog, S., Herden, C., Piepenbring, A., Neumann, D., Müller, H., Capelli, S., Müller, H., et al., 2011. Occurrence of avian bornavirus infection in captive psittacines in various European countries and its association with proventricular dilatation disease. Avian Pathol. 40 (4), 419–426.

Heffels-Redmann, U., Enderlein, D., Herzog, S., Piepenbring, A., Bürkle, M., Neumann, D., Herden, C., Lierz, M., 2012. Follow-up investigations on different courses of natural avian bornavirus infections in psittacines. Avian Dis. 56 (1), 153–159.

Heinig, A., 1969. Die Bornasche Krankheit der Pferde und Schafe. In: Röhrer, H. (Ed.), Handbuch der Virusinfektionen bei Tieren. VEB G. Fischer, Jena, p. 83.

Henkel, M., Planz, O., Fischer, T., Stitz, L., Rziha, H.-J., 2005. Prevention of virus persistence and protection against immunopathology after Borna disease virus infection of the brain by a novel Orf virus recombinant. J. Virol. 79 (1), 314–325.

Herden, C., Herzog, S., Wehner, T., Zink, C., Richt, J.A., Frese, K., 1999. Comparison of different methods of diagnosing Borna disease in horses post mortem. In: Wernery, U., Wade, J., Mumford, J.A., Kaaden, O.R. (Eds.), Equine Infectious Diseases VIII. R&W Publications, Newmarket, pp. 286–290.

Herden, C., Herzog, S., Richt, J.A., Nesseler, A., Christ, M., Failing, K., Frese, K., 2000. Distribution of Borna disease virus in the brain of rats infected with an obesity-inducing virus strain. Brain Pathol. 10 (1), 39–48.

Herden, C., Schluesener, H.J., Richt, J.A., 2005. Expression of allograft inflammatory factor-1 and haeme oxygenase-1 in brains of rats infected with the neurotropic Borna disease virus. Neuropathol. Appl. Neurobiol. 31 (5), 512–521.

Herden, C., Briese, T., Lipkin, W.I., Richt, J.A., 2013. Bornaviridae. In: Knipe, D.M., Howley, P.M. (Eds.), Fields Virology, sixth ed. Lippincott Williams & Wilkins, Philadelphia PA, pp. 1124–1150.

Herzog, S., Frese, K., Rott, R., 1991. Studies on the genetic control of resistance of black hooded rats to Borna disease. J. Gen. Virol. 72 (Pt. 3), 535–540.

Herzog, S., Pfeuffer, I., Haberzettl, K., Feldmann, H., Frese, K., Bechter, K., Richt, J.A., 1997. Molecular characterization of Borna disease virus from naturally infected animals and possible links to human disorders. Arch. Virol. Suppl. 13, 183–190.

Herzog, S., Enderlein, D., Heffels-Redmann, U., Piepenbring, A., Neumann, D., Kaleta, E.F., Müller, H., Lierz, M., Herden, C., 2010. Indirect immunofluorescence assay for intra vitam diagnosis of avian bornavirus infection in psittacine birds. J. Clin. Microbiol. 48 (6), 2282–2284.

Hilbe, M., Herrsche, R., Kolodziejek, J., Nowotny, N., Zlinszky, K., Ehrensperger, F., 2006. Shrews as reservoir hosts of Borna disease virus. Emerg. Infect. Dis. 12 (4), 675–677.

Hirano, N., Kao, M., Ludwig, H., 1983. Persistent, tolerant or subacute infection in Borna disease virus-infected rats. J. Gen. Virol. 64 (Pt. 7), 1521–1530.

Hock, M., Kraus, I., Schoehn, G., Jamin, M., Andrei-Selmer, C., Garten, W., Weissenhorn, W., 2010. RNA induced polymerization of the Borna disease virus nucleoprotein. Virology 397 (1), 64–72.

Hofer, M., Hausmann, J., Staeheli, P., Pagenstecher, A., 2004. Cerebral expression of interleukin-12 induces neurological disease via differential pathways and recruits antigen-specific T cells in virus-infected mice. Am. J. Pathol. 165 (3), 949–958.

Hoffmann, B., Tappe, D., Hoper, D., Herden, C., Boldt, A., Mawrin, C., Niederstrasser, O., Muller, T., Jenckel, M., et al., 2015. A variegated squirrel bornavirus associated with fatal human encephalitis. N. Engl. J. Med. 373 (2), 154–162.

Honda, T., 2017. Potential links between hepadnavirus and bornavirus sequences in the host genome and cancer. Front. Microbiol. 8, 2537.

Honda, T., Tomonaga, K., 2013. Nucleocytoplasmic shuttling of viral proteins in Borna disease virus infection. Viruses 5 (8), 1978–1990.

Honda, T., Tomonaga, K., 2016. Endogenous non-retroviral RNA virus elements evidence a novel type of antiviral immunity. Mob. Genet. Elem. 6 (3), e1165785.

Honda, T., Sofuku, K., Kojima, S., Yamamoto, Y., Ohtaki, N., Tomonaga, K., 2017. Linkage between the leader sequence and leader RNA production in Borna disease virus-infected cells. Virology 510, 104–110.

Honkavuori, K.S., Shivaprasad, H.L., Williams, B.L., Quan, P.L., Hornig, M., Street, C., Palacios, G., Hutchison, S.K., Franca, M., et al., 2008. Novel Borna virus in psittacine birds with proventricular dilatation disease. Emerg. Infect. Dis. 14 (12), 1883–1886.

Hoppes, S.M., Shivaprasad, H.L., 2020. Update on avian bornavirus and proventricular dilatation disease: diagnostics, pathology, prevalence, and control. Vet. Clin. North Am. Exot. Anim. Pract. 23 (2), 337–351.

Hoppes, S., Gray, P.L., Payne, S., Shivaprasad, H.L., Tizard, I., 2010. The isolation, pathogenesis, diagnosis, transmission, and control of avian bornavirus and proventricular dilatation disease. Vet. Clin. North Am. Exot. Anim. Pract. 13 (3), 495–508.

Hoppes, S.M., Tizard, I., Shivaprasad, H.L., 2013. Avian bornavirus and proventricular dilatation disease: diagnostics, pathology, prevalence, and control. Vet. Clin. North Am. Exot. Anim. Pract. 16 (2), 339–355.

Horie, M., 2017. The biological significance of bornavirus-derived genes in mammals. Curr. Opin. Virol. 25, 1–6.

Horie, M., Tomonaga, K., 2018. Paleovirology of bornaviruses: what can be learned from molecular fossils of bornaviruses. Virus Res. 262, 2–9.

Horie, M., Honda, T., Suzuki, Y., Kobayashi, Y., Daito, T., Oshida, T., Ikuta, K., Jern, P., Gojobori, T., et al., 2010. Endogenous non-retroviral RNA virus elements in mammalian genomes. Nature 463 (7277), 84–87.

Horie, M., Ueda, K., Ueda, A., Honda, T., Tomonaga, K., 2012. Detection of Avian bornavirus 5 RNA in Eclectus roratus with feather picking disorder. Microbiol. Immunol. 56 (5), 346–349.

Horie, M., Kobayashi, Y., Suzuki, Y., Tomonaga, K., 2013. Comprehensive analysis of endogenous bornavirus-like elements in eukaryote genomes. Philos. Trans. R. Soc. Lond. Ser. B Biol. Sci. 368 (1626), 20120499.

Horie, M., Kobayashi, Y., Honda, T., Fujino, K., Akasaka, T., Kohl, C., Wibbelt, G., Mühldorfer, K., Kurth, A., et al., 2016. An RNA-dependent RNA polymerase gene in bat genomes derived from an ancient negative-strand RNA virus. Sci. Rep. 6, 25873.

Hornig, M., Weissenböck, H., Horscroft, N., Lipkin, W.I., 1999. An infection-based model of neurodevelopmental damage. Proc. Natl. Acad. Sci. U. S. A. 96 (21), 12102–12107.

Hornig, M., Briese, T., Licinio, J., Khabbaz, R.F., Altshuler, L.L., Potkin, S.G., Schwemmle, M., Siemetzki, U., Mintz, J., et al., 2012. Absence of evidence for bornavirus infection in schizophrenia, bipolar disorder and major depressive disorder. Mol. Psychiatry 17 (5), 486–493.

Hyndman, T.H., Shilton, C.M., Stenglein, M.D., Wellehan, J.F.X., 2018. Divergent bornaviruses from Australian carpet pythons with neurological disease date the origin of extant Bornaviridae prior to the end-Cretaceous extinction. PLoS Pathog. 14 (2), e1006881.

Ibrahim, M.S., Watanabe, M., Palacios, J.A., Kamitani, W., Komoto, S., Kobayashi, T., Tomonaga, K., Ikuta, K., 2002. Varied persistent life cycles of Borna disease virus in a human oligodendroglioma cell line. J. Virol. 76 (8), 3873–3880.

Jamali, T., Jamali, Y., Mehrbod, M., Mofrad, M.R.K., 2011. Nuclear pore complex: biochemistry and biophysics of nucleocytoplasmic transport in health and disease. Int. Rev. Cell Mol. Biol. 287, 233–286.

Jha, M.K., Lee, W.-H., Suk, K., 2016. Functional polarization of neuroglia: implications in neuroinflammation and neurological disorders. Biochem. Pharmacol. 103, 1–16.

Joest, E., Degen, K., 1909. Über eigentümliche Kerneinschlüsse der Ganglienzellen bei der enzootischen Gehirn-Rückenmarksentzündung der Pferde. Z. Infkrankh. Haustiere 6, 348–356.

Jordan, I., Briese, T., Averett, D.R., Lipkin, W.I., 1999. Inhibition of Borna disease virus replication by ribavirin. J. Virol. 73 (9), 7903–7906.

Kamitani, W., Shoya, Y., Kobayashi, T., Watanabe, M., Lee, B.J., Zhang, G., Tomonaga, K., Ikuta, K., 2001. Borna disease virus phosphoprotein binds a neurite outgrowth factor, amphoterin/HMG-1. J. Virol. 75 (18), 8742–8751.

Kamitani, W., Ono, E., Yoshino, S., Kobayashi, T., Taharaguchi, S., Lee, B.-J., Yamashita, M., Kobayashi, T., Okamoto, M., et al., 2003. Glial expression of Borna disease virus phosphoprotein induces behavioral and neurological abnormalities in transgenic mice. Proc. Natl. Acad. Sci. U. S. A. 100 (15), 8969–8974.

Kao, M., Ludwig, H., Gosztonyi, G., 1984. Adaptation of Borna disease virus to the mouse. J. Gen. Virol. 65 (Pt. 10), 1845–1849.

Katzourakis, A., Gifford, R.J., 2010. Endogenous viral elements in animal genomes. PLoS Genet. 6 (11), e1001191.

Keller, D.L., Honkavuori, K.S., Briese, T., Lipkin, W.I., Muthuswamy, A., Steinberg, H., Sladky, K.K., 2010. Proventricular dilatation disease associated with Avian bornavirus in a scarlet macaw (Ara macao). J. Vet. Diagn. Investig. 22 (6), 961–965.

Kerski, A., de Kloet, A.H., de Kloet, S.R., 2012. Vertical transmission of avian bornavirus in Psittaciformes: avian bornavirus RNA and anti-avian bornavirus antibodies in eggs, embryos, and hatchlings obtained from infected sun conures (Aratinga solstitialis). Avian Dis. 56 (3), 471–478.

Kessler, S., Heenemann, K., Krause, T., Twietmeyer, S., Fuchs, J., Lierz, M., Corman, V.M., Vahlenkamp, T.M., Rubbenstroth, D., 2020. Monitoring of free-ranging and captive Psittacula populations in Western Europe for avian bornaviruses, circoviruses and polyomaviruses. Avian Pathol. 49 (2), 119–130.

Kiermayer, S., Kraus, I., Richt, J.A., Garten, W., Eickmann, M., 2002. Identification of the amino terminal subunit of the glycoprotein of Borna disease virus. FEBS Lett. 531 (2), 255–258.

Kistler, A.L., Gancz, A., Clubb, S., Skewes-Cox, P., Fischer, K., Sorber, K., Chiu, C.Y., Lublin, A., Mechani, S., et al., 2008. Recovery of divergent avian bornaviruses from cases of proventricular dilatation disease: identification of a candidate etiologic agent. Virol. J. 5 (1), 88.

Kistler, A.L., Smith, J.M., Greninger, A.L., de Risi, J.L., Ganem, D., 2010. Analysis of naturally occurring avian bornavirus infection and transmission during an outbreak of proventricular dilatation disease among captive psittacine birds. J. Virol. 84 (4), 2176–2179.

Kloet, S.R.d., Dorrestein, G.M., 2009. Presence of avian bornavirus RNA and anti-avian bornavirus antibodies in apparently healthy macaws. Avian Dis. 53 (4), 568–573.

Kloet, A.H.d., Kerski, A., Kloet, S.R.d., 2011. Diagnosis of Avian bornavirus infection in psittaciformes by serum antibody detection and reverse transcription polymerase chain reaction assay using feather calami. J. Vet. Diagn. Investig. 23 (3), 421–429.

Kobayashi, Y., Horie, M., Nakano, A., Murata, K., Itou, T., Suzuki, Y., 2016. Exaptation of bornavirus-like nucleoprotein elements in afrotherians. PLoS Pathog. 12 (8), e1005785.

Kojima, S., Sato, R., Yanai, M., Komatsu, Y., Horie, M., Igarashi, M., Tomonaga, K., 2019. Splicing-dependent subcellular targeting of Borna disease virus nucleoprotein isoforms. J. Virol. 93 (5), e01621-18.

Kolodziejek, J., Dürrwald, R., Herzog, S., Ehrensperger, F., Lussy, H., Nowotny, N., 2005. Genetic clustering of Borna disease virus natural animal isolates, laboratory and vaccine strains strongly reflects their regional geographical origin. J. Gen. Virol. 86 (Pt. 2), 385–398.

Korn, K., Coras, R., Bobinger, T., Herzog, S.M., Lücking, H., Stöhr, R., Huttner, H.B., Hartmann, A., Ensser, A., 2018. Fatal encephalitis associated with Borna disease virus 1. N. Engl. J. Med. 379 (14), 1375–1377.

Kramer, K., Schaudien, D., Eisel, U.L.M., Herzog, S., Richt, J.A., Baumgärtner, W., Herden, C., 2012. TNF-overexpression in Borna disease virus-infected mouse brains triggers inflammatory reaction and epileptic seizures. PLoS One 7 (7), e41476.

Kraus, I., Eickmann, M., Kiermayer, S., Scheffczik, H., Fluess, M., Richt, J.A., Garten, W., 2001. Open reading frame III of Borna disease virus encodes a nonglycosylated matrix protein. J. Virol. 75 (24), 12098–12104.

Kraus, I., Bogner, E., Lilie, H., Eickmann, M., Garten, W., 2005. Oligomerization and assembly of the matrix protein of Borna disease virus. FEBS Lett. 579 (12), 2686–2692.

Krey, H.F., Ludwig, H., Boschek, C.B., 1979. Multifocal retinopathy in Borna disease virus infected rabbits. Am. J. Ophthalmol. 87 (2), 157–164.

Kuhn, J.H., Dürrwald, R., Bào, Y., Briese, T., Carbone, K.M., Clawson, A.N., deRisi, J.L., Garten, W., Jahrling, P.B., et al., 2015. Taxonomic reorganization of the family Bornaviridae. Arch. Virol. 160 (2), 621–632.

Kupke, A., Becker, S., Wewetzer, K., Ahlemeyer, B., Eickmann, M., Herden, C., 2019. Intranasal Borna disease virus (BoDV-1) infection: insights into initial steps and potential contagiosity. Int. J. Mol. Sci. 20 (6), 1318.

Lancaster, K., Dietz, D.M., Moran, T.H., Pletnikov, M.V., 2007. Abnormal social behaviors in young and adult rats neonatally infected with Borna disease virus. Behav. Brain Res. 176 (1), 141–148.

Lau, N.C., Seto, A.G., Kim, J., Kuramochi-Miyagawa, S., Nakano, T., Bartel, D.P., Kingston, R.E., 2006. Characterization of the piRNA complex from rat testes. Science 313 (5785), 363–367.

Leal de Araujo, J., Rech, R.R., Heatley, J.J., Guo, J., Giaretta, P.R., Tizard, I., Rodrigues-Hoffmann, A., 2017. From nerves to brain to gastrointestinal tract: a time-based study of parrot bornavirus 2 (PaBV-2) pathogenesis in cockatiels (Nymphicus hollandicus). PLoS One 12 (11), e0187797.

Leal de Araujo, J., Rodrigues-Hoffmann, A., Giaretta, P.R., Guo, J., Heatley, J., Tizard, I., Rech, R.R., 2019. Distribution of viral antigen and inflammatory lesions in the central nervous system of cockatiels (Nymphicus hollandicus) experimentally infected with parrot bornavirus 2. Vet. Pathol. 56 (1), 106–117.

Leal de Araujo, J., Hameed, S.S., Tizard, I., Escandon, P., Giaretta, P.R., Heatley, J.J., Hoppes, S., Rech, R.R., 2020. Cardiac lesions of natural and experimental infection by parrot bornaviruses. J. Comp. Pathol. 174, 104–112.

Lebelt, J., Hagenau, K., 1996. Die Verteilung des Bornavirus in natürlich infizierten Tieren mit klinischer Erkrankung. Berl. Munch. Tierarztl. Wochenschr. 109 (5), 178–183.

Lennartz, F., Bayer, K., Czerwonka, N., Lu, Y., Kehr, K., Hirz, M., Steinmetzer, T., Garten, W., Herden, C., 2016. Surface glycoprotein of Borna disease virus mediates virus spread from cell to cell. Cell. Microbiol. 18 (3), 340–354.

Lewis, A.J., Whitton, J.L., Hatalski, C.G., Weissenböck, H., Lipkin, W.I., 1999. Effect of immune priming on Borna disease. J. Virol. 73 (3), 2541–2546.

Li, D., Lei, Y., Deng, J., Zhou, C., Zhang, Y., Li, W., Huang, H., Cheng, S., Zhang, H., et al., 2013. Human but not laboratory Borna disease virus inhibits proliferation and induces apoptosis in human oligodendrocytes in vitro. PLoS One 8 (6), e66623.

Lieb, K., Staeheli, P., 2001. Borna disease virus—does it infect humans and cause psychiatric disorders? J. Clin. Virol. 21 (2), 119–127.

Lierz, M., Hafez, H.M., Honkavuori, K.S., Gruber, A.D., Olias, P., Abdelwhab, E.M., Kohls, A., Lipkin, W.I., Briese, T., et al., 2009. Anatomical distribution of avian bornavirus in parrots, its occurrence in clinically healthy birds and ABV-antibody detection. Avian Pathol. 38 (6), 491–496.

Lierz, M., Piepenbring, A., Herden, C., Oberhäuser, K., Heffels-Redmann, U., Enderlein, D., 2011. Vertical transmission of avian bornavirus in psittacines. Emerg. Infect. Dis. 17 (12), 2390–2391.

Liesche, F., Ruf, V., Zoubaa, S., Kaletka, G., Rosati, M., Rubbenstroth, D., Herden, C., Goehring, L., Wunderlich, S., et al., 2019. The neuropathology of fatal encephalomyelitis in human Borna virus infection. Acta Neuropathol. 138 (4), 653–665.

Lin, C.-C., Wu, Y.-J., Heimrich, B., Schwemmle, M., 2013. Absence of a robust innate immune response in rat neurons facilitates persistent infection of Borna disease virus in neuronal tissue. Cell. Mol. Life Sci. 70 (22), 4399–4410.

Lipkin, W.I., Briese, T., Hornig, M., 2011. Borna disease virus—fact and fantasy. Virus Res. 162 (1–2), 162–172.

Liu, X., Zhao, L., Yang, Y., Bode, L., Huang, H., Liu, C., Huang, R., Zhang, L., Wang, X., et al., 2014. Human Borna disease virus infection impacts host proteome and histone lysine acetylation in human oligodendroglia cells. Virology 464–465, 196–205.

Ludwig, H., Kraft, W., Kao, M., Gosztonyi, G., Dahme, E., Krey, H., 1985. Borna virus infection (Borna disease) in naturally and experimentally infected animals: its significance for research and practice. Tierarztl. Prax. 13 (4), 421–453.

Ludwig, H., Bode, L., Gosztonyi, G., 1988. Borna disease: a persistent virus infection of the central nervous system. Prog. Med. Virol. 35, 107–151.

Ludwig, H., Furuya, K., Bode, L., Klein, N., Dürrwald, R., Lee, D.S., 1993. Biology and neurobiology of Borna disease viruses (BDV), defined by antibodies, neutralizability and their pathogenic potential. Arch. Virol. Suppl. 7, 111–133.

Mackenzie, J.S., Jeggo, M., 2013. Reservoirs and vectors of emerging viruses. Curr. Opin. Virol. 3 (2), 170–179.

Makino, A., Fujino, K., Parrish, N.F., Honda, T., Tomonaga, K., 2015. Borna disease virus possesses an NF-κB inhibitory sequence in the nucleoprotein gene. Sci. Rep. 5 (1), 8696.

Mandl, J.N., Ahmed, R., Barreiro, L.B., Daszak, P., Epstein, J.H., Virgin, H.W., Feinberg, M.B., 2015. Reservoir host immune responses to emerging zoonotic viruses. Cell 160 (1–2), 20–35.

Mannl, A., Gerlach, H., Leipold, R., 1987. Neuropathic gastric dilatation in psittaciformes. Avian Dis. 31 (1), 214–221.

Marton, S., Bányai, K., Gál, J., Ihász, K., Kugler, R., Lengyel, G., Jakab, F., Bakonyi, T., Farkas, S.L., 2015. Coding-complete sequencing classifies parrot bornavirus 5 into a novel virus species. Arch. Virol. 160 (11), 2763–2768.

Matsumoto, Y., Hayashi, Y., Omori, H., Honda, T., Daito, T., Horie, M., Ikuta, K., Fujino, K., Nakamura, S., et al., 2012. Bornavirus closely associates and segregates with host chromosomes to ensure persistent intranuclear infection. Cell Host Microbe 11 (5), 492–503.

Mayer, D., Baginsky, S., Schwemmle, M., 2005. Isolation of viral ribonucleoprotein complexes from infected cells by tandem affinity purification. Proteomics 5 (17), 4483–4487.

Metzler, A., Ehrensperger, F., Wyler, R., 1978. Natural borna virus infection in rabbits. J. Veterinary Med. Ser. B 25 (2), 161–164.

Mizutani, T., Inagaki, H., Araki, K., Kariwa, H., Arikawa, J., Takashima, I., 1998. Inhibition of Borna disease virus replication by ribavirin in persistently infected cells. Arch. Virol. 143 (10), 2039–2044.

Monaco, E., Hoppes, S., Guo, J., Tizard, I., 2012. The detection of avian bornavirus within psittacine eggs. J. Avian Med. Surg. 26 (3), 144–148.

Morales, J.A., Herzog, S., Kompter, C., Frese, K., Rott, R., 1988. Axonal transport of Borna disease virus along olfactory pathways in spontaneously and experimentally infected rats. Med. Microbiol. Immunol. 177 (2), 51–68.

Morimoto, K., Hooper, D.C., Bornhorst, A., Corisdeo, S., Bette, M., Fu, Z.F., Schäfer, M.K., Koprowski, H., Weihe, E., et al., 1996. Intrinsic responses to Borna disease virus infection of the central nervous system. Proc. Natl. Acad. Sci. U. S. A. 93 (23), 13345–13350.

Mukai, Y., Horie, M., Tomonaga, K., 2018. Systematic estimation of insertion dates of endogenous bornavirus-like elements in vesper bats. J. Vet. Med. Sci. 80 (8), 1356–1363.

Musser, J.M.B., Heatley, J.J., Koinis, A.V., Suchodolski, P.F., Guo, J., Escandon, P., Tizard, I.R., 2015. Ribavirin inhibits parrot bornavirus 4 replication in cell culture. PLoS One 10 (7), e0134080.

Myers, K.N., Barone, G., Ganesh, A., Staples, C.J., Howard, A.E., Beveridge, R.D., Maslen, S., Skehel, J.M., Collis, S.J., 2016. The bornavirus-derived human protein EBLN1 promotes efficient cell cycle transit, microtubule organisation and genome stability. Sci. Rep. 6, 35548.

Narayan, O., Herzog, S., Frese, K., Scheefers, H., Rott, R., 1983a. Behavioral disease in rats caused by immunopathological responses to persistent Borna virus in the brain. Science 220 (4604), 1401–1403.

Narayan, O., Herzog, S., Frese, K., Scheefers, H., Rott, R., 1983b. Pathogenesis of Borna disease in rats: immune-mediated viral ophthalmoencephalopathy causing blindness and behavioral abnormalities. J Infect Dis 148 (2), 305–315.

Nielsen, A.M.W., Ojkic, D., Dutton, C.J., Smith, D.A., 2018. Aquatic bird bornavirus 1 infection in a captive Emu (Dromaius novaehollandiae): presumed natural transmission from free-ranging wild waterfowl. Avian Pathol. 47 (1), 58–62.

Niller, H.H., Angstwurm, K., Rubbenstroth, D., Schlottau, K., Ebinger, A., Giese, S., Wunderlich, S., Banas, B., Forth, L.F., et al., 2020. Zoonotic spillover infections with Borna disease virus 1 leading to fatal human encephalitis, 1999–2019: an epidemiological investigation. Lancet Infect. Dis. 20 (4), 467–477.

Nishino, Y., Kobasa, D., Rubin, S.A., Pletnikov, M.V., Carbone, K.M., 2002. Enhanced neurovirulence of Borna disease virus variants associated with nucleotide changes in the glycoprotein and L polymerase genes. J. Virol. 76 (17), 8650–8658.

Nishino, Y., Ooishi, R., Kurokawa, S., Fujino, K., Murakami, M., Madarame, H., Hashimoto, O., Sugiyama, K., Funaba, M., 2009. Gene expression of the TGF-beta family in rat brain infected with Borna disease virus. Microbes Infect. 11 (8–9), 737–743.

Nitzschke, E., 1963. Untersuchungen über die experimentelle Bornavirus-Infektion bei der Ratte. Zentralbl. Veterinarmed. B 10 (6), 505–527.

Nobach, D., Herden, C., 2020. No evidence for European bats serving as reservoir for Borna disease virus 1 or other known mammalian orthobornaviruses. Virol. J. 17 (1), 11.

Nobach, D., Bourg, M., Herzog, S., Lange-Herbst, H., Encarnação, J.A., Eickmann, M., Herden, C., 2015. Shedding of infectious Borna disease virus-1 in living bicolored white-toothed shrews. PLoS One 10 (8), e0137018.

Nöske, K., Bilzer, T., Planz, O., Stitz, L., 1998. Virus-specific CD4+ T cells eliminate Borna disease virus from the brain via induction of cytotoxic CD8+ T cells. J. Virol. 72 (5), 4387–4395.

Nouri, M., Bredberg, A., Weström, B., Lavasani, S., 2014. Intestinal barrier dysfunction develops at the onset of experimental autoimmune encephalomyelitis, and can be induced by adoptive transfer of auto-reactive T cells. PLoS One 9 (9), e106335.

Nowotny, N., Kolodziejek, J., Jehle, C.O., Suchy, A., Staeheli, P., Schwemmle, M., 2000. Isolation and characterization of a new subtype of Borna disease virus. J. Virol. 74 (12), 5655–5658.

Okayama, S., Miura, N., Murakami, M., Funaba, M., Nishino, Y., 2009. Changes in Borna disease virus genome with adaptation to host. Microbes Infect. 11 (6–7), 721–724.

Olbert, M., Römer-Oberdörfer, A., Herden, C., Malberg, S., Runge, S., Staeheli, P., Rubbenstroth, D., 2016. Viral vector vaccines expressing nucleoprotein and phosphoprotein genes of avian bornaviruses ameliorate homologous challenge infections in cockatiels and common canaries. Sci. Rep. 6, 36840.

Otta, J., Jentzsch, K.D., 1960. Spontane Infektion mit dem virus der Bornaschen Krankheit bei Kaninchen. Mhefte. Vet. Med 15, 127.

Ouyang, N., Storts, R., Tian, Y., Wigle, W., Villanueva, I., Mirhosseini, N., Payne, S., Gray, P., Tizard, I., 2009. Histopathology and the detection of avian bornavirus in the nervous system of birds diagnosed with proventricular dilatation disease. Avian Pathol. 38 (5), 393–401.

Ovanesov, M.V., Sauder, C., Rubin, S.A., Richt, J., Nath, A., Carbone, K.M., Pletnikov, M.V., 2006. Activation of microglia by Borna disease virus infection: in vitro study. J. Virol. 80 (24), 12141–12148.

Ovanesov, M.V., Vogel, M.W., Moran, T.H., Pletnikov, M.V., 2007. Neonatal Borna disease virus infection in rats is associated with increased extracellular levels of glutamate and neurodegeneration in the striatum. J. Neurovirol. 13 (3), 185–194.

Ovanesov, M.V., Ayhan, Y., Wolbert, C., Moldovan, K., Sauder, C., Pletnikov, M.V., 2008a. Astrocytes play a key role in activation of microglia by persistent Borna disease virus infection. J. Neuroinflammation 5 (1), 50.

Ovanesov, M.V., Moldovan, K., Smith, K., Vogel, M.W., Pletnikov, M.V., 2008b. Persistent Borna disease virus (BDV) infection activates microglia prior to a detectable loss of granule cells in the hippocampus. J. Neuroinflammation 5 (1), 16.

Pacheco, R., Riquelme, E., Kalergis, A.M., 2010. Emerging evidence for the role of neurotransmitters in the modulation of T cell responses to cognate ligands. Cent. Nerv. Syst. Agents Med. Chem. 10 (1), 65–83.

Parrish, N.F., Fujino, K., Shiromoto, Y., Iwasaki, Y.W., Ha, H., Xing, J., Makino, A., Kuramochi-Miyagawa, S., Nakano, T., et al., 2015. piRNAs derived from ancient viral processed pseudogenes as transgenerational sequence-specific immune memory in mammals. RNA 21 (10), 1691–1703.

Payne, S., Covaleda, L., Jianhua, G., Swafford, S., Baroch, J., Ferro, P.J., Lupiani, B., Heatley, J., Tizard, I., 2011a. Detection and characterization of a distinct bornavirus lineage from healthy Canada geese (Branta canadensis). J. Virol. 85 (22), 12053–12056.

Payne, S., Shivaprasad, H.L., Mirhosseini, N., Gray, P., Hoppes, S., Weissenböck, H., Tizard, I., 2011b. Unusual and severe lesions of proventricular dilatation disease in cockatiels (Nymphicus hollandicus) acting as healthy carriers of avian bornavirus (ABV) and subsequently infected with a virulent strain of ABV. Avian Pathol. 40 (1), 15–22.

Payne, S.L., Delnatte, P., Guo, J., Heatley, J.J., Tizard, I., Smith, D.A., 2012. Birds and bornaviruses. Anim. Health Res. Rev. 13 (2), 145–156.

Peng, G., Zhang, F., Zhang, Q., Wu, K., Zhu, F., Wu, J., 2007. Borna disease virus P protein inhibits nitric oxide synthase gene expression in astrocytes. Virology 366 (2), 446–452.

Peng, G., Yan, Y., Zhu, C., Wang, S., Yan, X., Lu, L., Li, W., Hu, J., Wei, W., et al., 2008. Borna disease virus P protein affects neural transmission through interactions with gamma-aminobutyric acid receptor-associated protein. J. Virol. 82 (24), 12487–12497.

Perez, M., de la Torre, J.C., 2005. Identification of the Borna disease virus (BDV) proteins required for the formation of BDV-like particles. J. Gen. Virol. 86 (Pt. 7), 1891–1895.

Perez, M., Watanabe, M., Whitt, M.A., de la Torre, J.C., 2001. N-terminal domain of Borna disease virus G (p56) protein is sufficient for virus receptor recognition and cell entry. J. Virol. 75 (15), 7078–7085.

Perez, M., Sanchez, A., Cubitt, B., Rosario, D., de la Torre, J.C., 2003. A reverse genetics system for Borna disease virus. J. Gen. Virol. 84 (Pt. 11), 3099–3104.

Petzold, J., van den Brand, J.M.A., Nobach, D., Hoffmann, B., Hoffmann, D., Fast, C., Reusken, C.B.E.M., van Run, P.R.W.A., Schlottau, K., et al., 2019. Distribution of zoonotic variegated squirrel bornavirus 1 in naturally infected variegated and Prevost's squirrels. Sci. Rep. 9 (1), 11402.

Philadelpho, N.A., Rubbenstroth, D., Guimarães, M.B., Piantino Ferreira, A.J., 2014. Survey of bornaviruses in pet psittacines in Brazil reveals a novel parrot bornavirus. Vet. Microbiol. 174 (3–4), 584–590.

Piepenbring, A.K., Enderlein, D., Herzog, S., Kaleta, E.F., Heffels-Redmann, U., Ressmeyer, S., Herden, C., Lierz, M., 2012. Pathogenesis of avian bornavirus in experimentally infected cockatiels. Emerg. Infect. Dis. 18 (2), 234–241.

Piepenbring, A.K., Enderlein, D., Herzog, S., Al-Ibadi, B.I.H., Heffels-Redmann, U., Heckmann, J., Lange-Herbst, H., Herden, C., Lierz, M., 2016. Parrot Bornavirus (PaBV)-2 isolate causes different disease patterns in cockatiels than PaBV-4. Avian Pathol. 45 (2), 156–168.

Pikor, N.B., Astarita, J.L., Summers-Deluca, L., Galicia, G., Qu, J., Ward, L.A., Armstrong, S., Dominguez, C.X., Malhotra, D., et al., 2015. Integration of Th17- and lymphotoxin-derived signals initiates meningeal-resident stromal cell remodeling to propagate neuroinflammation. Immunity 43 (6), 1160–1173.

Planz, O., Bilzer, T., Sobbe, M., Stitz, L., 1993. Lysis of major histocompatibility complex class I-bearing cells in Borna disease virus-induced degenerative encephalopathy. J. Exp. Med. 178 (1), 163–174.

Planz, O., Pleschka, S., Ludwig, S., 2001. MEK-specific inhibitor U0126 blocks spread of Borna disease virus in cultured cells. J. Virol. 75 (10), 4871–4877.

Planz, O., Pleschka, S., Wolff, T., 2009. Borna disease virus: a unique pathogen and its interaction with intracellular signalling pathways. Cell. Microbiol. 11 (6), 872–879.

Plata-Salamán, C.R., Ilyin, S.E., Gayle, D., Romanovitch, A., Carbone, K.M., 1999. Persistent Borna disease virus infection of neonatal rats causes brain regional changes of mRNAs for cytokines, cytokine receptor components and neuropeptides. Brain Res. Bull. 49 (6), 441–451.

Pleschka, S., 2008. RNA viruses and the mitogenic Raf/MEK/ERK signal transduction cascade. Biol. Chem. 389 (10), 1273–1282.

Pleschka, S., Staeheli, P., Kolodziejek, J., Richt, J.A., Nowotny, N., Schwemmle, M., 2001. Conservation of coding potential and terminal sequences in four different isolates of Borna disease virus. J. Gen. Virol. 82 (Pt. 11), 2681–2690.

Pletnikov, M.V., Rubin, S.A., Vasudevan, K., Moran, T.H., Carbone, K.M., 1999. Developmental brain injury associated with abnormal play behavior in neonatally Borna disease virus-infected Lewis rats: a model of autism. Behav. Brain Res. 100 (2), 43–50.

Pletnikov, M.V., Rubin, S.A., Carbone, K.M., Moran, T.H., Schwartz, G.J., 2001. Neonatal Borna disease virus infection (BDV)-induced damage to the cerebellum is associated with sensorimotor deficits in developing Lewis rats. Brain Res. Dev. Brain Res. 126 (1), 1–12.

Pletnikov, M.V., Rubin, S.A., Vogel, M.W., Moran, T.H., Carbone, K.M., 2002. Effects of genetic background on neonatal Borna disease virus infection-induced neurodevelopmental damage. Brain Res. 944 (1–2), 97–107.

Poenisch, M., Unterstab, G., Wolff, T., Staeheli, P., Schneider, U., 2004. The X protein of Borna disease virus regulates viral polymerase activity through interaction with the P protein. J. Gen. Virol. 85 (Pt. 7), 1895–1898.

Poenisch, M., Staeheli, P., Schneider, U., 2008. Viral accessory protein X stimulates the assembly of functional Borna disease virus polymerase complexes. J. Gen. Virol. 89 (Pt. 6), 1442–1445.

Poenisch, M., Burger, N., Staeheli, P., Bauer, G., Schneider, U., 2009. Protein X of Borna disease virus inhibits apoptosis and promotes viral persistence in the central nervous systems of newborn-infected rats. J. Virol. 83 (9), 4297–4307.

Prat, C.M.A., Schmid, S., Farrugia, F., Cenac, N., Le Masson, G., Schwemmle, M., Gonzalez-Dunia, D., 2009. Mutation of the protein kinase C site in Borna disease virus phosphoprotein abrogates viral interference with neuronal signaling and restores normal synaptic activity. PLoS Pathog. 5 (5), e1000425.

Puorger, M.E., Hilbe, M., Müller, J.-P., Kolodziejek, J., Nowotny, N., Zlinszky, K., Ehrensperger, F., 2010. Distribution of Borna disease virus antigen and RNA in tissues of naturally infected bicolored white-toothed shrews, Crocidura leucodon, supporting their role as reservoir host species. Vet. Pathol. 47 (2), 236–244.

Qian, J., Zhai, A., Kao, W., Li, Y., Song, W., Fu, Y., Chen, X., Zhang, Q., Wu, J., et al., 2010. Modulation of miR-122 on persistently Borna disease virus infected human oligodendroglial cells. Antiviral Res. 87 (2), 249–256.

Raghav, R., Taylor, M., DeLay, J., Ojkic, D., Pearl, D.L., Kistler, A.L., deRisi, J.L., Ganem, D., Smith, D.A., 2010. Avian bornavirus is present in many tissues of psittacine birds with histopathologic evidence of proventricular dilatation disease. J. Vet. Diagn. Investig. 22 (4), 495–508.

Rall, I., Amann, R., Malberg, S., Herden, C., Rubbenstroth, D., 2019. Recombinant modified Vaccinia Virus Ankara (MVA) vaccines efficiently protect cockatiels against parrot bornavirus infection and proventricular dilatation disease. Viruses 11 (12), 1130.

Randall, R.E., Goodbourn, S., 2008. Interferons and viruses: an interplay between induction, signalling, antiviral responses and virus countermeasures. J. Gen. Virol. 89 (Pt. 1), 1–47.

Rauer, M., Götz, J., Schuppli, D., Staeheli, P., Hausmann, J., 2004. Transgenic mice expressing the nucleoprotein of Borna disease virus in either neurons or astrocytes: decreased susceptibility to homotypic infection and disease. J. Virol. 78 (7), 3621–3632.

Reuter, A., Ackermann, A., Kothlow, S., Rinder, M., Kaspers, B., Staeheli, P., 2010. Avian bornaviruses escape recognition by the innate immune system. Viruses 2 (4), 927–938.

Reuter, A., Horie, M., Höper, D., Ohnemus, A., Narr, A., Rinder, M., Beer, M., Staeheli, P., Rubbenstroth, D., 2016. Synergistic antiviral activity of ribavirin and interferon-α against parrot bornaviruses in avian cells. J. Gen. Virol. 97 (9), 2096–2103.

Richt, J.A., Rott, R., 2001. Borna disease virus: a mystery as an emerging zoonotic pathogen. Vet. J. 161 (1), 24–40.

Richt, J.A., Stitz, L., 1992. Borna disease virus-infected astrocytes function in vitro as antigen-presenting and target cells for virus-specific CD4-bearing lymphocytes. Arch. Virol. 124 (1–2), 95–109.

Richt, J.A., Schmeel, A., Frese, K., Carbone, K.M., Narayan, O., Rott, R., 1994. Borna disease virus-specific T cells protect against or cause immunopathological Borna disease. J. Exp. Med. 179 (5), 1467–1473.

Richt, J.A., Alexander, R.C., Herzog, S., Hooper, D.C., Kean, R., Spitsin, S., Bechter, K., Schüttler, R., Feldmann, H., et al., 1997. Failure to detect Borna disease virus infection in peripheral blood leukocytes from humans with psychiatric disorders. J. Neurovirol. 3 (2), 174–178.

Richt, J.A., Fürbringer, T., Koch, A., Pfeuffer, I., Herden, C., Bause-Niedrig, I., Garten, W., 1998. Processing of the Borna disease virus glycoprotein gp94 by the subtilisin-like endoprotease furin. J. Virol. 72 (5), 4528–4533.

Richt, J.A., Grabner, A., Herzog, S., Garten, W., Herden, C., Sellon, D.C., 2007. Chapter 22—Borna disease. In: Long, M.T. (Ed.), Equine Infectious Diseases. W.B. Saunders, Saint Louis, pp. 207–213.

Richter, K., Baur, K., Ackermann, A., Schneider, U., Hausmann, J., Staeheli, P., 2007. Pathogenic potential of Borna disease virus lacking the immunodominant CD8 T-cell epitope. J. Virol. 81 (20), 11187–11194.

Richter, K., Hausmann, J., Staeheli, P., 2009. Interferon-gamma prevents death of bystander neurons during CD8 T cell responses in the brain. Am. J. Pathol. 174 (5), 1799–1807.

Rosario, D., Perez, M., de la Torre, J.C., 2005. Functional characterization of the genomic promoter of Borna disease virus (BDV): implications of 3'-terminal sequence heterogeneity for BDV persistence. J. Virol. 79 (10), 6544–6550.

Rossi, G., Dahlhausen, R.D., Galosi, L., Orosz, S.E., 2018. Avian ganglioneuritis in clinical practice. Vet. Clin. North Am. Exot. Anim. Pract. 21 (1), 33–67.

Rott, R., Becht, H., 1995. Natural and experimental Borna disease in animals. Curr. Top. Microbiol. Immunol. 190, 17–30.

Rott, R., Herzog, S., Fleischer, B., Winokur, A., Amsterdam, J., Dyson, W., Koprowski, H., 1985. Detection of serum antibodies to Borna disease virus in patients with psychiatric disorders. Science 228 (4700), 755–756.

Rubbenstroth, D., Rinder, M., Stein, M., Höper, D., Kaspers, B., Brosinski, K., Horie, M., Schmidt, V., Legler, M., et al., 2013. Avian bornaviruses are widely distributed in canary birds (Serinus canaria f. domestica). Vet. Microbiol. 165 (3–4), 287–295.

Rubbenstroth, D., Schmidt, V., Rinder, M., Legler, M., Corman, V.M., Staeheli, P., 2014. Discovery of a new avian bornavirus genotype in estrildid finches (Estrildidae) in Germany. Vet. Microbiol. 168 (2–4), 318–323.

Rubbenstroth, D., Schmidt, V., Rinder, M., Legler, M., Twietmeyer, S., Schwemmer, P., Corman, V.M., 2016. Phylogenetic analysis supports horizontal transmission as a driving force of the spread of avian bornaviruses. PLoS One 11 (8), e0160936.

Rubbenstroth, D., Schlottau, K., Schwemmle, M., Rissland, J., Beer, M., 2019. Human bornavirus research: back on track!. PLoS Pathog. 15 (8), e1007873.

Rubin, S.A., Waltrip, R.W., Bautista, J.R., Carbone, K.M., 1993. Borna disease virus in mice: host-specific differences in disease expression. J. Virol. 67 (1), 548–552.

Rubin, S.A., Sylves, P., Vogel, M., PLETNIKOV, M., Moran, T.H., Schwartz, G.J., Carbone, K.M., 1999. Borna disease virus-induced hippocampal dentate gyrus damage is associated with spatial learning and memory deficits. Brain Res. Bull. 48 (1), 23–30.

Rudolph, M.G., Kraus, I., Dickmanns, A., Eickmann, M., Garten, W., Ficner, R., 2003. Crystal structure of the Borna disease virus nucleoprotein. Structure 11 (10), 1219–1226.

Runge, S., Olbert, M., Herden, C., Malberg, S., Römer-Oberdörfer, A., Staeheli, P., Rubbenstroth, D., 2017. Viral vector vaccines protect cockatiels from inflammatory lesions after heterologous parrot bornavirus 2 challenge infection. Vaccine 35 (4), 557–563.

Sa-Ardta, P., Rinder, M., Sanyathitiseree, P., Weerakhun, S., Lertwatcharasarakul, P., Lorsunyaluck, B., Schmitz, A., Korbel, R., 2019. First detection and characterization of Psittaciform bornaviruses in naturally infected and diseased birds in Thailand. Vet. Microbiol. 230, 62–71.

Sauder, C., Hallensleben, W., Pagenstecher, A., Schneckenburger, S., Biro, L., Pertlik, D., Hausmann, J., Suter, M., Staeheli, P., 2000. Chemokine gene expression in astrocytes of Borna disease virus-infected rats and mice in the absence of inflammation. J. Virol. 74 (19), 9267–9280.

Schamel, K., Staeheli, P., Hausmann, J., 2001. Identification of the immunodominant H-2K (k)-restricted cytotoxic T-cell epitope in the Borna disease virus nucleoprotein. J. Virol. 75 (18), 8579–8588.

Schaudien, D., 2007. Experimentelle Infektion von TNF-transgenen Mäusen mit dem Virus der Bornaschen Krankheit—Einfluss auf Zytokinprofil, Neurodegeneration und Neuroprotektion. PhD These. Tierärztliche Hochschule Hannover, Hannover.

Schlottau, K., Hoffmann, B., Homeier-Bachmann, T., Fast, C., Ulrich, R.G., Beer, M., Hoffmann, D., 2017a. Multiple detection of zoonotic variegated squirrel bornavirus 1 RNA in different squirrel species suggests a possible unknown origin for the virus. Arch. Virol. 162 (9), 2747–2754.

Schlottau, K., Jenckel, M., van den Brand, J., Fast, C., Herden, C., Höper, D., Homeier-Bachmann, T., Thielebein, J., Mensing, N., et al., 2017b. Variegated squirrel bornavirus 1 in squirrels, Germany and the Netherlands. Emerg. Infect. Dis. 23 (3), 477–481.

Schlottau, K., Forth, L., Angstwurm, K., Höper, D., Zecher, D., Liesche, F., Hoffmann, B., Kegel, V., Seehofer, D., et al., 2018. Fatal encephalitic Borna disease virus 1 in solid-organ transplant recipients. N. Engl. J. Med. 379 (14), 1377–1379.

Schmidt, P., Herzog, S., Frese, K., Grabner, A., Hafner, A., Herden, C., Schmahl, W., Wernery, U., 1999. Pathological and virological studies of Borna disease in horses. In: Wade, J., Mumford, J.A., Kaaden, O.R. (Eds.), Equine Infectious Diseases VIII. R&W Publications, Newmarket, pp. 493–494.

Schneemann, A., Schneider, P.A., Kim, S., Lipkin, W.I., 1994. Identification of signal sequences that control transcription of Borna disease virus, a nonsegmented, negative-strand RNA virus. J. Virol. 68 (10), 6514–6522.

Schneider, P.A., Schneemann, A., Lipkin, W.I., 1994. RNA splicing in Borna disease virus, a nonsegmented, negative-strand RNA virus. J. Virol. 68 (8), 5007–5012.

Schneider, P.A., Hatalski, C.G., Lewis, A.J., Lipkin, W.I., 1997. Biochemical and functional analysis of the Borna disease virus G protein. J. Virol. 71 (1), 331–336.

Schneider, U., Naegele, M., Staeheli, P., Schwemmle, M., 2003. Active Borna disease virus polymerase complex requires a distinct nucleoprotein-to-phosphoprotein ratio but no viral x protein. J. Virol. 77 (21), 11781–11789.

Schneider, U., Naegele, M., Staeheli, P., 2004. Regulation of the Borna disease virus polymerase complex by the viral nucleoprotein p38 isoform. Brief report. Arch. Virol. 149 (7), 1409–1414.

Schneider, U., Schwemmle, M., Staeheli, P., 2005. Genome trimming: a unique strategy for replication control employed by Borna disease virus. Proc. Natl. Acad. Sci. U. S. A. 102 (9), 3441–3446.

Schneider, U., Martin, A., Schwemmle, M., Staeheli, P., 2007. Genome trimming by Borna disease viruses: viral replication control or escape from cellular surveillance? Cell. Mol. Life Sci. 64 (9), 1038–1042.

Schulze, V., Lurz, P.W.W., Ferrari, N., Romeo, C., Steele, M.A., Marino, S., Mazzamuto, M.V., Calvignac-Spencer, S., Schlottau, K., et al., 2020. Search for polyoma-, herpes-, and bornaviruses in squirrels of the family Sciuridae. Virol. J. 17 (1), 42.

Schwardt, M., Mayer, D., Frank, R., Schneider, U., Eickmann, M., Planz, O., Wolff, T., Schwemmle, M., 2005. The negative regulator of Borna disease virus polymerase is a non-structural protein. J. Gen. Virol. 86 (Pt. 11), 3163–3169.

Schwemmle, M., De, B., Shi, L., Banerjee, A., Lipkin, W.I., 1997. Borna disease virus P-protein is phosphorylated by protein kinase Cepsilon and casein kinase II. J. Biol. Chem. 272 (35), 21818–21823.

Schwemmle, M., Salvatore, M., Shi, L., Richt, J., Lee, C.H., Lipkin, W.I., 1998. Interactions of the Borna disease virus P, N, and X proteins and their functional implications. J. Biol. Chem. 273 (15), 9007–9012.

Schwemmle, M., Jehle, C., Formella, S., Staeheli, P., 1999. Sequence similarities between human bornavirus isolates and laboratory strains question human origin. Lancet 354 (9194), 1973–1974.

Scordel, C., Huttin, A., Cochet-Bernoin, M., Szelechowski, M., Poulet, A., Richardson, J., Benchoua, A., Gonzalez-Dunia, D., Eloit, M., et al., 2015. Borna disease virus phosphoprotein impairs the developmental program controlling neurogenesis and reduces human GABAergic neurogenesis. PLoS Pathog. 11 (4), e1004859.

Shankar, V., Kao, M., Hamir, A.N., Sheng, H., Koprowski, H., Dietzschold, B., 1992. Kinetics of virus spread and changes in levels of several cytokine mRNAs in the brain after intranasal infection of rats with Borna disease virus. J. Virol. 66 (2), 992–998.

Shi, M., Lin, X.-D., Chen, X., Tian, J.-H., Chen, L.-J., Li, K., Wang, W., Eden, J.-S., Shen, J.-J., et al., 2018. The evolutionary history of vertebrate RNA viruses. Nature 556 (7700), 197–202.

Shivaprasad, H.L., Barr, B.C., Woods, L.W., Daft, B.M., Association of Avian Veterinarians, 1995. Spectrum of lesions (pathology) of proventricular Dilation Syndrome. In: Main Conference Proceedings. Association of Avian Veterinarians. AAV, pp. 505–506.

Siemetzki, U., Ashok, M.S., Briese, T., Lipkin, W.I., 2009. Identification of RNA instability elements in Borna disease virus. Virus Res. 144 (1–2), 27–34.

Sind, J.B.v., 1767. Der im Feld und auf der Reise geschwind heilende Pferdearzt, welcher einen gründlichen Unterricht von den gewöhnlichsten Krankheiten der Pferde im Feld und auf der Reise wie auch einen auserlesenen Vorrath der nützlichsten und durch die Erfahrung bewährtesten Heilungsmitteln eröffnet. bey Heinrich Ludwig Brönner.

Skirmuntt, E.C., Escalera-Zamudio, M., Teeling, E.C., Smith, A., Katzourakis, A., 2020. The potential role of endogenous viral elements in the evolution of bats as reservoirs for zoonotic viruses. Annu. Rev. Virol.

Sobbe, M., Bilzer, T., Gommel, S., Nöske, K., Planz, O., Stitz, L., 1997. Induction of degenerative brain lesions after adoptive transfer of brain lymphocytes from Borna disease virus-infected rats: presence of CD8+ T cells and perforin mRNA. J. Virol. 71 (3), 2400–2407.

Sofuku, K., Parrish, N.F., Honda, T., Tomonaga, K., 2015. Transcription profiling demonstrates epigenetic control of non-retroviral RNA virus-derived elements in the human genome. Cell Rep. 12 (10), 1548–1554.

Solbrig, M.V., Koob, G.F., Joyce, J.N., Lipkin, W.I., 1996. A neural substrate of hyperactivity in Borna disease: changes in brain dopamine receptors. Virology 222 (2), 332–338.

Song, W., Kao, W., Zhai, A., Qian, J., Li, Y., Zhang, Q., Zhao, H., Hu, Y., Li, H., et al., 2013. Borna disease virus nucleoprotein inhibits type I interferon induction through the interferon regulatory factor 7 pathway. Biochem. Biophys. Res. Commun. 438 (4), 619–623.

Staeheli, P., 2002. Bornaviruses. Virus Res. 82 (1–2), 55–59.

Staeheli, P., Sauder, C., Hausmann, J., Ehrensperger, F., Schwemmle, M., 2000. Epidemiology of Borna disease virus. J. Gen. Virol. 81 (Pt. 9), 2123–2135.

Staeheli, P., Sentandreu, M., Pagenstecher, A., Hausmann, J., 2001. Alpha/beta interferon promotes transcription and inhibits replication of Borna disease virus in persistently infected cells. J. Virol. 75 (17), 8216–8223.

Steinmetz, A., Pees, M., Schmidt, V., Weber, M., Krautwald-Junghanns, M.-E., Oechtering, G., 2008. Blindness as a sign of proventricular dilatation disease in a grey parrot (Psittacus erithacus erithacus). J. Small Anim. Pract. 49 (12), 660–662.

Stitz, L., Schilken, D., Frese, K., 1991. Atypical dissemination of the highly neurotropic Borna disease virus during persistent infection in cyclosporine A-treated, immunosuppressed rats. J. Virol. 65 (1), 457–460.

Stitz, L., Sobbe, M., Bilzer, T., 1992. Preventive effects of early anti-CD4 or anti-CD8 treatment on Borna disease in rats. J. Virol. 66 (6), 3316–3323.

Stitz, L., Dietzschold, B., Carbone, K.M., 1995. Immunopathogenesis of Borna disease. Curr. Top. Microbiol. Immunol. 190, 75–92.

Stitz, L., Nöske, K., Planz, O., Furrer, E., Lipkin, W.I., Bilzer, T., 1998. A functional role for neutralizing antibodies in Borna disease: influence on virus tropism outside the central nervous system. J. Virol. 72 (11), 8884–8892.

Stitz, L., Bilzer, T., Planz, O., 2002. The immunopathogenesis of Borna disease virus infection. Front. Biosci. 7, 541–555.

Suberbielle, E., Stella, A., Pont, F., Monnet, C., Mouton, E., Lamouroux, L., Monsarrat, B., Gonzalez-Dunia, D., 2008. Proteomic analysis reveals selective impediment of neuronal remodeling upon Borna disease virus infection. J. Virol. 82 (24), 12265–12279.

Sun, L., Guo, Y., He, P., Xu, X., Zhang, X., Wang, H., Tang, T., Zhou, W., Xu, P., et al., 2019. Genome-wide profiling of long noncoding RNA expression patterns and CeRNA analysis in mouse cortical neurons infected with different strains of Borna disease virus. Genes Dis. 6 (2), 147–158.

Szelechowski, M., Bétourné, A., Monnet, Y., Ferré, C.A., Thouard, A., Foret, C., Peyrin, J.-M., Hunot, S., Gonzalez-Dunia, D., 2014. A viral peptide that targets mitochondria protects against neuronal degeneration in models of Parkinson's disease. Nat. Commun. 5, 5181.

Takeuchi, H., Jin, S., Wang, J., Zhang, G., Kawanokuchi, J., Kuno, R., Sonobe, Y., Mizuno, T., Suzumura, A., 2006. Tumor necrosis factor-alpha induces neurotoxicity via glutamate release from hemichannels of activated microglia in an autocrine manner. J. Biol. Chem. 281 (30), 21362–21368.

Tappe, D., Schlottau, K., Cadar, D., Hoffmann, B., Balke, L., Bewig, B., Hoffmann, D., Eisermann, P., Fickenscher, H., et al., 2018. Occupation-associated fatal limbic encephalitis caused by variegated squirrel bornavirus 1, Germany, 2013. Emerg. Infect. Dis. 24 (6), 978–987.

Tappe, D., Frank, C., Homeier-Bachmann, T., Wilking, H., Allendorf, V., Schlottau, K., Muñoz-Fontela, C., Rottstegge, M., Port, J.R., et al., 2019a. Analysis of exotic squirrel trade and detection of human infections with variegated squirrel bornavirus 1, Germany, 2005 to 2018. Euro Surveill. 24 (8), 1800483.

Tappe, D., Schmidt-Chanasit, J., Rauch, J., Allartz, P., Herden, C., 2019b. Immunopathology of fatal human variegated squirrel bornavirus 1 encephalitis, Germany, 2011–2013. Emerg. Infect. Dis. 25 (6), 1058–1065.

Teng, D., Obika, S., Ueda, K., Honda, T., 2019. A small interfering RNA cocktail targeting the nucleoprotein and large protein genes suppresses Borna disease virus infection. Front. Microbiol. 10, 2781.

Tizard, I., Ball, J., Stoica, G., Payne, S., 2016a. The pathogenesis of bornaviral diseases in mammals. Anim. Health Res. Rev. 17 (2), 92–109.

Tizard, I., Shivaprasad, H.L., Guo, J., Hameed, S., Ball, J., Payne, S., 2016b. The pathogenesis of proventricular dilatation disease. Anim. Health Res. Rev. 17 (2), 110–126.

Tokunaga, T., Yamamoto, Y., Sakai, M., Tomonaga, K., Honda, T., 2017. Antiviral activity of favipiravir (T-705) against mammalian and avian bornaviruses. Antiviral Res. 143, 237–245.

Unterstab, G., Ludwig, S., Anton, A., Planz, O., Dauber, B., Krappmann, D., Heins, G., Ehrhardt, C., Wolff, T., 2005. Viral targeting of the interferon-beta-inducing Traf family member-associated NF-kappaB activator (TANK)-binding kinase-1. Proc. Natl. Acad. Sci. U. S. A. 102 (38), 13640–13645.

Villanueva, I., Gray, P., Mirhosseini, N., Payne, S., Hoppes, S., Honkavuori, K.S., Briese, T., Turner, D., Tizard, I., 2010. The diagnosis of proventricular dilatation disease: use of a Western blot assay to detect antibodies against avian Borna virus. Vet. Microbiol. 143 (2–4), 196–201.

Volmer, R., Monnet, C., Gonzalez-Dunia, D., 2006. Borna disease virus blocks potentiation of presynaptic activity through inhibition of protein kinase C signaling. PLoS Pathog. 2 (3), e19.

Volmer, R., Prat, C.M.A., Le Masson, G., Garenne, A., Gonzalez-Dunia, D., 2007. Borna disease virus infection impairs synaptic plasticity. J. Virol. 81 (16), 8833–8837.

Walker, M.P., Jordan, I., Briese, T., Fischer, N., Lipkin, W.I., 2000. Expression and characterization of the Borna disease virus polymerase. J. Virol. 74 (9), 4425–4428.

Weissenböck, H., Hornig, M., Hickey, W.F., Lipkin, W.I., 2000. Microglial activation and neuronal apoptosis in bornavirus infected neonatal Lewis rats. Brain Pathol. 10 (2), 260–272.

Weissenböck, H., Bakonyi, T., Sekulin, K., Ehrensperger, F., Doneley, R.J.T., Dürrwald, R., Hoop, R., Erdelyi, K., Gal, J., et al., 2009a. Avian bornaviruses in psittacine birds from Europe and Australia with proventricular dilatation disease. Emerg. Infect. Dis. 15 (9), 1453–1459.

Weissenböck, H., Sekulin, K., Bakonyi, T., Högler, S., Nowotny, N., 2009b. Novel avian bornavirus in a nonpsittacine species (Canary; Serinus canaria) with enteric ganglioneuritis and encephalitis. J. Virol. 83 (21), 11367–11371.

Weissenböck, H., Bagó, Z., Kolodziejek, J., Hager, B., Palmetzhofer, G., Dürrwald, R., Nowotny, N., 2017. Infections of horses and shrews with Bornaviruses in Upper Austria: a novel endemic area of Borna disease: a novel endemic area of Borna disease. Emerg. Microb. Infect. 6 (6), e52.

Wensman, J.J., Munir, M., Thaduri, S., Hörnaeus, K., Rizwan, M., Blomström, A.-L., Briese, T., Lipkin, W.I., Berg, M., 2013. The X proteins of bornaviruses interfere with type I interferon signalling. J. Gen. Virol. 94 (Pt. 2), 263–269.

Werner-Keiss, N., Garten, W., Richt, J.A., Porombka, D., Algermissen, D., Herzog, S., Baumgärtner, W., Herden, C., 2008. Restricted expression of Borna disease virus glycoprotein in brains of experimentally infected Lewis rats. Neuropathol. Appl. Neurobiol. 34 (6), 590–602.

Williams, B.L., Hornig, M., Yaddanapudi, K., Lipkin, W.I., 2008. Hippocampal poly(ADP-Ribose) polymerase 1 and caspase 3 activation in neonatal bornavirus infection. J. Virol. 82 (4), 1748–1758.

Wolff, T., Heins, G., Pauli, G., Burger, R., Kurth, R., 2006. Failure to detect Borna disease virus antigen and RNA in human blood. J. Clin. Virol. 36 (4), 309–311.

Wu, Y.-J., Schulz, H., Lin, C.-C., Saar, K., Patone, G., Fischer, H., Hübner, N., Heimrich, B., Schwemmle, M., 2013. Borna disease virus-induced neuronal degeneration dependent on host genetic background and prevented by soluble factors. Proc. Natl. Acad. Sci. U. S. A. 110 (5), 1899–1904.

Wünschmann, A., Honkavuori, K., Briese, T., Lipkin, W.I., Shivers, J., Armien, A.G., 2011. Antigen tissue distribution of Avian bornavirus (ABV) in psittacine birds with natural spontaneous proventricular dilatation disease and ABV genotype 1 infection. J. Vet. Diagn. Invest. 23 (4), 716–726.

Wüst, E.M., 2017. Untersuchungen zur vertikalen Übertragung des Parrot Bornavirus bei Psittaziden. Diss. Justus-Liebig-Universität Gießen, Gießen.

Yanai, M., Sakai, M., Makino, A., Tomonaga, K., 2017. Dual function of the nuclear export signal of the Borna disease virus nucleoprotein in nuclear export activity and binding to viral phosphoprotein. Virol. J. 14 (1), 126.

Yanai, M., Kojima, S., Sakai, M., Komorizono, R., Tomonaga, K., Makino, A., 2020. ADAR2 is involved in self and nonself recognition of Borna disease virus genomic RNA in the nucleus. J. Virol. 94 (6), e01513-19.

Zhai, A., Qian, J., Kao, W., Li, A., Li, Y., He, J., Zhang, Q., Song, W., Fu, Y., et al., 2013. Borna disease virus encoded phosphoprotein inhibits host innate immunity by regulating miR-155. Antiviral Res. 98 (1), 66–75.

Zhang, G., Kobayashi, T., Kamitani, W., Komoto, S., Yamashita, M., Baba, S., Yanai, H., Ikuta, K., Tomonaga, K., 2003. Borna disease virus phosphoprotein represses p53-mediated transcriptional activity by interference with HMGB1. J. Virol. 77 (22), 12243–12251.

Zhang, L., Lei, Y., Liu, X., Wang, X., Liu, Z., Li, D., Zheng, P., Zhang, L., Chen, S., et al., 2014. Glutamate and lipid metabolic perturbation in the hippocampi of asymptomatic Borna disease virus-infected horses. PLoS One 9 (6), e99752.

Zhang, X., Guo, Y., Xu, X., Tang, T., Sun, L., Wang, H., Zhou, W., Fang, L., Li, Q., et al., 2019. miR-146a promotes Borna disease virus 1 replication through IRAK1/TRAF6/NF-κB signaling pathway. Virus Res. 271, 197671.

Zheng, Y.M., Schäfer, M.K., Weihe, E., Sheng, H., Corisdeo, S., Fu, Z.F., Koprowski, H., Dietzschold, B., 1993. Severity of neurological signs and degree of inflammatory lesions in the brains of rats with Borna disease correlate with the induction of nitric oxide synthase. J. Virol. 67 (10), 5786–5791.

Zwick, W., 1939. Bornasche Krankheit und Enzephalomyelitis der Tiere. In: Gildemeister, E., Haagen, E., Waldmann, O. (Eds.), Handbuch der Viruskrankheiten. Gustav Fischer Verlag, Jena.

CHAPTER SIX

Immunopathology of Zika virus infection

Derek J. Platt[b], Jonathan J. Miner[a,b,c,*]

[a]Department of Medicine, Washington University in St. Louis School of Medicine, Saint Louis, MO, United States
[b]Department of Molecular Microbiology, Washington University in St. Louis School of Medicine, Saint Louis, MO, United States
[c]Department of Pathology and Immunology, Washington University in St. Louis School of Medicine, Saint Louis, MO, United States
*Corresponding author: e-mail address: jonathan.miner@wustl.edu

Contents

Abstract

Zika virus (ZIKV) is a mosquito-borne virus of the flavivirus genus in the *Flaviviridae* family. Flaviviruses are single-stranded, positive-sense RNA viruses that have been responsible for numerous human epidemics. Notable flaviviruses include mosquito-borne viruses such as yellow fever virus (YFV), Dengue virus (DENV), West Nile virus (WNV), Japanese encephalitis virus (JEV), as well as tick-borne viruses including Powassan virus (POWV) and tick-borne encephalitis virus (TBEV). Despite having been relatively obscure until the past decade, ZIKV has become a major global health concern, and is a topic of active research following multiple outbreaks across the globe. Here, we discuss ZIKV pathogenesis and the associated immunopathology, as well as advances in research, therapies, and vaccines developed using models of ZIKV pathogenesis.

1. Zika virus: Discovery, outbreaks, and clinical manifestations of disease

In 1947, Zika virus (ZIKV) was discovered in a yellow fever virus research station in the Ziika Forest of Uganda (Dick et al., 1952). Dick

Advances in Virus Research, Volume 107
ISSN 0065-3527
https://doi.org/10.1016/bs.aivir.2020.06.007

et al. collected serum from a febrile rhesus macaque (Rhesus 766) and inoculated mice both intraperitoneally and intracranially. ZIKV was eventually isolated by performing filtration on the brains of sick mice. The virus was additionally isolated from *Aedes africanus* mosquitoes, identifying *Aedes* species mosquitoes as ZIKV vectors. Subsequent studies demonstrated that mouse-adapted ZIKV is neurotropic in mice and has the capacity to cause lethality and neurological sequelae such as paralysis when administered intracranially (Dick, 1952). Early ZIKV transmission studies included laborious work that involved infecting mosquitoes by providing them with a ZIKV-infected blood meal, and exposing rhesus monkeys and newborn mice to infected mosquitos. After multiple unsuccessful attempts over the course of 2 months, researchers were able to induce transmission of ZIKV from mosquitoes to 10 neonatal mice and 1 rhesus monkey. Five of the 10 infected mice died, whereas the rhesus monkey did not exhibit any signs of infection but did produce suspected anti-ZIKV antibodies (Boorman and Porterfield, 1956). Clinical symptoms caused by ZIKV infection in humans were not identified until years later, when scientist William Bearcroft inoculated himself with ZIKV and reported mild, self-limited symptoms such as fever, headache, and nausea (Bearcroft, 1956). Concurrently, other groups developed tools to propagate ZIKV and perform neutralization tests in vitro and identified cell types in which ZIKV could grow, including green monkey kidney epithelial cells (Vero cells), which are still widely used for in vitro ZIKV growth and assays (Way et al., 1976). ZIKV can also be grown in cells isolated from *Aedes albopictus* mosquitoes (C6/36 cells) (Deng et al., 2016).

The first report of human ZIKV infection was in 1952 in Nigeria, from which the first isolation of ZIKV from a human subject was described (Macnamara, 1954). ZIKV was first isolated from *Aedes aegypti* mosquitoes in Malaysia in 1966 (Marchette et al., 1969), but the first report of human ZIKV infection in Asia was in Indonesia in 1977 (Olson and Ksiazek, 1981). Overall, only sporadic cases of ZIKV infection were reported in the literature over the course of nearly 60 years. However, it is likely that numbers during this period were inaccurate due to antigenic similarity between ZIKV and co-circulating flaviviruses such as DENV and YFV. The majority of case reports during this time period described patients who presented with symptoms similar to those reported by Bearcroft (1956). Additionally, some studies mentioned other symptoms such as maculopapular rash (Simpson, 1964) and jaundice (Macnamara, 1954). Only ~20% of infected individuals present with any symptoms.

The first known large ZIKV outbreak was on Yap Island in Micronesia in 2007 (Duffy et al., 2009; Lanciotti et al., 2008). During this outbreak, there were 49 confirmed and 59 probable cases of ZIKV infection in humans. Additionally, over 5000 individuals—about three quarters of the total population over the age of 3—had anti-ZIKV IgM antibodies. The most commonly reported symptoms were rash, fever, arthralgia, and conjunctivitis (Duffy et al., 2009). Despite the size of the outbreak on Yap Island, it was not until the 2013–2014 ZIKV outbreak in French Polynesia that ZIKV infection began to be associated with serious neurological symptoms in humans. During the outbreak, over 31,000 individuals sought treatment for potential ZIKV infection, of which almost 9000 were recorded by public health officials and almost 400 were confirmed in the laboratory (Cauchemez et al., 2016). A case-control study later discovered that 42 patients were diagnosed with Guillain-Barré syndrome, an autoimmune neurological disorder caused by peripheral nervous system damage that can lead to ascending paralysis. Of the 42 patients, 41 tested positive for anti-ZIKV antibodies. ZIKV is typically less neuroinvasive in adults than other neurotropic flaviviruses such as JEV, TBEV, POWV, and WNV. However, ZIKV efficiently infects, alters cell cycle progression, and induces cell death in neural progenitor cells (Tang et al., 2016). The potential causes of immune-mediated pathology following ZIKV infection remain unknown. Studies performed on plasma samples from ZIKV-infected patients in Salvador, Brazil demonstrated a significant increase in IgM and IgG anti-ganglioside antibodies. Since neuronal membranes contain gangliosides, it is plausible that anti-ganglioside antibodies contribute to the development of Guillain-Barré syndrome by targeting gangliosides and subsequent damage (Rivera-Correa et al., 2019). Other reported neurological manifestations of disease in adults include acute disseminated encephalomyelitis (Niemeyer et al., 2017) and encephalitis (Soares et al., 2016). Recent mouse studies suggest that the cytokine-mediated recruitment of innate immune cells may have a role in the immunopathology of acute disseminated encephalomyelitis, encephalitis, and Guillain-Barré syndrome following ZIKV infection (Hayashida et al., 2019). The involvement of innate and adaptive immune cells in ZIKV-related central nervous system pathology is currently an active area of research (Fig. 1).

Perhaps the most striking and noteworthy clinical features caused by ZIKV infection are those associated with the developing fetus. ZIKV has the capacity for transplacental infection and can induce damage in the developing fetus. The ability of ZIKV to productively infect the fetal brain is

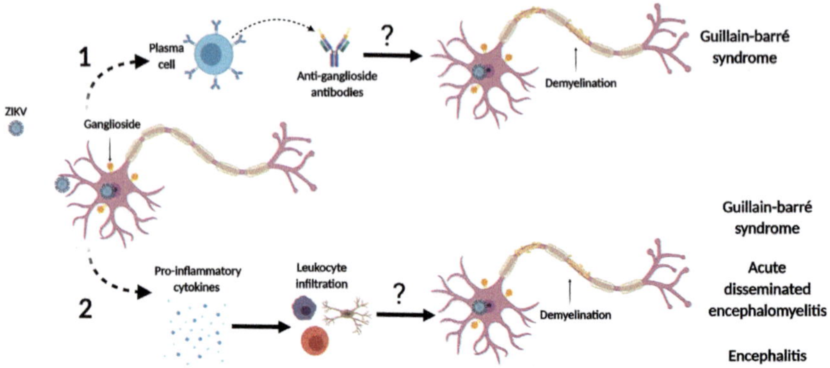

Fig. 1 Potential mechanisms of ZIKV-induced immunopathology. Neurological disease manifestations in a subset of ZIKV-infected adults include Guillain-Barré syndrome, acute disseminated encephalomyelitis, and encephalitis. Recent studies have suggested the involvement of both adaptive and innate immune cells in immunopathology. (1) Anti-self, anti-ganglioside antibodies could lead to immune activity against neurons and damage to myelin. (2) Alternatively, release of pro-inflammatory cytokines and recruitment of self-targeting innate immune cells may lead to demyelination and drive disease pathogenesis.

likely due to the fact that it preferentially infects neural progenitor cells that are instrumental in fetal brain development (Tang et al., 2016). ZIKV is not the only flavivirus that has been reported to cause transplacental congenital infection and disease phenotypes such as fetal demise, stillbirth, and spontaneous abortion, as there are reports of rare cases of JEV and WNV causing fetal damage in both humans and in animal models following vertical transmission (Alpert et al., 2003; Barnard and Voges, 1986; Centers for Disease Control and Prevention (CDC), 2002; Chaturvedi et al., 1980; Platt et al., 2018; Venter et al., 2011). However, the size of the 2015 ZIKV epidemic in the Americas made apparent the link between ZIKV infection and congenital outcomes (Icheku, 2017). Early reports of congenital ZIKV infection involved longitudinally monitoring infected pregnant women and performing ultrasonographic studies on the developing fetuses. These studies demonstrated abnormalities such as intrauterine growth restriction (IUGR), cerebral calcifications, microcephaly, and fetal demise in 42% of fetuses (Brasil et al., 2016). Other congenital outcomes include hearing defects and ocular defects (de Paula et al., 2016). To date, at least thousands of infants have been born with congenital abnormalities (Castro et al., 2018). Analysis of ZIKV-infected pregnant women in the United States concluded that of women who completed pregnancies and had evidence of recent ZIKV infection at any stage of pregnancy, 6% had ZIKV-related congenital abnormalities. Furthermore, 11% of fetuses or infants from women with confirmed first-trimester ZIKV infections demonstrated ZIKV-related congenital abnormalities (Honein et al., 2017). Another study demonstrated a similar trend showing a lower rate of complications at later stages, suggesting that ZIKV has a higher capacity to infect fetuses and induce congenital malformations at earlier stages in pregnancy (Shapiro-Mendoza et al., 2017).

ZIKV is unique among flaviviruses in that it has the capacity to be transmitted sexually. Though this ability first became widely known during the outbreak in the Americas, it was first reported years earlier by a scientist who contracted ZIKV in 2008 while working in Dakar, Senegal and infected his wife upon returning to his home in Colorado. His illness was characterized by maculopapular rash, headache, wrist and ankle arthralgia in addition to swollen ankles, and hematospermia. Days later, his wife began to exhibit similar symptoms. The presence of ZIKV in semen was later directly demonstrated in a patient during the 2013 French Polynesia outbreak (Foy et al., 2011). RT-PCR analysis of his blood was negative, but semen and urine samples contained 2.9×10^7 and 3.8×10^3 ZIKV RNA copies/mL, respectively, demonstrating that ZIKV RNA persists in semen and urine at high levels after it is no longer detectable in blood. ZIKV RNA is now known to be able to persist in semen up to a year after onset of symptoms, with infectious virus being able to be isolated up to 69 days after onset (Musso et al., 2015). While the majority of reported cases were male-to-female, there has also been a report of male-to-male transmission (Deckard et al., 2016), suggesting that ZIKV can be spread sexually by vaginal and anal intercourse. Due to the high probability of vector-borne-transmission in endemic areas, it is difficult to determine the contribution of sexual transmission to the total number of infections, although modeling approaches have suggested that about 3% of human cases are due to sexual transmission (Gao et al., 2016) (Fig. 2).

2. Zika virus structure and life cycle

Similar to other flaviviruses, ZIKV is single-stranded, positive-sense RNA virus. The genome is ~10.8 kb, consisting of a 5′ untranslated region (UTR), one open reading frame (ORF), and a 3′ UTR (Kuno and Chang, 2007). The ZIKV life cycle is maintained by propagation between both arthropod and mammalian vectors. In particular, ZIKV is transmitted most commonly to primates (including both human and non-human primates) by *Aedes* species mosquitoes (Boorman and Porterfield, 1956; Dick et al., 1952; Marchette et al., 1969). In a productive infectious cycle, an infected

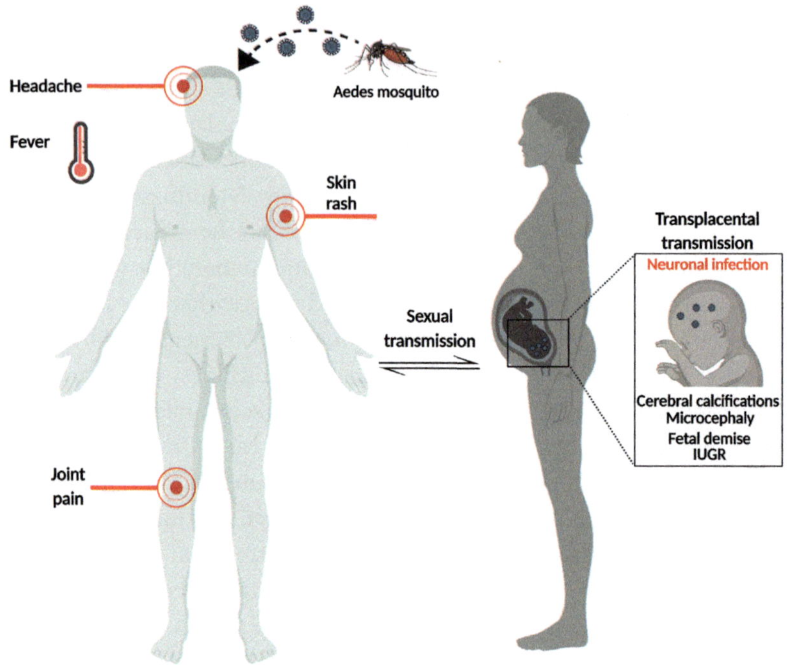

Fig. 2 Zika virus transmission and outcomes of congenital infection. ZIKV is transmitted from arthropod to human by infected Aedes species mosquitoes and human-to-human through sexual transmission. ZIKV infection is usually self-limiting and common symptoms of ZIKV infection include headache, fever, skin rash, and joint pain. Transplacental transmission of ZIKV during pregnancy results in often more severe clinical manifestations, including cerebral calcifications, microcephaly, fetal demise, and intrauterine growth restriction.

mosquito introduces ZIKV by injecting infectious virions into the dermis during a blood meal via insertion of its proboscis into the skin (Ha et al., 2017). Plaque assays on collected saliva from infected mosquitoes used to infect rhesus macaques measured $10^{1.5}$ to $10^{3.2}$ plaque-forming units of ZIKV (Dudley et al., 2017), although the viral load in studies with WNV demonstrating 10–10,000-fold more virus using an in vivo method may be more accurate for ZIKV as well (Styer et al., 2007). Once in the dermis, ZIKV infects host cells expressing the proper virus receptors. Receptors believed to be co-opted by ZIKV include C-type lectin receptors (CLRs) (Hamel et al., 2015), and certain lipid receptors such as T cell immunoglobulin mucin (TIM) (Hamel et al., 2015) and receptor kinases TYRO3, AXL, and MER (TAM) (Brouillette et al., 2018; Hamel et al., 2015; Meertens et al., 2017). However, TAM receptors are not required for ZIKV infection and disease in mice (Hastings et al., 2017).

Infection is achieved through a series of steps that begin with binding to the entry receptors, a process that is mediated by the viral envelope protein. Once bound, the infectious ZIKV virion is internalized by clathrin-mediated endocytosis, a process by which cells transport extracellular cargo molecules to the cell interior. During this process, clathrin-coated pits create pouches in the cell plasma membrane, eventually pinching off to form a clathrin-coated vesicle that later becomes an endosome (Takei and Haucke, 2001). The virion remains intact in the early endosome until the endosome matures into a late endosome, which has a low pH. The pH change promotes a structural change within the envelope protein to expose the fusion loop, which facilitates fusion to the endosomal membrane (Rawle et al., 2018). Following fusion to the endosomal membrane, the virion releases viral RNA into the host cytoplasm and uses host translation machinery to translate the ORF.

Similar to related flaviviruses, the ZIKV ORF encodes a total of 10 proteins (three structural, seven non-structural). The structural proteins include the capsid (C), pre-membrane/membrane (pr/M), and envelope (E) proteins. The non-structural proteins include NS1, NS2A, NS2B, NS3, NS4A, NS4B, and NS5. These proteins form a 3423 amino acid-long polyprotein. Collectively, these proteins form a viral particle with icosahedral symmetry, consisting of 60 repeating units (Kostyuchenko et al., 2016). Following translation, the polyprotein is cleaved by the viral serine protease NS2B-NS3. The other NS proteins, including NS1 (which becomes glycosylated) and the RNA-dependent RNA polymerase (RdRp) NS5, form a replication complex that is used to synthesize a full-length, (−)ssRNA intermediate that serves as a template for the production of subsequent (+)ssRNA encoding the polyproteins (Barnard et al., 2018). The translated C protein then binds viral RNA into a nucleocapsid to form the core of a mature virion, which is encapsulated by a lipid bilayer that contains the prM/M and E proteins in complex (Tan et al., 2020). The assembly of mature ZIKV virions, like other flavivirus virions, occurs at the endoplasmic reticulum (ER). At the ER, the envelope protein associates with the prM protein and a heterodimer complex is formed that effectively caps the fusion loop. Maturation occurs as the immature virus particle migrates through the trans-Golgi network, whose low pH induces separation of the prM-E heterodimer and proteolytic cleavage of the prM protein resulting in a pr peptide and a M protein (Tan et al., 2020). The mature virion then exits the host cell in a vesicle by exocytosis. In addition to serving replicative functions, several of the ZIKV proteins are also involved in viral immune evasion. This will be discussed later in the chapter.

3. Zika virus immune antagonism

During viral infection, recognition receptors (PRRs), including certain Toll-like receptors (TLRs) and retinoid-inducible gene I (RIG-I)-like receptors (RLRs) recognize pathogen-associated molecular patterns (PAMPs). This leads to activation of signaling pathways that induce antiviral and inflammatory responses (Cao, 2016). Since ZIKV is a positive-sense ssRNA virus that has a double stranded RNA (dsRNA) intermediate phase during replication (Paul and Bartenschlager, 2015), some of the most important TLRs for flavivirus recognition are TLR3, TLR7, and TLR8, which are all on the endosome and recognize viral RNA (Kawasaki and Kawai, 2014). Once activated, PRR pathways facilitate the phosphorylation and activation of IRF3 and/or IRF7, inducing the expression and secretion of type I IFNs. Type I IFNs then bind to the interferon alpha/beta receptor (IFNAR) and activate Janus kinase 1 (JAK1) and tyrosine kinase 2 (TYK2), which phosphorylate signal transducers and activators of transcription 1 (STAT1) and STAT2 (Rogatsky et al., 2014). STAT1 and STAT2 then dimerize and translocate to the nucleus, where they complex with IRF-9 to form IFN-stimulated gene factor 3 (ISGF3). After assembly, ISGF3 binds to IFN-stimulated response elements (ISREs), which induces the transcription of hundreds of interferon stimulated genes (ISGs), putting the cell in an antiviral state.

These signaling pathways are crucial to the restriction of several flaviviruses, including ZIKV. Accordingly, in order for ZIKV to be able to replicate efficiently in humans, it must antagonize the type I interferon IFN response or activation of PRRs. This is achieved mainly through non-structural proteins. One study demonstrated that in transfected HEK-293T cells, NS1 inhibits IFN-β production in a RIG-I expression-induced activation assay. Interestingly, two separate strains exhibited different levels of IFN-β production antagonism, with the African strain (FSS13025) reducing IFN-β expression by 30% and the Asian strain (PRVABC-59) reducing IFN-beta expression by 40%. The Asian strain was found to have a single amino acid change (A188V) that allows the virus to bind to and inhibit the phosphorylation of TBK1. This study utilized the same system in conjunction with expression of components from the RIG-I signaling pathway and ZIKV NS proteins, and found that expression of ZIKV NS2A, NS2B, NS4A, and NS5 also suppresses IFN-beta production, with activity not varying significantly between the African and Asian lineages (Xia et al., 2018). A common mechanism by which several flaviviruses inhibit the type I IFN response is through antagonism of JAK/STAT signaling. ZIKV prevents type I IFN-mediated phosphorylation of STAT1 and STAT2 in human DCs (Grant et al., 2016). Similar to DENV, ZIKV antagonizes human STAT2 by targeting the protein for proteasomal degradation by NS5. Although unlike DENV, ZIKV NS5 function does not utilize the E3 ubiquitin ligase, UBR4.

Another mechanism by which ZIKV antagonizes the type I IFN response is by inhibiting the cGMP-AMP synthase/stimulator of IFN genes (cGAS/STING) pathway. As suggested by its name, STING is a potent and essential stimulator of IFN genes that are essential in the inhibition of viral pathogens. In particular, STING is activated in response to cytosolic DNA or bacterial CDNs (Ishikawa et al., 2009; Ishikawa and Barber, 2008). Following cellular recognition of foreign cytosolic DNA (particularly viral DNA), cyclic GMP-AMP synthase (cGAS) catalyzes the formation of a type of CDN known as cyclic GMP-AMP, or cGAMP (Sun et al., 2013). Alternatively, CDNs produced by bacteria are also able to directly activate the STING signaling cascade, independently of cGAS (Ishikawa et al., 2009). Once bound to CDNs, STING forms a complex with tank-binding kinase 1 (TBK1). This complex then migrates through the Golgi apparatus to endolysosome compartments, where TBK1 phosphorylates and activates interferon regulatory factor 3 (IRF3) and nuclear factor-κB (NF-κB). Activated IRF3 and NF-κB translocate to the nucleus and activate the transcription of type I IFNs and IFN-stimulated genes (ISGs) (Barber, 2015). This pathway may seem unimportant in the restriction of RNA viruses such as ZIKV, but genetic ablation of cGAS in mice results in increased viral titers and lethality following WNV infection (Schoggins et al., 2014). Furthermore, ZIKV and other flaviviruses have been shown to effectively antagonize the pathway in non-human primate and human, but not mouse cells. This is accomplished by proteolytic cleavage of the STING dimer by the NS2B3 protease, a mechanism of antagonism also used by DENV NS2B3 (Aguirre et al., 2012). During flavivirus infection, mitochondrial damage leads to the release of mitochondrial DNA during infection. A study with DENV demonstrated that the NS2B protease targets cGAS for lysosomal degradation, avoiding detection of mitochondrial DNA (Aguirre et al., 2017). Thus, antagonizing the cGAS-STING pathway, even though it does not directly recognize RNA species, allows for higher infectivity among multiple flaviviruses.

4. Animal models of Zika virus infection

Animal models are an instrumental mechanism by which scientists learn about properties of viruses, such as tissue tropism, pathogenesis, and other aspects of infection that cannot be assessed in cell culture. Popular animal models include rodents such as mice, rats and guinea pigs, and non-human primates. Due to the ease of genetic manipulation, comparatively fast reproduction speed, and overall availability of laboratory mice, mouse models have become the most popular mechanism by which scientists study in vivo infection of ZIKV. Early mouse studies of ZIKV infection were accomplished by intracranial inoculation. While this method was useful in passaging viruses before cell culture techniques were readily available, the route of infection makes it difficult to distinguish between effects that are due to ZIKV and effects due to the immune response to the presence of foreign substances, include non-infectious viral antigens in the brain. Accordingly, many groups utilize subcutaneous routes of infection, which are more similar to mosquito bites. Interestingly, ZIKV is incapable of effectively antagonizing the type I IFN response in mice, preventing the virus from being able to efficiently replicate in mice and making it challenging to generate mouse models of ZIKV infection. Scientists have used a variety of different mechanisms to overcome this barrier by inhibiting the type I IFN response. This can be accomplished by genetic ablation of the IFNAR receptor (*Ifnar1*$^{-/-}$ mice) or administration of an *Ifnar1*$^{-/-}$-blocking antibody (MAR1-5A3). Inhibition of the type I IFN response can also be accomplished by using *Irf3*$^{-/-}$ *Irf5*$^{-/-}$ *Irf7*$^{-/-}$ triple knockout mice, which produce extremely low amounts of type I IFN, significantly blunting the type I IFN response (Lazear et al., 2016). Knockout of both the type II IFN receptor (IFNLR) and the IFNAR receptor (AG129 mice) also makes mice highly susceptible to ZIKV infection (Aliota et al., 2016). Mice in all three models exhibit significant weight loss and lethality starting around day seven after infection. Whereas *Irf3*$^{-/-}$ *Irf5*$^{-/-}$ *Irf7*$^{-/-}$ and *Ifnar1*$^{-/-}$ mice developed neurological disease and exhibited hindlimb weakness and subsequent paralysis before succumbing to lethality, AG129 mice did not. Furthermore, *Irf3*$^{-/-}$ *Irf5*$^{-/-}$ *Irf7*$^{-/-}$ and *Ifnar1*$^{-/-}$ mice exhibited viremia and high viral burden in the brain, spinal cord, and testes, while AG129 mice had high viral titers in the brain and visceral organs.

An early, commonly used mouse model to study vertical transmission of ZIKV involves crossing WT sires to *Ifnar1*$^{-/-}$ dams and subcutaneously inoculating dams with ZIKV at embryonic day six (E6), followed by harvesting 7 days later at E13.5. In this model, vertical transmission of ZIKV leads to high viral burden in the placenta and fetal heads, placental damage, IUGR, and fetal demise (Miner et al., 2016b). Using the *Ifnar1*$^{-/-}$ dam and WT sire model, additional studies have demonstrated that subcutaneous infection at later time points (E9.5 and E11.5) results in lower viral burden in the fetus without any noticeable placental or fetal abnormalities (Jagger et al., 2017). These data support the human studies that demonstrate higher rates of harmful fetal outcomes in fetuses from women infected at earlier stages in pregnancy (Honein et al., 2017). In order to assess the capacity of sexually transmitted ZIKV to affect fetuses, another study intravaginally inoculated *Ifnar1*$^{-/-}$ dams crossed to WT sires with ZIKV at E4.5 and E8.5 and assessed the fetuses at E18.5. Intravaginal inoculation led to fetal resorption and viral replication in the placenta and fetus on E18.5. While these models allow researchers to use more natural routes of infection and lower inoculation amounts, the approach makes it difficult to study the type I IFN response to ZIKV infection in mouse models. Other groups have utilized other methods that do not inhibit the type I IFN response, but which require extremely high amounts of virus and/or unnatural routes of infection (Cugola et al., 2016). Other methods to study ZIKV infection in mice that do not require genetic elimination of important elements of the type I IFN response involve a one-time administration of MAR1-5A3 and inoculation with a mouse-adapted, African strain of ZIKV (Dakar 41,519), which allows the virus to replicate more efficiently in mice, and infection of mice that express the human STAT2 protein instead of mouse STAT2 (Gorman et al., 2018).

Following the 2015 ZIKV outbreak in the Americas, ZIKV was found to be the first known flavivirus to be transmitted sexually in humans. Mouse models to evaluate potential impacts on reproductive health became a high priority. In order to determine the potential short- and long-term impacts of ZIKV infection on the reproductive system in male animals, one study administered MAR1-5A3 and inoculated male mice with Dakar 41,519 and a contemporary Asian ZIKV strain (H/PF/2013) (Govero et al., 2016). Evaluation of infected mice at varying time points revealed that ZIKV persisted in the testis and epididymis of male mice up to 28 days post infection. Infection occurred in spermatagonia, primary spermatocytes, and Sertoli cell in the testis, leading to cell death and subsequent destruction of the seminiferous tubules. Additionally, tissue injury caused decreased testosterone and inhibin B levels as well as oligospermia. Another study using

Ifnar1$^{-/-}$ mice produced findings consistent with the first, demonstrating infection and inflammation in the testis and epididymis, and damaged testes 60 days post infection (Ma et al., 2016). Assessment of the immune response in Leydig, Sertoli, and epididymal epithelial cells demonstrated the production of pro-inflammatory cytokines and chemokines in those cell types following ZIKV infection.

Congenital ZIKV infection can cause severe eye disease in the developing fetus such as optic nerve abnormalities and chorioretinal atrophy in newborns (de Paula et al., 2016). In adults, ZIKV infection was shown to lead to conjunctivitis and uveitis (Duffy et al., 2009; Furtado et al., 2016). In a mouse model, ZIKV infection of adult *Ifnar1*$^{-/-}$ and WT neonatal mice resulted in infection of the cornea, iris, optic nerve, retina. Infection also caused mice to develop conjunctivitis and panuveitis (Miner et al., 2016a). Though not an exhaustive list, these mouse models have been used to better understand many aspects of ZIKV pathogenesis and have helped scientists learn what may be possible pathologically.

While mice are invaluable for pathogenesis studies due to their ease of use and genetic tractability, there are benefits to research studies that use animals which are a natural host for ZIKV infection. For this reason, studies using non-human primates (NHPs) are highly relevant when available. Like mouse models, NHP models have been used to study ZIKV pathogenesis, tissue tropism, as well as to evaluate vaccines and therapeutics. Multiple groups have utilized rhesus, cynomolgus, and pigtail macaques to evaluate ZIKV infection in an NHP. Following subcutaneous inoculation with biologically relevant doses, one group demonstrated fever in four out of five animals and viremia in all animals, with the mean viremia duration being 6.6 days. Furthermore, they found that plasma titers ranged from $10^{4.16}$ to $10^{5.82}$ RNA copies/mL, which is similar to levels demonstrated in humans. ZIKV infection in rhesus macaques additionally led to infection of the central nervous system (CNS), and the intestines, liver, spleen, and several other visceral organs. Histopathological evaluation showed pathological changes in the CNS and damage to the liver, with correspondingly elevated levels of liver enzymes AST and ALT. ZIKV infection also induced a robust humoral immune response, with high levels of both IgM and IgG antibodies, including some neutralizing antibodies (Li et al., 2016). Another study also reported viremia and high viral burden in other bodily fluids including urine, saliva, cerebrospinal fluid, and vaginal fluid. Characterization of the immune response demonstrated proliferating NK cells and expanded $CD4^+$ and $CD8^+$ T cells. Inoculation of pregnant rhesus macaques with ZIKV led to viral RNA that persisted in plasma longer than in non-pregnant animals and 2×10^8 copies/mL of virus in the fetal brain, suggesting that the highly infected fetus may have been seeding the mother and leading to sustained viremia (Dudley et al., 2016). ZIKV RNA has also been reported in the semen of infected animals (Osuna et al., 2016). Several of the countries where ZIKV is endemic also have high rates of human immunodeficiency virus infection. Simian immunodeficiency virus and ZIKV co-infection studies demonstrated that delayed peak viremia, restricted NK cell proliferation, and a reduced humoral immune response to ZIKV (Vinton et al., 2019).

5. African vs Asian Zika virus strains and pathogenesis

ZIKV is closely related to DENV, with which it shares 55.6% amino acid identity in the viral envelope protein (Chang et al., 2017). Unlike DENV, which has four known serotypes, ZIKV only has one serotype. However, there are two genetically distinct lineages that are associated with different pathogenesis. ZIKV originated in Africa, but the Asian lineage was responsible for the 2007 outbreak on Yap Island, the 2013–2014 outbreak in French Polynesia that was associated with Guillain-Barre syndrome and congenital malformations, and the 2015–2016 outbreak in the Americas that highlighted the ability of ZIKV to cross the placenta and induce fetal damage. Although the cause of increased ZIKV virulence in the contemporary Asian lineage has yet to be determined, it is likely that sequence alterations have increased the pathogenicity of ZIKV. In support of this possibility, Yuan et al. demonstrated that an S139N substitution in the prM gene led to enhanced replication of ZIKV in both mouse and human neuronal progenitor cells, in addition to more severe microcephaly and increased lethality in neonatal mice (Yuan et al., 2017).

6. Development of neutralizing antibodies, diagnostic tests, and vaccines

Diagnosis of ZIKV infection can be accomplished by using molecular and serological methods. Molecular testing involves amplification of viral nucleic acids, RNA in the case of ZIKV infection. This is achieved by reverse transcription PCR (RT-PCR). Although RT-PCR is highly specific and thus reliable, ZIKV RNA in serum and plasma is relatively short-lived and is not reliably detectable more than 7 days after clinical onset of symptoms (Joguet et al., 2017). Since ZIKV symptoms can take up to several days to present due to the viral incubation period, there can be a relatively small window for RNA-based diagnosis. ZIKV RNA has been shown to remain in urine and semen for longer periods of time, but these sample types are typically not connected from patients who present with associated symptoms, and semen is restricted to individuals who produce it. Other studies have demonstrated that ZIKV RNA can be detected for up to 120 days in whole blood (Joguet et al., 2017), so some tests utilize quantitative RT-PCR (qRT-PCR) to detect ZIKV RNA (Voermans et al., 2019).

Serological methods utilize enzyme-based assays to detect ZIKV immunoglobulin M (IgM) in patient serum. Serological methods are useful because anti-ZIKV antibodies remain in circulation long after viral RNA levels have vanished. While serological assays have the benefit of detecting infection after the infection has passed, they are less accurate than RNA-based methods because of the structure of the ZIKV envelope (E) protein. The ZIKV E protein is divided into three separate domains, which include the beta-barrel domain (D1), the extended dimerization domain (DII), and the immunoglobulin-like domain (DIII) (Sirohi et al., 2016). DII is of special importance when discussing serological diagnostic development because it contains the fusion loop (FL) on the distal end (Dai et al., 2016). The fusion loop is indispensable for viral replication because of its role in fusion to the endosomal membrane (Rawle et al., 2018). Accordingly, it is highly conserved among related flaviviruses such as ZIKV, DENV, and WNV. Because of this fact, most antibodies developed against the fusion loop (which comprise a large fraction of the humoral response to flaviviruses) are cross-reactive and cannot accurately distinguish between ZIKV and other co-circulating flaviviruses.

While antibody-based tests can be used in the diagnosis of ZIKV, antibodies have also been shown to have therapeutic use in animal models and provide tremendous insight into vaccine development. As a part of the humoral immune response to ZIKV, a large portion of anti-ZIKV antibodies are developed against the FL (in DII). As mentioned earlier, the FL is highly conserved among flaviviruses. Despite anti-FL antibodies being broadly cross-reactive, they usually have poor neutralization activity against targeted viruses. Conversely, antibodies developed against DIII of the E protein are often highly effective at neutralizing ZIKV infection. Highly neutralizing monoclonal antibodies (mAbs) have been developed from both ZIKV-infected human patients and mice by flow cytometric sorting of antibody-secreting cells from ZIKV-infected humans or myeloma-splenocyte fusion of splenocytes from ZIKV-infected mice. Neutralizing, protective antibodies include those that are specific to ZIKV and also protect mice from lethal ZIKV challenge (Zhao et al., 2016), in addition to those that also cross-react with DENV and bind to a quaternary epitope on the E protein dimer-dimer interface (EDE mAbs) (Abbink et al., 2018; Barba-Spaeth et al., 2016; Dai et al., 2016; Fernandez et al., 2017) EDE mAbs have been demonstrated to reduce vertical transmission and protect against fetal demise in mice (Fernandez et al., 2017), protect severely immunocompromised mice from lethal ZIKV challenge (Dai et al., 2016), and provide therapeutic and prophylactic efficacy against ZIKV infection in rhesus monkeys (Abbink et al., 2018)

As with most viruses, the ultimate goal for flavivirus prevention is the development of a highly effective vaccine. However, for a variety of reasons, developing efficient and safe vaccines for flaviviruses has proved quite challenging. The first flavivirus vaccine developed was for YFV in the 1930s by generating a live, attenuated virus vaccine (French neurotropic vaccine strain, FNV) from passaging the wild-type French viscerotropic virus (FVV) strain 128 times in mouse brain. While the FNV vaccine led to a tremendous decline in YFV infections during the 40s and 50s, the neurotropism acquired by developing the vaccine strain led to an increase in postvaccine encephalitis cases in children. The FNV vaccine was soon discontinued and replaced with another live, attenuated virus strain. However, the attenuated 17D strain, which was developed by serial passage of the Asibi strain in chicken embryo tissue culture, has significantly reduced neurotropism (Ni et al., 2000). Although the 17D strain is effective at inducing neutralizing antibodies against YFV, transplacental transmission of YFV in humans occurred following administration of the vaccine (Tsai et al., 1993). The 17D strain is the only vaccine strain currently approved.

Another flavivirus to which an effective vaccine has been developed is TBEV. There are currently two vaccines being used in Western Europe, where TBEV strains are endemic. Though the vaccines were developed against the European serotype, immunity against all TBEV subtypes is induced by the vaccines (World Health Organization, 2020), which is a favorable outcome for viruses with different serotypes. Early TBEV vaccines were efficacious, but were quickly phased out due to adverse effects. The current TBEV vaccines occasionally cause mild adverse events, but none are serious or life-threatening (World Health Organization, 2020).

Another encephalitis virus for which scientists have developed a vaccine is JEV. As with early YFV and TBEV vaccines, early JEV vaccines used a live, attenuated virus strain and was associated with severe adverse events. A currently licensed live, attenuated vaccine (SA14–14-2) was derived from serial passaging of a wild-type virus (Eckels et al., 1988). An additional JEV vaccine utilizes the YFV 17D strain as a backbone and expresses the envelope proteins from SA14–14-2, generating a chimeric vaccine (Chambers et al., 1999).

Perhaps the most difficult flavivirus to develop vaccines against has been DENV, due to differences between effectiveness against different DENV serotypes and potential health complications due to the possibility of a phenomenon known as antibody-dependent enhancement (ADE). ADE occurs when either weakly neutralizing or non-neutralizing antibodies bind to the virion, but fail to induce neutralization, leading to increased infection of $Fc\gamma R^+$ cells and increased viral burden. ADE has been associated with increased disease severity in humans (Halstead, 1988) and demonstrated to increase infection and disease in mice and NHPs (Halstead, 1979). One vaccine that has been licensed in Brazil, El Salvador, Mexico, Paraguay, and the Philippines is CYD-TDV (Dengvaxia). This vaccine was generated using a combination of four YFV 17D-chimeric vaccines, each consisting of a prM and E sequence of a different DENV serotype. Though the vaccine has been shown to induce robust protection against serotypes 3 and 4, there is weak protection against DENV1 and none against DENV2. Furthermore, the vaccine increased the risk of hospitalization among children under the age of 9, and individuals with pre-existing immunity to DENV had more favorable outcomes than individuals without previous DENV immunity. Thus, the vaccine is approved for use in individuals over the age of 9 with laboratory-confirmed immunity to DENV.

The antigenic similarity between ZIKV and DENV is a potential concern when developing ZIKV vaccines. Although cross-reactive antibodies from DENV- and WNV-infected individuals can enhance ZIKV infection in culture and in mice (Bardina et al., 2017; Dejnirattisai et al., 2016; Priyamvada et al., 2016), studies in NHPs demonstrate no enhancement and there is no known increase in viral burden or disease severity in humans who had prior DENV exposure and were later infected with ZIKV. A variety of different strategies have been employed in developing vaccines for ZIKV. Live, attenuated viruses have been generated that express ZIKV prM-E on a DENV2 backbone (Xie et al., 2017) or have deletions in the 3′ UTR (Muruato et al., 2019; Shan et al., 2017a,b). Each of the vaccines reduced ZIKV titers in vitro and in vivo and protected mice from lethal ZIKV infection, with the vaccine developed by Shan et al. additionally preventing vertical transmission and testis damage in mice, and protecting rhesus macaques from ZIKV infection (Shan et al., 2017a). Other groups have utilized viral vectors such as vesicular stomatitis virus (VSV) expressing full-length prM-E or prM and soluble E with the Ebola virus glycoprotein (Emanuel et al., 2018) and measles vectors expressing prM and soluble E (MV-Zika-sE) (Nurnberger et al., 2019). The ZIKV vaccines on the VSV background protect *Ifnar1*$^{-/-}$ mice from lethal ZIKV challenge (Emanuel et al., 2018), and the MV-Zika-sE protected mice from lethality and viremia, in addition to induce robust neutralizing antibodies and T cell proliferation (Nurnberger et al., 2019). Vaccines can also use adenovirus vectors, including one using rhesus adenovirus 52 expressing ZIKV prM-E, which protects rhesus monkeys from infection and generated neutralizing antibodies (Xu et al., 2018), and a recently developed gorilla adenovirus-based vaccine that protects mice from vertical transmission of ZIKV and induces durable immunity in mice (Hassan et al., 2019).

Other highly pursued ZIKV vaccine platforms utilize the generation of both DNA and mRNA vaccines. Several groups concurrently developed DNA vaccines (Dowd et al., 2016; Gaudinski et al., 2018; Griffin et al., 2017; Tebas et al., 2017; Zou et al., 2018), with one (VRC5283) having advanced to phase II clinical trials. Multiple mRNA vaccines have performed well in pre-clinical studies, including a lipid nanoparticle (LNP) encapsulated mRNA vaccine that induces the production of highly neutralizing antibodies and protects from lethal ZIKV infection in mice (Richner et al., 2017) and elicits strong, durable neutralizing antibody responses in mice and NHPs (Pardi et al., 2017).

Although there are currently no licensed vaccines for ZIKV, the advances in technology that have built upon strategies employed to develop vaccines against other flaviviruses have been highly effective in producing vaccines that are protective and immunogenic in various cell and animal models. Many vaccines are currently undergoing clinical trial, but whether any of these vaccines will survive and be approved for use remains to be determined.

7. Summary

Since the explosive 2015 epidemic, ZIKV has become one of the most widely known and studied flaviviruses. Early ZIKV studies built a platform for the tremendous scientific advances that have occurred within the past

5 years. Numerous rodent and non-human primate models of infection have provided essential insight into ZIKV's neurotropism, capacity for transplacental congenital infection, and other aspects of immunopathology that follow infection. Furthermore, these models have allowed the testing of various therapeutics and vaccines. Although new cases have decreased since the most recent outbreak, the inability to currently treat ZIKV infection and the unique variety of clinical outcomes attributed to ZIKV cause it to remain a major topic of biomedical research.

References

Abbink, P., Larocca, R.A., Dejnirattisai, W., Peterson, R., Nkolola, J.P., Borducchi, E.N., Supasa, P., Mongkolsapaya, J., Screaton, G.R., Barouch, D.H., 2018. Therapeutic and protective efficacy of a dengue antibody against Zika infection in rhesus monkeys. Nat. Med. 24, 721–723.

Aguirre, S., Maestre, A.M., Pagni, S., Patel, J.R., Savage, T., Gutman, D., Maringer, K., Bernal-Rubio, D., Shabman, R.S., Simon, V., et al., 2012. DENV inhibits type I IFN production in infected cells by cleaving human STING. PLoS Pathog. 8, e1002934.

Aguirre, S., Luthra, P., Sanchez-Aparicio, M.T., Maestre, A.M., Patel, J., Lamothe, F., Fredericks, A.C., Tripathi, S., Zhu, T., Pintado-Silva, J., et al., 2017. Dengue virus NS2B protein targets cGAS for degradation and prevents mitochondrial DNA sensing during infection. Nat. Microbiol. 2, 17037.

Aliota, M.T., Caine, E.A., Walker, E.C., Larkin, K.E., Camacho, E., Osorio, J.E., 2016. Characterization of lethal Zika virus infection in AG129 mice. PLoS Negl. Trop. Dis. 10, e0004682.

Alpert, S.G., Fergerson, J., Noel, L.P., 2003. Intrauterine West Nile virus: ocular and systemic findings. Am. J. Ophthalmol. 136, 733–735.

Barba-Spaeth, G., Dejnirattisai, W., Rouvinski, A., Vaney, M.C., Medits, I., Sharma, A., Simon-Loriere, E., Sakuntabhai, A., Cao-Lormeau, V.M., Haouz, A., et al., 2016. Structural basis of potent Zika-dengue virus antibody cross-neutralization. Nature 536, 48–53.

Barber, G.N., 2015. STING: infection, inflammation and cancer. Nat. Rev. Immunol. 15, 760–770.

Bardina, S.V., Bunduc, P., Tripathi, S., Duehr, J., Frere, J.J., Brown, J.A., Nachbagauer, R., Foster, G.A., Krysztof, D., Tortorella, D., et al., 2017. Enhancement of Zika virus pathogenesis by preexisting antiflavivirus immunity. Science 356, 175–180.

Barnard, B.J., Voges, S.F., 1986. Flaviviruses in South Africa: pathogenicity for sheep. Onderstepoort J. Vet. Res. 53, 235–238.

Barnard, T.R., Rajah, M.M., Sagan, S.M., 2018. Contemporary Zika virus isolates induce more dsRNA and produce more negative-strand intermediate in human astrocytoma cells. Viruses 10.

Bearcroft, W.G., 1956. Zika virus infection experimentally induced in a human volunteer. Trans. R. Soc. Trop. Med. Hyg. 50, 442–448.

Boorman, J.P., Porterfield, J.S., 1956. A simple technique for infection of mosquitoes with viruses; transmission of Zika virus. Trans. R. Soc. Trop. Med. Hyg. 50, 238–242.

Brasil, P., Pereira Jr., J.P., Moreira, M.E., Ribeiro Nogueira, R.M., Damasceno, L., Wakimoto, M., Rabello, R.S., Valderramos, S.G., Halai, U.A., Salles, T.S., et al., 2016. Zika virus infection in pregnant women in Rio de Janeiro. N. Engl. J. Med. 375, 2321–2334.

Brouillette, R.B., Phillips, E.K., Patel, R., Mahauad-Fernandez, W., Moller-Tank, S., Rogers, K.J., Dillard, J.A., Cooney, A.L., Martinez-Sobrido, L., Okeoma, C., et al., 2018. TIM-1 mediates dystroglycan-independent entry of Lassa virus. J. Virol. 92.

Cao, X., 2016. Self-regulation and cross-regulation of pattern-recognition receptor signalling in health and disease. Nat. Rev. Immunol. 16, 35–50.

Castro, M.C., Han, Q.C., Carvalho, L.R., Victora, C.G., Franca, G.V.A., 2018. Implications of Zika virus and congenital Zika syndrome for the number of live births in Brazil. Proc. Natl. Acad. Sci. U. S. A. 115, 6177–6182.

Cauchemez, S., Besnard, M., Bompard, P., Dub, T., Guillemette-Artur, P., Eyrolle-Guignot, D., Salje, H., Van Kerkhove, M.D., Abadie, V., Garel, C., et al., 2016. Association between Zika virus and microcephaly in French Polynesia, 2013-15: a retrospective study. Lancet 387, 2125–2132.

Centers for Disease Control and Prevention (CDC), 2002. Intrauterine West Nile virus infection—New York. Morb. Mortal. Wkly Rep. 51, 1135–1136.

Chambers, T.J., Nestorowicz, A., Mason, P.W., Rice, C.M., 1999. Yellow fever/Japanese encephalitis chimeric viruses: construction and biological properties. J. Virol. 73, 3095–3101.

Chang, H.H., Huber, R.G., Bond, P.J., Grad, Y.H., Camerini, D., Maurer-Stroh, S., Lipsitch, M., 2017. Systematic analysis of protein identity between Zika virus and other arthropod-borne viruses. Bull. World Health Organ. 95, 517–525i.

Chaturvedi, U.C., Mathur, A., Chandra, A., Das, S.K., Tandon, H.O., Singh, U.K., 1980. Transplacental infection with Japanese encephalitis virus. J. Infect. Dis. 141, 712–715.

Cugola, F.R., Fernandes, I.R., Russo, F.B., Freitas, B.C., Dias, J.L., Guimaraes, K.P., Benazzato, C., Almeida, N., Pignatari, G.C., Romero, S., et al., 2016. The Brazilian Zika virus strain causes birth defects in experimental models. Nature 534, 267–271.

Dai, L., Song, J., Lu, X., Deng, Y.Q., Musyoki, A.M., Cheng, H., Zhang, Y., Yuan, Y., Song, H., Haywood, J., et al., 2016. Structures of the Zika virus envelope protein and its complex with a flavivirus broadly protective antibody. Cell Host Microbe 19, 696–704.

de Paula, F.B., de Oliveira Dias, J.R., Prazeres, J., Sacramento, G.A., Ko, A.I., Maia, M., Belfort Jr., R., 2016. Ocular findings in infants with microcephaly associated with presumed Zika virus congenital infection in Salvador, Brazil. JAMA Ophthalmol. 134, 529–535.

Deckard, D.T., Chung, W.M., Brooks, J.T., Smith, J.C., Woldai, S., Hennessey, M., Kwit, N., Mead, P., 2016. Male-to-male sexual transmission of Zika virus—Texas, January 2016. MMWR Morb. Mortal. Wkly Rep. 65, 372–374.

Dejnirattisai, W., Supasa, P., Wongwiwat, W., Rouvinski, A., Barba-Spaeth, G., Duangchinda, T., Sakuntabhai, A., Cao-Lormeau, V.M., Malasit, P., Rey, F.A., et al., 2016. Dengue virus sero-cross-reactivity drives antibody-dependent enhancement of infection with zika virus. Nat. Immunol. 17, 1102–1108.

Deng, C., Liu, S., Zhang, Q., Xu, M., Zhang, H., Gu, D., Shi, L., He, J., Xiao, G., Zhang, B., 2016. Isolation and characterization of Zika virus imported to China using C6/36 mosquito cells. Virol. Sin. 31, 176–179.

Dick, G.W., 1952. Zika virus. II. Pathogenicity and physical properties. Trans. R. Soc. Trop. Med. Hyg. 46, 521–534.

Dick, G.W., Kitchen, S.F., Haddow, A.J., 1952. Zika virus. I. Isolations and serological specificity. Trans. R. Soc. Trop. Med. Hyg. 46, 509–520.

Dowd, K.A., Ko, S.Y., Morabito, K.M., Yang, E.S., Pelc, R.S., DeMaso, C.R., Castilho, L.R., Abbink, P., Boyd, M., Nityanandam, R., et al., 2016. Rapid development of a DNA vaccine for Zika virus. Science 354, 237–240.

Dudley, D.M., Aliota, M.T., Mohr, E.L., Weiler, A.M., Lehrer-Brey, G., Weisgrau, K.L., Mohns, M.S., Breitbach, M.E., Rasheed, M.N., Newman, C.M., et al., 2016. A rhesus macaque model of Asian-lineage Zika virus infection. Nat. Commun. 7, 12204.

Dudley, D.M., Newman, C.M., Lalli, J., Stewart, L.M., Koenig, M.R., Weiler, A.M., Semler, M.R., Barry, G.L., Zarbock, K.R., Mohns, M.S., et al., 2017. Infection via mosquito bite alters Zika virus tissue tropism and replication kinetics in rhesus macaques. Nat. Commun. 8, 2096.

Duffy, M.R., Chen, T.H., Hancock, W.T., Powers, A.M., Kool, J.L., Lanciotti, R.S., Pretrick, M., Marfel, M., Holzbauer, S., Dubray, C., et al., 2009. Zika virus outbreak on Yap Island, Federated States of Micronesia. N. Engl. J. Med. 360, 2536–2543.

Eckels, K.H., Yu, Y.X., Dubois, D.R., Marchette, N.J., Trent, D.W., Johnson, A.J., 1988. Japanese encephalitis virus live-attenuated vaccine, Chinese strain SA14-14-2; adaptation to primary canine kidney cell cultures and preparation of a vaccine for human use. Vaccine 6, 513–518.

Emanuel, J., Callison, J., Dowd, K.A., Pierson, T.C., Feldmann, H., Marzi, A., 2018. A VSV-based Zika virus vaccine protects mice from lethal challenge. Sci. Rep. 8, 11043.

Fernandez, E., Dejnirattisai, W., Cao, B., Scheaffer, S.M., Supasa, P., Wongwiwat, W., Esakky, P., Drury, A., Mongkolsapaya, J., Moley, K.H., et al., 2017. Human antibodies to the dengue virus E-dimer epitope have therapeutic activity against Zika virus infection. Nat. Immunol. 18, 1261–1269.

Foy, B.D., Kobylinski, K.C., Chilson Foy, J.L., Blitvich, B.J., Travassos da Rosa, A., Haddow, A.D., Lanciotti, R.S., Tesh, R.B., 2011. Probable non-vector-borne transmission of Zika virus, Colorado, USA. Emerg. Infect. Dis. 17, 880–882.

Furtado, J.M., Esposito, D.L., Klein, T.M., Teixeira-Pinto, T., da Fonseca, B.A., 2016. Uveitis associated with Zika virus infection. N. Engl. J. Med. 375, 394–396.

Gao, D., Lou, Y., He, D., Porco, T.C., Kuang, Y., Chowell, G., Ruan, S., 2016. Prevention and control of Zika as a mosquito-borne and sexually transmitted disease: a mathematical modeling analysis. Sci. Rep. 6, 28070.

Gaudinski, M.R., Houser, K.V., Morabito, K.M., Hu, Z., Yamshchikov, G., Rothwell, R.S., Berkowitz, N., Mendoza, F., Saunders, J.G., Novik, L., et al., 2018. Safety, tolerability, and immunogenicity of two Zika virus DNA vaccine candidates in healthy adults: randomised, open-label, phase 1 clinical trials. Lancet 391, 552–562.

Gorman, M.J., Caine, E.A., Zaitsev, K., Begley, M.C., Weger-Lucarelli, J., Uccellini, M.B., Tripathi, S., Morrison, J., Yount, B.L., Dinnon 3rd, K.H., et al., 2018. An immunocompetent mouse model of Zika virus infection. Cell Host Microbe 23, 672–685.e676.

Govero, J., Esakky, P., Scheaffer, S.M., Fernandez, E., Drury, A., Platt, D.J., Gorman, M.J., Richner, J.M., Caine, E.A., Salazar, V., et al., 2016. Zika virus infection damages the testes in mice. Nature 540, 438–442.

Grant, A., Ponia, S.S., Tripathi, S., Balasubramaniam, V., Miorin, L., Sourisseau, M., Schwarz, M.C., Sanchez-Seco, M.P., Evans, M.J., Best, S.M., et al., 2016. Zika virus targets human STAT2 to inhibit type I interferon signaling. Cell Host Microbe 19, 882–890.

Griffin, B.D., Muthumani, K., Warner, B.M., Majer, A., Hagan, M., Audet, J., Stein, D.R., Ranadheera, C., Racine, T., De La Vega, M.A., et al., 2017. DNA vaccination protects mice against Zika virus-induced damage to the testes. Nat. Commun. 8, 15743.

Ha, Y.R., Kim, J.H., Ryu, J., Lee, S.J., 2017. Superb feeding behavior of Aedes albopictus transmitting Zika virus. PLoS One 12, e0184871.

Halstead, S.B., 1979. In vivo enhancement of dengue virus infection in rhesus monkeys by passively transferred antibody. J Infect Dis 140, 527–533.

Halstead, S.B., 1988. Pathogenesis of dengue: challenges to molecular biology. Science 239, 476–481.

Hamel, R., Dejarnac, O., Wichit, S., Ekchariyawat, P., Neyret, A., Luplertlop, N., Perera-Lecoin, M., Surasombatpattana, P., Talignani, L., Thomas, F., et al., 2015. Biology of Zika virus infection in human skin cells. J. Virol. 89, 8880–8896.

Hassan, A.O., Dmitriev, I.P., Kashentseva, E.A., Zhao, H., Brough, D.E., Fremont, D.H., Curiel, D.T., Diamond, M.S., 2019. A gorilla adenovirus-based vaccine against Zika virus induces durable immunity and confers protection in pregnancy. Cell Rep. 28, 2634–2646.e2634.

Hastings, A.K., Yockey, L.J., Jagger, B.W., Hwang, J., Uraki, R., Gaitsch, H.F., Parnell, L.A., Cao, B., Mysorekar, I.U., Rothlin, C.V., et al., 2017. TAM receptors are not required for Zika virus infection in mice. Cell Rep. 19, 558–568.

Hayashida, E., Ling, Z.L., Ashhurst, T.M., Viengkhou, B., Jung, S.R., Songkhunawej, P., West, P.K., King, N.J.C., Hofer, M.J., 2019. Zika virus encephalitis in immunocompetent mice is dominated by innate immune cells and does not require T or B cells. J. Neuroinflammation 16, 177.

Honein, M.A., Dawson, A.L., Petersen, E.E., Jones, A.M., Lee, E.H., Yazdy, M.M., Ahmad, N., Macdonald, J., Evert, N., Bingham, A., et al., 2017. Birth defects among fetuses and infants of US women with evidence of possible Zika virus infection during pregnancy. JAMA 317, 59–68.

Icheku, V., 2017. A review of the evidence linking Zika virus to the developmental abnormalities that lead to microcephaly in view of recent cases of birth defects in Africa. J. Mol. Pathol. Epidemiol. 2, 1–4.

Ishikawa, H., Barber, G.N., 2008. STING is an endoplasmic reticulum adaptor that facilitates innate immune signalling. Nature 455, 674–678.

Ishikawa, H., Ma, Z., Barber, G.N., 2009. STING regulates intracellular DNA-mediated, type I interferon-dependent innate immunity. Nature 461, 788–792.

Jagger, B.W., Miner, J.J., Cao, B., Arora, A., Smith, A.M., Kovacs, A., Mysorekar, I.U., Coyne, C.B., Diamond, M.S., 2017. Gestational stage and IFN-lambda signaling influence ZIKV infection in utero. Cell Host Microbe 22, 366–376.

Joguet, G., Mansuy, J.M., Matusali, G., Hamdi, S., Walschaerts, M., Pavili, L., Guyomard, S., Prisant, N., Lamarre, P., Dejucq-Rainsford, N., et al., 2017. Effect of acute Zika virus infection on sperm and virus clearance in body fluids: a prospective observational study. Lancet Infect. Dis. 17, 1200–1208.

Kawasaki, T., Kawai, T., 2014. Toll-like receptor signaling pathways. Front. Immunol. 5, 461.

Kostyuchenko, V.A., Lim, E.X., Zhang, S., Fibriansah, G., Ng, T.S., Ooi, J.S., Shi, J., Lok, S.M., 2016. Structure of the thermally stable Zika virus. Nature 533, 425–428.

Kuno, G., Chang, G.J., 2007. Full-length sequencing and genomic characterization of Bagaza, Kedougou, and Zika viruses. Arch. Virol. 152, 687–696.

Lanciotti, R.S., Kosoy, O.L., Laven, J.J., Velez, J.O., Lambert, A.J., Johnson, A.J., Stanfield, S.M., Duffy, M.R., 2008. Genetic and serologic properties of Zika virus associated with an epidemic, Yap State, Micronesia, 2007. Emerg. Infect. Dis. 14, 1232–1239.

Lazear, H.M., Govero, J., Smith, A.M., Platt, D.J., Fernandez, E., Miner, J.J., Diamond, M.S., 2016. A mouse model of Zika virus pathogenesis. Cell Host Microbe 19, 720–730.

Li, X.F., Dong, H.L., Huang, X.Y., Qiu, Y.F., Wang, H.J., Deng, Y.Q., Zhang, N.N., Ye, Q., Zhao, H., Liu, Z.Y., et al., 2016. Characterization of a 2016 clinical isolate of Zika virus in non-human primates. eBioMedicine 12, 170–177.

Ma, W., Li, S., Ma, S., Jia, L., Zhang, F., Zhang, Y., Zhang, J., Wong, G., Zhang, S., Lu, X., et al., 2016. Zika virus causes testis damage and leads to male infertility in mice. Cell 167, 1511–1524.e1510.

Macnamara, F.N., 1954. Zika virus: a report on three cases of human infection during an epidemic of jaundice in Nigeria. Trans. R. Soc. Trop. Med. Hyg. 48, 139–145.

Marchette, N.J., Garcia, R., Rudnick, A., 1969. Isolation of Zika virus from Aedes aegypti mosquitoes in Malaysia. Am. J. Trop. Med. Hyg. 18, 411–415.

Meertens, L., Labeau, A., Dejarnac, O., Cipriani, S., Sinigaglia, L., Bonnet-Madin, L., Le Charpentier, T., Hafirassou, M.L., Zamborlini, A., Cao-Lormeau, V.M., et al., 2017. Axl mediates ZIKA virus entry in human glial cells and modulates innate immune responses. Cell Rep. 18, 324–333.

Miner, J.J., Sene, A., Richner, J.M., Smith, A.M., Santeford, A., Ban, N., Weger-Lucarelli, J., Manzella, F., Ruckert, C., Govero, J., et al., 2016a. Zika virus infection in mice causes panuveitis with shedding of virus in tears. Cell Rep. 16, 3208–3218.

Miner, J.J., Cao, B., Govero, J., Smith, A.M., Fernandez, E., Cabrera, O.H., Garber, C., Noll, M., Klein, R.S., Noguchi, K.K., et al., 2016b. Zika virus infection during pregnancy in mice causes placental damage and fetal demise. Cell 165, 1081–1091.

Muruato, A.E., Shan, C., Fontes-Garfias, C.R., Liu, Y., Cao, Z., Gao, Q., Weaver, S.C., Shi, P.Y., 2019. Genetic stability of live-attenuated Zika vaccine candidates. Antiviral Res. 171, 104596.

Musso, D., Roche, C., Robin, E., Nhan, T., Teissier, A., Cao-Lormeau, V.M., 2015. Potential sexual transmission of Zika virus. Emerg. Infect. Dis. 21, 359–361.

Ni, H., Ryman, K.D., Wang, H., Saeed, M.F., Hull, R., Wood, D., Minor, P.D., Watowich, S.J., Barrett, A.D., 2000. Interaction of yellow fever virus French neurotropic vaccine strain with monkey brain: characterization of monkey brain membrane receptor escape variants. J. Virol. 74, 2903–2906.

Niemeyer, B., Niemeyer, R., Borges, R., Marchiori, E., 2017. Acute disseminated encephalomyelitis following Zika virus infection. Eur. Neurol. 77, 45–46.

Nurnberger, C., Bodmer, B.S., Fiedler, A.H., Gabriel, G., Muhlebach, M.D., 2019. A measles virus-based vaccine candidate mediates protection against Zika virus in an allogeneic mouse pregnancy model. J. Virol. 93.

Olson JG, Ksiazek TG, Suhandiman, Triwibowo: Zika virus, a cause of fever in Central Java, Indonesia. Trans. R. Soc. Trop. Med. Hyg. 1981, 75:389–393.

Osuna, C.E., Lim, S.Y., Deleage, C., Griffin, B.D., Stein, D., Schroeder, L.T., Omange, R.W., Best, K., Luo, M., Hraber, P.T., et al., 2016. Zika viral dynamics and shedding in rhesus and cynomolgus macaques. Nat. Med. 22, 1448–1455.

Pardi, N., Hogan, M.J., Pelc, R.S., Muramatsu, H., Andersen, H., DeMaso, C.R., Dowd, K.A., Sutherland, L.L., Scearce, R.M., Parks, R., et al., 2017. Zika virus protection by a single low-dose nucleoside-modified mRNA vaccination. Nature 543, 248–251.

Paul, D., Bartenschlager, R., 2015. Flaviviridae replication organelles: oh, what a tangled web we weave. Annu. Rev. Virol. 2, 289–310.

Platt, D.J., Smith, A.M., Arora, N., Diamond, M.S., Coyne, C.B., Miner, J.J., 2018. Zika virus–related neurotropic flaviviruses infect human placental explants and cause fetal demise in mice. Sci. Transl. Med. 10, eaao7090.

Priyamvada, L., Quicke, K.M., Hudson, W.H., Onlamoon, N., Sewatanon, J., Edupuganti, S., Pattanapanyasat, K., Chokephaibulkit, K., Mulligan, M.J., Wilson, P.C., et al., 2016. Human antibody responses after dengue virus infection are highly cross-reactive to Zika virus. Proc. Natl. Acad. Sci. U. S. A. 113, 7852–7857.

Rawle, R.J., Webster, E.R., Jelen, M., Kasson, P.M., Boxer, S.G., 2018. pH dependence of Zika membrane fusion kinetics reveals an off-pathway state. ACS Cent. Sci. 4, 1503–1510.

Richner, J.M., Himansu, S., Dowd, K.A., Butler, S.L., Salazar, V., Fox, J.M., Julander, J.G., Tang, W.W., Shresta, S., Pierson, T.C., et al., 2017. Modified mRNA vaccines protect against Zika virus infection. Cell 168, 1114–1125.e1110.

Rivera-Correa, J., de Siqueira, I.C., Mota, S., do Rosário, M.S., Pereira de Jesus, P.A., LCJ, A., Ernst, J.D., Rodriguez, A., 2019. Anti-ganglioside antibodies in patients with Zika virus infection-associated Guillain-Barré Syndrome in Brazil. PLoS Negl. Trop. Dis. 13, e0007695.

Rogatsky, I., Chandrasekaran, U., Manni, M., Yi, W., Pernis, A.B., 2014. Epigenetics and the IRFs: a complex interplay in the control of immunity and autoimmunity. Autoimmunity 47, 242–255.

Schoggins, J.W., MacDuff, D.A., Imanaka, N., Gainey, M.D., Shrestha, B., Eitson, J.L., Mar, K.B., Richardson, R.B., Ratushny, A.V., Litvak, V., et al., 2014. Pan-viral specificity of IFN-induced genes reveals new roles for cGAS in innate immunity. Nature 505, 691–695.

Shan, C., Muruato, A.E., Jagger, B.W., Richner, J., Nunes, B.T.D., Medeiros, D.B.A., Xie, X., Nunes, J.G.C., Morabito, K.M., Kong, W.P., et al., 2017a. A single-dose live-attenuated vaccine prevents Zika virus pregnancy transmission and testis damage. Nat. Commun. 8, 676.

Shan, C., Muruato, A.E., Nunes, B.T.D., Luo, H., Xie, X., Medeiros, D.B.A., Wakamiya, M., Tesh, R.B., Barrett, A.D., Wang, T., et al., 2017b. A live-attenuated Zika virus vaccine candidate induces sterilizing immunity in mouse models. Nat. Med. 23, 763–767.

Shapiro-Mendoza, C.K., Rice, M.E., Galang, R.R., Fulton, A.C., VanMaldeghem, K., Prado, M.V., Ellis, E., Anesi, M.S., Simeone, R.M., Petersen, E.E., et al., 2017. Pregnancy outcomes after maternal Zika virus infection during pregnancy—U.S. territories, January 1, 2016–April 25, 2017. Morb. Mortal. Wkly Rep. 66, 615–621.

Simpson, D.I., 1964. Zika virus infection in man. Trans. R. Soc. Trop. Med. Hyg. 58, 335–338.

Sirohi, D., Chen, Z., Sun, L., Klose, T., Pierson, T.C., Rossmann, M.G., Kuhn, R.J., 2016. The 3.8 A resolution cryo-EM structure of Zika virus. Science 352, 467–470.

Soares, C.N., Brasil, P., Carrera, R.M., Sequeira, P., de Filippis, A.B., Borges, V.A., Theophilo, F., Ellul, M.A., Solomon, T., 2016. Fatal encephalitis associated with Zika virus infection in an adult. J. Clin. Virol. 83, 63–65.

Styer, L.M., Kent, K.A., Albright, R.G., Bennett, C.J., Kramer, L.D., Bernard, K.A., 2007. Mosquitoes inoculate high doses of West Nile virus as they probe and feed on live hosts. PLoS Pathog. 3, 1262–1270.

Sun, L., Wu, J., Du, F., Chen, X., Chen, Z.J., 2013. Cyclic GMP-AMP synthase is a cytosolic DNA sensor that activates the type I interferon pathway. Science 339, 786–791.

Takei, K., Haucke, V., 2001. Clathrin-mediated endocytosis: membrane factors pull the trigger. Trends Cell Biol. 11, 385–391.

Tan, T.Y., Fibriansah, G., Kostyuchenko, V.A., Ng, T.S., Lim, X.X., Zhang, S., Lim, X.N., Wang, J., Shi, J., Morais, M.C., et al., 2020. Capsid protein structure in Zika virus reveals the flavivirus assembly process. Nat. Commun. 11, 895.

Tang, H., Hammack, C., Ogden, S.C., Wen, Z., Qian, X., Li, Y., Yao, B., Shin, J., Zhang, F., Lee, E.M., et al., 2016. Zika virus infects human cortical neural progenitors and attenuates their growth. Cell Stem Cell 18, 587–590.

Tebas, P., Roberts, C.C., Muthumani, K., Reuschel, E.L., Kudchodkar, S.B., Zaidi, F.I., White, S., Khan, A.S., Racine, T., Choi, H., et al., 2017. Safety and immunogenicity of an anti-Zika virus DNA vaccine—preliminary report. N. Engl. J. Med.

Tsai, T.F., Paul, R., Lynberg, M.C., Letson, G.W., 1993. Congenital yellow fever virus infection after immunization in pregnancy. J. Infect. Dis. 168, 1520–1523.

Venter, M., Human, S., van Niekerk, S., Williams, J., van Eeden, C., Freeman, F., 2011. Fatal neurologic disease and abortion in mare infected with lineage 1 West Nile virus, South Africa. Emerg. Infect. Dis. 17, 1534–1536.

Vinton, C.L., Magaziner, S.J., Dowd, K.A., Robertson, S.J., Amaro-Carambot, E., Karmele, E.P., Ortiz, A.M., Starke, C.E., Mudd, J.C., Whitehead, S.S., et al., 2019. Simian immunodeficiency virus infection of rhesus macaques results in delayed Zika virus clearance. mBio 10.

Voermans, J.J.C., Pas, S.D., van der Linden, A., GeurtsvanKessel, C., Koopmans, M., van der Eijk, A., Reusken, C., 2019. Whole-blood testing for diagnosis of acute Zika virus infections in routine diagnostic setting. Emerg. Infect. Dis. 25, 1394–1396.

Way, J.H., Bowen, E.T., Platt, G.S., 1976. Comparative studies of some African arboviruses in cell culture and in mice. J. Gen. Virol. 30, 123–130.

World Health Organization, 2020. Tick-Borne Encephalitis. WHO.

Xia, H., Luo, H., Shan, C., Muruato, A.E., Nunes, B.T.D., Medeiros, D.B.A., Zou, J., Xie, X., Giraldo, M.I., Vasconcelos, P.F.C., et al., 2018. An evolutionary NS1 mutation enhances Zika virus evasion of host interferon induction. Nat. Commun. 9, 414.

Xie, X., Yang, Y., Muruato, A.E., Zou, J., Shan, C., Nunes, B.T., Medeiros, D.B., Vasconcelos, P.F., Weaver, S.C., Rossi, S.L., et al., 2017. Understanding Zika virus stability and developing a chimeric vaccine through functional analysis. mBio 8.

Xu, K., Song, Y., Dai, L., Zhang, Y., Lu, X., Xie, Y., Zhang, H., Cheng, T., Wang, Q., Huang, Q., et al., 2018. Recombinant chimpanzee adenovirus vaccine AdC7-M/E protects against Zika virus infection and testis damage. J. Virol. 92.

Yuan, L., Huang, X.Y., Liu, Z.Y., Zhang, F., Zhu, X.L., Yu, J.Y., Ji, X., Xu, Y.P., Li, G., Li, C., et al., 2017. A single mutation in the prM protein of Zika virus contributes to fetal microcephaly. Science 358, 933–936.

Zhao, H., Fernandez, E., Dowd, K.A., Speer, S.D., Platt, D.J., Gorman, M.J., Govero, J., Nelson, C.A., Pierson, T.C., Diamond, M.S., et al., 2016. Structural basis of Zika virus-specific antibody protection. Cell 166, 1016–1027.

Zou, J., Xie, X., Luo, H., Shan, C., Muruato, A.E., Weaver, S.C., Wang, T., Shi, P.Y., 2018. A single-dose plasmid-launched live-attenuated Zika vaccine induces protective immunity. eBioMedicine 36, 92–102.

CHAPTER SEVEN

One hundred years of (influenza) immunopathology

David F. Boyd[a], Taylor L. Wilson[a,b], Paul G. Thomas[a,b,*]

[a]Department of Immunology, St. Jude Children's Research Hospital, Memphis, TN, United States
[b]Department of Microbiology, Immunology, and Biochemistry, College of Graduate Health Sciences, University of Tennessee Health Science Center, Memphis, TN, United States
*Corresponding author: e-mail address: paul.thomas@stjude.org

Contents

Abstract

It has been over 100 years since the 1918 influenza pandemic, one of the most infamous examples of viral immunopathology. Since that time, there has been an inevitable repetition of influenza pandemics every few decades and yearly influenza seasons, which have a significant impact on human health. Recently, noteworthy progress has been made in defining the cellular and molecular mechanisms underlying pathology induced by an exuberant host response to influenza virus infection. Infection with influenza viruses is associated with a wide spectrum of disease, from mild symptoms to severe complications including respiratory failure, and the severity of influenza disease is driven

Advances in Virus Research, Volume 107
ISSN 0065-3527
https://doi.org/10.1016/bs.aivir.2020.06.005

by a complex interplay of viral and host factors. This chapter will discuss mechanisms of infection severity using concepts of disease resistance and tolerance as a framework for understanding the balance between viral clearance and immunopathology. We review mechanistic studies in animal models of infection and correlational studies in humans that have begun to define these factors and discuss promising host therapeutic targets to improve outcomes from severe influenza disease.

1. Vaccination, treatment, and health burden of influenza

Although we have all become accustomed to the idea of an annual "flu season" preceded by reminders to get our flu vaccines, influenza virus infections remain a major cause of viral-related illness in the US and across the globe, placing a significant burden on healthcare systems. Each flu season, the US can expect several million symptomatic cases of the virus. The CDC estimates that there were almost 500,000 influenza-related hospitalizations and 34,000 deaths in the United States during the 2018–2019 season alone (CDC, 2020c). These numbers can become even larger during an outbreak of pandemic proportions. As a sobering example, the 1918 influenza pandemic strain of H1N1 infected almost one third of the world population (500 million) and caused an estimated 50 million deaths—675,000 of those deaths occurring within the US (CDC, 2020a). The more recent 2009 H1N1 pandemic caused 60 million symptomatic cases in the US with an estimate of 12,000 deaths (CDC, 2020b). Each year, the healthcare community braces for the impact of influenza by attempting to predict which strains will be most prevalent and educating the general public about preventative measures against the spread of the virus.

Healthcare providers do have a number of important tools to either prevent infection (vaccines) or limit disease severity once infection has occurred (antivirals). These tools, however, have their limitations. Vaccine efficacy varies year to year depending on the predominant circulating strain and prior infection/vaccination history of individuals in the population while antivirals have a narrow window of efficacy. Individuals that are hospitalized, currently experiencing a severe illness, or have risk factors for severe influenza are treated with antiviral therapies (Uyeki, 2014; Uyeki et al., 2019). Inhibition of the viral enzyme neuraminidase, responsible for cleaving virions from the surface of infected cells, has shown the greatest efficacy in slowing viral shedding. Recommended neuraminidase inhibitors are oral oseltamivir, inhaled zanamivir, or intravenous peramivir (Uyeki, 2014). The adamantanes, amantadine and rimantadine, were once the most popular treatment for severe influenza. This class of drug inhibits the M2 ion channel, preventing viral replication. Currently, the CDC does not recommend the use of adamantanes for treatment due to greater risk of developing resistance (CDC, 2019a). The proportion of adamantane-resistant strains in the US range from 25%–96%, depending on the season and strain most prevalent (Hussain et al., 2017). Resistance to the neuraminidase inhibitors is less common in the predominant influenza A strains at $<3.5\%$ for oseltamivir and $<1\%$ for zanamivir (Lee and Hurt, 2018). The endonuclease inhibitor, baloxivir, is also approved for the treatment of influenza. Recent studies, though, have confirmed that the predominant strains during the 2018–2019 season in Japan have reduced susceptibility to baloxivir (Imai et al., 2020).

There is a limited window within which antivirals can be effective at controlling symptom severity. Administration of neuraminidase inhibitors are recommended to begin early, within 48h after symptom onset (Cooper et al., 2003; Monto et al., 1999; Nicholson et al., 2000). Limited window of efficacy, delayed symptom alleviation, and the incidence of viral resistance to antiviral therapies, has led to a renewed interest in the use of therapeutics directed against the host. Emphasis on these host-derived therapeutics could significantly improve clinical outcomes for patients earlier and for longer periods than conventional antivirals without the potential for the virus to circumvent the medication's mechanism of action (Yip et al., 2018).

Despite vaccination efforts and the use of antivirals for treatment, hospitalization and death from influenza infection are common. Understanding mechanisms underlying both virus and host induced tissue damage will not only increase our ability to manage influenza effectively but will also offer greater understanding of other causes of pulmonary infection leading to acute lung injury. The initiating pathways that result in lung injury are unique to each infection type. Yet, these pathways likely coalesce and lead to the similar downstream effect of diffuse alveolar damage (Hendrickson and Matthay, 2013).

2. The spectrum of influenza disease

The disease associated with seasonal influenza spans a broad spectrum of symptoms. The rate of asymptomatic cases is difficult to determine, but studies have estimated that anywhere from 5% to 35% of infected individuals will display no detectable symptoms. Another subset of those infected will experience symptomatic disease of varying severity. These subclinical cases comprise 25%–61% of the total cases (Furuya-Kanamori et al., 2016). A smaller subset of patients will develop severe disease requiring hospitalization. Severity of infection is usually associated with viral replication and inflammation in the respiratory tract. The prevalence of these severe cases increases with a pandemic outbreak of the virus. Most adults have some level of protective immunity to influenza viruses due to prior infection or vaccination. However, antigenic shift, the mixing of genomic segments of different influenza viruses, can result in novel subtypes that the population has not encountered previously, leading to global spread of the virus and large numbers of severe cases (Glezen, 1996).

The most infamous example of the introduction of a novel influenza virus to humans following antigenic shift is the pandemic of 1918. Genetic analysis of the 1918 strain indicates that it contained hemagglutinin (HA) and neuraminidase (NA) genes most like that of avian influenza strains (Taubenberger, 2006). The 1918 pandemic was exceptional for many reasons, but one striking feature was a "W-shaped" mortality curve in which not only young children and the elderly exhibited high mortality rates, but also adults ages 20–40 years (Taubenberger and Morens, 2019). This "W-shaped" mortality curve has led to the hypothesis that the increased mortality in younger adults was due to an exuberant immune response that caused extensive lung damage in infected individuals. Other examples of severe influenza can be seen in nonpandemic avian strains of the virus. Avian H5N1 and H7N9 strains have caused the largest number of cases in humans. None of the avian strains of influenza spread readily in humans; however, once an infection is established, severe disease often follows. Individuals infected with avian strains usually require airway support through either nasal cannula or mechanical ventilation after development of acute respiratory distress syndrome (Chen et al., 2013; Gao et al., 2013). Pathological findings from H5N1-infected individuals display classical features of diffuse alveolar damage and severe viral pneumonia (Liem et al., 2009; Ng et al., 2006).

Severe influenza disease does not only occur with pandemic human viruses or rare cases of avian virus infection, but also with seasonal human viruses. Disease requiring hospitalization or producing significant mortality in the population occurs less often in seasonal strains of the virus. Nonetheless, seasonal influenza infections can cause hundreds of thousands of hospitalizations and tens of thousands of deaths each year in the United States alone. The vast majority of hospitalizations are seen in high-risk age groups (young children and adults >65 years of age), the immunocompromised, and individuals with genetic susceptibility or comorbidities (Mettelman and Thomas, 2020). A detailed understanding of the mechanisms leading to lung injury following influenza infection will provide clinicians an opportunity to improve the clinical management of hospitalized individuals.

3. Course of influenza virus infection in humans and mice

Influenza virus enters the body through mucosal surfaces, either through the nasal passages or the mouth where viral particles attach to the epithelial tissue of the upper airway. Within 48 h, shedding of newly formed virus reaches its peak (Carrat et al., 2008). Shedding continues for 5–10 days after initial infection (Lau et al., 2010). Symptoms began to present between 1 and 4 days after exposure. For most infections, the acute symptoms of influenza disease last for a period of 7–10 days (Ghebrehewet et al., 2016). However, it may take several weeks for full recovery. One of the critical determinants of viral tropism in the respiratory tract is the preference for modifications of the primary receptor for influenza viruses, sialic acid. On the surface of airway epithelial cells are glycosylated oligosaccharides ending in a sialic acid (SA) residue. The linkages between these glycans and the SA molecules are mediated through sialyltransferases that can form the bond in different conformations: α2,3, α2,6, and α2,8. These distinct linkages are species- and tissue-specific, and much remains to be understood about the distribution of these linkages across species. Human influenza viruses primarily target cells with the α2,6 linkage, whereas the avian viruses target the α2,3 linkage (Ibricevic et al., 2006; Mansfield, 2007). This SA preference contributes to the decreased transmission of avian viruses to humans, and viral tropism due to differences in distribution of α2,3, α2,6 linkages in the upper and lower respiratory tract may also influence the disease progression of more pathogenic strains of influenza (van Riel et al., 2007). In humans, α2,6 sialic acid linkages are more abundant within the upper respiratory tract (Kumlin et al., 2008; van Riel et al., 2010). Therefore, most seasonal human influenza virus infections are restricted to the upper airway.

Sensing of the virus initiates a series of coordinated host responses that together determine the outcome of infection. Mouse models of influenza have been invaluable in defining the course of influenza infection. The host response can be characterized by three general phases, the innate phase, the adaptive phase, and a late phase after viral clearance, which includes stages of tissue regeneration and repair in surviving hosts (Thomas et al., 2020). During the innate phase, epithelial cells act as a first line of defense against the virus and are armed with mechanisms to initiate the antiviral response. Within infected cells, several classes of pattern recognition receptors detect the presence of influenza virus. These pattern recognition receptors include toll-like receptors (TLRs), retinoic acid-inducible gene-1 like receptors (RLRs), NOD (nucleotide-binding oligomerization domain-containing) protein-like receptors or NLRs, and *Z*-DNA binding protein 1 (ZBP1). Pattern recognition receptor signaling by epithelial cells activates type I and type III IFNs. Type III IFNs are primarily produced by type II alveolar cells, whereas bronchial epithelial cells generate type I IFNs (Pulendran and Maddur, 2015). Interferons lead to the transcription of hundreds of downstream genes inducing an antiviral state in infected and bystander cells. Interferon-stimulated genes are responsible for impairing viral replication through degradation of viral RNA and inhibiting cellular transcription and translation required by the viral polymerase complex (Sanders et al., 2011). Some interferon-stimulated genes have also been found to lead to pyroptosis, a lytic, inflammatory form of cell death, and to apoptosis (Fujikura and Miyazaki, 2018). Interferon signaling is not the only downstream effect of pattern recognition receptor signaling. RIG-I recognizes the triphosphate group present at the 5′ end of viral dsRNA to disrupt the viral polymerase complex (Pichlmair et al., 2006). RIG-I can also directly activate the inflammasome, further contributing to early inflammatory signals (Pothlichet et al., 2013). After several activating signals are sensed by the inflammasome, of which NLRs are a significant portion, activation of caspase-1 occurs. Caspase-1 will cleave the inactive forms of IL-1β and IL-18 which leads to the recruitment of innate immune cells (Kanneganti et al., 2006).

This initial sensing of infection and production of pro-inflammatory cytokines and chemokines lead to an influx of innate immune cells into the lung. CCL2 secreted by infected and bystander parenchymal cells attracts alveolar macrophages and monocytes *via* their CCR2 receptors on the cell surface (Chen et al., 2018; Herold et al., 2006). Circulating monocytes differentiate into inflammatory macrophages. Both sets of macrophages are critical in early stages of infection by phagocytizing damaged and dying cells (Hashimoto et al., 2007). Another phagocytic cell type recruited in early time points is neutrophils. Neutrophils are an important source of inflammatory cytokines and assist in controlling viral spread within the respiratory tract. As will be discussed in more detail below, there is a potential negative consequence to their activity. Neutrophils produce neutrophil extracellular traps, or NETs, and reactive oxygen species (ROS) that may be further damaging to lung tissue (Narasaraju et al., 2011; Saffarzadeh et al., 2012). Natural killer (NK) cells are also important for early viral clearance by targeting and killing infected cells. NK cells are a source of IFNγ which may impair viral replication kinetics ((Biondo et al., 2019; Weizman et al., 2017). Lastly, dendritic cells serve as a bridge between the initial innate immune responses and adaptive immunity through the presentation of viral antigens and activation of T lymphocytes. Once primed with antigen, dendritic cells are recruited to the lymph node through CCR7 and CCL19/21 and present antigenic peptides *via* MHC class II molecules on the cell surface (Chen et al., 2018; Hintzen et al., 2006). This coordinated innate phase of the host response is important for limiting viral spread, controlling the infection until adaptive lymphocytes are recruited to the lung starting around day five of the infection (Biondo et al., 2019).

In the adaptive phase of the host response, lymphocytes are critical for clearance and resolution of influenza infection. Both CD4+ and CD8+ T cells are involved in this process. CD8+ T cells are activated in the lymph nodes and are induced to differentiate into cytotoxic T lymphocytes (CTLs) through the production of type I IFNs, IFNγ, IL-2, and IL-12 (Pipkin et al., 2010; Whitmire et al., 2005). CTLs encounter MHC class I molecules presenting viral antigens on the surface of infected cells and induce cell death through two mechanisms: release of cytotoxic granules, granzymes and perforins, or expression of apoptotic markers. As their name suggests, perforins form pores in the cell membrane that allow the entry of granzymes to induce apoptosis. FAS-L and TRAIL expression by CTLs can also induce cellular death (Mirandola et al., 2004; Topham et al., 1997). CD4+ T cells function to assist in the response of other cell types through expression of cell surface markers and elaboration of cytokines. After activation by lymph node dendritic cells, CD4+ lymphocytes migrate to the lung where they differentiate into Th1 cells. These cells produce many pro-inflammatory cytokines that activate macrophages and induce CD8+ T cell differentiation. CD4+ lymphocytes can also provide additional activating signals to B cells for antibody production and even differentiate into other T cell types, like regulatory T cells, Th17 cells, or CD4+ lymphocytes with cytotoxic capabilities (Brown et al., 2012).

T lymphocytes are not the only adaptive cells involved in the coordinated attack against influenza. Early in infection, B cells can mount a

T-cell independent response by producing low affinity IgM and IgD immunoglobulins. Once T cells join the scene, they can further activate B cells resident in germinal centers of lymphoid tissue where B lymphocytes undergo maturation of their B cell receptors (BCRs) to produce receptors of greater affinity and effectiveness against the virus (Hermans et al., 2018). Switching of B cell receptors to different classes of immunoglobulins also occurs. Secretory IgA can prevent viral entry at mucosal surfaces through neutralizing of viral particles and could contribute to early antiviral responses (van Riet et al., 2012). However, a robust antibody response is not generated until days 7–14 following infection, so antibodies provide the greatest protection in secondary exposures to similar influenza viral strains (Lam and Baumgarth, 2019). One of the chief benefits of the adaptive branch of the immune system is the development of protective memory against subsequent influenza infection.

The final phase of the host response against infection is resolution of immunopathology-mediated tissue damage. In uncomplicated influenza cases, there are signs of recovery even at early time points (Kuiken and Taubenberger, 2008). Successful repair requires resolution of inflammation and proliferation of progenitor cell types, deposition of matrix by the stromal cell compartment, and regeneration of the intricate capillary system surrounding the alveoli (Herold et al., 2015). This process involves coordination of anti-inflammatory cytokines that stimulate tissue regeneration. Together, immune and parenchymal cells define the tissue environments within which resolution of inflammation and repair take place. Reparative mediators are produced naturally as negative regulators of activating signals by many immune cell populations. For example, IL-10 released by effector T cells and regulatory T cells (Tregs) serves as a "brake" against the proinflammatory response by impairing effector T cell function and inhibiting antigen presenting cell activity, effectively halting the continuation of immunopathology (Rojas et al., 2017). A number of immune cell types actively participate in the repair process, including innate lymphoid cells (ILCs), Tregs, and macrophages (Herold et al., 2015). Overall, resolving lung pathology is a complex process that involves the crosstalk of multiple cellular pathways among immune and parenchymal cells.

4. Course of disease in severe infections

Seasonal influenza virus infections are self-limiting, often leading to a complete resolution of disease and a return to homeostasis without medical intervention. In certain instances, though, severe pathology occurs as a result of infection—either due to a particularly pathogenic strain or a seasonal strain in individuals that are highly vulnerable. Severe influenza disease is characterized by persistent inflammation in the lower respiratory tract and dysregulated host responses that lead to excessive immunopathology. Severe disease is not only seen in cases of avian influenza. Seasonal influenza viruses can also lead to severe infections in certain high-risk populations. One of the most vulnerable populations are individuals that are immunosuppressed. Immunocompromised states include chronic immunosuppressive conditions like HIV/AIDS and solid organ transplantation, transient immunosuppressive conditions such as hematopoietic stem cell transplant, and chronic or transient use of corticosteroids (Kunisaki and Janoff, 2009).

Other populations particularly susceptible to severe disease from seasonal influenza are children and the elderly. According to the CDC, 7000 to 26,000 flu-related hospitalizations of children have occurred since 2010 (CDC, 2019). Some symptoms are similar to what is seen in adults such as fever, cough, headache, and body aches (Silvennoinen et al., 2009). However, children are more likely to present with extrapulmonary symptoms. Vomiting and diarrhea is common in children of all age groups, especially in hospitalized cases (Al Subaie and Al Saadi, 2012). Otitis media, inflammation of the middle ear, can occur after influenza infection in young children, leading to additional visits to outpatient healthcare facilities (Ruf and Knuf, 2014). Pneumonia is another complication of influenza, especially in infants under the age of 2 years (Neuzil et al., 2002). Neurological dysfunction, including convulsions, febrile seizures, and encephalitis also occur in severely ill children (Ruf and Knuf, 2014).

Elderly patients (> 65 years) account for most of the morbidity and 90% of influenza-related deaths (Thompson et al., 2003; Wilhelm, 2018). Older patients usually do not present with traditional influenza symptoms, but with worsening of comorbidities, particularly cardio-pulmonary chronic illnesses like chronic obstructive pulmonary disease (COPD) and heart disease (Neuzil et al., 2002, 2003). In these patients, dyspnea or labored breathing, cough, and sputum production are exacerbated. Immunosenescence may contribute to the increased infection rates within elderly patients. Aging causes a decline in the overall immune response. This can leave patients with greater susceptibility to infection (Doherty et al., 2016). As summarized previously, an overly active immune response leading to increased mortality in younger patients is presumed to have occurred in the 1918 pandemic influenza. Any benefit of a dampened immune response due to immunosenescence is likely offset by the increased susceptibility to infection and the incidence of comorbidities. Other chronic illnesses associated with disease severity are asthma, blood disorders (sickle cell anemia), endocrine disorders (diabetes most commonly), cancer, and obesity (CDC, 2019b).

Another interesting complication of influenza is secondary bacterial pneumonia. Most influenza-related deaths in the 1918 pandemic were due to a secondary bacterial infection. Secondary bacterial infections were also common during pandemics in 1957 (H2N2), 1968 (H3N2) and 2009 (H1N1) (Morens et al., 2008). *Staphylococcus aureus*, *Streptococcus pneumoniae*, and *Streptococcus pyogenes* are the most common bacteria leading to secondary infection (Chertow and Memoli, 2013). A compromised physical barrier of the upper airway epithelium likely leads to bacterial colonization, and loss of mucociliary clearance, deposition of fibrin, and tissue regeneration can expose sites for bacterial attachment (McCullers, 2006). After colonization in the lung, blockage of the small airways by cell debris and edematous fluid along with inflammation of the tissues surrounding the lung occurs with significant loss of lung function (Kostrzewska et al., 1974).

5. Lung pathology in influenza

Influenza causes major changes to the lung parenchyma, contributing to disease symptoms and potential complications. In uncomplicated influenza, the virus readily attaches to ciliated epithelial cells within the trachea. Attachment and infection within the lower bronchial epithelium decreases with depth (van Riel et al., 2007). Histological analysis from biopsies of the trachea and bronchus reveals diffuse inflammation of tissues from both sites. These lesions are present within 1 day of symptom onset. Swelling or edema, loss of cilia, and desquamation of epithelial cells occurs in lung lesions (Kuiken and Taubenberger, 2008). Infiltration of inflammatory immune cells is also evident. These changes are transient with evidence of reparative pathways beginning at 2 days after symptoms development (Kuiken and Taubenberger, 2008). Within 1–6 weeks after disease onset, thickened surface epithelium and the presence of lymphocytes within the lamina propria were the only histological differences between infected and healthy individuals (Camner et al., 1973).

One of the most common complications of severe influenza cases is viral pneumonia. The progression to pneumonia requires involvement of the lower respiratory tract. Human influenza viruses primarily target ciliated epithelium of the bronchus and type I pneumocytes, but nonciliated epithelium, type II pneumocytes, and alveolar macrophages can also be permissive to the virus (van Riel et al., 2010). Lesions of viral pneumonia show diffuse alveolar damage within the distal bronchial tissue. This damage includes necrotic changes to alveolar epithelium and flooding of edematous fluid within the lumen of the alveoli (Kuiken and Taubenberger, 2008). This fluid contains remnants of necrotic epithelial cells and fibrin. The walls of the alveoli, or septa, are widened due to edema, neutrophilic infiltrates, and hyperemia of the alveolar capillaries. In later stages, interstitial fibrosis develops as well as proliferation of type II pneumocytes. Although massive changes occur within the lower bronchial tree, lesions of the upper airway are much like uncomplicated cases. If bacteria colonize the compromised lung epithelium, purulent inflammation occurs with neutrophil infiltration and production of pus (Kuiken and Taubenberger, 2008).

Avian influenza viruses can also cause diffuse alveolar damage. Pulmonary hemorrhaging is also a pathological finding (Korteweg and Gu, 2008). This is due, in part, to the localization of infection within the lower respiratory tract. Avian strains preferentially bind the type II pneumocytes and alveolar macrophages of the bronchioles with decreasing attachment in the trachea and beyond (van Riel et al., 2010). Type II pneumocytes are essential to lung function by producing surfactant that decreases surface tension to aid in breathing. Type II pneumocytes also have advanced regenerative capabilities which are important for re-epithelization following tissue erosion. Alveolar macrophages are also significant in the early control of the antiviral response through phagocytosis and production of cytokines. Impaired activity of these two cell types might contribute to the heightened pathogenicity seen in avian viruses (Kuiken and Taubenberger, 2008).

6. Role of the virus in tissue damage/pathology

Although the host response to influenza virus infection is the primary driver of pathology in the respiratory tract, the virus does induce important changes to infected cells and in turn alterations in lung structure and physiology. From an evolutionary perspective, it is not advantageous to the virus to induce debilitating disease in the host. Severe influenza disease is almost always the result of infection of the lower respiratory tract, and transmission occurs less efficiently from this compartment compared to the upper airways. Avian influenza viruses that have not adapted to humans (*e.g.* H5N1 and H7N9 viruses) infect the lower respiratory tract with high rates of severe disease and mortality in confirmed cases, but are poorly transmissible to humans (Herfst et al., 2014). Mechanistically, replication of pathogenic avian viruses in the lower respiratory tract has been attributed to the receptor binding preference of avian hemagglutinin (HA) for a 2,3-linked sialic acid, which is enriched in the lower respiratory tract in humans. Human seasonal viruses can also cause severe disease associated with lower respiratory tract infection as a result of uncontrolled viral replication, which typically occurs in people that are genetically susceptible, belong to a high-risk age group, or are immunocompromised (Mettelman and Thomas, 2020).

Cells that are infected by influenza viruses are not simply passive vessels for viral replication. Instead, these cells harbor a hostile environment within which the virus must efficiently utilize the host cellular machinery for its replication while also avoiding antiviral proteins. Both parenchymal and immune cells maintain sophisticated sets of machinery that can sense the presence of the virus and respond with dramatic changes to the cell, including a diverse array of cell death and inflammatory pathways. Cell death often occurs following infection, although not always (Heaton et al., 2014). Influenza viruses are able to assemble new virions and egress from the cell without causing cell lysis. Thus, the majority of cells that die following influenza virus infection do so in a coordinated fashion, whether the manner of death is inflammatory or not.

Over the last two decades, a number of host proteins that bind to and sense the presence of influenza virus infection have been identified. These viral sensors include members of four main classes of proteins—toll-like receptors (TLRs3, 7, 8), retinoic acid-inducible gene I (RIG-I), NOD-LRR and pyrin domain-containing 3 (NLRP3), and Z-DNA-binding protein 1 (ZBP1). TLRs 3/7/8, RIG-I, and ZBP1 recognize various forms of nucleic acid associated with influenza virus infection, while NLRP3 activation is nucleic acid dependent, following recognition by RIG-I and ZBP1 (Iwasaki and Pillai, 2014; Thomas et al., 2020). Detection of viral infection initiates a cascade of carefully coordinated events that lead to upregulation of inflammatory pathways that alert other cells to the presence of infection or to the death of the cell. Cell death can be viewed as a defense strategy as much as a consequence of viral infection (Vaux et al., 1994; Yatim and Albert, 2011). Influenza viruses, like all viruses, are obligate, intracellular pathogens, and without the appropriate cellular machinery cannot complete their life cycle. Death of the infected cell can be an efficient way to halt viral replication and also alert neighboring cells to the presence of infection.

While there are cell extrinsic signals, *e.g.* cytokines such as TNFa and type-I interferons and death ligands such as TRAIL, that can induce cell death in uninfected cells, this section will focus on cell-intrinsic mechanisms of cell death following sensing of various viral nucleic acids and proteins.

An influenza virus-infected cell can generally die by one of three well-defined cell death pathways, apoptosis, necroptosis, or pyroptosis, with the first being non-inflammatory and the latter two inflammatory (Thomas et al., 2020). Although the manner of death tends to be cell-type dependent, the precise mechanisms underlying cell death decisions within a cell type, for example epithelial cells, remain unclear (Thomas et al., 2020).

Cell death in target epithelial cells in the lower respiratory tract and alveolar regions is of particular concern because of their critical role in lung function. Disruption of the epithelial barrier in the airways and in the alveoli are hallmarks of severe influenza disease in humans and animal models of infection (Beigel et al., 2005; Shieh et al., 2010; Short et al., 2014). In the alveoli, maintenance of the epithelial-endothelial barrier is essential for blood oxygenation. Although the contribution of each of these cell death pathways to disease during human influenza virus infections remains to be defined, studies in the mouse model have identified the important molecules involved in each pathway and the contribution of each form of cell death to the outcome of infection.

6.1 Sensing of influenza nucleic acid and associated cellular changes

An interesting recent finding has been that a single viral sensor, Z-DNA binding protein 1 (ZBP1), appears to be a critical upstream regulator of all three forms of cell death (Kuriakose et al., 2016; Thapa et al., 2016). Despite its name suggesting DNA binding, ZBP1 recognizes a left-handed, helical form of RNA, Z-RNA, produced by influenza viruses in infected cells (Zhang et al., 2020). Downstream of ZBP1 recognition of influenza Z-RNAs, apoptotic and necroptotic cell death pathways share some of the same molecular machinery, including the receptor interacting serine/threonine protein kinase (RIPK) family members 1 and 3 (Nogusa et al., 2016; Thapa et al., 2016). Studies in mice have demonstrated that RIPK3-dependent apoptosis and necroptosis are parallel pathways that contribute to viral clearance and reduction in lung damage (Nogusa et al., 2016). In the absence of RIPK3-mediated apoptosis, mice with intact necroptosis machinery are able to clear infection similar to wild-type animals. Likewise, mice deficient for MLKL, the key executioner of necroptosis, are able to effectively clear the virus and survive the infection. There may, however, be disadvantages to inflammatory forms of cell death, like necroptosis, that can contribute to excessive lung inflammation following influenza infection. Experiments inoculating with a high dose of influenza virus indicate that mice sufficient for necroptosis exhibit higher levels of neutrophil infiltration and higher mortality compared to wild-type animals (Zhang et al., 2020). This survival phenotype was independent of the viral burden in the lungs and likely a result of widespread lung involvement in the immune response initially driven by inflammatory signals released upon necroptosis of infected cells.

Another form of inflammatory cell death, pyroptosis, is also regulated by ZBP1 (Kesavardhana et al., 2017; Kuriakose et al., 2016). Pyroptosis in response to influenza virus infection is mediated by the NLRP3 inflammasome leading to activation of caspase-1 and subsequently IL1β and Gasdermin D, which forms pores in the cellular membrane (Allen et al., 2009; Ichinohe et al., 2010; Man et al., 2017; Thomas et al., 2009). The NLRP3 inflammasome does not appear to play an important role in controlling viral infection at early stages. Instead, the inflammasome and associated pyroptosis appear to help coordinate innate immune responses to limit the extent of lung damage. In addition to ZBP1 recognition of influenza nucleic acid, the viral protein M2 can activate the NLRP3 inflammasome through its ion channel activity in the Golgi apparatus (Ichinohe et al., 2010). Thus, in addition to the presence of viral nucleic acid in a target cell, cellular changes (in this case changes in intracellular ion concentrations) can also alert the cell to the presence of infection and lead to cell death.

Together, these studies define the cell death pathways that can influence the progression of the infection and host response. Mercifully, a target cell can only die one way following viral infection. It is still unclear what factors determine the way in which a cell dies, or—if we give the cell more agency—how it decides its manner of death from the several options. These decisions have important consequences for downstream adaptive immune responses and more directly for lung function and mortality as there is a threshold of type I pneumocyte cell death beyond which the host is likely not to survive (Sanders et al., 2013a,b). The fate of infected, or virus-stimulated cells, is not always death, and other phenotypic changes can occur that can contribute to virus-induced alterations of lung function. Influenza virus can directly inhibit epithelial sodium ion channels, which are responsible for maintaining an osmotic gradient and preventing fluid accumulation in the alveoli (Chen et al., 2004; Kunzelmann et al., 2000; Lazrak et al., 2009; Short et al., 2014; Wolk et al., 2008). There is also some evidence that influenza viruses can disrupt epithelial cell tight junctions and lead to epithelial-endothelial barrier permeability (Short et al., 2014).

The presence of virus in the host and virus-induced damage or changes to cells of the respiratory tract initiates host responses that are often more damaging to the tissue than the infection itself. Therefore, the extent of viral spread in the respiratory tract and compartmentalization in either the upper or lower airways (or both) are important determinants of the extent and localization of the host response. The mechanisms and consequences of the host response will be discussed below.

7. Role of the host response in tissue damage

7.1 Immune cells and disease resistance

There is a broad consensus that the main driver of morbidity and mortality during symptomatic influenza virus infection is immunopathology. Several lines of evidence point to the host response, and not the virus, as the primary cause of tissue damage and compromised lung function. First, many studies have found that the severity of influenza disease and the outcome of infection in humans do not correlate with the viral load in the respiratory tract. This lack of correlation between the amount of virus measured and the severity of disease is consistent in animal models of infection (Granados et al., 2017; Lee et al., 2011; Meschi et al., 2011; Oshansky et al., 2014). As previously mentioned, compartmentalization of the virus in the upper respiratory tract typically leads to only mild symptoms, while infection in the lower respiratory tract often leads to more severe disease. The inability to sample from deeper in the respiratory tract (*i.e.* the lung tissue) in patients with mild and severe disease could be a limitation in identifying such an association. Nonetheless, well-controlled studies in animal models of infection have also found that the viral burden in the lungs is a poor predictor of outcome (Sanders et al., 2013a,b). Second, antivirals that directly interfere with viral replication have a narrow therapeutic window of efficacy, typically within the first 48–72 h following infection (Govorkova and McCullers, 2013). The lack of efficacy beyond the first several days of the infection suggest that the critical immunological decisions that will determine the course of disease have already been made and will continue whether or not there are high levels of viral replication. A third line of evidence related to viral burden is that in very severe cases of influenza, mortality often occurs well after viral clearance and following a long period of hospitalization. This extended period of inflammation and compromised lung function in the absence of ongoing viral replication indicate that once the infection has been initiated and involves enough of the respiratory tract, the host response continues to sustain the inflammation and associated lung damage. Importantly, damage to the epithelial barriers in the respiratory tract also makes the host more susceptible to secondary bacterial infections, which can be lethal.

Given the contributions of both virus- and host-induced damage to the respiratory tract, a careful balance between clearing the infection and limiting tissue damage to preserve lung function is required for host survival. A useful framework for thinking about this balance is in terms of disease resistance and tolerance as strategies to defend against infection (Medzhitov et al., 2012). Disease resistance refers to immune mechanisms that actively decrease the viral burden and eventually clear the pathogen. The cost of resistance mechanisms is often damage to the infected tissue and loss of at least some organ function. Disease tolerance refers to mechanisms that preserve the integrity of or repair tissue to promote survival without directly affecting viral burden (Ayres and Schneider, 2012; Medzhitov et al., 2012). For decades the field of influenza immunology has overwhelmingly focused on mechanisms of disease resistance that prevent infection or limit viral spread in the host. In particular, there has been a strong bias for the study of professional immune cells that mobilize to recognize and kill infected cells or produce proteins that neutralize the virus or otherwise interfere with its replication. Few studies have focused on mechanisms of disease tolerance (Jamieson et al., 2010, 2013; Wang et al., 2016) and the role of stromal cells in regulating the host response to infection (Teijaro et al., 2011). An important knowledge gap in the field is how inflammatory or damage signals generated during the infection are translated into structural and physiological changes to the respiratory tract, which ultimately determine the outcome of infection. Given the compartmentalization of function in different parts of the respiratory tract, the most effective host strategy for survival may depend on where the infection and inflammation are localized. For example, the upper respiratory tract, including the nasal passages and trachea may benefit from a resistant strategy to limit viral spread, while the lower respiratory, where gas exchange occurs, may benefit from a tolerant strategy that favors preservation of lung function (Fig. 1).

The term "cytokine storm" is frequently invoked to describe a systemic response to infection with excessive levels of proinflammatory cytokines and chemokines that drive immunopathology (Clark, 2007; Guo and Thomas, 2017; Tisoncik et al., 2012). This characterization has become popular in reference to the exuberant immune responses observed in human cases of avian influenza and the 1918 pandemic H1N1 virus. This systemic inflammation could be thought of an extreme form of resistance to a pathogen.

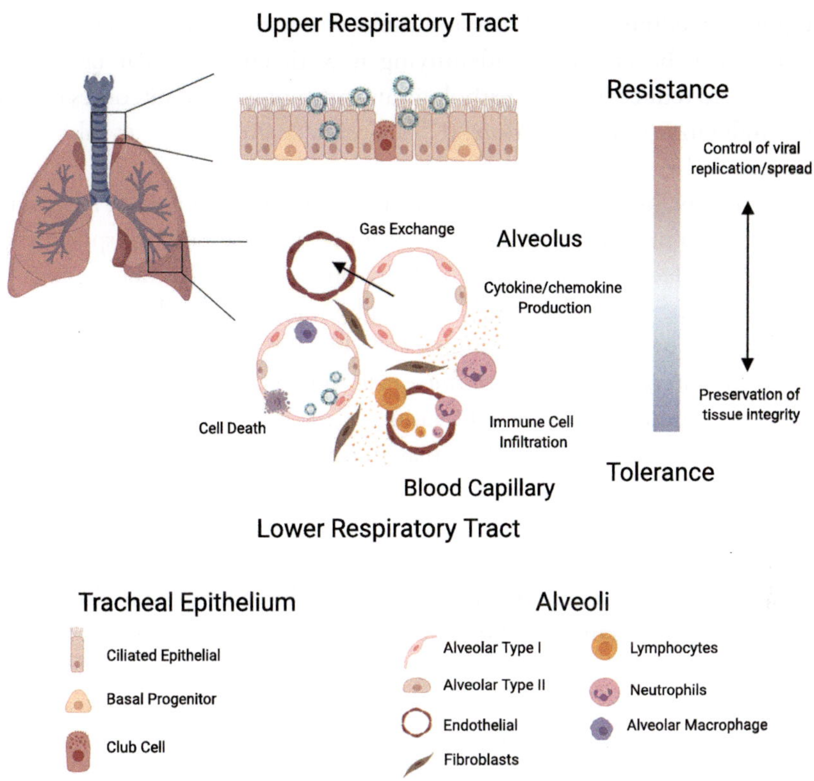

Fig. 1 Disease resistance and tolerance strategies in the upper vs. lower respiratory tract during influenza virus infection. During mild cases of influenza, viral replication is typically restricted to the upper respiratory tract, including the nasal and tracheal epithelium. In contrast, severe infections are characterized by viral replication and inflammation in the lower respiratory tract, involving the distal airways and terminal alveoli, where gas exchange occurs. Given the differences in structure and function of the upper and lower respiratory tracts, the effectiveness of different defense strategies may depend on the tissue compartment. For example, a resistance strategy emphasizing control of viral replication and spread may be more beneficial in the upper respiratory tract. In the lower respiratory tract, a tolerance strategy promoting preservation of tissue integrity and function and limiting immune cell infiltration, cytokine/chemokine production, and inflammatory cell death may be more beneficial for host survival. Illustration created with BioRender.com.

While high overall levels of cytokines and chemokines certainly correlate with severe disease in many cases, it remains unclear which factors are most important for promoting the tissue damage and loss of organ function. "Storm" seems to imply a certain chaos and betrays what is likely an ordered, although complex, sequence of events resulting in an overactive host response. Defining the cellular and molecular mechanisms underlying this process will be critical for identifying new therapeutics that target host response to reduce immunopathology and promote better outcomes following influenza virus infection (D'Elia et al., 2013; Liu et al., 2016; Teijaro, 2015).

In this section, we will attempt to move beyond the idea of a cytokine storm and summarize findings from recent studies that point to specific cellular and molecular mechanisms in the respiratory tract that may be responsible for changes in tissue structure and physiology. We will emphasize not only the role of professional immune cells, but also of stromal cells, and studies on the interactions between these cells in determining disease outcome. Given the immense volume of literature on the role of professional immune cells during influenza infection, it may not be surprising to know that many immune cell types have been implicated in both protective and pathogenic responses. Each immune cell type can play a protective or pathogenic role depending on the magnitude, timing, or localization of the response. Thus, seemingly conflicting reports may not actually be incompatible. In general, a well-timed, restrained response is helpful in clearing infection without inflicting too much damage on the tissue. Studies from mouse models, in which regular, timed tissue sampling is possible, suggest that immunological decisions leading to extensive immunopathology and severe disease are made very early in the infection (Askovich et al., 2013; Brandes et al., 2013). This section will focus on three immune cell types—neutrophils, monocytes, and CD8+ T cells—that have been consistently reported as drivers of immunopathology during severe infections in both animal models of infection and in human cases.

7.2 Neutrophils

One of the more consistent findings in the influenza literature is that excessive and sustained neutrophil infiltration is associated with severe disease and poor clinical outcomes (Camp and Jonsson, 2017). Numerous studies have described immune factors that promote or modulate neutrophil infiltration into the lung and associated tissue damage. High levels of neutrophil-associated cytokines and chemokines are associated with disease severity (Fiore-Gartland et al., 2017; Guan et al., 2018; Tang et al., 2019). Excessive neutrophil infiltration is also a key feature of severe complications of influenza infections, including acute respiratory distress syndrome (Matthay et al., 2019). Neutrophils may contribute to tissue damage through a number of direct mechanisms, including generation of reactive oxygen species, release of neutrophil extracellular traps (Narasaraju et al., 2011), and production of extracellular matrix degrading enzymes that facilitate migration into the lung (Bradley et al., 2012; Rojas-Quintero et al., 2018). Neutrophils can have a major impact on epithelial and endothelial barrier integrity. Neutrophils also have a beneficial effect for the host and the protective or pathogenic potential of neutrophils depends on the timing and magnitude of the response (Bradley et al., 2012; Brandes et al., 2013). In addition to early effects on limiting viral replication, apoptotic neutrophils may also play a role in resolving lung inflammation and promoting tissue repair following uptake by macrophages (Bosurgi et al., 2017; Fadok et al., 1998).

7.3 Monocytes

Another innate immune cell type that has been consistently associated with excessive immunopathology is the monocyte. Definition of this general cell type is complicated by the fact that monocytes can differentiate into a number of different states upon reaching the lung, including dendritic cells and macrophages. One common feature of monocytes is that these cells express the chemokine receptor (CCR2) for CCL2/MCP-1. Genetic deletion of CCR2 in mice and antibodies antagonizing CCR2 have been important tools for defining the role of these cells in influenza pathogenesis.

At least two distinct classes of circulating monocytes have been defined in humans based on the surface markers CD14 and CD16 (FcγRIII) and on Ly6C and CX3CR1 in mice (Murray, 2018)—patrolling monocytes ($CD14_{lo}CD16+$ or $Ly6C_{lo}CX3CR1+$) and inflammatory monocytes (CD14+CD16− or $Ly6C_{hi}CX3CR1-$). While the precise role of patrolling monocytes during influenza infection is unclear, these cells do monitor the vascular endothelium and can promote macrophage-mediated tissue repair (Carlin et al., 2013; Murray, 2018). The role of inflammatory monocytes, previously known as TNF/iNOS-producing dendritic cells. (Serbina et al., 2003), has been much better studied in influenza infection. In mouse models of infection, these cells accumulate in high abundance in the respiratory tract starting early in infection and persist until later stages when an adaptive response emerges (Aldridge et al., 2009; Cole et al., 2017; Lin et al., 2008). The activity of these cells reflect the need for balance in pathways that promote disease resistance or tolerance. On the one hand, inflammatory monocytes contribute to immunopathology by trafficking to the respiratory tract and producing inflammatory mediators. Dampening the magnitude of the inflammatory monocyte response using pharmacological inhibitors of chemokine production or using an anti-CCR2 antibody can promote improved outcomes in both juvenile (Coates et al., 2018) and adult mice (Aldridge et al., 2009). On the other hand, these cells regulate the magnitude of the cytotoxic CD8+ T cell responses that are necessary for effective clearance of virally infected cells (Aldridge et al., 2009). This highly abundant cell type is an effective antigen presenting cell that promotes CD8+ T cell responses during the acute phase of primary infection and also contributes to the establishment of tissue-resident immunological memory (Aldridge et al., 2009; Dunbar et al., 2020).

Findings on the association of inflammatory monocyte responses with severity of influenza disease have translated well from mice to humans. Several studies have reported an association between frequency of inflammatory monocytes in the nasal washes (Oshansky et al., 2014) and in the blood (Cole et al., 2017) with disease severity. These associations with disease severity are consistently independent of the viral burden in the respiratory tract.

7.4 CD8 T cells

In addition to the innate immune cells described above that respond rapidly to the infection, adaptive immune cells have also been implicated in driving immunopathology and severe disease. Decades of research have identified cytotoxic CD8+ T cells (CTLs) as both critical mediators of killing of infection cells and viral clearance and of potentially lethal immunopathology. CTLs mediate immunopathology through direct lysis of infected cells presenting viral peptides (Enelow et al., 1998; Xu et al., 2004) and by production of cytokines that can kill bystander cells in a non-specific way and recruit and activate other immune cells (Small et al., 2001; Zhao et al., 2000).

In mouse models of severe influenza virus infection, the accumulation of cytokine producing CTLs, many of which are influenza-specific, in the lungs coincides with a decrease in lung function and the presentation of acute respiratory distress. A series of studies have established that influenza-specific CD8+ T cells are able to mediate acute lung damage and compromise respiratory function by killing epithelial cells expressing influenza antigen. These studies were performed in an elegant, nonviral model of inflammation in which type II pneumocytes endogenously express influenza HA antigen in a transgenic mouse model. Transfer of stimulated CD8+ T cells into transgenic mice lead to a decrease in type

II cells, an increase in myeloid inflammation, and a decrease in respiratory function (Enelow et al., 1998). CD8+ T cells led to increased chemokine expression in type II cells (Zhao et al., 2000) and lung inflammation continued in the absence of the transferred cells (Small et al., 2001). Lung damage was also in part due to apoptosis of bystander type II cells induced by CD8+ T cell-derived TNFα (Liu et al., 1999).

Correlational studies in humans are consistent with mouse studies demonstrating a role for CD8+ T cells in promoting lethal immunopathology. In humans, severe or lethal cases of influenza exhibit extensive lymphocyte infiltration in the respiratory tract at late stages of infection (Mauad et al., 2010; Wang et al., 2014, 2015). Studies of CD8+ T cell responses in severe avian H7N9 infections demonstrated that a delayed T cell response and persistent lymphocyte infiltration into the respiratory tract along with elevated inflammatory cytokine level were associated with the most severe infections (Wang et al., 2014, 2015).

7.5 Immune cells and disease tolerance

In addition to these cellular mechanisms of disease resistance, several lineages of professional immune cells promote disease tolerance through their interaction with stromal cells in the respiratory tract. Regulatory T cells (Tregs) are well known for their ability to restrain inflammatory responses from other immune cells. However, these cells also play a direct role in promoting tissue repair following influenza virus infection through the production of amphigregulin (AREG), an EGF-like molecule that is important for maintaining epithelial barrier integrity (Arpaia et al., 2015). Interestingly, a number of different immune cell responses related to disease tolerance converge on the production and activity of AREG. Type 2 innate lymphoid cells (ILC2s) are also an important source of AREG and contribute to lung repair following infection (Monticelli et al., 2011). Production of AREG in both Tregs and ILC2s is regulated by the alarmin IL-33, released by damaged epithelial cells, through its engagement of the receptor ST2/IL1RL1.

Upstream of the IL-33-AREG pathway, IL-17A produced by γδ T cells, a subset of lymphocytes with innate tissue sensing ability and rapid kinetics (Hayday, 2019), can stimulate the production of IL-33 in several different stromal cells, including epithelial cells and fibroblasts (Guo et al., 2018). IL-17A production by γδ T cells in neonatal mice and subsequent stimulation of AREG is a protective mechanism to promote tissue repair independent of viral clearance (Guo et al., 2018). The rapid kinetics and production of IL17-A indicate that mechanisms of disease tolerance can be active even at very early stages of the infection without affecting antiviral pathways required for limiting viral spread and mediating clearance of the infection.

Alveolar macrophages, a type of self-renewing macrophage that is resident in the lung alveoli and critical for maintaining homeostasis, have also been implicated in protection from infection sequelae. Several studies in mice have demonstrated that reduction or ablation of alveolar macrophages leads to increased lung inflammation and tissue damage by decreasing the spread of infection (Cardani et al., 2017; Purnama et al., 2014; Schneider et al., 2014). In contrast to other mechanisms of disease tolerance mentioned above, alveolar macrophage protection of lung integrity during infection does appear to depend on the extent of viral infection as these cells are able to protect alveolar type I cells from infection. Mechanistically, this protection may be linked to the ability of alveolar macrophages to suppress the cysteinyl leukotriene pathway in alveolar type I cells, which confers susceptibility to productive infection (Cardani et al., 2017).

All of the responses of immune cells described in this section are dependent upon recruitment to the site of infection in the respiratory tract or stimulation of the cells by inflammatory signals in the tissue environment. Studies of the lung environment during infection have heavily focused on the quantification of inflammatory cytokines and chemokines. However, there are a number of other tissue derived factors, including those involved in lung remodeling, that may also control the balance between disease resistance and tolerance. The following section will highlight recent studies defining the responses of parenchymal cells in the lung, including epithelial cells, endothelial cells, and fibroblasts, to influenza virus infection and how they might also regulate this essential balance. This tissue level control of inflammation and repair may open new opportunities for preserving lung tissue integrity and dampening excessive immune responses.

7.6 Parenchymal cells and the balance between resistance and tolerance

Studies of immunopathology in influenza infection have overwhelmingly focused on the contribution of professional, migrating immune cells such as neutrophils, monocytes, and T cells described above. Much less attention has been given to parenchymal cells, such as epithelial cells, endothelial cells, and fibroblasts in coordinating and determining the trajectory of the host response. This lack of attention is somewhat surprising given that these cells are by definition tissue-resident and present in the respiratory tract at the very earliest stages of infection. It is becoming increasingly clear that these cell types have many of the same functions as migrating immune cells, including the ability to sense viral infection, produce inflammatory cytokines and chemokines, and generate a type of immune memory (Ordovas-Montanes et al., 2020). As structural, non-immune cells take on the dual roles of directly supporting lung function and participating in the host inflammatory response, they are critical cell types in determining the balance between disease resistance and tolerance and ultimately the outcome of influenza infection.

7.7 Epithelial cells

As the primary target cells of influenza viruses and sites of viral replication, epithelial cells in the respiratory tract act as sentinel cells of infection. There are many different types of epithelial cells in the respiratory tract from nasal passages of the upper airways down to the distal alveoli of the lung. These distinct cell types include ciliated cells, mucus-secreting goblet cells, secretory club cells, basal cells, and type I and II alveolar epithelial cells. As described earlier, epithelial cells express sensors of viral nucleic acids, viral proteins, and virus-induced cellular changes, that lead to cytokine/chemokine production and cell death by several mechanisms.

Epithelial cells produce cytokines that induce antiviral responses (type I and III interferons), bystander cell death (TNF), and proinflammatory chemokines (IP-10/CXCL10, MCP-1, IL8, IL6, CCL5) that attract immune cells. The extent and magnitude of these responses represent critical early immunological decisions that determine subsequent responses (Askovich et al., 2013; Sanders et al., 2011).

Epithelial cells also play important roles in disease tolerance mechanisms related to barrier maintenance and tissue repair. AREG derived from Tregs was previously mentioned as an important regulator of epithelial barrier integrity and was shown to be protective against severe influenza disease in numerous studies (Arpaia et al., 2015; Guo et al., 2018; Hall et al., 2016; Jamieson et al., 2013; Monticelli et al., 2011). Epithelial cells not only respond to AREG, but also produce this molecule in response to infection (Hall et al., 2016; Vermillion et al., 2018). Although the precise mechanism by which AREG promotes tolerance during influenza virus infection has not been clearly defined, it likely involves both promotion of epithelial cell proliferation and local suppression of inflammation (Zaiss et al., 2015).

Recently, there have been exciting advances identifying epithelial progenitor cells that mediate lung repair and regeneration following severe influenza infection. While alveolar type II cells have the capacity to proliferate and differentiate into alveolar type I cells, in cases of severe infection when type II cells are ablated in a region area of the lung, facultative progenitors mobilize to regenerate damaged epithelia (Kumar et al., 2011; Vaughan et al., 2014; Xi et al., 2017; Zacharias et al., 2018; Zuo et al., 2014). A cumulative finding from these studies is that the nature of alveolar regeneration and the effectiveness of that process depends on compartmentalized severity of the infection (Kanegai et al., 2016). Alveolar type II cell regeneration of damaged lung following influenza infection, from a specialized alveolar epithelial progenitor that is responsive to Wnt, is more effective than repair mediated by basal progenitor cells that migrate from the proximal airways when alveolar type II cells are completely ablated (Xi et al., 2017; Zacharias et al., 2018). Ineffective tissue repair that results in areas of lung epithelia with persistent pathology is driven by localized hypoxia and NOTCH signaling that reinforce a cytokeratin 5 program (Kanegai et al., 2016; Xi et al., 2017). In mice, these areas of persistent pathology, consisting of both immune and stromal cells, can last for at least 200 days (Kanegai et al., 2016) and perhaps indefinitely.

7.8 Endothelial cells

Preservation of the barrier between alveolar epithelial cells and endothelial cells is essential for blood oxygenation and lung function. Endothelial cell activation is a feature of severe influenza virus infections, in particular those caused by highly pathogenic avian viruses (Short et al., 2014). Although human endothelial cells can be infected with influenza A viruses *in vitro* (Armstrong et al., 2012; Chan et al., 2009, 2019; Ocaña-Macchi et al., 2009; Zeng et al., 2012), it remains controversial the extent to which endothelial cells get infected, either productively or abortively, in humans (Short et al., 2019).

Nearly 10 years ago, Teijaro and colleagues made the important finding that lung endothelial cells are critical amplifiers of cytokine and chemokine production during influenza virus infection and push the host towards hypercytokinemia. Using pharmacological agonists of the sphingosine-1-phosphate receptor 1 ($S1P_1R$), highly expressed on lymphatic and vascular endothelial cells, the authors demonstrated a blunting of cytokine production, a reduction in innate immune cell infiltration, and improved survival in mice without affecting the viral burden (Teijaro et al., 2011). In a subsequent study, the amplification of cytokine production in endothelial cells was found to be independent of innate sensors of viral infection (TLR3 and TLR7) and dependent on IL-1 receptor signaling (Teijaro et al., 2014). Together, these studies suggest that this endothelial contribution to hypercytokinemia is in response to damage signals from infected cells and not from the cell intrinsic sensing of influenza virus.

Defining specific cell types that contribute to hypercytokinemia was an important step in deconvoluting the "cytokine storm." These studies have opened the door for manipulating specific cellular targets to modulate excessive inflammation while leaving intact a host response sufficient to clear infection.

7.9 Other mesenchymal cells

In addition to epithelial and endothelial cells, the other main group of cells in the respiratory tract are mesenchymal cells, including fibroblasts, smooth muscle cells, and pericytes, which make up about 30% of cells interstitium (Jordana et al., 1994). Mesenchymal cells coordinate lung development and define tissue structure and function during homeostasis through their interactions and communication with epithelial and endothelial cells (Nowarski et al., 2017). Mesenchymal cells play a critical role in defining tissue environments through their ability to both synthesize new extracellular matrix (ECM) proteins and produce matrix remodeling enzymes that modify or degrade these structural proteins (Bonnans et al., 2014). Through this ECM remodeling activity, mesenchymal cells help define the tissue microenvironments that direct cell proliferation, differentiation, and migration.

To date, mesenchymal cells have received surprisingly little attention in the influenza field. Studies of the role of fibroblasts in lung injury have primarily focused on the late stages of tissue repair and long-term sequelae, including lung fibrosis (Duffield et al., 2013). Very few studies have addressed the role of mesenchymal cells during the acute phase of influenza virus infection and the contribution of these cells to the initial inflammatory response. At least one study, using a viral mimetic (poly I:C), has demonstrated that lung fibroblasts upregulate inflammatory programs in response to infected or damaged epithelial cells that release IL-1α (Suwara et al., 2014). Several recent studies in mice have identified ECM proteases, from both the matrix metalloproteinase (MMP) and A disintegrin and metalloproteinase with thrombospondin motifs (ADAMTS), that may determine the balance between disease resistance and tolerance and affect the outcome of of infection. These matrix proteases include MMP-9 (Bradley et al., 2012; Rojas-Quintero et al., 2018), MMP-14 (Talmi-Frank et al., 2016), and ADAMTS-5 (McMahon et al., 2016). Although these studies have primarily attributed activity of these proteases to immune cells (neutrophils, macrophages, and T cells), each of these enzymes is also expressed by parenchymal cells, including lung fibroblasts.

In the coming years, more studies on the role of lung stromal cells in regulating the balance between influenza disease resistance and tolerance will hopefully define the mechanisms by which inflammatory signals from infection are translated into changes in lung structure and function. Defining these mechanisms will inform the development of new host therapeutics that preserve lung tissue integrity and function to improve severe disease outcomes.

8. Future questions and potential therapeutic targets to reduce immunopathology

Currently, antivirals are the only therapeutics approved for treatment of influenza in the US. These therapeutics are efficacious in minimizing the spread of virions within the host and shortening symptomatic disease. However, in order to be effective, they must be administered early in infection, often within the first 1–2 days after symptom onset. Furthermore, antiviral drug resistance poses a significant challenge to their use and efficacy. As influenza viruses continue to adapt, we must be armed with newer and better drugs to limit the replication of the virus and/or combat the symptoms and potential complications of disease. There are a few candidates for novel therapeutics that either "pump-the-breaks" on pathways leading to lung pathology or enhance responses contributing to disease resolution.

As discussed above, severe influenza disease is often associated with the excessive production of cytokines that lead to not only lung pathology but systemic disease, such as organ-failure and neurological dysfunction. Hypercytokinemia has been demonstrated in H5N1-infected patients as well as those infected with pandemic H1N1 (de Jong et al., 2006; Tisoncik et al., 2012). Interleukin-6 (IL-6) is consistently reported as associated with severity of infection (Hagau et al., 2010; Kaiser et al., 2001; Oshansky et al., 2014) and has been suggested as a potential biomarker for severe H1N1 infection in humans (Paquette et al., 2012). It has also been demonstrated that the development of symptoms coincides with rising serum IL-6 levels (Kaiser et al., 2001). IL-6 is produced by many innate immune cells, but also other non-immune cells like epithelial cells, fibroblasts, and endothelial cells

(Dienz et al., 2012; Kishimoto, 2010). Thus, IL-6 might be a potential therapeutic target to ameliorate severe disease. The IL-6 receptor monoclonal antibody, tocilizumab, is used commonly for the treatment of chronic inflammatory disease such as rheumatoid arthritis and juvenile idiopathic arthritis (Alten, 2011; Turnier and Brunner, 2016). In a 2013 study, looking at the effects of tocilizumab in juvenile idiopathic arthritis patients infected with influenza, tocilizumab significantly reduced clinical symptoms in all 10 patients participating in the study (Kawada et al., 2013).This study suggests that IL-6 blockage could be beneficial for clinical management of influenza; however, there is evidence that IL-6 is essential for tissue protection in influenza infection. IL-6 may promote neutrophil recruitment to the lung and activate neutrophil-mediated viral clearance (Dienz et al., 2012). Viral clearance *via* macrophage-mediated phagocytosis has been shown to be modulated by IL-6 signaling (Yang et al., 2017). This study also found that mice deficient in IL-6 had enhanced fibroblast proliferation, an indicator of fibrotic changes following acute lung injury. Lastly, another group showed that IL-6 deficient mice had increased pro-inflammatory cytokine production and severe lung damage (Lauder et al., 2013). The pleiotropic effects of IL-6 that have been defined in mouse models of influenza will be important to consider as decisions about clinical trials and treatment in humans are being made.

Corticosteroids have also gained interest as a treatment modality for severe influenza. Corticosteroids are often the first-line defense against chronic inflammatory diseases, such as asthma. These drugs serve as master regulators of the inflammation through the modulation of inflammatory factors, including cytokines, chemokines, their corresponding receptors, and cell-surface adhesion markers (Barnes, 2006). Animal models have demonstrated that corticosteroid use can be beneficial in alleviating lung injury (Li et al., 2012; Ottolini et al., 2003). However, there is significant uncertainty as to efficacy and safety of adjunctive corticosteroids for influenza treatment in humans. Two studies found an association between corticosteroid use and risk of secondary bacterial infection (Li et al., 2017; Moreno et al., 2018). Several meta-analysis studies suggest a link between corticosteroid use and increased length of hospital stay, greater dependency on mechanical ventilation, and increased mortality following influenza infection. However, there is a need for more robust studies to conclusively determine the efficacy (Lansbury et al., 2019; Nedel et al., 2016; Ni et al., 2019; Zhou et al., 2020). There is concern that impaired host antiviral responses could be a reason for limited clinical efficacy and increased mortality. Consistent with this idea, a recent study found that treatment of human airway epithelial cells and mice with glucocorticoids decreases the expression of antiviral interferon-stimulated genes (ISGs), and, thus, leads to enhanced viral replication (Thomas et al., 2014).

The contribution of lipid molecules to either pathogenicity or tissue recovery has been a recent area of research that has revealed some potential therapeutic targets (Teijaro, 2015). Cyclooxygenease (COX) inhibitors prevent the conversion of arachidonic acid into inflammatory prostaglandins. Some early studies show that when administered in combination with antivirals, a COX-2 inhibitor, celecoxib, improved survival in mice infected with avian H5N1 (Zheng et al., 2008). More recently, lipidomic profiling discovered lipoxygenase metabolites that correlate with resolution of disease as well as induction of lung pathology (Tam et al., 2013). Hepoxylin and 12-HETE are generated from activity of 12/15-lipoxygenase and shown to be elevated during the recovery phase following infection. These two molecules are known to inhibit macrophage influx into the lung, limit fibrosis, and pro-inflammatory cytokine production (Jankov et al., 2002; Krönke et al., 2009). Alternatively, metabolites of 5-lipoxygenase were shown to be elevated in the initial pathogenic phase of viral spread (Tam et al., 2013). Downstream TLR4 signaling activates oxidized phospholipid, OxPAPC, to induce cytokine production by macrophages leading to acute lung injury in H5N1-infected mice (Imai et al., 2008). Shirey and colleagues administered the TLR4-antagonist, Eritoran, to influenza-infected mice. Eritoran was able to limit symptom progression, lung pathology, cytokine production, oxidized phospholipid expression, and viral titers (Shirey et al., 2013).

Sphingosine-1-phosphate (S1P) pathways also have potential as a therapeutic target. S1P receptors (S1PRs) are present on many cell types, including endothelial and immune cells. These receptors influence several immune responses including cytokine production and cellular activity, as well as trafficking of immune cells across tissues (Rivera et al., 2008). The drug Fingolimod (FTY720) is a S1PR pro-agonist that modulates TCR-dependent and independent T-cell activation (Baer et al., 2018). Another S1PR agonist, AAL-R, protects against influenza-related mortality through inhibition of aberrant cytokine production by innate cells and T-cell responses leading to lung immunopathology while preserving host antiviral responses (Walsh et al., 2011). These studies suggest there is significant potential in lipid signaling as therapeutic targets for influenza.

Another opportunity for therapeutic intervention during influenza lies in the extracellular matrix (ECM). The ECM is important for disease pathogenesis and progression because it serves as both a reservoir and modulator of immune pathways (Boyd and Thomas, 2017). Matrix proteases, including the MMP and ADAMTS families, modify or degrade specific components of the lung ECM. Several recent studies have identified both pathogenic and protective roles for matrix proteases in influenza infection (Bradley et al., 2012; McMahon et al., 2016; Rojas-Quintero et al., 2018; Talmi-Frank et al., 2016) suggesting that specific proteases may play important roles in determining resistant or tolerant tissue environments in the lung. For example, the membrane-bound MT1-MMP (MMP14), produced by both myeloid and parenchymal cells, has been shown to have a pathogenic role in influenza infection by restructuring the collagen matrix. Inhibition of MT1-MMP in mice protected the lung from matrix degradation, and blockage of MT1-MMP, along with oseltamivir, led to protection from secondary bacterial infection (Talmi-Frank et al., 2016). Given the diversity of ECM structural proteins and matrix remodeling enzymes there are likely other therapeutic targets within the lung ECM.

Severe influenza infection has gained popularity as a model of acute lung injury and acute respiratory distress syndrome and it has been an exciting decade for research defining the mechanisms of influenza immunopathology. These mechanisms now extend beyond the concept of a 'cytokine storm' and have begun to link specific inflammatory signals or cell types to changes in lung structure and function, which ultimately determine whether a host survives the infection. Recent work has highlighted the importance of lung parenchymal cells in regulating states of disease resistance and tolerance. The mechanisms of immunopathology defined in influenza models may also translate to other respiratory infections and lung diseases. Hopefully, these findings will help us avoid the next hundred years of (influenza) immunopathology.

Acknowledgments

This work was funded by the National Institute of Allergy and Infectious Diseases under HHS contract HHSN272201400006C for the St. Jude Center of Excellence for Influenza Research and Surveillance (P.G.T.) and ALSAC. The content is solely the responsibility of the authors anddoes not necessarily represent the official views of the National Institutes of Health.

References

Al Subaie, S.S., Al Saadi, M.A., 2012. Features associated with severe disease in hospitalized children with 2009 influenza A (H1N1) infection at a university hospital in Riyadh, Saudi Arabia. Ann. Saudi Med. 32 (1), 53–58.

Aldridge, J.R., et al., 2009. TNF/iNOS-producing dendritic cells are the necessary evil of lethal influenza virus infection. Proc. Natl. Acad. Sci. U. S. A. 106 (13), 5306–5311.

Allen, I.C., et al., 2009. The NLRP3 inflammasome mediates in vivo innate immunity to influenza A virus through recognition of viral RNA. Immunity 30 (4), 556–565.

Alten, R., 2011. Tocilizumab: a novel humanized anti-interleukin 6 receptor antibody for the treatment of patients with rheumatoid arthritis. Ther. Adv. Musculoskelet. Dis. 3 (3), 133–149.

Armstrong, S.M., et al., 2012. Influenza infects lung microvascular endothelium leading to microvascular leak: role of apoptosis and claudin-5. PLoS One 7 (10), e47323.

Arpaia, N., et al., 2015. A distinct function of regulatory T cells in tissue protection. Cell 162 (5), 1078–1089.

Askovich, P.S., et al., 2013. Differential host response, rather than early viral replication efficiency, correlates with pathogenicity caused by influenza viruses. PLoS One 8 (9), e74863. Edited by P. Digard.

Ayres, J.S., Schneider, D.S., 2012. Tolerance of infections. Annu. Rev. Immunol. 30, 271–294.

Baer, A., Colon-Moran, W., Bhattarai, N., 2018. Characterization of the effects of immunomodulatory drug fingolimod (FTY720) on human T cell receptor signaling pathways. Sci. Rep. 8 (1), 10910.

Barnes, P.J., 2006. How corticosteroids control inflammation: quintiles prize lecture 2005. Br. J. Pharmacol. 148 (3), 245–254.

Beigel, J.H., et al., 2005. Avian influenza A (H5N1) infection in humans. N. Engl. J. Med. 353 (13), 1374–1385.

Biondo, C., et al., 2019. The dual role of innate immunity during influenza. Biomed. J. 42 (1), 8–18.

Bonnans, C., Chou, J., Werb, Z., 2014. Remodelling the extracellular matrix in development and disease. Nat. Rev. Mol. Cell Biol. 15 (12), 786–801.

Bosurgi, L., et al., 2017. Macrophage function in tissue repair and remodeling requires IL-4 or IL-13 with apoptotic cells. Science 356 (6342), 1072–1076.

Boyd, D.F., Thomas, P.G., 2017. Towards integrating extracellular matrix and immunological pathways. Cytokine 98, 79–86.

Bradley, L.M., et al., 2012. Matrix metalloprotease 9 mediates neutrophil migration into the airways in response to influenza virus-induced toll-like receptor signaling. PLoS Pathog. 8 (4), e1002641. Edited by Y. Kawaoka.

Brandes, M., et al., 2013. A systems analysis identifies a feedforward inflammatory circuit leading to lethal influenza infection. Cell 154 (1), 197–212.

Brown, D.M., et al., 2012. Multifunctional CD4 cells expressing gamma interferon and perforin mediate protection against lethal influenza virus infection. J. Virol. 86 (12), 6792–6803.

Camp, J.V., Jonsson, C.B., 2017. A role for neutrophils in viral respiratory disease. Front. Immunol. 8, 550.

Camner, P., Jarstrand, C., Philipson, K., 1973. Tracheobronchial clearance in patients with influenza. Am. Rev. Respir. Dis. 108, 131–135. https://doi.org/10.1164/arrd.1973.108.1.131.

Cardani, A., et al., 2017. Alveolar macrophages prevent lethal influenza pneumonia by inhibiting infection of type-1 alveolar epithelial cells. PLoS Pathog. 13 (1), e1006140.

Carlin, L.M., et al., 2013. Nr4a1-dependent Ly6C(low) monocytes monitor endothelial cells and orchestrate their disposal. Cell 153 (2), 362–375.

Carrat, F., et al., 2008. Time lines of infection and disease in human influenza: a review of volunteer challenge studies. Am. J. Epidemiol. 167 (7), 775–785.
CDC, 2019. Estimated Influenza Illnesses, Medical visits, Hospitalizations, and Deaths in the United States—2017–2018 Influenza Season. CDC. Available at: https://www.cdc.gov/flu/about/burden/2017-2018.htm (Accessed: 10 February 2020).
CDC, 2019a. Influenza Antiviral Drug Resistance. CDC. Available at: https://www.cdc.gov/flu/treatment/antiviralresistance.htm (Accessed: 25 May 2020).
CDC, 2019b. People at High Risk of Flu. Centers for Disease Control and Prevention. Available at: https://www.cdc.gov/flu/highrisk/index.htm (Accessed: 25 May 2020).
CDC, 2020a. 1918 Pandemic (H1N1 virus) | Pandemic Influenza (Flu). CDC. Available at: https://www.cdc.gov/flu/pandemic-resources/1918-pandemic-h1n1.html (Accessed: 25 May 2020).
CDC, 2020b. 2009 H1N1 Pandemic. Centers for Disease Control and Prevention. Available at: https://www.cdc.gov/flu/pandemic-resources/2009-h1n1-pandemic.html (Accessed: 25 May 2020).
CDC, 2020c. Burden of Influenza. Centers for Disease Control and Prevention. Available at: https://www.cdc.gov/flu/about/burden/index.html (Accessed: 25 May 2020).
Chan, M.C.W., et al., 2009. Influenza H5N1 virus infection of polarized human alveolar epithelial cells and lung microvascular endothelial cells. Respir. Res. 10, 102.
Chan, L.L.Y., et al., 2019. Risk assessment of the tropism and pathogenesis of the highly pathogenic avian influenza A/H7N9 virus using ex vivo and in vitro cultures of human respiratory tract. J. Infect. Dis. 220 (4), 578–588.
Chen, X.-J., et al., 2004. Influenza virus inhibits ENaC and lung fluid clearance. Am. J. Physiol. Lung Cell. Mol. Physiol. 287 (2), L366–L373.
Chen, Y., et al., 2013. Human infections with the emerging avian influenza A H7N9 virus from wet market poultry: clinical analysis and characterisation of viral genome. Lancet 381 (9881), 1916–1925.
Chen, X., et al., 2018. Host immune response to influenza A virus infection. Front. Immunol. 9, 320.
Chertow, D.S., Memoli, M.J., 2013. Bacterial coinfection in influenza: a grand rounds review. JAMA 309 (3), 275–282.
Clark, I.A., 2007. The advent of the cytokine storm. Immunol. Cell Biol. 85 (4), 271–273.
Coates, B.M., et al., 2018. Inflammatory monocytes drive influenza A virus–mediated lung injury in juvenile mice. J. Immunol. 200 (7), ji1701543.
Cole, S.L., et al., 2017. M1-like monocytes are a major immunological determinant of severity in previously healthy adults with life-threatening influenza. JCI Insight 2 (7), e91868. https://doi.org/10.1172/jci.insight.91868.
Cooper, N.J., et al., 2003. Effectiveness of neuraminidase inhibitors in treatment and prevention of influenza A and B: systematic review and meta-analyses of randomised controlled trials. BMJ 326 (7401), 1235.
de Jong, M.D., et al., 2006. Fatal outcome of human influenza A (H5N1) is associated with high viral load and hypercytokinemia. Nat. Med. 12 (10), 1203–1207.
D'Elia, R.V., et al., 2013. Targeting the "cytokine storm" for therapeutic benefit. Clin. Vaccine Immunol. 20 (3), 319–327.
Dienz, O., et al., 2012. Essential role of IL-6 in protection against H1N1 influenza virus by promoting neutrophil survival in the lung. Mucosal Immunol. 5 (3), 258–266.
Doherty, M., et al., 2016. Vaccination of special populations: protecting the vulnerable. Vaccine 34 (52), 6681–6690.
Duffield, J.S., et al., 2013. Host responses in tissue repair and fibrosis. Annu. Rev. Pathol. Mech. Dis. 8 (1), 241–276.
Dunbar, P.R., et al., 2020. Pulmonary monocytes interact with effector T cells in the lung tissue to drive TRM differentiation following viral infection. Mucosal Immunol. 13 (1), 161–171.

Enelow, R.I., et al., 1998. Structural and functional consequences of alveolar cell recognition by CD8(+) T lymphocytes in experimental lung disease. J. Clin. Invest. 102 (9), 1653–1661.

Fadok, V.A., et al., 1998. Macrophages that have ingested apoptotic cells in vitro inhibit proinflammatory cytokine production through autocrine/paracrine mechanisms involving TGF-beta, PGE2, and PAF. J. Clin. Invest. 101 (4), 890–898.

Fiore-Gartland, A., et al., 2017. Cytokine profiles of severe influenza virus-related complications in children. Front. Immunol. 8, 1423.

Fujikura, D., Miyazaki, T., 2018. Programmed cell death in the pathogenesis of influenza. Int. J. Mol. Sci. 19 (7), 2065. https://doi.org/10.3390/ijms19072065.

Furuya-Kanamori, L., et al., 2016. Heterogeneous and dynamic prevalence of asymptomatic influenza virus infections. Emerg. Infect. Dis. J. 22 (6), 1052.

Gao, R., et al., 2013. Human infection with a novel avian-origin influenza A (H7N9) virus. N. Engl. J. Med. 368 (20), 1888–1897.

Ghebrehewet, S., MacPherson, P., Ho, A., 2016. Influenza. BMJ 355, i6258.

Glezen, W.P., 1996. Emerging infections: pandemic influenza. Epidemiol. Rev. 18 (1), 64–76.

Govorkova, E.A., McCullers, J.A., 2013. Therapeutics against influenza. Curr. Top. Microbiol. Immunol. 370, 273–300.

Granados, A., et al., 2017. Influenza and rhinovirus viral load and disease severity in upper respiratory tract infections. J. Clin. Virol. 86, 14–19.

Guan, W., et al., 2018. Clinical correlations of transcriptional profile in patients infected with avian influenza H7N9 virus. J. Infect. Dis. 218 (8), 1238–1248.

Guo, X.-Z.J., Thomas, P.G., 2017. New fronts emerge in the influenza cytokine storm. Semin. Immunopathol. 39 (5), 541–550.

Guo, X.-Z.J., et al., 2018. Lung γδ T cells mediate protective responses during neonatal influenza infection that are associated with type 2 immunity. Immunity 49 (3), 531–544.e6.

Hagau, N., et al., 2010. Clinical aspects and cytokine response in severe H1N1 influenza A virus infection. Crit. Care 14 (6), R203.

Hall, O.J., et al., 2016. Progesterone-based therapy protects against influenza by promoting lung repair and recovery in females. PLoS Pathog. 12 (9), e1005840.

Hashimoto, Y., et al., 2007. Evidence for phagocytosis of influenza virus-infected, apoptotic cells by neutrophils and macrophages in mice. J. Immunol. 178 (4), 2448–2457.

Hayday, A.C., 2019. γδ T cell update: adaptate orchestrators of immune surveillance. J. Immunol. 203 (2), 311–320.

Heaton, N.S., et al., 2014. Long-term survival of influenza virus infected club cells drives immunopathology. J. Exp. Med. 211 (9), 1707–1714.

Hendrickson, C.M., Matthay, M.A., 2013. Viral pathogens and acute lung injury: investigations inspired by the SARS epidemic and the 2009 H1N1 influenza pandemic. Semin. Respir. Crit. Care Med. 34 (4), 475–486.

Herfst, S., et al., 2014. Avian influenza virus transmission to mammals. In: Compans, R.W., Oldstone, M.B.A. (Eds.), Influenza Pathogenesis and Control. In: vol. I. Springer International Publishing, Cham, pp. 137–155.

Hermans, D., Webby, R.J., Wong, S.-S., 2018. Atypical antibody responses to influenza. J. Thorac. Dis. 10 (Suppl. 19), S2238–S2247.

Herold, S., et al., 2006. Alveolar epithelial cells direct monocyte transepithelial migration upon influenza virus infection: impact of chemokines and adhesion molecules. J. Immunol. 177 (3), 1817–1824.

Herold, S., et al., 2015. Influenza virus-induced lung injury: pathogenesis and implications for treatment. Eur. Respir. J. 45 (5), 1463–1478.

Hintzen, G., et al., 2006. Induction of tolerance to innocuous inhaled antigen relies on a CCR7-dependent dendritic cell-mediated antigen transport to the bronchial lymph node. J. Immunol. 177 (10), 7346–7354.

Hussain, M., et al., 2017. Drug resistance in influenza A virus: the epidemiology and management. Infect. Drug Resist. 10, 121–134.

Ibricevic, A., et al., 2006. Influenza virus receptor specificity and cell tropism in mouse and human airway epithelial cells. J. Virol. 80 (15), 7469–7480.

Ichinohe, T., Pang, I.K., Iwasaki, A., 2010. Influenza virus activates inflammasomes via its intracellular M2 ion channel. Nat. Immunol. 11 (5), 404–410.

Imai, Y., et al., 2008. Identification of oxidative stress and toll-like receptor 4 signaling as a key pathway of acute lung injury. Cell 133 (2), 235–249.

Imai, M., et al., 2020. Influenza A variants with reduced susceptibility to baloxavir isolated from Japanese patients are fit and transmit through respiratory droplets. Nat. Microbiol. 5 (1), 27–33.

Iwasaki, A., Pillai, P.S., 2014. Innate immunity to influenza virus infection. Nat. Rev. Immunol. 14 (5), 315–328.

Jamieson, A.M., et al., 2010. Influenza virus-induced glucocorticoids compromise innate host defense against a secondary bacterial infection. Cell Host Microbe 7 (2), 103–114.

Jamieson, A.M., et al., 2013. Role of tissue protection in lethal respiratory viral-bacterial coinfection. Science 340 (6137), 1230–1234.

Jankov, R.P., et al., 2002. Hepoxilin analogs inhibit bleomycin-induced pulmonary fibrosis in the mouse. J. Pharmacol. Exp. Ther. 301 (2), 435–440.

Jordana, M., et al., 1994. Immune-inflammatory functions of fibroblasts. Eur. Respir. J. 7 (12), 2212–2222.

Kaiser, L., et al., 2001. Symptom pathogenesis during acute influenza: interleukin-6 and other cytokine responses. J. Med. Virol. 64 (3), 262–268.

Kanegai, C.M., et al., 2016. Persistent pathology in influenza-infected mouse lungs. Am. J. Respir. Cell Mol. Biol. 55 (4), 613–615.

Kanneganti, T.-D., et al., 2006. Bacterial RNA and small antiviral compounds activate caspase-1 through cryopyrin/Nalp3. Nature 440 (7081), 233–236.

Kawada, J.-I., et al., 2013. Clinical characteristics of influenza virus infection in juvenile idiopathic arthritis patients treated with tocilizumab. Mod. Rheumatol. 23 (5), 972–976.

Kesavardhana, S., et al., 2017. ZBP1/DAI ubiquitination and sensing of influenza vRNPs activate programmed cell death. J. Exp. Med. 214 (8), jem.20170550.

Kishimoto, T., 2010. IL-6: from its discovery to clinical applications. Int. Immunol. 22 (5), 347–352.

Korteweg, C., Gu, J., 2008. Pathology, molecular biology, and pathogenesis of avian influenza A (H5N1) infection in humans. Am. J. Pathol. 172 (5), 1155–1170.

Kostrzewska, K., et al., 1974. Pulmonary staphylococcal complications in patients during the influenza epidemic in 1971–1972. Mater. Med. Pol. 6 (3), 207–212.

Krönke, G., et al., 2009. 12/15-lipoxygenase counteracts inflammation and tissue damage in arthritis. J. Immunol. 183 (5), 3383–3389.

Kuiken, T., Taubenberger, J.K., 2008. Pathology of human influenza revisited. Vaccine 26 (Suppl. 4), D59–D66.

Kumar, P.A., et al., 2011. Distal airway stem cells yield alveoli in vitro and during lung regeneration following H1N1 influenza infection. Cell 147 (3), 525–538.

Kumlin, U., et al., 2008. Sialic acid tissue distribution and influenza virus tropism. Influenza Other Respir. Viruses 2 (5), 147–154.

Kunisaki, K.M., Janoff, E.N., 2009. Influenza in immunosuppressed populations: a review of infection frequency, morbidity, mortality, and vaccine responses. Lancet Infect. Dis. 9 (8), 493–504.

Kunzelmann, K., et al., 2000. Influenza virus inhibits amiloride-sensitive Na+ channels in respiratory epithelia. Proc. Natl. Acad. Sci. U. S. A. 97 (18), 10282–10287.

Kuriakose, T., et al., 2016. ZBP1/DAI is an innate sensor of influenza virus triggering the NLRP3 inflammasome and programmed cell death pathways. Sci. Immunol. 1 (2). https://doi.org/10.1126/sciimmunol.aag2045.

Lam, J.H., Baumgarth, N., 2019. The multifaceted B cell response to influenza virus. J. Immunol. 202 (2), 351–359.

Lansbury, L., et al., 2019. Corticosteroids as adjunctive therapy in the treatment of influenza. Cochrane Database Syst. Rev. 2, CD010406.

Lau, L.L.H., et al., 2010. Viral shedding and clinical illness in naturally acquired influenza virus infections. J. Infect. Dis. 201 (10), 1509–1516.

Lauder, S.N., et al., 2013. Interleukin-6 limits influenza-induced inflammation and protects against fatal lung pathology. Eur. J. Immunol. 43 (10), 2613–2625.

Lazrak, A., et al., 2009. Influenza virus M2 protein inhibits epithelial sodium channels by increasing reactive oxygen species. FASEB J. 23 (11), 3829–3842.

Lee, N., Hurt, A.C., 2018. Neuraminidase inhibitor resistance in influenza: a clinical perspective. Curr. Opin. Infect. Dis. 31 (6), 520–526.

Lee, C.K., et al., 2011. Comparison of pandemic (H1N1) 2009 and seasonal influenza viral loads, Singapore. Emerg. Infect. Dis. 17 (2), 287–291.

Li, C., et al., 2012. Corticosteroid treatment ameliorates acute lung injury induced by 2009 swine origin influenza A (H1N1) virus in mice. PLoS One 7 (8), e44110.

Li, H., et al., 2017. Effect of low-to-moderate-dose corticosteroids on mortality of hospitalized adolescents and adults with influenza A(H1N1)pdm09 viral pneumonia. Influenza Other Respi. Viruses 11 (4), 345–354.

Liem, N.T., et al., 2009. Clinical features of human influenza A (H5N1) infection in Vietnam: 2004–2006. Clin. Infect. Dis. 48 (12), 1639–1646.

Lin, K.L., et al., 2008. CCR2+ monocyte-derived dendritic cells and exudate macrophages produce influenza-induced pulmonary immune pathology and mortality. J. Immunol. 180 (4), 2562–2572.

Liu, A.N., et al., 1999. Perforin-independent CD8(+) T-cell-mediated cytotoxicity of alveolar epithelial cells is preferentially mediated by tumor necrosis factor-alpha: relative insensitivity to Fas ligand. Am. J. Respir. Cell Mol. Biol. 20 (5), 849–858.

Liu, Q., Zhou, Y.-H., Yang, Z.-Q., 2016. The cytokine storm of severe influenza and development of immunomodulatory therapy. Cell. Mol. Immunol. 13 (1), 3–10.

Man, S.M., Karki, R., Kanneganti, T.-D., 2017. Molecular mechanisms and functions of pyroptosis, inflammatory caspases and inflammasomes in infectious diseases. Immunol. Rev. 277 (1), 61–75.

Mansfield, K.G., 2007. Viral tropism and the pathogenesis of influenza in the mammalian host. Am. J. Pathol. 171 (4), 1089–1092.

Matthay, M.A., et al., 2019. Acute respiratory distress syndrome. Nat. Rev. Dis. Primers 5 (1), 18.

Mauad, T., et al., 2010. Lung pathology in fatal novel human influenza A (H1N1) infection. Am. J. Respir. Crit. Care Med. 181 (1), 72–79.

McCullers, J.A., 2006. Insights into the interaction between influenza virus and pneumococcus. Clin. Microbiol. Rev. 19 (3), 571–582.

McMahon, M., et al., 2016. ADAMTS5 is a critical regulator of virus-specific T cell immunity. PLoS Biol. 14 (11), e1002580. Edited by P. Marrack.

Medzhitov, R., Schneider, D.S., Soares, M.P., 2012. Disease tolerance as a defense strategy. Science 335 (6071), 936–941.

Meschi, S., et al., 2011. Duration of viral shedding in hospitalized patients infected with pandemic H1N1. BMC Infect. Dis. 11, 140.

Mettelman, R.C., Thomas, P.G., 2020. Human susceptibility to influenza infection and severe disease. Cold Spring Harb. Perspect. Med. https://doi.org/10.1101/cshperspect.a038711.

Mirandola, P., et al., 2004. Activated human NK and CD8+ T cells express both TNF-related apoptosis-inducing ligand (TRAIL) and TRAIL receptors but are resistant to TRAIL-mediated cytotoxicity. Blood 104 (8), 2418–2424.

Monticelli, L.A., et al., 2011. Innate lymphoid cells promote lung-tissue homeostasis after infection with influenza virus. Nat. Immunol. 12 (11), 1045–1054.

Monto, A.S., et al., 1999. Efficacy and safety of the neuraminidase inhibitor zanamivirin the treatment of influenza A and B virus infections. J. Infect. Dis. 180 (2), 254–261.

Moreno, G., et al., 2018. Corticosteroid treatment in critically ill patients with severe influenza pneumonia: a propensity score matching study. Intensive Care Med. 44 (9), 1470–1482.

Morens, D.M., Taubenberger, J.K., Fauci, A.S., 2008. Predominant role of bacterial pneumonia as a cause of death in pandemic influenza: implications for pandemic influenza preparedness. J. Infect. Dis. 198 (7), 962–970.

Murray, P.J., 2018. Immune regulation by monocytes. Semin. Immunol. 35, 12–18.

Narasaraju, T., et al., 2011. Excessive neutrophils and neutrophil extracellular traps contribute to acute lung injury of influenza pneumonitis. Am. J. Pathol. 179 (1), 199–210.

Nedel, W.L., et al., 2016. Corticosteroids for severe influenza pneumonia: a critical appraisal. Pediatr. Crit. Care Med. 5 (1), 89–95.

Neuzil, K.M., et al., 2002. Burden of interpandemic influenza in children younger than 5 years: a 25-year prospective study. J. Infect. Dis. 185 (2), 147–152.

Neuzil, K.M., et al., 2003. Recognizing influenza in older patients with chronic obstructive pulmonary disease who have received influenza vaccine. Clin. Infect. Dis. 36 (2), 169–174.

Ng, W.-F., et al., 2006. The comparative pathology of severe acute respiratory syndrome and avian influenza A subtype H5N1—a review. Hum. Pathol. 37 (4), 381–390.

Ni, Y.-N., et al., 2019. The effect of corticosteroids on mortality of patients with influenza pneumonia: a systematic review and meta-analysis. Crit. Care 23 (1), 99.

Nicholson, K.G., et al., 2000. Efficacy and safety of oseltamivir in treatment of acute influenza: a randomised controlled trial. Neuraminidase inhibitor flu treatment investigator group. Lancet 355 (9218), 1845–1850.

Nogusa, S., et al., 2016. RIPK3 activates parallel pathways of MLKL-driven necroptosis and FADD-mediated apoptosis to protect against influenza A virus. Cell Host Microbe 20 (1), 13–24.

Nowarski, R., Jackson, R., Flavell, R.A., 2017. The stromal intervention: regulation of immunity and inflammation at the epithelial-mesenchymal barrier. Cell 168 (3), 362–375.

Ocaña-Macchi, M., et al., 2009. Hemagglutinin-dependent tropism of H5N1 avian influenza virus for human endothelial cells. J. Virol. 83 (24), 12947–12955.

Ordovas-Montanes, J., et al., 2020. Distribution and storage of inflammatory memory in barrier tissues. Nat. Rev. Immunol. 20 (5), 1–13. Nature Publishing Group.

Oshansky, C.M., et al., 2014. Mucosal immune responses predict clinical outcomes during influenza infection independently of age and viral load. Am. J. Respir. Crit. Care Med. 189 (4), 449–462.

Ottolini, M., et al., 2003. Combination anti-inflammatory and antiviral therapy of influenza in a cotton rat model. Pediatr. Pulmonol. 36 (4), 290–294.

Paquette, S.G., et al., 2012. Interleukin-6 is a potential biomarker for severe pandemic H1N1 influenza A infection. PLoS One 7 (6), e38214.

Pichlmair, A., et al., 2006. RIG-I-mediated antiviral responses to single-stranded RNA bearing 5'-phosphates. Science 314 (5801), 997–1001.

Pipkin, M.E., et al., 2010. Interleukin-2 and inflammation induce distinct transcriptional programs that promote the differentiation of effector cytolytic T cells. Immunity 32 (1), 79–90.

Pothlichet, J., et al., 2013. Type I IFN triggers RIG-I/TLR3/NLRP3-dependent inflammasome activation in influenza A virus infected cells. PLoS Pathog. 9 (4), e1003256.

Pulendran, B., Maddur, M.S., 2015. Innate immune sensing and response to influenza. Curr. Top. Microbiol. Immunol. 386, 23–71.

Purnama, C., et al., 2014. Transient ablation of alveolar macrophages leads to massive pathology of influenza infection without affecting cellular adaptive immunity. Eur. J. Immunol. 44 (7), 2003–2012.

Rivera, J., Proia, R.L., Olivera, A., 2008. The alliance of sphingosine-1-phosphate and its receptors in immunity. Nat. Rev. Immunol. 8 (10), 753–763.

Rojas, J.M., et al., 2017. IL-10: a multifunctional cytokine in viral infections. J. Immunol. Res. 2017, 6104054.

Rojas-Quintero, J., et al., 2018. Matrix metalloproteinase-9 deficiency protects mice from severe influenza A viral infection. JCI Insight 3 (24), 21.

Ruf, B.R., Knuf, M., 2014. The burden of seasonal and pandemic influenza in infants and children. Eur. J. Pediatr. 173 (3), 265–276.

Saffarzadeh, M., et al., 2012. Neutrophil extracellular traps directly induce epithelial and endothelial cell death: a predominant role of histones. PLoS One 7 (2), e32366.

Sanders, C.J., Doherty, P.C., Thomas, P.G., 2011. Respiratory epithelial cells in innate immunity to influenza virus infection. Cell Tissue Res. 343 (1), 13–21.

Sanders, C.J., et al., 2013a. Compromised respiratory function in lethal influenza infection is characterized by the depletion of type I alveolar epithelial cells beyond threshold levels. Am. J. Physiol. Lung Cell. Mol. Physiol. 304 (7), L481–L488.

Sanders, C.J., et al., 2013b. Intranasal influenza infection of mice and methods to evaluate progression and outcome. In: Allen, I.C. (Ed.), Mouse Models of Innate Immunity: Methods and Protocols. Humana Press, Totowa, NJ, pp. 177–188.

Schneider, C., et al., 2014. Alveolar macrophages are essential for protection from respiratory failure and associated morbidity following influenza virus infection. PLoS Pathog. 10 (4), e1004053.

Serbina, N.V., et al., 2003. TNF/iNOS-producing dendritic cells mediate innate immune defense against bacterial infection. Immunity 19 (1), 59–70.

Shieh, W.-J., et al., 2010. 2009 pandemic influenza A (H1N1): pathology and pathogenesis of 100 fatal cases in the United States. Am. J. Pathol. 177 (1), 166–175.

Shirey, K.A., et al., 2013. The TLR4 antagonist Eritoran protects mice from lethal influenza infection. Nature 497 (7450), 498–502.

Short, K.R., et al., 2014. Pathogenesis of influenza-induced acute respiratory distress syndrome. Lancet Infect. Dis. 14 (1), 57–69.

Short, K.R., Kuiken, T., Van Riel, D., 2019. Role of endothelial cells in the pathogenesis of Influenza in humans. J. Infect. Dis. 220 (11), 1859–1860.

Silvennoinen, H., et al., 2009. Clinical presentation of influenza in unselected children treated as outpatients. Pediatr. Infect. Dis. J. 28 (5), 372–375.

Small, B.A., et al., 2001. CD8(+) T cell-mediated injury in vivo progresses in the absence of effector T cells. J. Exp. Med. 194 (12), 1835–1846.

Suwara, M.I., et al., 2014. IL-1α released from damaged epithelial cells is sufficient and essential to trigger inflammatory responses in human lung fibroblasts. Mucosal Immunol. 7 (3), 684–693.

Talmi-Frank, D., et al., 2016. Extracellular matrix proteolysis by MT1-MMP contributes to influenza-related tissue damage and mortality. Cell Host Microbe 20 (4), 458–470.

Tam, V.C., et al., 2013. Lipidomic profiling of influenza infection identifies mediators that induce and resolve inflammation. Cell 154 (1), 213–227.

Tang, B.M., et al., 2019. Neutrophils-related host factors associated with severe disease and fatality in patients with influenza infection. Nat. Commun. 10 (1), 3422.

Taubenberger, J.K., 2006. The origin and virulence of the 1918 "Spanish" influenza virus. Proc. Am. Philos. Soc. 150 (1), 86–112.

Taubenberger, J.K., Morens, D.M., 2019. The 1918 influenza pandemic and its legacy. Cold Spring Harb. Perspect. Med, a038695. https://doi.org/10.1101/cshperspect.a038695.

Teijaro, J.R., 2015. The role of cytokine responses during influenza virus pathogenesis and potential therapeutic options. In: Oldstone, M.B.A., Compans, R.W. (Eds.), Influenza Pathogenesis and Control. In: vol. II. Springer International Publishing, Cham, pp. 3–22.

Teijaro, J.R., et al., 2011. Endothelial cells are central orchestrators of cytokine amplification during influenza virus infection. Cell 146 (6), 980–991.

Teijaro, J.R., et al., 2014. Mapping the innate signaling cascade essential for cytokine storm during influenza virus infection. Proc. Natl. Acad. Sci. U. S. A. 111 (10), 3799–3804.

Thapa, R.J., et al., 2016. DAI senses influenza A virus genomic RNA and activates RIPK3-dependent cell death. Cell Host Microbe 20 (5), 674–681.

Thomas, P.G., et al., 2009. The intracellular sensor NLRP3 mediates key innate and healing responses to influenza A virus via the regulation of caspase-1. Immunity 30 (4), 566–575.

Thomas, B.J., et al., 2014. Glucocorticosteroids enhance replication of respiratory viruses: effect of adjuvant interferon. Sci. Rep. 4, 7176.

Thomas, P.G., Shubina, M., Balachandran, S., 2020. ZBP1/DAI-dependent cell death pathways in influenza A virus immunity and pathogenesis. Curr. Top. Microbiol. Immunol. https://doi.org/10.1007/82_2019_190.

Thompson, W.W., et al., 2003. Mortality associated with influenza and respiratory syncytial virus in the United States. JAMA 289 (2), 179–186.

Tisoncik, J.R., et al., 2012. Into the eye of the cytokine storm. Microbiol. Mol. Biol. Rev. 76 (1), 16–32.

Topham, D.J., Tripp, R.A., Doherty, P.C., 1997. CD8+ T cells clear influenza virus by perforin or Fas-dependent processes. J. Immunol. 159 (11), 5197–5200.

Turnier, J.L., Brunner, H.I., 2016. Tocilizumab for treating juvenile idiopathic arthritis. Expert Opin. Biol. Ther. 16 (4), 559–566.

Uyeki, T.M., 2014. Preventing and controlling influenza with available interventions. N. Engl. J. Med. 370 (9), 789–791.

Uyeki, T.M., et al., 2019. Clinical practice guidelines by the infectious diseases society of America: 2018 update on diagnosis, treatment, chemoprophylaxis, and institutional outbreak management of seasonal influenzaa. Clin. Infect. Dis. 68 (6), e1–e47.

van Riel, D., et al., 2007. Human and avian influenza viruses target different cells in the lower respiratory tract of humans and other mammals. Am. J. Pathol. 171 (4), 1215–1223.

van Riel, D., et al., 2010. Seasonal and pandemic human influenza viruses attach better to human upper respiratory tract epithelium than avian influenza viruses. Am. J. Pathol. 176 (4), 1614–1618.

van Riet, E., et al., 2012. Mucosal IgA responses in influenza virus infections; thoughts for vaccine design. Vaccine 30 (40), 5893–5900.

Vaughan, A.E., et al., 2014. Lineage-negative progenitors mobilize to regenerate lung epithelium after major injury. Nature 517 (7536), 621–625.

Vaux, D.L., Haecker, G., Strasser, A., 1994. An evolutionary perspective on apoptosis. Cell 76 (5), 777–779.

Vermillion, M.S., et al., 2018. Production of amphiregulin and recovery from influenza is greater in males than females. Biol. Sex Differ. 9 (1), 24.

Walsh, K.B., et al., 2011. Suppression of cytokine storm with a sphingosine analog provides protection against pathogenic influenza virus. Proc. Natl. Acad. Sci. U. S. A. 108 (29), 12018–12023.

Wang, Z., et al., 2014. Early hypercytokinemia is associated with interferon-induced transmembrane protein-3 dysfunction and predictive of fatal H7N9 infection. Proc. Natl. Acad. Sci. U. S. A. 111 (2), 769–774.

Wang, Z., et al., 2015. Recovery from severe H7N9 disease is associated with diverse response mechanisms dominated by $CD8^+$ T cells. Nat. Commun. 6 (1), 6833.

Wang, A., et al., 2016. Opposing effects of fasting metabolism on tissue tolerance in bacterial and viral inflammation. Cell 166 (6), 1512–1525.e12.

Weizman, O.-E., et al., 2017. ILC1 confer early host protection at initial sites of viral infection. Cell 171 (4), 795–808.e12.

Whitmire, J.K., Tan, J.T., Whitton, J.L., 2005. Interferon-gamma acts directly on CD8+ T cells to increase their abundance during virus infection. J. Exp. Med. 201 (7), 1053–1059.

Wilhelm, M., 2018. Influenza in older patients: a call to action and recent updates for vaccinations. Am. J. Manag. Care 24 (2 Suppl), S15–S24.

Wolk, K.E., et al., 2008. Influenza A virus inhibits alveolar fluid clearance in BALB/c mice. Am. J. Respir. Crit. Care Med. 178 (9), 969–976.

Xi, Y., et al., 2017. Local lung hypoxia determines epithelial fate decisions during alveolar regeneration. Nat. Cell Biol. 19 (8), 904–914.

Xu, L., et al., 2004. Cutting edge: pulmonary immunopathology mediated by antigen-specific expression of TNF-alpha by antiviral CD8+ T cells. J. Immunol. 173 (2), 721–725.

Yang, M.-L., et al., 2017. IL-6 ameliorates acute lung injury in influenza virus infection. Sci. Rep. 7, 43829.

Yatim, N., Albert, M.L., 2011. Dying to replicate: the orchestration of the viral life cycle, cell death pathways, and immunity. Immunity 35 (4), 478–490.

Yip, T.-F., et al., 2018. Advancements in host-based interventions for influenza treatment. Front. Immunol. 9, 1547.

Zacharias, W.J., et al., 2018. Regeneration of the lung alveolus by an evolutionarily conserved epithelial progenitor. Nature 555 (7695), 251–255.

Zaiss, D.M.W., et al., 2015. Emerging functions of amphiregulin in orchestrating immunity, inflammation, and tissue repair. Immunity 42 (2), 216–226.

Zeng, H., et al., 2012. Human pulmonary microvascular endothelial cells support productive replication of highly pathogenic avian influenza viruses: possible involvement in the pathogenesis of human H5N1 virus infection. J. Virol. 86 (2), 667–678.

Zhang, T., et al., 2020. Influenza virus Z-RNAs induce ZBP1-mediated necroptosis. Cell 180 (6), 1115–1129.e13.

Zhao, M.Q., et al., 2000. Alveolar epithelial cell chemokine expression triggered by antigen-specific cytolytic CD8(+) T cell recognition. J. Clin. Invest. 106 (6), R49–R58.

Zheng, B.-J., et al., 2008. Delayed antiviral plus immunomodulator treatment still reduces mortality in mice infected by high inoculum of influenza A/H5N1 virus. Proc. Natl. Acad. Sci. U. S. A. 105 (23), 8091–8096.

Zhou, Y., et al., 2020. Use of corticosteroids in influenza-associated acute respiratory distress syndrome and severe pneumonia: a systemic review and meta-analysis. Sci. Rep. 10 (1), 3044.

Zuo, W., et al., 2014. p63+Krt5+ distal airway stem cells are essential for lung regeneration. Nature 517 (7536), 616–620.

CHAPTER EIGHT

Immunopathology and immune homeostasis during viral infection in insects

Andrea González-González*, Marta L. Wayne
Department of Biology, University of Florida, Gainesville, FL, United States
*Corresponding author: e-mail address: andreagg@gmail.com

Contents

Abstract

Organisms clear infections by mounting an immune response that is normally turned off once the pathogens have been cleared. However, sometimes this immune response is not properly or timely arrested, resulting in the host damaging itself. This immune dysregulation may be referred to as immunopathology. While our knowledge of immune and metabolic pathways in insects, particularly in response to viral infections, is growing, little is known about the mechanisms that regulate this immune response and hence little is known about immunopathology in this important and diverse group of organisms. In this chapter we focus both on documenting the molecular mechanisms described involved in restoring immune homeostasis in insects after viral infections

Advances in Virus Research, Volume 107
ISSN 0065-3527
https://doi.org/10.1016/bs.aivir.2020.06.001

and on identifying potential mechanisms for future investigation. We argue that learning about the immunopathological consequences of an improperly regulated immune response in insects will benefit both insect and human health.

1. Introduction

Insects represent the most abundant and diverse group of organisms in the world (Stork, 2018). They play an important role in natural habitats by performing ecosystem services such as pollination, seed dispersal, pest control, and conferring soil fertility by nutrient and waste recycling (Losey and Vaughan, 2006). In addition, insects are of direct economic importance (honey, wax, and silk production) and represent a human and livestock concern due to their impacts on crops as well as their disease transmission capacity (World Health Organization, 2020).

Insect population densities are regulated in part by diverse pathogens such as bacteria, fungi, nematodes, parasitoids and viruses (Marchioro and Foerster, 2016). Viral infections are distinctive. In contrast to bacterial or fungal infections, viruses hijack the host machinery (ribosome and nucleus) to replicate themselves and to produce new infective virions ready to be released and infect new healthy host cells (Wasik and Turner, 2013). To counteract and clear viral infections, insects use different strategies such as the RNA interference (RNAi) mechanism as well as cellular (apoptosis, phagocytosis, and autophagy) and humoral processes (production of antimicrobial peptides) (reviewed by Palmer et al., 2018). In addition, host-related processes such as nutritional status, heat-shock response and metabolic changes, as well as interactions with the host's resident microbiota, can impact the antiviral response in insects (Dolezal et al., 2019; Merkling et al., 2015b, 2019; Neu and Mainou, 2020; Xu et al., 2013).

Normally the antiviral immune response is tuned such that the response is appropriate to clear an infection, and turns off once the infection has been cleared. However, sometimes the immune system mounts an improper or over-reactive response causing self-damage or immunopathology (Graham et al., 2005). The study of immunopathology caused by viral infections in insects is an understudied field, especially relative to the amount known about immunopathologies in humans and other animals, and in plants. We argue that, besides gaining valuable knowledge about the mechanisms and evolutionary consequences of immunopathology in insects, studying how defective immune responses arise and evolve can contribute to a better understanding (and the potential manipulation) of how viral diseases affect vector competency, pollinator health, and food production.

Given both the potential importance and the relative paucity of knowledge about insect immunopathology, in this chapter we focus both on documenting the molecular mechanisms described so far involved in restoring homeostasis in insects after viral infections and on identifying potential mechanisms for future investigation. We begin by briefly describing the immune response pathways elicited by viral infections in insects in order to present an overview of the complexity of this response. We then describe the loss of homeostasis and explain the negative regulatory feedbacks described during viral infections. Finally, we suggest future research directions to be addressed in order to move the field forward.

2. Antiviral immune pathways in insects

Three main mechanisms have been described to be in charge of controlling and clearing viral infections in insects: RNAi, humoral and cellular responses.

2.1 RNA interference pathways

RNAi refers to a group of mechanisms implicated in the regulation of gene expression that uses small non-coding (nc)RNAs associated with proteins from the Argonaute family to degrade or silence target RNA sequences. RNAi is present across animals, plants and fungi, where it also plays a role in controlling viral infections, as well as in development and other processes.

In insects, *Drosophila melanogaster* and mosquitoes have been used extensively as model systems to study and describe the architecture and functioning of RNAi. However, RNAi has also been investigated in other organisms like bees, bumble bees and lepidopterans (reviewed by Gammon and Mello, 2015; Jayachandran et al., 2012; Kolliopoulou and Swevers, 2014; Lee et al., 2019; McMenamin et al., 2018; Sánchez-Vargas et al., 2009). Three RNAi pathways have been described in insects: small interfering (siRNA), microRNA (miRNA), and PIWI-interacting (piRNA) pathways (see Mongelli and Saleh, 2016; Schuster et al., 2019 for comprehensive reviews). Although all three pathways share this basic principle of gene silencing, they differ in the biogenesis of the small RNAs, the nature and processing of the sequence targets, and their biological functions (Carthew and Sontheimer, 2009; Ha and Kim, 2014; Schuster et al., 2019; Théron et al., 2014).

2.1.1 siRNA pathway

The siRNA pathway is the classical antiviral response in insects, is expressed in both germline and somatic tissues, (Gammon and Mello, 2015) and is triggered when endogenous or exogenous double strand RNA (dsRNA) is sensed in the host cytoplasm (Paro et al., 2015). This dsRNA is recognized and sliced by Dicer 2 (Dcr-2) and finally degraded by Argonaute 2 (Ago-2) inside the RNA-induced silencing complex (RISC) following a series of well described molecular steps in association with some cofactors (reviewed by Mongelli and Saleh, 2016). Typical endogenous dsRNAs arise from folding within transcripts, from sense/antisense interactions between transcripts (Czech et al., 2008; Okamura et al., 2008) and from transposable elements (Biryukova and Ye, 2015; van Rij and Berezikov, 2009); while virus-derived dsRNA—such as replication intermediates of single-stranded RNA viruses (ssRNA)—are considered to be common exogenous molecules (Bernstein et al., 2001). The siRNA response is also effective against dsDNA viruses by targeting dsRNA produced by bidirectional overlapping transcription during dsDNA virus infection (Bronkhorst et al., 2012; Kemp et al., 2013).

2.1.2 miRNA pathway

The miRNA pathway in insects has been primarily associated with the regulation of gene expression at a post-transcriptional level in germline and somatic cells, particularly during developmental processes such as tissue differentiation and cell proliferation (Carthew et al., 2017; Chandra et al., 2017; Jones and Newbury, 2010). However, it is also involved in regulating other biological processes such as metabolism and innate immune response (Li et al., 2017; Lucas and Raikhel, 2013). This RNAi pathway is found across the tree of life: in insects as well as other animals, plants, protists, and even within viruses themselves (Ghildiyal and Zamore, 2009). The miRNAs produced by DNA viruses suppress host defense responses promoting their replication, and therefore are an important part of host-parasite interactions (Hussain and Asgari, 2014; Skalsky and Cullen, 2010).

2.1.3 piRNA pathway

In insects, the piRNA pathway is in control of silencing the expression of transposable elements in germline and somatic tissues by distinguishing between host protein-coding genes and from mobile elements thus avoiding their deleterious effects (Lewis et al., 2018, reviewed in Iwasaki et al., 2015). This function can be considered as analogous to the immune response's role in discriminating "self" from "non-self" followed by the elimination of "parasitic" elements (Malone and Hannon, 2009). Several studies have assessed the antiviral role of the piRNA pathway in different insect model systems such as fruit flies, different mosquito species, midges and the silkworm both *in vivo* and *in vitro*, however, the results are contradictory and inconclusive (reviewed in Kolliopoulou et al., 2019).

2.2 Humoral response

Besides having a predominant role in the antibacterial and antifungal response (Boman et al., 1972), signal transduction pathways such as Jak-STAT and the nuclear factor κB (Nf-κB) pathways, Toll and Imd, also play a role in the defense against viruses in insects (reviewed by Liu et al., 2018; Palmer et al., 2018). Generally speaking, upon infection, the insect activates pattern recognition receptors (PRRs) that recognize and bind pathogen-associated molecular patterns (PAMPs) that are specific to the pathogen (i.e., not present in the host). Dicer-2 has been proposed to function as a PRR during viral infection given its phylogenetic similarity to mammalian retinoic acid-inducible gene-I-like receptors (also known as RIG-I-like receptors or RLRs), which induce an antiviral interferon response (Deddouche et al., 2008). The following candidate PAMPs have been suggested: dsRNA generated during viral replication, tissue damage, and cell debris (reviewed in Lamiable and Imler, 2014). Whatever the PAMP may be, we know that once the host receptor binds to it, a cascade of signal transduction reactions take place in the cytoplasm ending in the transcriptional induction of effector genes or antimicrobial peptides (AMPs) in the nucleus capable of inhibiting pathogen replication (Kingsolver et al., 2013). These AMPs are mainly synthetized in fat body cells and then secreted into the hemolymph where they reach high concentrations being still detected in few days or weeks after the pathogen infection (Haine et al., 2008; Lemaitre and Hoffmann, 2007).

2.3 Cellular response

Apoptosis or programmed cell death is a process of self-destruction important for maintaining tissue homeostasis and proper immune response in multicellular organisms (Elmore, 2007; Kerr et al., 1972). In insects, apoptosis has been shown to participate in the antiviral response against different RNA and DNA viruses (Clem et al., 1991; Lamiable et al., 2016c; Liu et al., 2013; Nainu et al., 2015; Vaidyanathan and Scott, 2006). When mammalian cells recognize viruses or intracellular debris produced by changes in the host physiology after viral infection, they undergo apoptotic death to prevent the damage from spreading. This is followed by their phagocytosis, which is mediated by hemocytes or blood cells. In insects, the process is similar except that plasmatocytes (functionally similar to mammalian macrophages) are in charge of engulfing and degrading dead cells, debris and infecting viruses (Nainu et al., 2015). These patrolling cells are abundant: in *D. melanogaster*, plasmatocytes represent about 95% of the circulating hemocytes, the other 5% is composed of crystal cells that mediate melanization and wound healing.

Another antiviral cellular response is autophagy. Besides being associated with maintaining cellular homeostasis during nutrient deprivation by engulfing and recycling damaged organelles or proteins, autophagy is involved in the control of viral infection caused by Vesicular stomatitis virus (VSV) and Rift valley fever virus (RVFV) in *D. melanogaster* (Lamiable et al., 2016c; Moy et al., 2014).

2.4 Nutritional signaling and host metabolism

Immune response involves a readjustment of nutrient allocation toward immune-related tasks (Dolezal et al., 2019; Varanasi and Rouse, 2018). Perhaps, then, it is unsurprising that diet and immune response are intricately intertwined. For example, one recently described antiviral mechanism is associated with proper lipid metabolism: *D. melanogaster* flies lacking the gene *Sgroppino*, which localizes to peroxisomes (organelles involved in lipid metabolism), are hypersensitive to infection with different ssRNA viruses such as Drosophila C virus (DCV), Cricket paralysis virus (CrPV), Flock house virus (FHV) (Merkling et al., 2019). Similarly, the extracellular signal-regulated kinase (ERK) pathway links nutritional status with antiviral response. Components of this pathway limit VSV, Sindbis virus (SINV) and DCV infection in Drosophila S2 cells and adult flies. The ERK response is active in the gut epithelium, which enhances the role of the gut as an immune barrier to avoid further systemic infections (Xu et al., 2013) for viruses acquired by feeding.

ERK also plays a role in antiviral priming via the gut microbiome. Here, the gut microbiome somehow senses viral infection and then stimulates Imd signaling which in turns induces the ERK pathway (Sansone et al., 2015). More evidence supporting the role of the gut microbiome in antiviral activity is provided by the observation that antibiotic-treated fruit flies present higher DCV titers compared to untreated flies; however, this differential viral load does not induce differential mortality (Ferreira et al., 2014). Similarly, in *Aedes aegypti*, the presence of different Gram-negative bacterial species protects the midgut against Dengue virus (DENV) infection (Ramirez et al., 2012). The interplay between nutritional status, the gut microbiome, and antiviral response is complex and bears further investigation.

3. Immunopathology in insects

Immunopathology, which we define as an inappropriate immune response to an infection, can cause harm to the host in different ways. If the response is weak (immunodeficiency), immunopathology could take the form of pathogen proliferation. Alternatively, illness in the host can be caused by an overactive response to a pathogen, or even by the inability to distinguish self from invader (autoimmunity) (Fig. 1) (Marshall et al., 2018). Many of the diseases caused by viruses in humans such as dengue, malaria, Zika, and influenza have pernicious consequences in some individuals not necessarily because of the presence of high viral levels, but due to the series of immunopathological responses triggered during the process of clearing the infection (Culshaw et al., 2017; Damjanovic et al., 2012; Voller, 1974).

One of the most important factors that leads to immunopathology of the overactive sort is inflammation. While inflammation plays a central role in immunity, its proper resolution is essential for the organism to return to its homeostatic state (Serhan and Savill, 2005). Widely recognized as a physiological response to tissue injury or infection, inflammation can be localized initially and then become systemic. It causes an increase in temperature, pain, swelling and redness. In mammals the inflammatory response is orchestrated by innate and adaptive immune mechanisms (Panayidou and Apidianakis, 2013; Rouse and Sehrawat, 2010). In insects, the Toll, Imd and Jak-STAT humoral pathways share some similarities with the inflammatory pathways of mammals such as the common ancestry shared by certain components of each pathway although there are structural and functional differences as well (reviewed in Sheehan et al., 2018).

In general, the immune response in insects is efficient against a variety of pathogens due to the non-specificity of its inflammatory component. It is also fast acting since it involves the constitutive expression of enzymatic cascades (Khan et al., 2017; Sadd and Siva-Jothy, 2006). Upon viral infection, the host deals both with the infectious process itself and with the toxic environment produced by its own immune response, which in turn induces

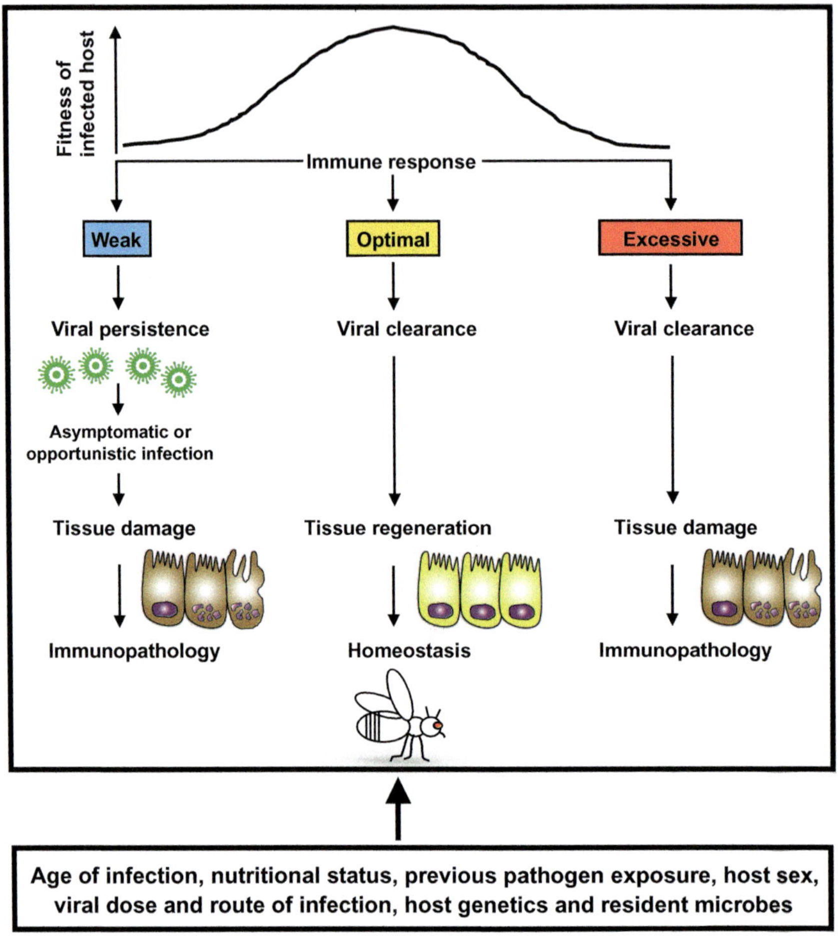

Fig. 1 Relationship between host fitness and magnitude of immune response to viral infection in insects. Following viral infection, the host immune system elicits a response that involves the activation, induction and effects of different signaling pathways, cellular components, and molecules that cause tissue damage as part of pathogen elimination. If the immune response is not turned off in a timely manner, it will cause severe tissue damage, i.e., immunopathology. On the other hand, if the response is not robust enough, it will either allow the establishment of a fatal viral load, or a persistent infection. The low viral levels common in persistent infections can themselves be the cause of tissue damage. An optimal response is one that will resolve the infection and limit tissue damage, thus maintaining the host homeostasis and maximizing its fitness. Different host factors such as the age of the host, nutritional status, sex, previous exposure to the same or other pathogens, viral dose and route of infection, genetics and resident microbiota can influence the outcome of the infection.

tissue remodeling and damage to eliminate the infected cells. If prolonged or exaggerated, cytotoxicity and tissue damage can lead to immunopathology and threaten the life of the infected host (Graham et al., 2005; Rouse and Sehrawat, 2010). What mechanisms allow the host to sense the moment to resolve the infection and reestablish homeostasis, that threshold where antiviral function ceases and unnecessary host tissue damage starts?

4. Restoring immune homeostasis

The immune response to viral and other type of infections is an energetically intensive response. It includes, of course, the up- and down-regulation of genes belonging to different immune pathways (see Section 2 "Antiviral immune pathways in insects") in addition to other physiological changes, such as stress response and tissue repair. The appropriate deployment of this response is achieved by different interconnected layers of regulation that ensure the arrest of the pathogen's proliferation or its successful clearance (Murray and Smale, 2012). Once the infection is resolved or controlled, turning the immune response off is key to avoid further self-inflicted damage to the host. Thus, re-establishing the immune homeostasis is accomplished by several layers of negative regulatory feedbacks within and among immune pathways (Kim et al., 2007; Schneider, 2007; Wang and Xia, 2018).

As described previously, there is an increasing body of knowledge about the mechanisms involved in how the infecting viruses are eliminated by insect hosts; however, studies about the regulation of this response in order to avoid the potential detrimental effects of the immune system are fewer (Lamiable et al., 2016a,b). Fortunately, several studies have shed some light into some regulatory mechanisms during bacterial infections that can be used as starting points to understand how the immune response turns off during viral infections in insects (Brandt et al., 2004; Gordon et al., 2005, 2008; Kim et al., 2007; Schneider, 2007); accordingly we include some bacterial examples below.

4.1 Negative feed-back loops and the regulation of the immune response in insects

Feedback loops play an important role in the amplification and attenuation of intracellular signaling pathways. Everything starts with an extracellular signal sensed by transmembrane receptors that convert this extracellular signal into an intracellular one. Next, this intracellular signal is amplified by positive-feedback loops such that it activates specific signaling routes that control different functions inside the cell. Positive feedbacks are regulated by negative feedbacks that ensure the cell initiates an intracellular signal once the "appropriate" level of the extracellular response has been reached (Reth and Brummer, 2004).

In the context of the immune response, extracellular signaling refers to the sensing and recognition of a pathogenic organism. The resulting intracellular signaling activates specific pathways such as Imd, Toll or Jak-Stat, which in turn trigger the transcription of effector molecules such as antimicrobial peptides that will get rid of the pathogens. To restore homeostasis, negative-feedback loops to restrict the amplification of the intracellular signal are activated once the signal has reached a particular duration, strength or concentration. Loss of function mutations of any of the signaling elements involved in negative-feedback loops would be expected to trigger a pathological hyperactivation of the signaling pathway causing disease. Similarly, hyperexpression of such genes might be expected to cause under activation of immune response.

A clear example of the importance of these negative feedback loops as modulators of the immune response can be found in mammalian macrophages infected with *Mycobacterium tuberculosis* (Mtb). The inhibition of negative regulatory pathways by synthetic chemical inhibitors during Mtb infection promotes exaggerated tissue damage that leads to an immunopathological response that actually facilitates the transmission of this bacterial pathogen (Brace et al., 2017).

Some negative-feedback loops have been described in insects, in particular in fruit flies in the context of bacterial infections. During infection with *Listeria monocytogenes*, peptidoglycans stimulate the Imd pathway by inducing the translocation of the NF-κB transcription factor *Relish* to the nucleus where it will activate the transcription of *edin* (*elevated during infection*) as well as some antimicrobial peptides (AMPs) such as Diptericin B (*DiptB*) and different Attacins (*AttA*, *AttB*, and *AttD*). This Imd signaling is restricted by the cytokine WntD, a component of the negative-feedback loop that lessens the activity of *Relish* by potentially blocking with the help of its receptor Frizzled (Fz)4, the extracellular domain of the transmembrane receptor that activates the Imd pathway. Upon infection with *L. monocytogenes*, *wntD* loss-of-function mutants die faster than wild-type flies. Importantly, the bacterial loads of both groups are similar, suggesting *wntD* mutant flies die because of the exacerbated inflammatory response elicited by the overproduction of *edin* and other AMPs rather than as a result of the bacteria per se (Gordon et al., 2005, 2008; Rahimi et al., 2016).

Signaling events that trigger negative-feedback loops during the immune response to bacterial or fungal infections in insects are diverse and relatively well-characterized: (1) ubiquitination status of the signaling molecules, (2) conformational changes of transmembrane domains by cytokines, (3) inactivation of serine proteases activity by serpins, (4) interactions between different signaling pathways, and (5) the disruption of interactions between signaling molecules (Schneider, 2007; reviewed by Wang and Xia, 2018). In contrast, little is known about the negative feedbacks loops that regulate the immune system during viral infections. Some of the current described mechanisms involve the suppression of signal transduction by cytokines, epigenetic regulators and by some components of the RNAi machinery. One might imagine that disruption of any of these processes could result in immunopathology as a result of infection, but without further characterization, it would be difficult to know where to start to systematically search for sources of virally induced immunopathology.

4.1.1 Cytokines and AMPs in the humoral antiviral response

In both insects and mammals, the activation of the Toll, Imd and Jak-STAT signal transduction pathways involves the production of cytokines, a diverse group of small proteins produced after different stimuli. Cytokines play an important role in a vast number of biological processes such as hematopoiesis, cell proliferation and differentiation. They are particularly important because of their role as mediators of cell-cell communication, which allows them to activate and regulate the immune and inflammatory responses by acting as either transcription factors or extracellular ligands (reviewed in Shears and Hayakawa, 2019; Vanha-aho et al., 2016). Insects and mammals also both produce antimicrobial peptides (AMPs), small cationic peptides best known for their role in the innate immunity response against pathogens, which are also involved in controlling tissue damage, mediating inflammation, and maintaining gut homeostasis (Hanson and Lemaitre, 2020; Lee and Lee, 2014).

Improper regulation of the immune response leads to immunopathology. In mammals, an exacerbated or prolonged activation of the signaling pathways by pro-inflammatory cytokines such as IL-2 and IL-6 has been associated with chronic inflammation, cancers or autoimmunity (Belkaid, 2007; Kuraishy et al., 2011; Lee and Lee, 2014). In contrast, anti-inflammatory cytokines such as IL-10 and TGFβ function as immunomodulators that control the intensity and resolution of the immune response by negatively regulating some of the components of the immune transduction pathways (Banchereau et al., 2012; Couper et al., 2008; Li and Flavell, 2008; Murray and Smale, 2012).

In insects, Spätzle and products of the Unpaired gene family (1, 2, 3) have been described as pro-inflammatory cytokines that activate the Toll and Jak-STAT pathways in fruit flies and mosquitoes, respectively, following bacterial, fungal and viral infection (Harrison et al., 1998; Liu et al., 2018; Mizuguchi et al., 1998; Shin et al., 2006; Weber et al., 2003). The honeybee genome encodes two putative Spätzle orthologues that bind the Toll receptor (McMenamin et al., 2018). The antiviral gene *Vago* has been suggested to act as a cytokine in mosquitoes and bumble bees by interacting with the Jak-STAT pathway demonstrating the crosstalk between RNAi and an induced response (Niu et al., 2016; Paradkar et al., 2012). Diedel, WntD, and Eiger are among the identified cytokines that suppress the immune response in insects to bacterial and viral infections in *D. melanogaster* (Lamiable et al., 2016a,b).

4.1.1.1 Regulation of Imd by cytokines

The immune gene *diedel* (*die*) encodes a small (115 amino acid long gene) cysteine-rich protein and was first described as a negative regulator of the Jack/STAT pathway (Boutros et al., 2002; Müller et al., 2005). The gene *die* is transcribed in the fat body of fruit flies where it is strongly expressed in flies infected with the enveloped viruses SINV and VSV. This overexpression of *die* is regulated by the Nf-κB transcription factor Dif; however, is not activated by the canonical Toll pathway, since *die* is also induced in mutants lacking the Toll adapter *MyD88*. Furthermore, null *die*$^{-}$ fly mutants have a reduced lifespan when infected with a sublethal dose of SINV compared to wild-type flies. Genome wide analysis suggests that these higher mortality levels are due to an overactivation of different Imd regulated genes and not due to the presence of high viral load since the viral levels present in both wild-type and mutant flies are similar. Overall these findings suggest that Diedel is in charge of maintaining the homeostasis of the flies by down-regulating the expression of the Imd pathway (Lamiable et al., 2016a,b).

Another potential Imd-related cytokine is WntD. Besides being involved in the embryonic development of *D. melanogaster* via the Toll pathway (Rahimi et al., 2016), the gene *wntD* has been described to regulate other biological processes such as immune defense and tissue homeostasis (Ganguly et al., 2005; Gordon et al., 2005, 2008). However, in contrast to its role in modulating the Imd pathway during infection with *Listeria monocytogenes*, WntD does not form a negative-feedback loop during the antiviral response in fruit flies. Lamiable et al. (2016b) infected *wntD* null mutant flies with SINV and found that they present similar survival rates and viral loads compared to the control mutants (injected with buffer).

The cytokine Eiger and its receptor Wengen are members of the Drosophila tumor necrosis factor (TNF). Along with Traf1 and Traf2, two TNF receptor-associated factors, Eiger/Wengen activate the apoptosis pathway and the Imd pathway respectively (Cha et al., 2003; Igaki, 2002; Kanda et al., 2002). When injected with *Salmonella typhimurium* or *Mycobacterium marinum*, *eiger*—knockdown mutants outlive wild-type flies, suggesting the activity of *eiger* is deleterious for the flies and can cause immunopathology that resembles TNF-induced metabolic collapse in vertebrates when infected by intracellular pathogens. In contrast, *eiger* mutants infected with extracellular bacteria such as *Enterococcus faecalis* or *Staphylococcus aureus* (among others) die faster than wild-type flies. One explanation to this different outcome is that *eiger* leads to apoptosis-phagocytosis via the c-Jun N-terminal kinase (JNK) pathway when the host is infected by extracellular bacteria. However, when infected with intracellular bacteria, the apoptotic/phagocytic activity elicited by *eiger* is deleterious because to kill intracellular bacteria, one must kill the host cell in which they are residing (Brandt et al., 2004; Péan and Dionne, 2014; Schneider et al., 2007).

In the context of viral infections, upregulaton of *eiger* and *wengen* has been detected 10 days following infection with the picorna-like Nora virus, a virus that produces a persistent infection in insects (Lopez et al., 2018). Given that viruses are intracellular pathogens, it would be worthwhile to assess the role of *eiger* as immunomodulator during viral infections by performing similar survival and pathogen studies in $eiger^{-}$ mutants infected with different type of viruses.

4.1.1.2 Regulation of Toll

Similarly, overactivation of the Toll pathway has shown to be detrimental in fruit flies during infection with the non-enveloped virus Drosophila X Virus (DXV), leading to damage potentially linked to an immunopathological response rather than to a high viral burden. In particular, gain of function Tl^{10b} fly mutants show lower lifespan associated to reduced viral titers compared to wild-type flies when infected with DXV virus (Zambon et al., 2005). Tl^{10b} fly mutants constitutively activate the Toll pathway. Since these mutants harbor a point mutation in the Toll transmembrane receptor, we speculate that this point mutation might have changed the conformation of the transmembrane protein domains, making it difficult to regulate the excessive activity of this transmembrane receptor and thus changing the appropriate response. Currently no target cytokine has been identified. Regardless, the behavior of the Tl^{10b} mutants suggests that Toll would be an interesting candidate to explore with respect to immunopathology, and that control of Toll is key to return to homeostasis.

4.1.2 Epigenetic regulation

In mammals, epigenetic regulation also mediates the onset of the inflammatory response by altering gene expression (Higashijima et al., 2020; Wilson, 2008). The degree of DNA coiling determines the accessibility or inaccessibility of a gene to the transcriptional machinery, thereby determining the level of expression. DNA coiling is in turn determined by chemical modifications of the histones: acetylation or demethylation renders transcriptionally active chromatin, while histone deacetylation or methylation is associated with gene repression.

In *D. melanogaster*, histone H3 lysine 9 methyltransferase G9a acts as an immunomodulator by dampening the overactive response of a subset of genes belonging to the Jak-STAT pathway during DCV infection in the fat body of adult flies. G9a mutants showed decreased levels of histone H3 lysine 9 dimethylation, which altered the transcriptional repression in fruit flies (Kramer et al., 2011). G9a loss of function mutants also show a reduced lifespan after DCV infection compared to wild-type flies and this sensitivity is not due by an increase in viral titers since both, mutants and wild-type flies present similar levels of RNA viral (Merkling et al., 2015a).

4.1.3 RNA interference pathways

Although the siRNA pathway is considered to be a major immune response against viral infections in insects, with a robust and growing body of work depicting the way it functions, little is known about the mechanisms underlying its regulation (Torri et al., 2020). In humans, regulation of Dicer has been studied extensively given its role in the generation of small regulatory RNAs, inflammation, chromatin structure remodeling and degradation of apoptotic DNA (reviewed in Kurzynska-Kokorniak et al., 2015). TRPB (transactivation response element RNA-binding protein), PACT (protein activator of protein kinase R) and ADAR1 (adenosine deaminase acting on RNA enzyme) are three different proteins described to modulate Dicer catalytic activity (Kurzynska-Kokorniak et al., 2015). In particular, ADAR1 catalyzes the adenosine-to-inosine editing of dsRNA, pri- and pre-mRNAs and increases the rate of substrate processing by direct protein-protein interaction with Dicer (Ota et al., 2013).

The regulation of the siRNA pathway is poorly understood (Kurzynska-Kokorniak et al., 2015; Lim et al., 2011; Torri et al., 2020). In insects, recent work has shown that Dcr-2 and Ago-2, two core components of the siRNA pathway, are regulated at the protein level instead of at the transcriptional level through an unknown mechanism. In addition, besides being involved in the antiviral, antibacterial and, antifungal response in fruit flies by modulating Toll signaling, Dcr-2 has been shown to play a role as a general gene regulatory mechanism (Torri et al., 2020; Wang et al., 2015). Furthermore, in a system developed by Harsh et al. (2018) to study the host immune response to Zika virus infections using *D. melanogaster*, they found that Dcr-2 confers resistance to Zika virus and also controls some other general functions related to keeping the general homeostasis since Dcr-2 fly mutants show lower midgut epithelial recovery and decreased locomotor activity after infection with Zika virus. Recently Deng et al. (2020) demonstrated that the enzyme ADAR suppresses the innate immune induction of uninfected *D. melanogaster* flies. This work demonstrated that mutant Adar ($Adar^{E374A}$) flies that have a catalytically inactive Adar enzyme present an aberrant upregulation of the antiviral signaling protein *Vago*. These mutants also have upregulation of different AMPs that are normally activated by IMD or Toll due to the loss of RNA editing, and this immune induction is mediated by Dcr-2.

In addition, recent studies have shown that some host miRNAs are involved in maintaining homeostasis in fruit flies by maintaining basal levels of AMPs such as *Drosomycin* and *Diptericin* in the absence of infection (Choi and Hyun, 2012) or by negatively regulating the Toll and Imd pathways during *Escherichia coli* infections (Li et al., 2017, 2019).

4.2 Persistent infections in insect vectors

While we often think of acute infections, which are either quickly cleared or result in death, some infections persist over longer periods. Persistent infections may be seen as an alternative homeostasis to the uninfected state. However, when infections persist, the inflammatory response and other immune mechanisms remain active for long periods, thereby creating another scenario for immunopathology. Insect vectors represent a fascinating and medically important example of the establishment of viral persistent infections. Perhaps the best known examples of insects that transmit viruses (also called arboviruses) to animal hosts such as humans and livestock are mosquitoes and ticks. However, some plant viruses are also vectored by phytophagous insects. The cycle starts when the vector ingests a meal. If the meal is viremic, there is a proliferation and spread of the virus from the mosquito, tick or aphid midgut to their salivary glands through the hemocoel via hemolymph circulation (ticks: Nuttall, 2014; mosquitoes: Salazar et al., 2007; plant virus vectors Whitfield et al., 2015). The hallmark of this process is the establishment of a persistent infection: an infection that is long lasting, without mortality as a result of the virus or obvious deleterious effects of a continuous immune response. Understanding such persistent infections is important to explain the persistence of disease, as the fitness of the vector affects the transmission of virus to new vertebrate or plant hosts (i.e., vector competency) (Lee et al., 2019). Thus, the mechanisms limiting immunopathology during persistent infection could be interesting candidates for control of vector competence.

One of the mechanisms explaining this persistence, curiously, is the production of viral-derived DNA (vDNA) by the host via reverse transcription using fragments of infecting RNA viruses as templates. It has been shown that these vDNAs enhance the RNAi response mounted by mosquitoes from the genus *Aedes* against dengue or chikungunya infections. When the synthesis of these vDNA is inhibited, the survival of infected mosquitoes decreases however the viral loads are similar to those present in untreated mosquitoes (Goic et al., 2016). Thus, while vDNA and RNAi are required for high host fitness during infection, these processes are not sufficient to clear infection and thus seem to be required for maintaining immune homeostasis during persistent infections.

Defective viral genomes (DVGs) are preferential templates of vDNAs (Poirier et al., 2018). These DVGs are shortened versions of viral genomes incapable of replication, and are produced by point mutations or by recombination during the replication of RNA viruses. The production of vDNAs is regulated by Dicer-2 using an alternative mechanism to the one used in the canonical siRNA pathway. The helicase domain of Dicer-2 recognizes and modulates the reverse transcription of DVGs into vDNAs. In addition, these vDNAs might integrate into the host and become endogenous viral elements (EVEs) (Poirier et al., 2018). EVEs are considered to be a type of fossil record of previous viral infections present in a wide range of eukaryotic species and in mosquitoes and other arthropods, and constitute a potential mechanism of immune memory as well (ter Horst et al. 2019; Whitfield et al., 2017).

The best known example of DVGs are defective interfering particles (DIPs). These are particles containing reduced or truncated viral genomes that interfere with viral replication. DIPs have been detected during viral infection *in vivo* and *in vitro*, and they proliferate in cell lines persistently infected with a wide range of RNA viruses. In some cases, they prevent the development of cytopathic effects in newly infected cells, possibly by acting as templates to produce vDNAs (reviewed in Rezelj et al., 2018).

Other mechanisms that promote persistence are linked to cellular renewal, where apoptosis plays an important role in controlling the proliferation of virus upon infection (O'Neill et al., 2015). In addition, apoptosis can limit the neuronal damage of mosquito brains caused by some viruses thus preventing the development of lethal infections and reducing the deleterious effects of viral infections (Xiao et al., 2015).

5. Concluding remarks and future directions

Mounting an immune response is energetically demanding and involves a series of molecular and cellular reactions devoted to maintaining the health of the host by getting rid of pathogenic intruders. However, the immune response is a double-edged sword because if not carefully controlled, it can itself be detrimental and cause disease. The study of the pathological consequences inflicted by the immune response has long been studied in humans (and in mammalian models). Far less is known about immunopathology in insects. The fruit fly *D. melanogaster*, along with mosquitoes and honeybees, is being developed as a model system to decipher the activation of diverse immune pathways to combat pathogenic infections caused by bacteria, virus and fungi. Furthermore, studies in *D. melanogaster* have shed some light on the regulatory feedbacks governing the proper functioning of this immune response and the mechanisms to turn the response off, in particular against bacterial infections, and, to a lesser extent, against viral infections. More recently, metabolic and stress host responses to infections, along with tissue repair, have gained more attention and have been suggested to play an important role in maintaining the physiological homeostasis of fruit flies.

We argue that studying how viral infections could cause immunopathology in insects deserves attention not just because insects are good model systems, but also because of the impact this information might have on insect-related issues such as human health and agriculture. In the previous sections, we have given a general overview of the immune pathways and other molecular mechanisms that insects use to clear viral infections. We have also documented the negative regulatory feedbacks involved in turning this immune response off. And although this chapter is about viral infections, some mechanisms elicited during bacterial infections are also involved in the antiviral response and therefore, they constitute a promising way to understand and study the regulation of the immune response during viral infections. Next, we discuss the study of particular immune pathways and candidate genes using both experimental and modeling approaches that will allow us to expand our knowledge about how an immunopathological response is elicited in insects during viral infections.

To start, we need to better characterize the immunopathology phenotype caused by different types of viruses (DNA and RNA viruses; enveloped and non-enveloped) at a physiological level by using different read-outs (other than mortality rates) that can tell us about the general condition of the host. For example, measuring the RNA levels of vitellogenin can be used to evaluate the "biological age" of the host, where increased levels of vitellogenin would tell us the host has been subjected to high levels of stress either caused by infection, oxidative stress or nutritional fluctuations (Grozinger and Zayed, 2020).

Similarly, a detailed description of the molecular mechanisms underlying the pro- and anti-inflammatory responses, as well as tissue remodeling after viral infection is lacking as well. It would be useful to study the basal antimicrobial and cytokine activity in the absence of infection in order to depict the nature of homeostatic gene expression and regulation by using wild type, AMP deficient, and cytokine deficient lines. Considering that there is no universal antiviral response to all viruses, we can expect a plethora of regulatory mechanisms controlling the proper return to normal functioning not just of the immune response but also the whole organismic physiology. One might thus begin with a particular family of arboviruses, for example flaviviruses, and describe commonalities and variation in response to these important viruses.

5.1 Candidate pathways and genes during the humoral response

As we explained before, three different cytokines have been described as modulators of the immune response to bacterial infections in insects. From these, the role of wntD and Diedel has been studied in the context of viral infections; Diedel was required to maintain immune homeostasis in fruit flies, while wntD did not seem to be important. We encourage further studies of the role of the cytokine Eiger as well as the possible protein interactions established among the Toll transmembrane receptor and other molecules as potential regulatory mechanisms during viral infections.

5.2 Candidate genes during apoptosis, phagocytosis and autophagy

After infection, clearing apoptotic cells is key to maintaining host homeostasis. Incorrect regulation of apoptosis and phagocytosis is the cause of inflammatory or autoimmune immunopathologies in humans including but not limited to Parkinson's, Alzheimer's and Celiac disease (reviewed by Elmore, 2007; Green et al., 2016). While the exact mechanisms for how viral infected cells in adult insects are targeted and engulfed by plasmatocytes is not clear, there is strong evidence that phagocytosis of dead cells plays an important role in protecting insects including *D. melanogaster*, the mosquito *Culex pipiens pipiens* and, honey bees from the effects of different RNA viruses (Annoscia et al., 2019; Lamiable et al., 2016c; Nainu et al., 2015, 2017; Vaidyanathan and Scott, 2006). We propose the characterization of the negative-feedback loop governing the regulation of the apoptotic response in insects as a strong candidate to maintain immune homeostasis. In *Drosophila* homeostasis, the caspase Dronc, which is responsible for triggering apoptosis, is inhibited (Steller, 2008). During viral infection, Dronc is activated after its repressor (DIAP1) is depleted. Continuous activation of Dronc, for example due to the inability of the host to increase the levels of DIAP1 and/or by continuous transcription of the pro-apoptotic RHG genes (*reaper*, *hid*, *grim*, and *sickle*), could lead to an overactive immune response and thus immunopathology. Prolonged apoptotic response would be expected to result in higher mortality of fruit flies without an increase in viral titers.

Similarly, the mechanisms underlying the interaction of plasmatocytes during viral infection in adult flies is not well understood, however, a study using Drosophila S2 cells found that the plasmatocyte engulfment receptors Draper and integrin $\beta\nu$ are involved in targeting apoptotic cells that contain particular markers on their cell surface after DCV infection (Nainu et al., 2015). An impaired activation or inhibition of this recognition could again lead to an immunopathological response. How this interaction is regulated is unknown. What we do know is that priming of naive plasmatocytes by natural phagocytosis of apoptotic cells is required to trigger pro-inflammatory and tissue repair response during *Escherichia coli* infection in fruit flies (Weavers et al., 2016).

As mentioned before, autophagy has been associated with maintaining the cellular homeostasis when there is a lack of nutrients or other cell stress. During viral infections, autophagy responds to the cell stress caused by the viral replication and degrades viral components; however RNA viruses can use autophagy to enhance their own replication by using the double-membrane organelles as places for viral replication and to protect themselves from detection by the immune response and further degradation (Choi et al., 2018). In mammals, autophagy has been demonstrated to protect the host from immunopathological consequences during acute infection by controlling the pro-inflammatory response (Pfeiler et al., 2016).

During infection with SVS, autophagy is induced by the recognition of the transmembrane receptor Toll-7 and the subsequent repression of the PI3K (phosphatidylinositol 3-kinase)-Akt-Tor pathway normally involved in nutrient signaling (Nakamoto et al., 2012; Shelly et al., 2009). It is tempting to speculate that the activation of the PI3K-Akt-Tor pathway will limit autophagy once the viral infection is contained, thus avoiding further self-inflicted damage as can occur in mammals.

5.3 Heat-shock response

Heat-shock response is a cellular protective mechanism activated by different stressors such as high temperature or infections where chaperon proteins prevent cellular stress response by protein folding and re-folding and by activating autophagy (reviewed in Lindquist, 1986; Richter et al., 2010). Transcriptomic and functional studies demonstrated that the heat-shock response is an antiviral response induced upon DCV infection in Drosophila S2 cells and adult fruit flies (Merkling et al., 2015b). A similar antiviral role was detected in honeybees when infected with the model virus Sindbis-GFP, (Brutscher et al., 2017; McMenamin et al., 2020). In mammals, heat-shock response has been proposed to modulate chronic inflammatory disease such rheumatoid arthritis or atherosclerosis by promoting the production of anti-inflammatory cytokines (Res et al., 1991; Van Eden et al., 2005). Due the overarching and vital role of the heat-shock response, there are regulatory pathways described in mammals that ensure the proper activation of this response (described in Merkling et al., 2015b); however, little is known about these regulatory mechanisms during viral infections in insects.

5.4 Experimental and modeling approaches

We encourage hypothesis-driven perturbation of feedback loops to investigate and describe the regulation of the immune response in insects to viral infections. In particular, negative-feedback loops play an essential role in regulating the response that determines the immunity does not turn into an immunopathological response. The advantage of using these feedback loops is that they allow us to understand the amplification and repression mechanisms of the intracellular signaling by considering at the same time, the upstream and downstream elements in the immune response. The use of chemical inhibitors, the reconstruction of a particular signal pathway and the identification of new protein interactions have been shown to be useful to discover new feedback loops (Reth and Brummer, 2004).

We also encourage the use of mathematical modeling to quantify the dynamics of feedback loops and to describe under which conditions negative-feedback loops are activated. These loops alternately amplify or attenuate intracellular signals once these signals reach particular strength and/or duration thresholds, all of which are quantifiable variables. In an elegant study, Rahimi et al. (2016) used reaction-diffusion equations to predict a feedback loop involved in the Toll-dependent dorsoventral patterning of fruit flies embryos using the kinetic parameters associated with the spread of morphogens. Luckily, the integration of mathematical modeling into the study of the immune response has gained interest in order to give a quantitative and predictive touch to the field of immunology and go beyond being descriptive (Eberhardt et al., 2016).

Given the crosstalk between immune pathways such as the induction of effector genes by components of the siRNA pathway (see Ahlers et al., 2019; Deddouche et al., 2008; Paradkar et al., 2012; West and Silverman, 2018 for descriptions of pathway interactions in the context of viral infections in insects), one way to study the underlying mechanisms that regulate the fate of viral infections (host death, viral clearance, or establishment of a persistent infection) from beginning to end is by using systems biology approaches. For example, time series transcriptomic data will allow us to understand temporal patterns of expression and coexpression, and has the potential to help us finding biological markers useful to predict infection outcomes and potential candidates to develop therapeutic strategies to treat persistent diseases as in (Argilaguet et al., 2019). We understand the challenges of performing time series experiments in insects given the destructive nature of the sampling, as well as potential tissue specific transcription, etc. However, we can start by performing cell culture studies that can give us insights into molecular processes going on in an organismal context; and single-cell technologies may be usefully brought to bear on these studies.

5.5 Final considerations

In this work we emphasize the importance of studying the mechanisms that regulate the immune response to viral infections in insects as a way to lay out a foundation on which to build our understanding of insect health. In addition, as previously suggested by Oliveira et al. (2020), this information is relevant to control of arboviruses, particularly those vectored by mosquitoes. Learning about the regulatory pathways that promote immunopathology during persistent infections will identify potential mechanisms to decrease vector competence by reducing mosquito fitness following infection, and thereby viral transmission. As we described, there are several pathways involved and many of these act in a host- and virus-specific manner. Thus, we should expect a diversity of regulatory mechanisms that reduce the immune response once the infection has been controlled or cleared. Likewise, we should keep in mind that the immune response is energetically costly to the host and different factors such as host sex, age, viral dose, route of infection, previous pathogen exposure (immune priming), coinfection, environment, nutritional status, age of infection, host genotype and host microbiome are all factors that influence the immune response and immunopathologies in insects.

References

Ahlers, L.R.H., Trammell, C.E., Carrell, G.F., Mackinnon, S., Torrevillas, B.K., Chow, C.Y., Luckhart, S., Goodman, A.G., 2019. Insulin potentiates JAK/STAT signaling to broadly inhibit flavivirus replication in insect vectors. Cell Rep. 29, 1946–1960.e5.

Annoscia, D., Brown, S.P., Di Prisco, G., De Paoli, E., Del Fabbro, S., Frizzera, D., Zanni, V., Galbraith, D.A., Caprio, E., Grozinger, C.M., Pennacchio, F., Nazzi, F., 2019. Haemolymph removal by varroa mite destabilizes the dynamical interaction between immune effectors and virus in bees, as predicted by Volterra's model. Proc. R. Soc. Lond. Ser. B Biol. Sci. 286, 20190331.

Argilaguet, J., Pedragosa, M., Esteve-Codina, A., Riera, G., Vidal, E., Peligero-Cruz, C., Casella, V., Andreu, D., Kaisho, T., Bocharov, G., Ludewig, B., Heath, S., Meyerhans, A., 2019. Systems analysis reveals complex biological processes during virus infection fate decisions. Genome Res. 29, 907–919.

Banchereau, J., Pascual, V., O'Garra, A., 2012. From IL-2 to IL-37: the expanding spectrum of anti-inflammatory cytokines. Nat. Immunol. 13, 925–931.

Belkaid, Y., 2007. Regulatory T cells and infection: a dangerous necessity. Nat. Rev. Immunol. 7, 875–888.

Bernstein, E., Caudy, A.A., Hammond, S.M., Hannon, G.J., 2001. Role for a bidentate ribonuclease in the initiation step of RNA interference. Nature 409, 363–366.

Biryukova, I., Ye, T., 2015. Endogenous siRNAs and piRNAs derived from transposable elements and genes in the malaria vector mosquito *Anopheles gambiae*. BMC Genomics 16, 278.

Boman, H.G., Nilsson, I., Rasmuson, B., 1972. Inducible antibacterial defence system in *Drosophila*. Nature 237, 232–235.

Boutros, M., Agaisse, H., Perrimon, N., 2002. Sequential activation of signaling pathways during innate immune responses in *Drosophila*. Dev. Cell 3, 711–722.

Brace, P.T., Tezera, L.B., Bielecka, M.K., Mellows, T., Garay, D., Tian, S., Rand, L., Green, J., Jogai, S., Steele, A.J., Millar, T.M., Sanchez-Elsner, T., Friedland, J.S., Proud, C.G., Elkington, P.T., 2017. *Mycobacterium tuberculosis* subverts negative regulatory pathways in human macrophages to drive immunopathology. PLoS Pathog. 13, e1006367.

Brandt, S.M., Dionne, M.S., Khush, R.S., Pham, L.N., Vigdal, T.J., Schneider, D.S., 2004. Secreted bacterial effectors and host-produced Eiger/TNF drive death in a *Salmonella*-infected fruit fly. PLoS Biol. 2, e418.

Bronkhorst, A.W., van Cleef, K.W.R., Vodovar, N., İnce, İ.A., Blanc, H., Vlak, J.M., Saleh, M.-C., Van Rij, R.P., 2012. The DNA virus invertebrate iridescent virus 6 is a target of the *Drosophila* RNAi machinery. Proc. Natl. Acad. Sci. U. S. A. 109, E3604–E3613.

Brutscher, L.M., Daughenbaugh, K.F., Flenniken, M.L., 2017. Virus and dsRNA-triggered transcriptional responses reveal key components of honeybee antiviral defense. Sci. Rep. 7, 6448.

Carthew, R.W., Sontheimer, E.J., 2009. Origins and mechanisms of miRNAs and siRNAs. Cell 136, 642–655.

Carthew, R.W., Agbu, P., Giri, R., 2017. MicroRNA function in *Drosophila melanogaster*. Semin. Cell Dev. Biol. 65, 29–37.

Cha, G.-H., Cho, K.S., Lee, J.H., Kim, M., Kim, E., Park, J., Lee, S.B., Chung, J., 2003. Discrete functions of TRAF1 and TRAF2 in *Drosophila melanogaster* mediated by c-Jun N-Terminal kinase and NF-kappaB-dependent signaling pathways. Mol. Cell. Biol. 23, 7982–7991.

Chandra, S., Vimal, D., Sharma, D., Rai, V., Gupta, S.C., Chowdhuri, D.K., 2017. Role of miRNAs in development and disease: lessons learnt from small organisms. Life Sci. 185, 8–14.

Choi, I.K., Hyun, S., 2012. Conserved microRNA miR-8 in fat body regulates innate immune homeostasis in *Drosophila*. Dev. Comp. Immunol. 37, 50–54.

Choi, Y., Bowman, J.W., Jung, J.U., 2018. Autophagy during viral infection—a double-edged sword. Nat. Rev. Microbiol. 16, 341–354.

Clem, R.J., Fechheimer, M., Miller, L.K., 1991. Prevention of apoptosis by a baculovirus gene during infection of insect cells. Science 254, 1388–1390.

Couper, K.N., Blount, D.G., Riley, E.M., 2008. IL-10: the master regulator of immunity to infection. J. Immunol. 180, 5771–5777.

Culshaw, A., Mongkolsapaya, J., Screaton, G.R., 2017. The immunopathology of dengue and Zika virus infections. Curr. Opin. Immunol. 48, 1–6.

Czech, B., Malone, C.D., Zhou, R., Stark, A., Schlingeheyde, C., Dus, M., Perrimon, N., Kellis, M., Wohlschlegel, J.A., Sachidanandam, R., Hannon, G.J., Brennecke, J., 2008. An endogenous small interfering RNA pathway in *Drosophila*. Nature 453, 798–802.

Damjanovic, D., Small, C.-L., Jeyananthan, M., McCormick, S., Xing, Z., 2012. Immunopathology in influenza virus infection: uncoupling the friend from foe. Clin. Immunol. 144, 57–69.

Deddouche, S., Matt, N., Budd, A., Mueller, S., Kemp, C., Galiana-Arnoux, D., Dostert, C., Antoniewski, C., Hoffmann, J.A., Imler, J.-L., 2008. The DExD/H-box helicase Dicer-2 mediates the induction of antiviral activity in *Drosophila*. Nat. Immunol. 9, 1425–1432.

Deng, P., Khan, A., Jacobson, D., Sambrani, N., McGurk, L., Li, X., Jayasree, A., Hejatko, J., Shohat-Ophir, G., O'Connell, M.A., Li, J.B., Keegan, L.P., 2020. Adar RNA editing-dependent and -independent effects are required for brain and innate immune functions in *Drosophila*. Nat. Commun. 11, 1580.

Dolezal, T., Krejcova, G., Bajgar, A., Nedbalova, P., Strasser, P., 2019. Molecular regulations of metabolism during immune response in insects. Insect Biochem. Mol. Biol. 109, 31–42.

Eberhardt, M., Lai, X., Tomar, N., Gupta, S., Schmeck, B., Steinkasserer, A., Schuler, G., Vera, J., 2016. Third-kind encounters in biomedicine: immunology meets mathematics and informatics to become quantitative and predictive. Methods Mol. Biol. 1386, 135–179.

Elmore, S., 2007. Apoptosis: a review of programmed cell death. Toxicol. Pathol. 35, 495–516.

Ferreira, Á.G., Naylor, H., Esteves, S.S., Pais, I.S., Martins, N.E., Teixeira, L., 2014. The Toll-Dorsal pathway is required for resistance to viral oral infection in *Drosophila*. PLoS Pathog. 10, e1004507.

Gammon, D.B., Mello, C.C., 2015. RNA interference-mediated antiviral defense in insects. Curr. Opin. Insect Sci. 8, 111–120.

Ganguly, A., Jiang, J., Ip, Y.T., 2005. *Drosophila* WntD is a target and an inhibitor of the Dorsal/Twist/Snail network in the gastrulating embryo. Development 132, 3419–3429.

Ghildiyal, M., Zamore, P.D., 2009. Small silencing RNAs: an expanding universe. Nat. Rev. Genet. 10, 94–108.

Goic, B., Stapleford, K.A., Frangeul, L., Doucet, A.J., Gausson, V., Blanc, H., Schemmel-Jofre, N., Cristofari, G., Lambrechts, L., Vignuzzi, M., Saleh, M.-C., 2016. Virus-derived DNA drives mosquito vector tolerance to arboviral infection. Nat. Commun. 7, 12410.

Gordon, M.D., Dionne, M.S., Schneider, D.S., Nusse, R., 2005. WntD is a feedback inhibitor of Dorsal/NF-κB in *Drosophila* development and immunity. Nature 437, 746–749.

Gordon, M.D., Ayres, J.S., Schneider, D.S., Nusse, R., 2008. Pathogenesis of *Listeria*-infected *Drosophila wntD* mutants is associated with elevated levels of the novel immunity gene editing. PLoS Pathog. 4, e1000111.

Graham, A.L., Allen, J.E., Read, A.F., 2005. Evolutionary causes and consequences of immunopathology. Annu. Rev. Ecol. Evol. Syst. 36, 373–397.

Green, D.R., Oguin, T.H., Martinez, J., 2016. The clearance of dying cells: table for two. Cell Death Differ. 23, 915–926.

Grozinger, C.M., Zayed, A., 2020. Improving bee health through genomics. Nat. Rev. Genet. 21, 277–291.

Ha, M., Kim, V.N., 2014. Regulation of microRNA biogenesis. Nat. Rev. Mol. Cell Biol. 15, 509–524.

Haine, E.R., Moret, Y., Siva-Jothy, M.T., Rolff, J., 2008. Antimicrobial defense and persistent infection in insects. Science 322, 1257–1259.

Hanson, M.A., Lemaitre, B., 2020. New insights on *Drosophila* antimicrobial peptide function in host defense and beyond. Curr. Opin. Immunol. 62, 22–30.

Harrison, D.A., McCoon, P.E., Binari, R., Gilman, M., Perrimon, 1998. *Drosophila* unpaired encodes a secreted protein that activates the JAK signaling pathway. Genes Dev. 12, 3252–3263.

Harsh, S., Ozakman, Y., Kitchen, S.M., Paquin-Proulx, D., Nixon, D.F., Eleftherianos, I., 2018. Dicer-2 regulates resistance and maintains homeostasis against zika virus infection in *Drosophila*. J. Immunol. 201, 3058–3072.

Higashijima, Y., Matsui, Y., Shimamura, T., Nakaki, R., Nagai, N., Tsutsumi, S., Abe, Y., Link, V.M., Osaka, M., Yoshida, M., Watanabe, R., Tanaka, T., Taguchi, A., Miura, M., Ruan, X., Li, G., Inoue, T., Nangaku, M., Kimura, H., Furukawa, T., Aburatani, H., Wada, Y., Ruan, Y., Glass, C.K., Kanki, Y., 2020. Coordinated demethylation of H3K9 and H3K27 is required for rapid inflammatory responses of endothelial cells. EMBO J. 39, e103949.

Hussain, M., Asgari, S., 2014. MicroRNAs as mediators of insect host–pathogen interactions and immunity. J. Insect Physiol. 70, 151–158.

Igaki, T., 2002. Eiger, a TNF superfamily ligand that triggers the *Drosophila* JNK pathway. EMBO J. 21, 3009–3018.

Iwasaki, Y.W., Siomi, M.C., Siomi, H., 2015. PIWI-interacting RNA: its biogenesis and functions. Annu. Rev. Biochem. 84, 405–433.

Jayachandran, B., Hussain, M., Asgari, S., 2012. RNA interference as a cellular defense mechanism against the DNA virus baculovirus. J. Virol. 86, 13729.

Jones, C.I., Newbury, S.F., 2010. Functions of microRNAs in *Drosophila* development. Biochem. Soc. Trans. 38, 1137–1143.

Kanda, H., Igaki, T., Kanuka, H., Yagi, T., Miura, M., 2002. Wengen, a member of the *Drosophila* tumor necrosis factor receptor superfamily, is required for Eiger signaling. J. Biol. Chem. 277, 28372–28375.

Kemp, C., Mueller, S., Goto, A., Barbier, V., Paro, S., Bonnay, F., Dostert, C., Troxler, L., Hetru, C., Meignin, C., Pfeffer, S., Hoffmann, J.A., Imler, J.-L., 2013. Broad RNA interference–mediated antiviral immunity and virus-specific inducible responses in Drosophila. J. Immunol. 190, 650–658.

Kerr, J.F., Wyllie, A.H., Currie, A.R., 1972. Apoptosis: a basic biological phenomenon with wide-ranging implications in tissue kinetics. Br. J. Cancer 26, 239–257.

Khan, I., Agashe, D., Rolff, J., 2017. Early-life inflammation, immune response and ageing. Proc. R. Soc. Lond. Ser. B Biol. Sci. 284, 20170125.

Kim, L.K., Choi, U.Y., Cho, H.S., Lee, J.S., Lee, W.-B., Kim, J., Jeong, K., Shim, J., Kim-Ha, J., Kim, Y.-J., 2007. Down-regulation of NF-κB target genes by the AP-1 and STAT complex during the innate immune response in *Drosophila*. PLoS Biol. 5, e238.

Kingsolver, M.B., Huang, Z., Hardy, R.W., 2013. Insect antiviral innate immunity: pathways, effectors, and connections. J. Mol. Biol. 425, 4921–4936.

Kolliopoulou, A., Swevers, L., 2014. Recent progress in RNAi research in *Lepidoptera*: intracellular machinery, antiviral immune response and prospects for insect pest control. Curr. Opin. Insect Sci. 6, 28–34.

Kolliopoulou, A., Santos, D., Taning, C.N.T., Wynant, N., Vanden Broeck, J., Smagghe, G., Swevers, L., 2019. PIWI pathway against viruses in insects. Wiley Interdiscip. Rev. RNA 10, e1555.

Kramer, J.M., Kochinke, K., Oortveld, M.A.W., Marks, H., Kramer, D., de Jong, E.K., Asztalos, Z., Westwood, J.T., Stunnenberg, H.G., Sokolowski, M.B., Keleman, K., Zhou, H., van Bokhoven, H., Schenck, A., 2011. Epigenetic regulation of learning and memory by *Drosophila* EHMT/G9a. PLoS Biol. 9, e1000569.

Kuraishy, A., Karin, M., Grivennikov, S.I., 2011. Tumor promotion via injury- and death-induced inflammation. Immunity 35, 467–477.

Kurzynska-Kokorniak, A., Koralewska, N., Pokornowska, M., Urbanowicz, A., Tworak, A., Mickiewicz, A., Figlerowicz, M., 2015. The many faces of dicer: the complexity of the mechanisms regulating dicer gene expression and enzyme activities. Nucleic Acids Res. 43, 4365–4380.

Lamiable, O., Imler, J.-L., 2014. Induced antiviral innate immunity in *Drosophila*. Curr. Opin. Microbiol. 20, 62–68.

Lamiable, O., Kellenberger, C., Kemp, C., Troxler, L., Pelte, N., Boutros, M., Marques, J.T., Daeffler, L., Hoffmann, J.A., Roussel, A., Imler, J.-L., 2016a. Cytokine Diedel and a viral homologue suppress the IMD pathway in *Drosophila*. Proc. Natl. Acad. Sci. U. S. A. 113, 698–703.

Lamiable, O., Meignin, C., Imler, J.-L., 2016b. WntD and Diedel: two immunomodulatory cytokines in *Drosophila* immunity. Fly (Austin) 10, 187–194.

Lamiable, O., Arnold, J., de Faria, I.J.D.S., Olmo, R.P., Bergami, F., Meignin, C., Hoffmann, J.A., Marques, J.T., Imler, J.-L., 2016c. Analysis of the contribution of hemocytes and autophagy to *Drosophila* antiviral immunity. J. Virol. 90, 5415–5426.

Lee, K.-A., Lee, W.-J., 2014. *Drosophila* as a model for intestinal dysbiosis and chronic inflammatory diseases. Dev. Comp. Immunol. 42, 102–110.

Lee, W.-S., Webster, J.A., Madzokere, E.T., Stephenson, E.B., Herrero, L.J., 2019. Mosquito antiviral defense mechanisms: a delicate balance between innate immunity and persistent viral infection. Parasit. Vectors 12, 165.

Lemaitre, B., Hoffmann, J., 2007. The host defense of *Drosophila melanogaster*. Annu. Rev. Immunol. 25, 697–743.

Lewis, S.H., Quarles, K.A., Yang, Y., Tanguy, M., Frézal, L., Smith, S.A., Sharma, P.P., Cordaux, R., Gilbert, C., Giraud, I., Collins, D.H., Zamore, P.D., Miska, E.A., Sarkies, P., Jiggins, F.M., 2018. Pan-arthropod analysis reveals somatic piRNAs as an ancestral defence against transposable elements. Nat. Ecol. Evol. 2, 174–181.

Li, M.O., Flavell, R.A., 2008. Contextual regulation of inflammation: a duet by transforming growth factor-beta and interleukin-10. Immunity 28, 468–476.

Li, S., Shen, L., Sun, L., Xu, J., Jin, P., Chen, L., Ma, F., 2017. Small RNA-Seq analysis reveals microRNA-regulation of the Imd pathway during *Escherichia coli* infection in *Drosophila*. Dev. Comp. Immunol. 70, 80–87.

Li, R., Huang, Y., Zhang, Q., Zhou, H., Jin, P., Ma, F., 2019. The miR-317 functions as a negative regulator of Toll immune response and influences *Drosophila* survival. Dev. Comp. Immunol. 95, 19–27.

Lim, D.-H., Oh, C.-T., Lee, L., Hong, J.-S., Noh, S.-H., Hwang, S., Kim, S., Han, S.-J., Lee, Y.S., 2011. The endogenous siRNA pathway in *Drosophila* impacts stress resistance and lifespan by regulating metabolic homeostasis. FEBS Lett. 585, 3079–3085.

Lindquist, S., 1986. The heat-shock response. Annu. Rev. Biochem. 55, 1151–1191.

Liu, B., Behura, S.K., Clem, R.J., Schneemann, A., Becnel, J., Severson, D.W., Zhou, L., 2013. P53-mediated rapid induction of apoptosis conveys resistance to viral infection in *Drosophila melanogaster*. PLoS Pathog. 9, e1003137.

Liu, T., Xu, Y., Wang, X., Gu, J., Yan, G., Chen, X.-G., 2018. Antiviral systems in vector mosquitoes. Dev. Comp. Immunol. 83, 34–43.

Lopez, W., Page, A.M., Carlson, D.J., Ericson, B.L., Cserhati, M.F., Guda, C., Carlson, K.A., 2018. Analysis of immune-related genes during Nora virus infection of *Drosophila melanogaster* using next generation sequencing. AIMS Microbiol. 4, 123–139.

Losey, J.E., Vaughan, M., 2006. The economic value of ecological services provided by insects. Bioscience 56, 311–323.

Lucas, K., Raikhel, A.S., 2013. Insect MicroRNAs: biogenesis, expression profiling and biological functions. Insect Biochem. Mol. Biol. 43, 24–38.

Malone, C.D., Hannon, G.J., 2009. Small RNAs as guardians of the genome. Cell 136, 656–668.

Marchioro, C.A., Foerster, L.A., 2016. Biotic factors are more important than abiotic factors in regulating the abundance of *Plutella xylostella* L., in Southern Brazil. Rev. Bras. Entomol. 60, 328–333.

Marshall, J.S., Warrington, R., Watson, W., Kim, H.L., 2018. An introduction to immunology and immunopathology. Allergy Asthma Clin. Immunol. 14, 49.

McMenamin, A.J., Daughenbaugh, K.F., Parekh, F., Pizzorno, M.C., Flenniken, M.L., 2018. Honeybee and bumble bee antiviral defense. Viruses 10, 395.

McMenamin, A.J., Daughenbaugh, K.F., Flenniken, M.L., 2020. The heat shock response in the western honeybee (*Apis mellifera*) is antiviral. Viruses 12, 245.

Merkling, S.H., Bronkhorst, A.W., Kramer, J.M., Overheul, G.J., Schenck, A., Van Rij, R.P., 2015a. The epigenetic regulator G9a mediates tolerance to RNA virus infection in *Drosophila*. PLoS Pathog. 11, e1004692.

Merkling, S.H., Overheul, G.J., van Mierlo, J.T., Arends, D., Gilissen, C., Van Rij, R.P., 2015b. The heat shock response restricts virus infection in *Drosophila*. Sci. Rep. 5, 12758.

Merkling, S.H., Riahi, H., Overheul, G.J., Schenck, A., Van Rij, R.P., 2019. Peroxisome-associated *Sgroppino* links fat metabolism with survival after RNA virus infection in *Drosophila*. Sci. Rep. 9, 2065.

Mizuguchi, K., Parker, J.S., Blundell, T.L., Gay, N.J., 1998. Getting knotted: a model for the structure and activation of spätzle. Trends Biochem. Sci. 23, 239–242.

Mongelli, V., Saleh, M.-C., 2016. Bugs are not to be silenced: small RNA pathways and antiviral responses in insects. Annu. Rev. Virol. 3, 573–589.

Moy, R.H., Gold, B., Molleston, J.M., Schad, V., Yanger, K., Salzano, M.-V., Yagi, Y., Fitzgerald, K.A., Stanger, B.Z., Soldan, S.S., Cherry, S., 2014. Antiviral autophagy restricts rift valley fever virus infection and is conserved from flies to mammals. Immunity 40, 51–65.

Müller, P., Kuttenkeuler, D., Gesellchen, V., Zeidler, M.P., Boutros, M., 2005. Identification of JAK/STAT signalling components by genome-wide RNA interference. Nature 436, 871–875.

Murray, P.J., Smale, S.T., 2012. Restraint of inflammatory signaling by interdependent strata of negative regulatory pathways. Nat. Immunol. 13, 916–924.

Nainu, F., Tanaka, Y., Shiratsuchi, A., Nakanishi, Y., 2015. Protection of insects against viral infection by apoptosis-dependent phagocytosis. J. Immunol. 195, 5696–5706.

Nainu, F., Shiratsuchi, A., Nakanishi, Y., 2017. Induction of apoptosis and subsequent phagocytosis of virus-infected cells as an antiviral mechanism. Front. Immunol. 8, 1220.

Nakamoto, M., Moy, R.H., Xu, J., Bambina, S., Yasunaga, A., Shelly, S.S., Gold, B., Cherry, S., 2012. Virus recognition by Toll-7 activates antiviral autophagy in *Drosophila*. Immunity 36, 658–667.

Neu, U., Mainou, B.A., 2020. Virus interactions with bacteria: partners in the infectious dance. PLoS Pathog. 16, e1008234.

Niu, J., Meeus, I., Smagghe, G., 2016. Differential expression pattern of Vago in bumblebee (*Bombus terrestris*), induced by virulent and avirulent virus infections. Sci. Rep. 6, 34200.

Nuttall, P.A., 2014. Tick-borne viruses. In: Sonenshine, D.E., Roe, R.M. (Eds.), Biology of Ticks. In: vol. 2. Oxford University Press, New York, pp. 180–210.

Okamura, K., Chung, W.-J., Ruby, J.G., Guo, H., Bartel, D.P., Lai, E.C., 2008. The *Drosophila* hairpin RNA pathway generates endogenous short interfering RNAs. Nature 453, 803–806.

Oliveira, J.H., Bahia, A.C., Vale, P.F., 2020. How are arbovirus vectors able to tolerate infection? Dev. Comp. Immunol. 103, 103514.

O'Neill, K., Olson, B.J.S.C., Huang, N., Unis, D., Clem, R.J., 2015. Rapid selection against arbovirus-induced apoptosis during infection of a mosquito vector. Proc. Natl. Acad. Sci. U. S. A. 112, E1152–E1161.

Ota, H., Sakurai, M., Gupta, R., Valente, L., Wulff, B.-E., Ariyoshi, K., Iizasa, H., Davuluri, R.V., Nishikura, K., 2013. ADAR1 forms a complex with dicer to promote microRNA processing and RNA-induced gene silencing. Cell 153, 575–589.

Palmer, W.H., Varghese, F.S., Van Rij, R.P., 2018. Natural variation in resistance to virus infection in dipteran insects. Viruses 10, 118.

Panayidou, S., Apidianakis, Y., 2013. Regenerative inflammation: lessons from *Drosophila* intestinal epithelium in health and disease. Pathogens 2, 209–231.

Paradkar, P.N., Trinidad, L., Voysey, R., Duchemin, J.-B., Walker, P.J., 2012. Secreted Vago restricts West Nile virus infection in Culex mosquito cells by activating the Jak-STAT pathway. Proc. Natl. Acad. Sci. U. S. A. 109, 18915–18920.

Paro, S., Imler, J.-L., Meignin, C., 2015. Sensing viral RNAs by Dicer/RIG-I like ATPases across species. Curr. Opin. Immunol. 32, 106–113.

Péan, C.B., Dionne, M.S., 2014. Intracellular infections in *Drosophila melanogaster*: host defense and mechanisms of pathogenesis. Dev. Comp. Immunol. 42, 57–66.

Pfeiler, S., Khandagale, A.B., Magenau, A., Nichols, M., Heijnen, H.F.G., Rinninger, F., Ziegler, T., Seveau, S., Schubert, S., Zahler, S., Verschoor, A., Latz, E., Massberg, S., Gaus, K., Engelmann, B., 2016. Distinct surveillance pathway for immunopathology during acute infection via autophagy and SR-BI. Sci. Rep. 6, 34440.

Poirier, E.Z., Goic, B., Tomé-Poderti, L., Frangeul, L., Boussier, J., Gausson, V., Blanc, H., Vallet, T., Loyd, H., Levi, L.I., Lanciano, S., Baron, C., Merkling, S.H., Lambrechts, L., Mirouze, M., Carpenter, S., Vignuzzi, M., Saleh, M.-C., 2018. Dicer-2-dependent

generation of viral DNA from defective genomes of RNA viruses modulates antiviral immunity in insects. Cell Host Microbe 23, 353–365.e8.
Rahimi, N., Averbukh, I., Haskel-Ittah, M., Degani, N., Schejter, E.D., Barkai, N., Shilo, B.-Z., 2016. A WntD-dependent integral feedback loop attenuates variability in *Drosophila* Toll Signaling. Dev. Cell 36, 401–414.
Ramirez, J.L., Souza-Neto, J., Torres Cosme, R., Rovira, J., Ortiz, A., Pascale, J.M., Dimopoulos, G., 2012. Reciprocal tripartite interactions between the *Aedes aegypti* midgut microbiota, innate immune system and dengue virus influences vector competence. PLoS Negl. Trop. Dis. 6, e1561.
Res, P.C.M., Thole, J.E.R., de Vries, R.R.P., 1991. Heat shock proteins in immunopathology. Curr. Opin. Immunol. 3, 924–929.
Reth, M., Brummer, T., 2004. Feedback regulation of lymphocyte signalling. Nat. Rev. Immunol. 4, 269–277.
Rezelj, V.V., Levi, L.I., Vignuzzi, M., 2018. The defective component of viral populations. Curr. Opin. Virol. 33, 74–80.
Richter, K., Haslbeck, M., Buchner, J., 2010. The heat shock response: life on the verge of death. Mol. Cell 40, 253–266.
Rouse, B.T., Sehrawat, S., 2010. Immunity and immunopathology to viruses: what decides the outcome? Nat. Rev. Immunol. 10, 514–526.
Sadd, B.M., Siva-Jothy, M.T., 2006. Self-harm caused by an insect's innate immunity. Proc. R. Soc. Lond. Ser. B Biol. Sci. 273, 2571–2574.
Salazar, M.I., Richardson, J.H., Sánchez-Vargas, I., Olson, K.E., Beaty, B.J., 2007. Dengue virus type 2: replication and tropisms in orally infected *Aedes aegypti* mosquitoes. BMC Microbiol. 7, 9.
Sánchez-Vargas, I., Scott, J.C., Poole-Smith, B.K., Franz, A.W.E., Barbosa-Solomieu, V., Wilusz, J., Olson, K.E., Blair, C.D., 2009. Dengue virus type 2 infections of *Aedes aegypti* are modulated by the mosquito's RNA interference pathway. PLoS Pathog. 5, e1000299.
Sansone, C.L., Cohen, J., Yasunaga, A., Xu, J., Osborn, G., Subramanian, H., Gold, B., Buchon, N., Cherry, S., 2015. Microbiota-dependent priming of antiviral intestinal immunity in *Drosophila*. Cell Host Microbe 18, 571–581.
Schneider, D.S., 2007. How and why does a fly turn its immune system off? PLoS Biol. 5, e247.
Schneider, D.S., Ayres, J.S., Brandt, S.M., Costa, A., Dionne, M.S., Gordon, M.D., Mabery, E.M., Moule, M.G., Pham, L.N., Shirasu-Hiza, M.M., 2007. *Drosophila* eiger mutants are sensitive to extracellular pathogens. PLoS Pathog. 3, e41.
Schuster, S., Miesen, P., Van Rij, R.P., 2019. Antiviral RNAi in insects and mammals: parallels and differences. Viruses 11, 448.
Serhan, C.N., Savill, J., 2005. Resolution of inflammation: the beginning programs the end. Nat. Immunol. 6, 1191–1197.
Shears, S.B., Hayakawa, Y., 2019. Functional multiplicity of an insect cytokine family assists defense against environmental stress. Front. Physiol. 10, 222.
Sheehan, G., Garvey, A., Croke, M., Kavanagh, K., 2018. Innate humoral immune defenses in mammals and insects: the same, with differences? Virulence 9, 1625–1639.
Shelly, S., Lukinova, N., Bambina, S., Berman, A., Cherry, S., 2009. Autophagy is an essential component of *Drosophila* immunity against vesicular stomatitis virus. Immunity 30, 588–598.
Shin, S.W., Bian, G., Raikhel, A.S., 2006. A toll receptor and a cytokine, Toll5A and Spz1C, are involved in toll antifungal immune signaling in the mosquito *Aedes aegypti*. J. Biol. Chem. 281, 39388–39395.
Skalsky, R.L., Cullen, B.R., 2010. Viruses, microRNAs, and host interactions. Annu. Rev. Microbiol. 64, 123–141.

Steller, H., 2008. Regulation of apoptosis in *Drosophila*. Cell Death Differ. 15, 1132–1138.

Stork, N.E., 2018. How many species of insects and other terrestrial arthropods are there on Earth? Annu. Rev. Entomol. 63, 31–45.

Ter Horst, A.M., Nigg, J.C., Dekker, F.M., Falk, B.W., 2019. Endogenous viral elements are widespread in arthropod genomes and commonly give rise to PIWI-interacting RNAs. J. Virol. 93, e02124–18.

Théron, E., Dennis, C., Brasset, E., Vaury, C., 2014. Distinct features of the piRNA pathway in somatic and germ cells: from piRNA cluster transcription to piRNA processing and amplification. Mobile DNA 5, 28.

Torri, A., Mongelli, V., Mondotte, J.A., Saleh, M.-C., 2020. Viral infection and stress affect protein levels of dicer 2 and *Argonaute* 2 in *Drosophila melanogaster*. Front. Immunol. 11, 338.

Vaidyanathan, R., Scott, T.W., 2006. Apoptosis in mosquito midgut epithelia associated with west nile virus infection. Apoptosis 11, 1643–1651.

Van Eden, W., van der Zee, R., Prakken, B., 2005. Heat-shock proteins induce T-cell regulation of chronic inflammation. Nat. Rev. Immunol. 5, 318–330.

Van Rij, R.P., Berezikov, E., 2009. Small RNAs and the control of transposons and viruses in *Drosophila*. Trends Microbiol. 17, 163–171.

Vanha-aho, L.-M., Valanne, S., Rämet, M., 2016. Cytokines in *Drosophila* immunity. Immunol. Lett. 170, 42–51.

Varanasi, S.K., Rouse, B.T., 2018. How host metabolism impacts on virus pathogenesis. Curr. Opin. Virol. 28, 37–42.

Voller, A., 1974. Immunopathology of malaria. Bull. World Health Organ. 50, 177–186.

Wang, F., Xia, Q., 2018. Back to homeostasis: negative regulation of NF-κB immune signaling in insects. Dev. Comp. Immunol. 87, 216–223.

Wang, Z., Wu, D., Liu, Y., Xia, X., Gong, W., Qiu, Y., Yang, J., Zheng, Y., Li, J., Wang, Y.-F., Xiang, Y., Hu, Y., Zhou, X., 2015. *Drosophila* Dicer-2 has an RNA interference-independent function that modulates Toll immune signaling. Sci. Adv. 1, e1500228.

Wasik, B.R., Turner, P.E., 2013. On the biological success of viruses. Annu. Rev. Microbiol. 67, 519–541.

Weavers, H., Evans, I.R., Martin, P., Wood, W., 2016. Corpse engulfment generates a molecular memory that primes the macrophage inflammatory response. Cell 165, 1658–1671.

Weber, A.N.R., Tauszig-Delamasure, S., Hoffmann, J.A., Lelievre, E., Gascan, H., Ray, K.P., Morse, M.A., Imler, J.-L., Gay, N.J., 2003. Binding of the *Drosophila* cytokine Spätzle to Toll is direct and establishes signaling. Nat. Immunol. 4, 794–800.

West, C., Silverman, N., 2018. p38b and JAK-STAT signaling protect against invertebrate iridescent virus 6 infection in *Drosophila*. PLoS Pathog. 14, e1007020.

Whitfield, A.E., Falk, B.W., Rotenberg, D., 2015. Insect vector-mediated transmission of plant viruses. Virology 479-480, 278–289.

Whitfield, Z.J., Dolan, P.T., Kunitomi, M., Tassetto, M., Seetin, M.G., Oh, S., Heiner, C., Paxinos, E., Andino, R., 2017. The diversity, structure, and function of heritable adaptive immunity sequences in the Aedes aegypti genome. Curr Biol. 27, 3511–3519.e7.

Wilson, A.G., 2008. Epigenetic regulation of gene expression in the inflammatory response and relevance to common diseases. J. Periodontol. 79, 1514–1519.

World Health Organization, 2020. Vector-Borne Diseases. https://www.who.int/newsroom/fact-sheets/detail/vector-borne-diseases.

Xiao, X., Zhang, R., Pang, X., Liang, G., Wang, P., Cheng, G., 2015. A neuron-specific antiviral mechanism prevents lethal flaviviral infection of mosquitoes. PLoS Pathog. 11, e1004848.

Xu, J., Hopkins, K., Sabin, L., Yasunaga, A., Subramanian, H., Lamborn, I., Gordesky-Gold, B., Cherry, S., 2013. ERK signaling couples nutrient status to antiviral defense in the insect gut. Proc. Natl. Acad. Sci. U. S. A. 110, 15025–15030.
Zambon, R.A., Nandakumar, M., Vakharia, V.N., Wu, L.P., 2005. The toll pathway is important for an antiviral response in *Drosophila*. Proc. Natl. Acad. Sci. U. S. A. 102, 7257–7262.

CHAPTER NINE

Immunopathogenesis of alphaviruses

Victoria K. Baxter[a], Mark T. Heise[b,c,*]

[a]Department of Pathology and Laboratory Medicine, University of North Carolina at Chapel Hill, Chapel Hill, NC, United States
[b]Department of Genetics, University of North Carolina at Chapel Hill, Chapel Hill, NC, United States
[c]Department of Microbiology and Immunology, University of North Carolina at Chapel Hill, Chapel Hill, NC, United States
[*]Corresponding author: e-mail address: mark_heisem@med.unc.edu

Contents

Abstract

Alphaviruses, members of the enveloped, positive-sense, single-stranded RNA *Togaviridae* family, represent a reemerging public health threat as mosquito vectors expand into new geographic territories. The Old World alphaviruses, which include chikungunya virus, Ross River virus, and Sindbis virus, tend to cause a clinical syndrome characterized by fever, rash, and arthritis, whereas the New World alphaviruses, which consist of Venezuelan equine encephalitis virus, eastern equine encephalitis virus, and western equine encephalitis virus, induce encephalomyelitis. Following recovery from the acute phase of infection, many patients are left with debilitating persistent joint and neurological complications that can last for years. Clues from human cases and studies using animal models strongly suggest that much of the disease and pathology induced by alphavirus infection, particularly atypical and chronic manifestations, is mediated by

Advances in Virus Research, Volume 107
ISSN 0065-3527
https://doi.org/10.1016/bs.aivir.2020.06.002

the immune system rather than directly by the virus. This review discusses the current understanding of the immunopathogenesis of the arthritogenic and neurotropic alphaviruses accumulated through both natural infection of humans and experimental infection of animals, particularly mice. As treatment following alphavirus infection is currently limited to supportive care, understanding the contribution of the immune system to the disease process is critical to developing safe and effective therapies.

1. Introduction to alphaviruses

The Alphaviruses are a genus in the *Togaviridae* family, a group of enveloped, positive-sense, single-stranded RNA viruses that are found worldwide (Griffin, 2013) (Fig. 1). Alphaviruses of clinical importance in humans are transmitted by mosquitos, typically species from the *Culex* and *Aedes* genera, and represent a re-emerging public health threat as the insect vectors expand into new territories. The viruses circulate in sylvatic cycle between mosquitoes and either birds, nonhuman primates (NHP), or small mammals (Lim et al., 2018). However, promiscuous mosquitoes harboring an alphavirus sometimes feed on humans, creating an urban cycle that can result in massive widespread epidemics. Alphaviruses are divided into two major groups, Old World, and New World, which is based on their geographic distribution and typical clinical disease syndrome induced (Table 1).

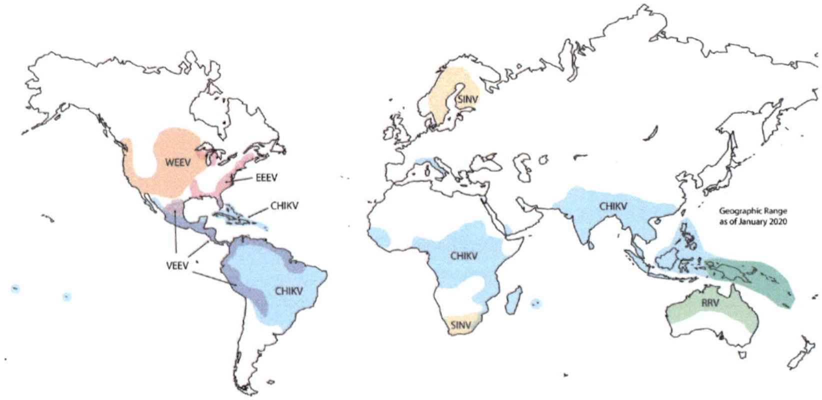

Fig. 1 Geographic range of cases of human disease induced by select Old World (CHIKV, RRV, and SINV) and New World (VEEV, EEEV, and WEEV) alphaviruses. CHIKV = chikungunya virus, RRV = Ross River virus, SINV = Sindbis virus, VEEV = Venezuelan equine encephalitis virus, EEEV = eastern equine encephalitis virus, WEEV = western equine encephalitis virus.

Table 1 Clinical disease syndromes induced by natural alphavirus infection in humans.

Disease syndrome	Clinical signs and symptoms	Associated alphaviruses
Arthritogenic disease	• Fever • Headache • Rash • Polyarthritis (inflammation in multiple joints) • Polyarthralgia (pain in multiple joints) • Myalgia (muscle pain) • Synovitis (inflammation of a synovial membrane) • Tendonitis (tendon inflammation)	Old World alphaviruses: • CHIKV • RRV • MAYV • ONNV • SINV
Neurologic disease	• Fever • Headache • Stupor or coma • Seizures • Encephalitis (brain inflammation) • Myelitis (spinal cord inflammation) • Meningitis (inflammation of the meninges) • Radiculitis (inflammation of the spinal nerve roots)	New world alphaviruses • VEEV • EEEV • WEEV Old World alphaviruses • CHIKV (atypical disease manifestation)

The Old World alphaviruses, which include chikungunya virus (CHIKV), Ross River virus (RRV), o'nyong-nyong virus, (ONNV), Mayaro virus (MAYV), and Sindbis virus (SINV), are typically found in Africa, Asia, Europe, and Oceania and induce an arthritogenic disease characterized by fever, rash, and arthralgia (reviewed in Suhrbier et al., 2012). CHIKV was initially identified in 1952 in present-day Tanzania but has been responsible for sporadic epidemics of fever and arthralgia, commonly known as chikungunya fever, across Africa and Asia since the 1960s (Robinson, 1955; Ross, 1956; Zeller et al., 2016). The virus expanded throughout the Indian Ocean region in the mid-2000s, and in a separate outbreak was introduced into the Caribbean in 2013. The virus resulted in over 1 million new cases within a year of arrival in the Americas and has become endemic in Central and South America and the Caribbean (Yactayo et al., 2016; Zeller et al., 2016). Because of ease of human travel, adaptive mutations in CHIKV, and spread of *Aedes* spp. mosquito vectors, cases of CHIKV infection have now been reported in over 40 countries (Tsetsarkin et al., 2007; Yactayo et al., 2016; Zeller et al., 2016). Closely related to CHIKV, ONNV is responsible for periodic outbreaks of fever and arthralgia across Central and East Africa (Forrester et al., 2012; Rezza et al., 2017). First isolated from mosquitos in Queensland, Australia in 1959, RRV is now endemic in Australia and several Pacific Islands and is responsible for approximately 4000–8000 cases each year in Northern Australia (Endy, 2020; Mylonas et al., 2002). Though part of the Semliki Forest phylogenetic complex that includes RRV and therefore classified as an Old World alphavirus, MAYV is native to and responsible for increasingly frequent outbreaks of arthritogenic disease in Latin America, having first been isolated in Trinidad in 1954 (Anderson et al., 1957; Forrester et al., 2012; Ganjian and Riviere-Cinnamond, 2020). The SINV strain group is the most widely distributed alphavirus, being found across Europe, Asia, Africa and Oceania (Lundström and Pfeffer, 2010). However, most cases of arthritis and arthralgia are reported in Scandinavia, caused by the Ockelbo, Pogosta, or Karelian strains of SINV, or South Africa (Laine et al., 2004; Storm et al., 2014). While clinical disease induced by the Old World alphaviruses is typically self-limiting and rarely causes death, persistent chronic arthralgia and arthritis lasting months after recovery from acute illness is common and can be debilitating (Alla and Combe, 2011).

The New World alphaviruses, which include Venezuelan equine encephalitis virus (VEEV), Eastern equine encephalitis virus (EEEV), and Western equine encephalitis virus (WEEV), are found in North, Central, and South America and typically induce neurological disease (Hollidge et al., 2010). EEEV used to be subdivided into North American and South American variants, but recent ecologic and genetic studies have determined the variants belong to distinct separate lineages (Arrigo et al., 2010). South American EEEV, which readily infects a wide variety of mammals but has only been attributed to about a dozen human cases of encephalitis, has since been renamed Madariaga virus, while members of the North American EEEV lineage have retained the EEEV name (Carrera et al., 2013; Luciani et al., 2015). The sequences of the capsid and 3′ untranslated region of WEEV is similar to EEEV, but the E1 and E2 glycoprotein sequences are more closely related to the Old World alphavirus SINV, indicating that WEEV resulted from a recombination event (Hahn et al., 1988). VEEV, EEEV, and WEEV typically circulate between mosquitoes and passerine birds in the wild; however, promiscuity by permissive mosquitoes occasionally results in infection of other species, particularly humans and horses (Go et al., 2014). These large mammals are typically dead-end hosts, though VEEV is capable of replicating at high enough titers in the blood to permit continued transmission (Weaver et al., 2004).

Human cases of Venezuelan equine encephalitis (VEE), eastern equine encephalitis (EEE), and western equine encephalitis (WEE) are only sporadically reported in the United States each year, though outbreaks of EEE and VEE over the last few decades have increased and are being diagnosed in previously unreported locations. An outbreak of VEE in the mid-1990s in Venezuela and Columbia affected an estimated 75,000–100,000 people (Weaver et al., 1996), and the number of human cases of EEE in the northeastern United States has markedly increased in the last decade, with states such as Vermont and Maine reporting locally acquired human cases for the first time (Lubelczyk et al., 2013; Molaei et al., 2015). In 2019, the CDC confirmed 38 cases of EEE resulting in 15 deaths were diagnosed in 10 states, more than double the highest number of annual EEE cases in the past decade (Centers for Disease Control and Prevention et al., 2019). Unlike the Old World alphaviruses, mortality is not uncommon following infection with the New World alphaviruses. While most people infected with WEEV remain asymptomatic, EEE carries a substantial death rate, ranging from 30% to 70% of symptomatic individuals, depending on the report (Calisher, 1994; Griffin, 2013; Steele et al., 2007). People who survive the initial clinical illness, especially those infected as infants or children, tend to develop chronic persistent neurological deficits (Bruyn and Lennette, 1953; Earnest et al., 1971; León, 1975; Palmer and Finley, 1956; Villari et al., 1995).

Currently, no treatments beyond symptomatic care are available for patients infected with the alphaviruses (Go et al., 2014). Licensed combination vaccines for EEEV, WEEV, and VEEV are available for horses, but no effective vaccine is approved for non-military use in humans (Bartelloni et al., 1970; Steele et al., 2007). While the explosive increase in CHIKV outbreaks and expansion to the Americas over the past 15 years have created a renewed interest in developing effective vaccines and therapies against the arthritogenic alphaviruses, no candidates have yet been successfully developed (reviewed in Powers, 2018). The reemerging public health concern and propensity to cause chronic debilitating effects presented by the alphaviruses necessitate better understanding of the mechanisms mediating the pathogenesis of infection so that safe and effective vaccines and therapies may be developed.

2. Potential outcomes following alphavirus infection

Cellular infection by an alphavirus can ultimately result in one of three outcomes. First, the virus directly induces death of the cell, typically through a mechanism such as apoptosis. For example, infection of neonatal mouse pups with SINV results in extensive apoptosis of multiple cells in the central nervous system (CNS), including neurons (Labrada et al., 2002). Second, the virus induces death through a secondary, rather than direct, mechanism, such as through effects of the immune system or excitotoxicity. And third, which is the optimal outcome, viral infection is cleared, and the cell survives. However, viral persistence following recovery is becoming increasingly recognized with alphaviruses (Fragkoudis et al., 2018; Hoarau et al., 2010; Levine and Griffin, 1992; Morrison et al., 2011), and reactivation of viral replication and relapse of disease must be controlled by the immune system, re-opening the possibility of pathology being induced as a bystander effect. For example, Semliki Forest virus (SFV) RNA persists in the brains of immunocompetent mice following recovery from acute infection, and suppression of the antibody response several months later results in reactivation of infectious virus production (Fragkoudis et al., 2018). The outcome of infection is dependent on a myriad of factors, including viral species and strain, type of cell infected, and host age, genetics, and immune response.

3. Protective effects of the immune system during Alphavirus infection

While the focus of this review is on virus-induced immune pathology, it is also important to remember that the host immune response plays an essential role in both controlling and clearing alphaviruses, while also mediating long-term protective immunity following either primary infection or vaccination. The type I interferon (IFN) system is essential for protecting against alphavirus replication, and mice that lack a functional IFN system or even specific antiviral interferon stimulated genes exhibit enhanced susceptibility to virus replication and disease (Burdeinick-Kerr et al., 2007; Byrnes et al., 2000; Couderc et al., 2008; Gardner et al., 2012; Partidos et al., 2011; Ryman et al., 2000; Schilte et al., 2010; Wollish et al., 2013). Other components of the innate immune response, such as the complement cascade and Toll-like receptor (TLR) signaling also protect from alphaviruses such as SINV, VEEV, and CHIKV (Brooke et al., 2012; Her et al., 2015;

Hirsch et al., 1980; Neighbours et al., 2012; Rudd et al., 2012; Schilte et al., 2010; Wollish et al., 2013).

Adaptive immunity is essential for clearance of alphaviruses from infected tissues and mediating long term protective immunity. While glutamine antagonist-mediated inhibition of lymphocyte proliferation and CNS infiltration results in mitigated clinical disease and pathology in SINV infected mice, viral clearance is also delayed (Baxter et al., 2017; Potter et al., 2015). Virus-specific antibody is responsible for viral clearance from the serum and is a major mediator of protective immunity following either primary alphavirus infection or vaccination (Lum et al., 2013; Mallilankaraman et al., 2011). Virus specific antibody directed against the E2 glycoprotein also mediates non-cytolytic clearance of alphaviruses from the CNS (Levine et al., 1991). T cells also make important contributions to alphavirus protective immunity, with local IFN-γ production by CD4+ and CD8+ T cells playing a cooperative role with anti-SINV antibody in facilitating clearance of SINV from the CNS (Baxter and Griffin, 2020; Binder and Griffin, 2001; Burdeinick-Kerr et al., 2007). CD4+ T cells promote recovery from VEEV-induced CNS disease (Brooke et al., 2010), and CD8+ T cells mediate clearance of RRV from infected muscle tissues (Burrack et al., 2015). CD8+ T cell responses can also mediate vaccine-induced protection against CHIKV (Broeckel et al., 2019), and γδ T cells protect from CHIKV-induced joint disease (Long et al., 2015). Mice lacking functional adaptive immune systems are also prone to developing chronic alphavirus infection (Burrack et al., 2015; Hawman et al., 2013; Lum et al., 2013; Nilaratanakul et al., 2018; Seymour et al., 2015; Teo et al., 2013). Therefore, while understanding how alphavirus-interactions with the immune system promote virus-induced disease, it is also important to keep in mind the beneficial aspect of the immune response when considering immune suppressive strategies for treating alphavirus-induced disease. Table 2 provides a summary of the pathologic and protective effects of different components of the immune response seen during infection with different alphaviruses.

4. Natural infection with alphaviruses

4.1 Clinical disease, pathology, and immune response in human patients

4.1.1 Old World alphaviruses

Clinical disease induced by the Old World alphaviruses is variable but most commonly characterized by fever, rash, and polyarthritis (inflammation of

Table 2 Pathogenic and protective roles of the immune response during alphaviral infection.

	Pathogenic	Protective
Alphavirus-induced arthralgia and myalgia		
CHIKV	*Cytokines* • TNF-α (Chen et al., 2017; Gardner et al., 2010; Malvy et al., 2009; Ninla-aesong et al., 2019; Thanapati et al., 2017) • IFN-γ (Gardner et al., 2010; Lee et al., 2015; Malvy et al., 2009; Ninla-aesong et al., 2019; Teo et al., 2013; Thanapati et al., 2017; Venugopalan et al., 2014) • IL-1β (Chen et al., 2017; Malvy et al., 2009; Ninla-aesong et al., 2019; Ng et al., 2009; Poo et al., 2014) • IL-2 (Lee et al., 2015) • IL-6 (Chaaitanya et al., 2011; Chen et al., 2015, 2017; Chirathaworn et al., 2013; Chow et al., 2011; Gardner et al., 2010; Kelvin et al., 2011; Lee et al., 2015; Lohachanakul et al., 2012; Malvy et al., 2009; Ninla-aesong et al., 2019; Ng et al., 2009; Sepúlveda-Delgado et al., 2017) • IL-8 (Chaaitanya et al., 2011; Malvy et al., 2009; Ninla-aesong et al., 2019; Theilacker et al., 2013) • IL-10 (Lee et al., 2015; Malvy et al., 2009; Poo et al., 2014) • IL-12 (Ninla-aesong et al., 2019) • IL-13 (Venugopalan et al., 2014) • IL-17 (Chow et al., 2011; Dias et al., 2018; Gualberto Cavalcanti et al., 2019) • IL-27 (Gualberto Cavalcanti et al., 2019)	*Cytokines* • IFN-α/β (Couderc et al., 2008; Gardner et al., 2012; Partidos et al., 2011; Schilte et al., 2010) • IFN-α (Hoarau et al., 2010; Teng et al., 2015) • IFN-γ (Teo et al., 2013) • IL-8 (Lohachanakul et al., 2012) • IL-10 (Chaaitanya et al., 2011; Dias et al., 2018) • IL-12 (Hoarau et al., 2010) • TGF-β (Ninla-aesong et al., 2019) *Immune receptors/signaling molecules* • TLR3 (Her et al., 2015) • MyD88 (Schilte et al., 2010) • IRF3 (Rudd et al., 2012) • IRF7 (Rudd et al., 2012) *Chemokines* • CCL2 (Poo et al., 2014) • CCL11 (Chow et al., 2011) • HGF (Chow et al., 2011) *Immune cells* • Macrophages (Gardner et al., 2010; Poo et al., 2014) • NK cells (Hoarau et al., 2010) • CD8+ T cells (Broeckel et al., 2019; Hoarau et al., 2010) • Tregs (Lee et al., 2015) • γδ T cells (Long et al., 2015) *Antibodies* • Anti-CHIKV antibody (Gardner et al., 2010; Hawman et al., 2013; Kam et al., 2012; Lum et al., 2013; Mallilankaraman et al., 2011; Miner et al., 2017; Venugopalan et al., 2014)

Table 2 Pathogenic and protective roles of the immune response during alphaviral infection.—cont'd

	Pathogenic	Protective
	Inflammasome • NLRP3 (Chen et al., 2017) *Chemokines* • CCL2 (Chaaitanya et al., 2011; Chen et al., 2015; Chen et al., 2017; Chirathaworn et al., 2013; Gardner et al., 2010; Lohachanakul et al., 2012; Miner et al., 2017; Ninla-aesong et al., 2019; Reddy et al., 2014; Teng et al., 2015) • CCL3 (Chaaitanya et al., 2011) • CCL4 (Chaaitanya et al., 2011; Miner et al., 2017) • CCL5 (Miner et al., 2015) • CCL7 (Chen et al., 2015) • CCL8 (Chen et al., 2015) • CXCL1 (Poo et al., 2014) • CXCL2 (Poo et al., 2014) • CXCL9 (Kelvin et al., 2011; Reddy et al., 2014) • CXCL10 (Kelvin et al., 2011; Lee et al., 2015; Miner et al., 2017) *Growth factors/cell migration/tissue remodeling* • G-CSF (Poo et al., 2014) • GM-CSF (Chow et al., 2011) • CTLA4 (Miner et al., 2017) • MMP-1 (Ninla-aesong et al., 2019) • MMP-3 (Ninla-aesong et al., 2019) *Immune cells* • Macrophages (Gardner et al., 2010; Hoarau et al., 2010) • T cells (Hawman et al., 2013) • CD4+ T cells (Hoarau et al., 2010; Lee et al., 2015; Miner et al., 2017; Teo et al., 2013) • CD8+ T cells (Dias et al., 2018)	

Continued

Table 2 Pathogenic and protective roles of the immune response during alphaviral infection.—cont'd

	Pathogenic	Protective
	Antibodies • RF (Amaral et al., 2020b; Amaral and Schoen, 2018; Bouquillard and Combe, 2009; Ganu and Ganu, 2011) • ACPA (Amaral et al., 2020b; Bouquillard and Combe, 2009)	
RRV	*Cytokines* • TNF-α (Chen et al., 2014; Rulli et al., 2009) • IL-1β (Chen et al., 2014) • IL-6 (Chen et al., 2014) *Chemokines* • CCL2 (Chen et al., 2014; Rulli et al., 2009) *Complement* • C3 (Gunn et al., 2012; Morrison et al. 2007, 2008) • MBL (Gunn et al., 2012, 2018) *Immune cells* • Macrophages (Clarris et al., 1975; Fraser et al., 1981; Fraser and Becker, 1984; Lidbury et al., 2000; Rulli et al., 2009)	*Cytokines* • IFN-α/β (Chen et al., 2014) • TLR7 (Neighbours et al., 2012) • MyD88 (Neighbours et al., 2012) *Immune cells* • CD8+ T cells (Burrack et al., 2015) *Antibodies* • Anti-RRV antibody (Lidbury et al., 2000)
MAYV	*Cytokines* • TNF-α (Santiago et al., 2015) • IFN-γ (Santiago et al., 2015; Tappe et al., 2016) • IL-1β (de Castro-Jorge et al., 2019) • IL-5 (Tappe et al., 2016) • IL-6 (Tappe et al., 2016) • IL-7 (Santiago et al., 2015; Tappe et al., 2016) • IL-8 (Santiago et al., 2015; Tappe et al., 2016) • IL-9 (Santiago et al., 2015; Tappe et al., 2016) • IL-10 (Tappe et al., 2016) • IL-12 (Tappe et al., 2016)	*Cytokines* • IL-1Ra (Santiago et al., 2015) *Growth factors/cell migration/tissue remodeling* • PDGF-BB (Santiago et al., 2015)

Table 2 Pathogenic and protective roles of the immune response during alphaviral infection.—cont'd

	Pathogenic	Protective
	• IL-13 (Santiago et al., 2015; Tappe et al., 2016) • IL-17 (Santiago et al., 2015; Tappe et al., 2016) • IL-18 (de Castro-Jorge et al., 2019) *Inflammasome* • NLRP3 (de Castro-Jorge et al., 2019) *Chemokines* • CCL2 (Santiago et al., 2015) • CCL3 (Tappe et al., 2016) • CCL4 (Tappe et al., 2016) • CCL5 (Tappe et al., 2016) • CXCL10 (Tappe et al., 2016) *Growth factors/cell migration/tissue remodeling* • G-CSF (Santiago et al., 2015) • GM-CSF (Tappe et al., 2016) • VEGF (Santiago et al., 2015)	
Alphavirus-induced encephalomyelitis		
VEEV, EEEV, and WEEV	*Cytokines* • TNF-α (Lukaszewski and Brooks, 2000; Schoneboom et al. 1999, 2000a) • IFN-γ (Grieder et al., 1997; Honnold et al., 2015a) • IL-2 (Gupta et al., 2017) • IL-4 (Gupta et al., 2017) • IL-6 (Grieder et al., 1997) • IL-8 (Gupta et al., 2017) • IL-9 (Gupta et al., 2017) • IL-12 (Grieder et al., 1997; Gupta et al., 2017; Sharma and Maheshwari, 2009) • IL-17A (Gupta et al., 2017; Sharma et al., 2008) *Chemokines* • CCL2 (Gupta et al., 2017; Phillips et al., 2013; Sharma and Maheshwari, 2009)	*Cytokines* • IFN-α/β (Grieder et al., 1997; Lukaszewski and Brooks, 2000; Schoneboom et al., 2000b) • TNF-α (Logue et al., 2010) • IFN-γ (Logue et al., 2010) • IL-12 (Logue et al., 2010) *Immune receptors/signaling molecules* • IRF2 (Schoneboom et al., 2000b) *Growth factors/cell migration/tissue remodeling* • iNOS (Schoneboom et al., 2000b *Complement* • C3 (Brooke et al., 2012) *Immune cells* • CD4+ T cells (Brooke et al., 2010)

Continued

Table 2 Pathogenic and protective roles of the immune response during alphaviral infection.—cont'd

	Pathogenic	Protective
	• CCL3 (Sharma et al., 2008) • CCL4 (Honnold et al., 2015a) • CCL5 (Gupta et al., 2017; Honnold et al., 2015a; Sharma et al., 2008) • CCL6 (Gupta et al., 2017) • CCL12 (Sharma et al., 2008) • CXCL9 (Honnold et al., 2015a; Sharma et al., 2008) • CXCL10 (Phillips et al., 2013; Sharma et al., 2008) • CXCL11 (Sharma et al., 2008) • CXCL13 (Sharma et al., 2008) *Growth factors/cell migration/tissue remodeling* • G-CSF (Honnold et al., 2015a) • LY6 (Gupta et al., 2017) • MIF (Sharma et al., 2008) • iNOS (Schoneboom et al., 1999, 2000a) *Immune cells* • T cells (Charles et al., 2001; Dal Canto and Rabinowitz, 1981; Woodman et al., 1975)	
SINV	*Cytokines* • TNF-α (Baxter and Griffin, 2016, 2020; Carmen et al., 2009; Kulcsar et al., 2015; Lee et al., 2013; Sarid et al., 2001; Schultz et al., 2019) • IFN-γ (Baxter and Griffin, 2016, 2020; Kulcsar et al., 2015; Lee et al., 2013; Wesselingh et al., 1994) • IL-1β (Baxter and Griffin, 2016; Kulcsar et al., 2015; Prow and Irani, 2007) • IL-2 (Wesselingh et al., 1994)	*Cytokines* • IFN-α/β (Byrnes et al., 2000; Ryman et al., 2000) • IFN-β (Burdeinick-Kerr et al., 2007) • IFN-γ (Baxter and Griffin, 2016, 2020; Binder and Griffin, 2001; Burdeinick-Kerr et al., 2007; Lee et al., 2013) • IL-10 (Kulcsar et al., 2014, 2015; Martin and Griffin, 2018) *Immune receptors/signaling molecules* • TLR7 (Wollish et al., 2013) • MyD88 (Wollish et al., 2013) • TRIF (Wollish et al., 2013)

Table 2 Pathogenic and protective roles of the immune response during alphaviral infection.—cont'd

	Pathogenic	Protective
	• IL-6 (Baxter and Griffin, 2016; Kulcsar et al., 2015; Lee et al., 2013; Schultz et al., 2019) • IL-10 (Rowell and Griffin, 1999) • IL-12 (Kulcsar et al., 2015) • IL-17 (Lee et al., 2013; Martin and Griffin, 2018) • IL-23A (Martin and Griffin, 2018) *Chemokines* • CCL2 (Baxter and Griffin, 2016; Kulcsar et al., 2015; Schultz et al., 2019) • CCL5 (Baxter and Griffin, 2016; Schultz et al., 2019) • CXCL9 (Baxter and Griffin, 2016) • CXCL10 (Baxter and Griffin, 2016; Schultz et al., 2019) *Growth factors/cell migration/tissue remodeling* • GM-CSF (Baxter and Griffin, 2016, 2020; Kulcsar et al., 2014, 2015) • LIF (Baxter et al., 2018; Wesselingh et al., 1994) *Immune cells* • T cells (Baxter et al., 2017; Greene et al., 2008; Kimura and Griffin, 2003; Kulcsar et al., 2014; Levine et al., 1991; Manivannan et al., 2016; Potter et al., 2015; Rowell and Griffin, 2002; Wesselingh et al., 1994) • CD4+ T cells (Lee et al., 2013; Martin and Griffin, 2018; Rowell and Griffin, 1999, 2002; Schultz et al., 2019) • Th17 cells (Kulcsar et al., 2014, 2015)	• STAT1 (Byrnes et al., 2000) • IRF3 (Schultz et al., 2019) • IRF7 (Schultz et al., 2019) *Complement* • C5 (Hirsch et al., 1980) *Immune cells* • CD8+ T cells (Baxter and Griffin, 2020; Kimura and Griffin, 2000; Rowell and Griffin, 1999) • Tregs (Kulcsar et al., 2015; Martin and Griffin, 2018) • B cells (Baxter et al., 2017; Baxter and Griffin, 2016) *Antibodies* • Anti-SINV antibody (Baxter et al., 2018; Baxter and Griffin, 2016; Burdeinick-Kerr et al., 2007; Byrnes et al., 2000; Kulcsar et al., 2015; Levine et al., 1991; Nilaratanakul et al., 2018)

Continued

Table 2 Pathogenic and protective roles of the immune response during alphaviral infection.—cont'd

	Pathogenic	Protective
SFV	*Cytokines* • TNF-α (McKimmie et al., 2005; Morris et al., 1997) • IFN-γ (Blackman and Morris, 1984; Morris et al., 1997) • IL-6 (Morris et al., 1997) *Immune cells* • T cells (Amor et al., 1996; Berger, 1980; (Amor et al., 1996; Fazakerley et al., 1988, 1993, 2006; Fazakerley and Webb, 1987a, 1987b; Gates et al., 1984; Pathak et al., 1983) • CD8+ T cells (Mokhtarian et al., 2003; Subak-Sharpe et al., 1993) • B cells (Mokhtarian et al., 2003) • Microglia (Mokhtarian et al., 2003) *Antibodies* • Anti-SFV antibody (Fazakerley and Webb, 1987b; Mokhtarian et al., 1999)	*Cytokines* • IFN-α/β (McKimmie et al., 2005) • TNF-α (Morris et al., 1997) • IFN-γ (Morris et al., 1997; Keogh et al., 2003) *Immune cells* • γδ T cells (Mokhtarian et al., 2012; Safavi et al., 2011) • B cells (Amor et al., 1996) *Antibodies* • Anti-SFV antibody (Amor et al., 1996; (Amor et al., 1996; Fazakerley et al., 2006; Fazakerley and Webb, 1987b; Fragkoudis et al., 2008, 2018; Mokhtarian et al., 2012; Safavi et al., 2011)

multiple joints), and polyarthralgia (pain in multiple joints) (Table 1, reviewed in Suhrbier et al., 2012). The frequency of clinical disease development varies by virus, with most RRV and SINV infections being asymptomatic but as high as 80% of CHIKV and MAYV infections resulting in clinical symptoms (Azevedo et al., 2009; Brummer-Korvenkontio et al., 2002; Endy, 2020; Harley et al., 2001; Suhrbier et al., 2012; Vijayakumar et al., 2011). Following an incubation period ranging from 2 to 10 days, patients generally first present with a fever (Suhrbier et al., 2012). Soon after, polyarthralgia most commonly affecting small joints, such as hands/fingers, wrists, ankles, and knees, develops. Synovial fluid sampled early in the RRV disease process contains high numbers of monocytes and activated macrophages with RRV antigen variably detected by immunofluorescence (Clarris et al., 1975; Fraser et al., 1981). Other clinical symptoms commonly seen during acute infection include headache, myalgia (muscle pain), and a maculopapular rash extending over the trunk, limbs, and face (Suhrbier et al., 2012). Histologically, the CHIKV-induced rash is characterized by edema of the epidermis and dermis and perivascular inflammation consisting of lymphocytes and melanophages in the dermis (Riyaz et al., 2010).

Multiple case reports and series have examined the immune response throughout the course of Old World alphavirus infection, particularly with CHIKV. Numerous pro-inflammatory cytokines and chemokines are reported to be elevated in patients' serum during the acute phase of infection and include tumor necrosis factor alpha (TNF-α), IFN-α, IFN-γ, interleukin (IL) 1 beta (IL-1β), IL-5, IL-6, IL-7, IL-8, IL-9, IL-10, IL-12, IL-13, IL-15, IL-17, IL-18, IL-17A, IL-18, IL-27, IL-29, granulocyte-colony stimulating factor (G-CSF), granulocyte-macrophage colony-stimulating factor (GM-CSF), CC chemokine ligand 2 (CCL2), CCL3, CCL4, CCL5, CXC chemokine ligand 9 (CXCL9), and CXCL10 (Chaaitanya et al., 2011; Chen et al., 2017; Chirathaworn et al., 2013; Chopra et al., 2014; Chow et al., 2011; Gualberto Cavalcanti et al., 2019; Kelvin et al., 2011; Lohachanakul et al., 2012; Ng et al., 2009; Reddy et al., 2014; Santiago et al., 2015; Tappe et al., 2016; Teng et al., 2015; Venugopalan et al., 2014; Wauquier et al., 2011). Vascular endothelial growth factor and platelet endothelial growth factor, potent angiogenic factors, have also been found elevated in serum starting early in MAYV infection (Santiago et al., 2015; Tappe et al., 2016). During the acute phase of CHIKV infection, peripheral blood mononuclear cells (pBMCs) show elevated NOD-, LRR- and pyrin domain-containing protein 3 (NLRP3) and caspase-1 mRNA expression (Chen et al., 2017), and circulating CD8+ T cells express high levels of cytotoxic markers such as granzyme B, perforin, and the degranulation marker CD107A (Dias et al., 2018). Increased levels of pro-inflammatory cytokines, lymphopenia, monocytopenia, and neutrophilia are all associated with more severe CHIKV viremia (Chow et al., 2011; Reddy et al., 2014; Teng et al., 2015). Patients with more severe RRV-induced disease have higher levels of serum mannose binding lectin (MBL), a pattern recognition receptor involved in activation of the complement system (Gunn et al., 2012). During the recovery or convalescent phase of alphavirus infection, multiple cytokines and chemokines in the serum have been reported to stay elevated or even increase further compared to the acute phase, including TNF-α, IFN-γ, IL-1β, IL, 2, IL-5 IL-6, IL-7, IL-8, IL-9, IL-12, IL-10, Il-13, IL-17A, IL-18, GM-CSF, CCL2, CCL4, and CXCL10 (Chirathaworn et al., 2010, 2013; Chopra et al., 2014; Kelvin et al., 2011; Santiago et al., 2015). This indicates that even in individuals who are considered recovered from alphavirus infection, the immune response remains activated.

Death following Old World alphavirus infection is rare but can occur, especially in neonatal or elderly patients with comorbidities (Economopoulou et al., 2009; Tandale et al., 2009). Fatal cases have been associated with respiratory, renal, hepatic and neurological disease, and hypertension and underlying respiratory and cardiovascular disease increase the risk of severe and fatal disease (Chua et al., 2010; Economopoulou et al., 2009; Hoz et al., 2015; Mercado et al., 2018). Postmortem histopathological reports are rarely published due to the low mortality rate associated with arthritogenic alphavirus infection and lack of medical infrastructure in the geographical locations where transmission and disease tend to occur. However, postmortem CHIKV-induced pathology described in case reports includes acute tubulointerstitial nephritis with lymphocytic intertubular inflammation and tubular necrosis, viral pneumonia with pulmonary edema and congestion, hepatocellular necrosis with mixed inflammation, and acute pericarditis with mild mononuclear inflammation (Mercado et al., 2018; Mercado et al., 2016; Reilly et al., 2020). Neurological disease associated with CHIKV infection has the highest mortality rate of potential atypical disease manifestations (Economopoulou et al., 2009; Tandale et al., 2009). Patients who develop neurological complications following CHIKV infection have shown elevated TNF-α, IFN-α, IL-6, IL-8, CCL2, CCL5, CCL17, and CXCL9 in cerebrospinal fluid (CSF) samples compared with patients with non-CHIKV-induced neurological disease (Kashyap et al., 2014). Autopsy of a 73-year-old man who died of CHIKV-associated encephalomyeloradiculitis (inflammation of the brain, spinal cord, and spinal nerve roots) revealed a grossly swollen brain with severe edema and multifocal ischemic changes (Ganesan et al., 2008). Microscopically, demyelination in the subcortical white matter with edema and myelin pallor was noted, and mild perivascular lymphocytic inflammation with phagocytic microglia (gitter cells) was found in the basal ganglia. Pooled risk of neonatal death following maternal-fetal transmission of CHIKV has been reported to be 2.8% (Contopoulos-Ioannidis et al., 2018). Fetal death due to CHIKV infection during early gestation results in no gross malformations, and while babies infected during late gestation are born without any clinical symptoms, within a week of birth, they develop joint edema, petechiae (small red or purple spots due to bleeding in the skin), rash, thrombocytopenia, disseminated intravascular coagulopathy (small blood clots throughout the bloodstream that deplete platelets and clotting factors and block small blood vessels), and encephalopathy with a high rate of persistent disabilities (Gérardin et al., 2008; Touret et al., 2006).

Acute symptoms usually resolve within a week or two, but the accompanying arthralgia and arthritis can last months to years following initial infection, creating significant physical, emotional, mental, and financial disability in affected patients. Multiple reports have described the incidence of persistent arthralgia and arthritis following alphavirus infection, especially with CHIKV, but also RRV, MAYV, and ONNV. Up to 81% of patients develop persistent symptoms, including joint tenderness, swelling, and pain, musculoskeletal stiffness and pain, and fatigue (Bonifay et al., 2018; Borgherini et al., 2008; Chang et al., 2017; Couzigou et al., 2018; Duvignaud et al., 2018; Forechi et al., 2018; Gauri et al., 2016; Gérardin et al., 2011; Huits et al., 2018; Lima et al., 2019; Llagonne-Barets et al., 2016; Murillo-Zamora et al., 2018; Ninla-aesong et al., 2020; Panato et al., 2019; Sánchez et al., 2019; Sissoko et al., 2009; Sosa-Martínez et al., 2018; Theilacker et al., 2013; Tritsch et al., 2019; Watson et al., 2020). Similar to acute clinical disease, small joints such as fingers, hands, wrists, ankles, and knees are more likely to be chronically affected (de Souza et al., 2019; Ninla-aesong et al., 2020; Sánchez et al., 2019). Radiographic and ultrasonographic findings include joint effusions, bone erosions, bone marrow edema, and inflammation of the subcutaneous connective tissues (cellulitis), synovial membranes of a joint (synovitis), and tendons (tendonitis) (Amaral et al., 2020a; Mogami et al., 2017). On synovial biopsy of patients with chronic polyarthritis from CHIKV infection, synovitis with inflammation consisting of macrophages, natural killer (NK) cells, CD4+ T cells, and plasma cells, synovial lining hyperplasia, and vascular proliferation is found (Ganu and Ganu, 2011; Hoarau et al., 2010). Synovial cytology in chronic CHIKV-induced arthritis consists mostly of CD14+ monocytes/macrophages, less than 5% of which stain positive for CHIKV, with smaller numbers of activated NK cells and CD4+ T cells (Hoarau et al., 2010). Mononuclear cells, mostly consisting of CD4+ T cells, have also been detected in synovial effusions from chronic RRV disease cases (Fraser and Becker, 1984). Risk factors that are associated with development of chronic arthralgia or arthritis include being female, older age, smoking, more severe acute disease with involvement of multiple joints, persistent viremia, preexisting co-morbidities such as diabetes mellitus, cardiovascular disease, or joint disease, and dengue virus (DENV) co-infection (Amaral et al., 2020a,b; Badawi et al., 2018; Delgado Enciso et al., 2018; Elsinga et al., 2017, 2018; Heath et al., 2018; Huits et al., 2018; Murillo-Zamora et al., 2018; Ninla-aesong et al., 2020; Patel et al., 2019; Sosa-Martínez et al., 2018; Tritsch et al., 2019). Chronic CHIKV arthritis has also been determined to be a risk factor for acute coronary syndrome (OR = 3.0), possibly because the chronic arthritis may trigger plaque destabilization in patients with concomitant cardiovascular disease (Patel et al., 2019).

Overall, persistent arthralgia and arthritis following alphavirus infection is associated with a pro-inflammatory state. Patients who develop chronic arthralgia or arthritis following alphavirus infection tend to have elevated serum levels of pro-inflammatory cytokines and chemokines, including IFN-α, IFN-γ, TNF-α, IL-1β, IL-6, IL-7, IL-8, IL-10, IL-12, IL-15, IL-17, IL-27, GM-CSF, CCL2, CCL3, CCL4, CXCL9, and CXCL10 (Amaral et al., 2020a; Chaaitanya et al., 2011; Chow et al., 2011; Gualberto Cavalcanti et al., 2019; Hoarau et al., 2010; Kelvin et al., 2011; Ninla-aesong et al., 2019; Reddy et al., 2014; Santiago et al., 2015; Sepúlveda-Delgado et al., 2017). Chronic chikungunya patients with tender joints and arthritis show significantly elevated IL-27 and IL-17A serum levels, respectively (Gualberto Cavalcanti et al., 2019). Compared to patients who recover from CHIKV infection, individuals with persistent arthralgia and arthritis tend to show anemia, elevated platelets (thrombocytosis), and elevated C-reactive protein (Amaral et al., 2020a). pBMCs from chikungunya patients with chronic arthralgia possess more IFN-α mRNA and more TNF-α- and IFN-γ-secreting NKT cells with lower cytotoxic activity compared to recovered patients (Hoarau et al., 2010; Thanapati et al., 2017). Chronic CHIKV patients have a delayed IgG3 response and lower antibody avidity index despite mounting an adequate neutralizing antibody response (Anfasa et al., 2019; Kam et al., 2012). Intensity of T cells responses in chronic chikungunya patients versus recovered chikungunya patients has not been found to differ (Hoarau et al., 2013). Osteoblasts from RRV patients with osteoarthritis show enhanced RRV infectivity and replication associated with increased expression of cytokines that favor bone resorption, including TNF-α, IL-1β, IL-6, and CCL2, and delayed type I IFN signaling (Chen et al., 2014).

4.1.2 New World alphaviruses

Neurological manifestations following VEEV, EEEV, or WEEV infection mainly consist of encephalitis or encephalomyelitis (Table 1). The likelihood of encephalitis development following infection with the New World alphaviruses depends on the viral species but is overall the minority of infections, especially with VEEV and WEEV (Greenlee, 2014). Neurological disease, when present, is similar for VEE, EEE, and WEE, though EEEV infections tend to be more severe. Following a prodromal phase lasting approximately 2–7 days consisting of fever and headache sometimes accompanied by myalgia, signs of irritability, and abdominal pain, onset of encephalitis is typically abrupt and progression rapid (Deresiewicz et al., 1997; Greenlee, 2014; Steele et al., 2007). Neurological manifestations include headache, stupor or coma, seizures, focal weakness, and cranial-nerve palsies. Blood chemistries reveal leukocytosis with a left shift (increased immature cells) and low blood sodium (hyponatremia), and CSF analysis shows elevated protein and increased cells (pleocytosis) initially consisting primarily of neutrophils but evolving primarily to lymphocytes (Deresiewicz et al., 1997; Przelomski et al., 1988; Rozdilsky et al., 1968; Silverman et al., 2013). Changes on brain MRI include symmetric or asymmetric lesions in multiple brain regions, including the basal ganglia, hippocampi, thalami, insular cortex, midbrain, and brainstem (Deresiewicz et al., 1997; Nickerson et al., 2016). Edema and periventricular white matter changes similar to those seen in multiple sclerosis patients have also been seen (Deresiewicz et al., 1997). Prognostic indicators for a poorer outcome include young age, particularly infants and children, shorter prodromal phase (<5 days), severe CSF pleocytosis, and coma (Deresiewicz et al., 1997; Finley et al., 1955; Przelomski et al., 1988; Rozdilsky et al., 1968).

Death, mostly associated with EEE, has mainly been reported to occur during second to third week of hospitalization (Deresiewicz et al., 1997). On postmortem examination, brains tend to be grossly congested and edematous, with diffuse meningoencephalitis found microscopically (Greenlee, 2014). Histopathological changes in the central nervous system (CNS) include perivascular and parenchymal inflammation consisting of mononuclear cells and neutrophils accompanying neuronal loss, especially in the basal ganglia and thalamus (Booss and Esiri, 2003; Lindenberg and Haymaker, 1982; Przelomski et al., 1988; Rozdilsky et al., 1968). Demyelinating lesions and oligodendrocyte death have also been reported, and inflammation of the spinal cord (myelitis) and blood vessels (vasculitis) are frequently present with EEE. Intranuclear or intracytoplasmic viral inclusion bodies are conspicuous by their absence (Rozdilsky et al., 1968; Steele et al., 2007).

Patients who survive the initial course of infection with VEEV, EEEV, and WEEV, especially those infected as infants or children, are often left with persistent neurological sequalae (Silverman et al., 2013). Reports of neurological sequelae are best described in patients who have recovered from WEE and include decreased motor skills and altered gait, intellectual and/or learning disability, facial paralysis, impaired speech, and Parkinson-like symptoms (Bruyn and Lennette, 1953; Earnest et al., 1971; Herzon et al., 1957;

Mulder et al., 1951; Palmer and Finley, 1956). Neuropsychological symptoms, including anxiety, depression, paranoia, attention deficit disorder, and schizophrenia-like personality changes have also been associated with WEE (Deaton et al., 1986; Fulton and Burton, 1953; Herzon et al., 1957; Palmer and Finley, 1956). VEE has been associated with recurrent headaches, fatigue, depression, seizures, paralysis, and intellectual disability (Bowen et al., 1976; León, 1975; Rivas et al., 1997). The course of neurological sequelae has been reported to range from a static nonprogressive condition to relapsing-remitting disease to complete recovery and reversion to normal function (Bowen et al., 1976; Palmer and Finley, 1956; Silverman et al., 2013). Regardless, neurological sequalae induced by New World alphavirus infection represent a significant source of physical, emotional, mental, and financial disability for affected patients and their caretakers (Villari et al., 1995).

4.2 Immune-mediated disease induced by alphaviruses

As discussed above, alphaviruses can be directly cytolytic, and direct virus-induced cell killing is likely to contribute to alphavirus-induced disease processes. However, there is also abundant evidence that alphaviruses can interact with the host immune system to cause virus-induced disease. These interactions include evidence that alphavirus-induced inflammation contributes to the pathogenesis of arthralgia and/or myalgia that are characteristic of Old World alphaviruses such as CHIKV and RRV, as well as the role of immune cells in driving pathologic aspects of alphavirus-induced CNS disease. However, alphavirus infections are also associated with autoimmune or inflammatory syndromes, and we will discuss each of these alphavirus-induced disease outcomes in the following section (Table 3).

Persistent arthralgia and arthritis have been reported with high incidence following infection with many of the Old World alphaviruses. Multiple characteristics and findings of alphavirus-induced chronic arthralgia and arthritis suggest the pathogenesis is mediated by the immune system rather than directly by the virus. While some studies have found evidence of CHIKV persistence during chronic infection (Hoarau et al., 2010; Ozden et al., 2007), several other studies evaluating synovial fluid in patients with chronic CHIKV- or RRV-induced arthritis found no evidence of CHIKV RNA via PCR, viral proteins via immunohistochemistry or mass spectrometry, or infectious virus via culture (Chang et al., 2018; Fraser and Becker, 1984; Ganu and Ganu, 2011). Regardless of whether chronic viral replication is

Table 3 Immune-mediated conditions induced by alphavirus infection.

Organ system	Condition	Description	Associated alphaviruses
Musculoskeletal	Chronic polyarthritis	Persistent pain, stiffness, and swelling in and around multiple joints	CHIKV, RRV, MAYV, SINV
Integumentary	Psoriasis	Chronic disease characterized by raised, red, scaly patches on the skin	CHIKV
	Lichen planus	Inflammatory skin condition characterized by bumps, sores, and rough scaly patches on the skin and mouth that often itch	
	Type I leprosy reaction	Delayed hypersensitivity to *Mycobacterium leprae* antigens characterized by increased redness and hardness of existing lesions	
Nervous	Guillain Barre Syndrome	Autoimmune disorder where the immune system attacks the myelin sheaths or axons of peripheral nerves	CHIKV
	Acute disseminated encephalomyelitis	Autoimmune disorder characterized by sudden widespread inflammation that damages brain and spinal cord white matter	
	Transverse myelitis	Autoimmune disorder characterized by sudden transient inflammation of the spinal cord	
Ocular	Anterior uveitis	Inflammation of the middle layer of the eye, which includes the iris and ciliary body	CHIKV
	Choroiditis	Inflammation of the choroid, the pigmented vascular layer between the retina and sclera	
	Panuveitis	Inflammation of all layers of the uvea, which includes the iris, ciliary body, and choroid	
	Retinitis	Inflammation of the retina, the thin layer of tissue at the back of the eye that converts light to neural signal	
	Neuroretinitis	Inflammation of the retina defined by optic disc edema and stellate macular exudates	
	Optic neuritis	Inflammation of the optic nerve	

driving disease pathogenesis, these results suggest that chronic alphavirus-induced arthralgia may be driven by indirect mechanisms. Patients infected with CHIKV have shown flare-ups of preexisting immune-mediated dermatoses, such as psoriasis, lichen planus, and type I leprosy reactions, indicating viral infection induces strong activation of the immune system (Riyaz et al., 2010). Most patients who develop persistent arthralgia following alphavirus infection are females, who are also far more prone to developing autoimmunities than males (Amaral et al., 2020a,b; de Souza et al., 2019; Heath et al., 2018; Huits et al., 2018; Sosa-Martínez et al., 2018; Tritsch et al., 2019). While treatment of chronic alphavirus-induced arthralgia and arthritis is primarily limited to supportive care anchored by nonsteroidal anti-inflammatory drugs, positive outcomes have been reported following treatment with disease modifying anti-rheumatic drugs. Both sulfasalazine and methotrexate, immunosuppressive drugs used to treat autoimmune conditions such as psoriasis, Crohn's disease, and rheumatoid arthritis, have been used to inhibit proinflammatory cytokine production and reduce joint swelling, tenderness, and pain in patients with CHIKV- or MAYV-induced persistent arthralgia and arthritis with some success (Amaral et al., 2020a,b; Ganu and Ganu, 2011; Malvy et al., 2009; Poon et al., 2019; Theilacker et al., 2013; Zaid et al., 2018). Together, these findings strongly suggest that persistent arthralgia and arthritis induced by alphaviruses are immune-mediated.

Parallels have been drawn between chronic chikungunya arthritis and rheumatoid arthritis. The two conditions share numerous similarities: (1) both conditions are characterized by symmetrical polyarthritis of the small joints, (2) both conditions are most commonly diagnosed in middle-aged females, (3) symptoms for both conditions include arthralgia, arthritis, myalgia, stiffness, and fatigue, (4) both conditions have similar bloodwork and radiographic findings, and (5) many of the same pro-inflammatory cytokines are elevated in the serum and synovial fluid (Amaral et al., 2020a). Multiple case studies have found some chronic chikungunya arthritis patients meet the diagnostic criteria for rheumatoid arthritis according to the American College of Rheumatology/European League Against Rheumatism 2010 Rheumatoid Arthritis Classification Criteria; this classification system evaluates the joints involved, serologic test results for rheumatoid factor and/or anti-cycle citrullinated peptide antibody, elevated acute phase response as evidenced by abnormal C-reactive protein or erythrocyte sedimentation rate, and symptom duration (Aletaha et al., 2010; Amaral et al., 2020b; Amaral and Schoen, 2018; Bouquillard and Combe, 2009; Huits et al., 2018; Miner et al., 2015; Sepúlveda-Delgado et al., 2017). Using the

American College of Rheumatology/European League Against Rheumatism 2010 Rheumatoid Arthritis Classification Criteria, a study found that in North Kerala, India, where chikungunya epidemics are frequent, CHIKV infection is significantly related to rheumatoid arthritis development (OR = 8.73) (Paul and Pariyapurath, 2018). Matrix metalloproteinases 1 and 3, which have been shown to correlate with rheumatoid arthritis disease activity and progression, are elevated in chronic chikungunya arthritis patients compared to healthy controls (Green, 2003; Ninla-aesong et al., 2019). NK cell and T cell phenotypes are similar between rheumatoid arthritis and chronic chikungunya arthritis, with pBMCs from both rheumatoid arthritis and chronic chikungunya arthritis patients having higher percentages of activated and effector CD4+ and CD8+ T cells than control patients (Miner et al., 2015). The frequency of human leukocyte antigen (HLA) alleles HLA-DRB1*01 and HLA-DRB1*04, which are associated with increased risk for developing rheumatoid arthritis (MacGregor et al., 1995; Weyand et al., 1992), is significantly higher in patients with chronic joint manifestations following CHIKV and SINV infection than the general population (Bouquillard and Combe, 2009; Sane et al., 2012). The structural polyproteins of multiple Old World alphaviruses share homology with numerous putative human proteins involved in rheumatoid arthritis and demonstrate similar binding patterns to HLA-DRB1 (Venigalla et al., 2020). These potential mimics are predicted to elicit an efficient B cell response, providing a possible autoimmune mechanism by which alphavirus infections are inducing chronic arthralgia and arthritis.

Instances of the immune response inducing pathology following alphavirus infection in human patients has been demonstrated with atypical neurological manifestations associated with the arthritogenic alphaviruses. In particular, multiple case reports of Guillain-Barre syndrome (GBS) have been reported in the literature. GBS is an autoimmune demyelinating polyneuropathy or axonal neuropathy characterized by acute onset of limb or muscle weakness usually progressing to flaccid paralysis (reviewed in Hughes and Cornblath, 2005). The general pathophysiology of GBS consists of an antibody-mediated attack of the myelin surrounding peripheral nerve axons (acute inflammatory demyelinating polyneuropathy subtype) or the axon itself (acute motor axonal neuropathy and acute motor and sensory axonal neuropathy subtypes) due to molecular mimicry between microbial antigens and axolemmal surface molecules (Bourque et al., 2015; Willison et al., 2016). Most cases are preceded by an infection less than 1 month prior, and the type of infection influences the clinical phenotype and prognosis.

While GBS is most commonly associated with *Campylobacter jejuni* infection (Jacobs et al., 1998), infection with arboviruses, including alphaviruses, are becoming increasingly recognized as contributors to GBS development (Davis et al., 2006; Lebrun et al., 2009; Oehler et al., 2015; Sahu et al., 2014).

The first published association was a 1998 report out of Australia suggesting a connection between Barmah Forest virus infection and GBS (Phan et al., 1998). Since that report though, almost all of the associations between GBS and alphavirus infection have involved the naturally arthritogenic alphavirus CHIKV. The association between GBS and CHIKV was first proposed following an outbreak on Reunion Island in the Indian Ocean in 2006. Four reports collectively described eight patients admitted to hospitals with GBS during the 2006 CHIKV outbreak who tested negative for the typical infectious agents associated with GBS (*C. jejuni*, cytomegalovirus, *Mycoplasma pneumoniae*, Epstein-Barr virus, dengue virus) and had serum IgM against CHIKV (Lebrun et al., 2009; Lémant et al., 2008; Tournebize et al., 2009; Wielanek et al., 2007). More cases of GBS were reported concurrent with a CHIKV outbreak in French Polynesia from October 2014 to March 2015, where an estimated 25% of the population became infected (Aubry and Cao-Lormeau, 2019; Koeltz et al., 2018; Oehler et al., 2015). Antibodies against GM2 and GD1a, gangliosides present in nodes of Ranvier along nerve axons, were each found in two patients from the outbreak, further supporting the hypothesis that CHIKV infection can trigger GBS (Oehler et al., 2015). The French Polynesia outbreak was determined to have originated from an infected individual traveling from the Caribbean (Nhan et al., 2014), where CHIKV had recently spread from Southeast Asia in 2013. Retrospective analysis of severe disease manifestations of CHIKV infections during the initial outbreak in Martinique and Guadeloupe in the French West Indies revealed six patients with GBS (Crosby et al., 2016). A case series evaluation of CHIKV-related GBS in 13 patients in the Caribbean found that 54% of cases were classified as the acute inflammatory demyelinating polyneuropathy subtype and 15% of cases were classified as the acute motor and sensory axonal neuropathy subtype, with the remaining cases classified as more uncommon variants (Balavoine et al., 2017). A case-controlled study evaluating the association between CHIKV infection and GBS in the French West Indies found an odds ratio of 8.3, indicating CHIKV infection was a risk factor for GBS (Stegmann-Planchard et al., 2019).

The association of GBS with arboviral infection became especially apparent during the Zika virus (ZIKV) outbreak in Central and South America from 2015 to 2016. Though humans cases were sporadically described in Africa, Asia, and the Pacific Islands starting in the 1950's, ZIKV, a flavivirus, was first detected in Brazil in early 2015; by March 2016, over 50,000 suspected cases had been reported, and the approximately 5000 cases of ZIKV-associated congenital microcephaly incited public outcry (Kindhauser et al., 2016; Pinheiro et al., 2016). ZIKV, the closely related flavivirus DENV, and CHIKV, for which local infections have been reported since 2014 (Azevedo et al., 2015), are all transmitted by the same endemic *Aedes* spp. mosquito vectors (Weaver and Lecuit, 2015). Infection with all three viruses induces fever and rash, making it very difficult to provide an accurate diagnosis via clinical symptoms alone, and many laboratory tests are cross-reactive among the viruses, particularly the flaviviruses ZIKV and DENV (Braga et al., 2017). Reports of GBS in the Americas during the 2015–2016 Zika virus outbreak increased markedly (Capasso et al., 2019), with the inciting agent mostly thought to be ZIKV, though many patients showed evidence of recent CHIKV or DENV infection as well (Acevedo et al., 2017; Anaya et al., 2017; de Azevedo et al., 2018; Dirlikov et al., 2016; Lima et al., 2019; Malta et al., 2017; Mancera-Páez et al., 2018; Mehta et al., 2018; Styczynski et al., 2017; Vieira et al., 2018; Zambrano et al., 2016). During the ZIKV outbreak, other cases of GBS were reported following CHIKV infection, with no evidence of prior ZIKV infection (Del Carpio-Orantes et al., 2018; Villamil-Gómez et al., 2016). Wachira et al. performed a systematic review evaluating factors reported to trigger GBS prior to and after 2007 and found that both CHIKV and ZIKV were identified in the literature as new etiological agents associated with the autoimmune condition (Wachira et al., 2019). Circulation of multiple arboviruses, particularly CHIKV, ZIKV, and DENV, in affected areas and high frequency of co-infections among patients complicate doctors' and scientists' ability to determine which pathogen(s) is responsible for triggering GBS. Further studies evaluating the exact mechanism by which arboviruses trigger GBS and how co-infections contribute to development of the condition are warranted.

Other immune-mediated neurological conditions have been reported as atypical manifestations of CHIKV infection in human patients. Acute disseminated encephalomyelitis (ADEM) is multifocal neurologic and demyelinating syndrome usually triggered by a prior infection whose pathophysiology is thought to be due to T cell activation following formation of antibodies against myelin and other self-antigens in the CNS (reviewed in Gray and Gorelick, 2016). Several cases of ADEM have been reported following CHIKV infection (Anand et al., 2019; Carvalho et al., 2020; Maity et al., 2014;

Musthafa et al., 2008; Taraphdar et al., 2015; Teixeira et al., 2019), with a cross-sectional study of CHIKV cases during an outbreak in Pakistan showing an ADEM incidence rate of 5% (Barr et al., 2018). Transverse myelitis, another immune-mediated neurologic condition with variable pathological lesions involving the spinal cord, including demyelination, axonal degeneration, and perivascular infiltration of monocytes and lymphocytes (reviewed in Krishnan et al., 2006), has also been recently documented following CHIKV infection (Choudhary et al., 2016; Farias et al., 2018; González-Galván et al., 2017; Hameed et al., 2019; Kumar et al., 2019; Neri et al., 2018). Multiple case reports and series have described atypical ocular manifestations following CHIKV infection, including anterior uveitis, choroiditis, panuveitis, retinitis, neuroretinitis, and optic neuritis (Chanana et al., 2007; Deeba et al., 2019; Lalitha et al., 2007; Lin et al., 2018; Mahendradas et al., 2008, 2010; Mahesh et al., 2009; Mittal et al., 2007; Murthy et al., 2008; Nair et al., 2012; Neri et al., 2018; Rocha et al., 2018; Rose et al., 2011; Salceanu and Raman, 2018; Scripsema et al., 2015; Ulloa-Padilla et al., 2018). Whether the pathological changes for these ocular conditions are due to a direct viral mechanism or an immune-mediated mechanism are not currently known. However, in most of the reported cases, ocular disease was bilateral, developed after a period of latency following initiation of classical CHIKV symptoms (fever, arthralgia, etc.), and responded to immunosuppressive therapy such as corticosteroids, all suggesting an immune-mediated, rather than direct viral, mechanism (Merle et al., 2018; Rose et al., 2011).

An in-depth systematic review that effectively describes the range of neurological disease induced by CHIKV infection in human patients from CHIKV outbreaks in Reunion Island, Italy, Thailand, the Caribbean, and Pacific Islands was recently published by Cerny et al. (2017). Of 64 neurological cases described, encephalitis was the most common neurological complication diagnosed (27 cases), followed by optic neuropathy (18 cases), neuroretinitis (5 cases), GBS (5 cases), and ADEM (3 cases). Encephalitis cases were more commonly diagnosed in infants and children or elderly patients with co-morbidities, but in contrast, most of the cases involving the peripheral nervous system (PNS) (optic neuropathy, neuroretinitis, GBS) were diagnosed in previously healthy young to middle-aged adults. While encephalitis cases were mainly treated with supportive care and had a high rate of moderate to severe complications (100%) and severe outcomes (47%), cases involving the brainstem, spinal cord, or PNS were treated with immunosuppressive therapy (e.g., steroids, intravenous immunoglobulin), resulting in a good (66%) or partial (33%) response to therapy, and much lower complication (29%) and severe outcome (18%) rates. These findings caused the authors to suggest that based on the treating physicians' assumptions, CHIKV-induced encephalitis has a direct viral pathomechanism, while neurological conditions involving the brainstem, spinal cord, or PNS has an underlying autoimmune or immunopathogenic mechanism. This hypothesis is further supported by the finding that the mean latency from development of classical CHIKV infection systems (fever, arthralgia, rash, etc.) to development of neurological complications was 3.3 days for encephalitis cases but 16.2 days for brainstem, spinal cord, or PNS cases; furthermore, for several patients who developed PNS disease, the latency period was marked by a complete lack of symptoms. Evaluation of the location of different diagnoses showed that while most cases from the Reunion Island outbreak were thought to have a direct viral pathomechanism (88%), most cases from outbreaks in India were autoimmune conditions involving the PNS (82%). This suggests that genetic factors, whether viral or host, are likely contributing to the development and pathogenesis of immune-mediated neurological disease during CHIKV infection.

5. Alphavirus-induced arthralgia and myalgia

As noted above, arthritogenic alphaviruses, and especially CHIKV, have been associated with several autoimmune neurologic disease states in humans, including GBS. However, the major disease signs associated with arthralgia-associated alphaviruses, which include CHIKV, RRV, and MAYV, are severe acute arthralgia and myalgia, that can progress to chronic disease in a subset of patients (reviewed in Burt et al., 2017). As discussed above, there is evidence for the host inflammatory response contributing to alphavirus-induced arthralgia and myalgia in humans. However, much of the evidence for immune pathology in driving these disease states comes from animal models. CHIKV causes inflammatory arthritis and myositis both in mice and in non-human primates, and these models have been useful systems both for studying the pathogenesis of CHIKV-induced disease and platforms for evaluating CHIKV therapeutics and vaccines (reviewed in Haese et al., 2016). Although most analysis has focuses on CHIKV, mouse models of RRV and MAYV also indicate that host inflammatory responses play a pathologic role in virus-induced myositis and arthritis (de Castro-Jorge et al., 2019; Lidbury et al., 2000).

Components of both the innate and adaptive immune system contribute to alphavirus-induced arthritis and myositis. Monocytes and/or macrophages are a prominent feature of both CHIKV and RRV-induced arthritis or myositis in mouse models, and depletion of these cells results in decreased disease signs (Gardner et al., 2010; Lidbury et al., 2000). Given the importance of monocyte and macrophages in the pathogenesis of CHIKV and RRV, additional studies have evaluated the potential for inhibitors of monocyte recruitment as therapies for alphavirus-induced arthritis. Bindarit, an inhibitor of monocyte chemotactic proteins (CCL2, CCL8), protects against both CHIKV and RRV-induced disease (Chen et al., 2015; Rulli et al., 2009), which suggest that targeting monocyte chemotactic proteins may be of therapeutic benefit. However, mice that are genetically deficient for CCR2, which acts as a receptor for several chemokines, including CCL2 led to enhanced CHIKV-induced arthritis characterized by neutrophil infiltration into joint tissues (Poo et al., 2014). This suggests that the inflammatory pathways that regulate monocyte recruitment into CHIKV infected tissues is complex and requires further study.

The complement cascade, which plays a protective role in the context of alphavirus-induced CNS disease (Brooke et al., 2012; Hirsch et al., 1980), has also been shown to play a role in alphavirus-induced arthritis and/or myositis. RRV infected mice show high levels of complement deposition in infected skeletal muscle (Morrison et al., 2007). Mice lacking the C3 component of complement, which is essential for activation and amplification of the pathways, show significantly reduced disease signs and muscle damage, yet viral loads and inflammatory cell recruitment into the infected muscle are unchanged. Further analysis found that complement receptor 3 deficiency led to a similar decrease in virus-induced disease and tissue damage, which suggests that complement receptor 3 expression on macrophages or other inflammatory cells may be contributing to inflammatory cell-mediated tissue damage in RRV-infected tissues (Morrison et al., 2008). Studies by Gunn, et al. went on to show that RRV-induced complement activation occurred via the MBL dependent pathway of complement activation and provided evidence that differences in MBL levels may affect RRV-induced disease severity in humans (Gunn et al., 2012). In a follow-up study, Gunn, et al., demonstrated that N-linked glycans on the RRV E2 protein were required for MBL binding, complement activation, and RRV-induced muscle damage (Gunn et al., 2018). It remains to be determined whether the complement cascade's role in driving virus-induced tissue damage is specific to RRV or if complement plays a larger role in the pathogenesis of other arthralgia associated alphaviruses.

As noted above, adaptive immunity plays an important role in protecting from arthralgia associated alphaviruses, with both antiviral antibody and CD8+ T cells contributing to protection from virus-induced disease. However, there is also evidence that CD4+ T cells contribute to CHIKV-induced disease, since mice lacking CD4+ T cells show a reduction in CHIKV-induced footpad swelling and joint pathology (Teo et al., 2013). Follow up studies by this group showed that expansion of regulatory T cells (Tregs) ameliorates CHIKV-induced disease in the mouse by inhibiting recruitment of pathologic CD4+ T cells into sites of CHIKV infection (Lee et al., 2015). These findings not only provide important insights into the pathogenesis of CHIKV-induced disease, but also have important implications for both the development of CHIKV vaccines and for potential therapeutic approaches for treating CHIKV-induced arthritis. This latter point is illustrated by the findings of Miner et al., where the disease-modifying anti-rheumatic drug cytotoxic T-lymphocyte-associated protein 4 (CTLA4) immunoglobulin, which blocks T cell co-stimulation, had benefit in treating CHIKV-induced arthritis and joint swelling in a mouse model, and this activity was enhanced when CTLA4 immunoglobulin was combined with a potent CHIKV specific monoclonal antibody (Miner et al., 2017).

6. Alphavirus-induced encephalomyelitis

6.1 Venezuelan equine encephalitis virus (VEEV), eastern equine encephalitis virus (EEEV), and western equine encephalitis virus (WEEV)

The neurotropic alphaviruses primarily consist of VEEV, EEEV, and WEEV, and animal models have helped shed light on the pathogenesis and immune response to infection with these viruses. Horses represent a highly permissive aberrant host for natural infection with the equine encephalitides but have also been used to study experimental infection. During infection with VEEV, severity of disease ranges from asymptomatic to rapid death, with clinically affected equids demonstrating fever, lethargy, weakness, and convulsions (Dietz et al., 1978; Gleiser et al., 1962; Monlux and Luedke, 1973; Roberts et al., 1970; Sahu et al., 2003; Walton et al., 1973). Pathological changes are consistent with meningoencephalitis and include perivascular and parenchymal infiltration of lymphocytes, mononuclear cells, and neutrophils with variable presence of glial cell proliferation and/or hypertrophy (gliosis) and clustering around neurons (satellitosis), neuronal damage, and vasculitis in the CNS. EEEV infection has a much higher mortality rate in horses than VEEV, with pathological lesions in the CNS characterized by perivascular cuffing with mononuclear cells, neutrophilic infiltration in the parenchyma, gliosis, neuronophagia, and vascular damage, consistent with diffuse polioencephalomyelitis (inflammation of the gray matter of the brain and spinal cord) (Del Piero et al., 2001; Franklin et al., 2002). WEEV appears to be the least virulent of the equine encephalitides, with most horses developing non-clinical infections (Potter et al., 1977). However, the large size and expense of housing horses precludes most in depth experimental studies evaluating the pathogenesis of VEEV, EEEV, and WEEV infection, necessitating the use of smaller animal models.

While experimental infections have been performed in other animals such as hamsters, rabbits, and guinea pigs, mice and nonhuman primates are considered to be the most relevant animal models for evaluating the pathology and immune response induced by VEEV, EEEV, and WEEV. Experimental infection of mice with virulent strains of VEEV by subcutaneous or footpad inoculation (considered to mimic natural infection by mosquito bite) results in CNS invasion via axonal transport up olfactory neurons following initial infection of dermal dendritic cells, replication in the draining lymph nodes, and viremic dissemination throughout the body (Charles et al., 1995; Davis et al., 1994; MacDonald and Johnston, 2000; Ryzhikov et al., 1991; Steele et al., 1998, 2006; Vogel et al., 1996). Aerosol or intranasal infection leads to a direct infection of olfactory neuroepithelium, resulting in infection of the brain as early as 16 h post infection (HPI) (Bocan et al., 2019; Cain et al., 2017; Ryzhikov et al., 1991; Steele et al., 1998; Vogel et al., 1996). Once in the CNS, VEEV rapidly disseminates, primarily targeting neurons in both the brain and spinal cord, though macrophages may also be infected (Jackson et al., 1991; Steele et al., 1998; Vogel et al., 1996). Histopathological changes are consistent with encephalomyelitis and characterized by thick perivascular cuffs and parenchymal infiltration, initially of neutrophils but then evolving to mostly mononuclear cells, as well as an increase in astrocytes (astrocytosis) (Ryzhikov et al., 1991; Schoneboom et al., 2000a; Sharma et al., 2008; Sharma and Maheshwari, 2009; Steele et al., 1998, 2007). Neuronal damage has been reported to occur through both apoptosis and necrosis (Bocan et al., 2019; Jackson and Rossiter, 1997; Schoneboom et al., 2000a; Steele et al., 1998; Steele and Twenhafel, 2010).

Infection of NHPs with VEEV via the intraperitoneal route results in transient fever and viremia within the first 3 days post infection (DPI) and infection of the brain by 6 DPI (Gleiser et al., 1962). Histopathological changes are similar but generally less severe than those seen in mouse models and include multifocal perivascular cuffs composed mostly of lymphocytes, gliosis, and rare evidence of neuronal damage associated with microglial clusters (glial nodules). Experimental intranasal infection with VEEV results in lesions in the olfactory bulbs by 48 HPI but is otherwise results in similar, though slightly more severe, pathological lesions in the CNS (Danes et al., 1973). Most experimental VEEV infection studies in NHP's are conducted in the context of vaccine research using aerosol exposure. Infection of the CNS following challenge by this route is variable but induces lesions similar to those seen in humans when present, such as perivascular cuffing and gliosis (Fine et al., 2008; Steele and Twenhafel, 2010).

Unlike VEEV, EEEV does not initially replicate in dendritic cells or macrophages (Gardner et al., 2008; Levitt et al., 1979) but instead in fibroblasts, skeletal muscle, and osteoblasts (Vogel et al., 2005). Also unlike with the mouse model of VEEV infection, neuroinvasion by EEEV occurs via the hematogenous route rather than via axonal transport into the olfactory bulb (Vogel et al., 2005). EEEV can be detected in the brains of mice infected subcutaneously as early as 24 HPI, and multifocal positive EEEV antigen staining seen in multiple regions of the brain supports hematogenous spread. Similar to VEEV, neurons represent the main infected cell, though glial cells may also be infected (Roy et al., 2009; Vogel et al., 2005). Pathological changes in the brain include diffuse neuronal necrosis but fairly mild infiltration of neutrophils and eosinophils. Though not a known natural route of infection, aerosol challenge with EEEV has been studied mainly in the context of biodefense, as many strains of EEEV are classified as Select Agents. In this challenge model, infection has been demonstrated to first occur in the olfactory bulb and then disperse throughout the brain via transneuronal spread, similar to aerosol models of VEEV infection (Honnold et al., 2015a; Roy et al., 2009; Steele and Twenhafel, 2010). Histopathological findings include fulminant meningoencephalitis with abundant neuronal vacuolation, degeneration, and necrosis but few inflammatory infiltrates (Honnold et al., 2015b; Phelps et al., 2019). An aerosol challenge guinea pig model of EEE has shown a similar route of viral entry into the CNS, and pathological changes in the brain consist of marked neuronal necrosis and both perivascular and parenchymal inflammation (Roy et al., 2009). When inoculated with EEEV subcutaneously, hamsters have been shown to be exquisitely sensitive to infection, with virus detected in lung, liver, muscle, heart, kidney, spleen early in infection and titers increasing in the brain from 3 DPI to death at 5 or 6 DPI (Paessler et al., 2004). Similar to humans infected with EEEV, hamsters demonstrate fulminant vasculitis and subependymal hemorrhages with mixed inflammation in the brain. Neuronal death is mainly attributed to virus-mediated apoptosis (Honnold et al., 2015b; Vogel et al., 2005).

NHPs have also been used to evaluate the pathology induced by EEEV infection. Direct inoculation of virus into the brains of juvenile rhesus macaques results in severe encephalitis characterized by neuronal necrosis and loss with relatively mild inflammation (Nathanson et al., 1969). Aerosol challenge of cynomolgus macaques with virulent EEEV results in severe meningoencephalomyelitis, characterized by diffuse neuronal necrosis, perivascular cuffing with mononuclear cells and neutrophils, gliosis, satellitosis, and vasculitis (Reed et al., 2007; Roy et al., 2013; Steele and Twenhafel, 2010). Perivascular spaces are often expanded by edema or hemorrhage, and viral antigen is detected in multiple brain regions, primarily targeting neurons. Marmosets, small callitrichid monkeys native to Central and South America, have also been evaluated as potential models for EEE, and infection with EEEV induces meningoencephalitis with perivascular hemorrhages, similar to what is seen in human patients (Adams et al., 2008; Porter et al., 2017). When comparing the pathology induced by VEEV versus EEEV in different animal models, despite both viruses primarily targeting neurons, EEEV infection induces markedly more severe neuronal damage and death. However, these pathological changes are associated with relatively little inflammation, whereas VEEV infection of the brain results in massive immune cell infiltration, but only minimal damage to neurons.

Generally considered the least neurovirulent of the New World alphaviruses, animal studies of WEEV infection have shown variable pathology. Mice infected with more virulent WEEV strains demonstrate high mortality via multiple inoculation routes and develop severe meningoencephalitis characterized by neuronal degeneration and necrosis, edema, and mild infiltration of mononuclear cells (Aguilar, 1970; Logue et al., 2009; Phelps et al., 2017). Infection with the less virulent strains demonstrate more perivascular inflammation but minimal neuronal damage (Logue et al., 2009). Following recovery from WEEV, CD-1 mice demonstrate persistent gliosis (Bantle et al., 2019). High mortality in WEEV-infected neonatal mice due to severe inflammation and/or necrosis in peripheral tissues such as bone marrow, skeletal muscle, and peripheral nerves without any CNS involvement (Aguilar, 1970) is similar to disease seen in some reports of hamsters infected with VEEV (Gorelkin and Jahrling, 1975; Jackson et al., 1991; Jahrling and Scherer, 1973; Walker et al., 1976). Challenge experiments via subcutaneous, intranasal, and aerosol routes in mice indicate WEEV invades the brain via axonal transport up the olfactory tract similar to VEEV (Bantle et al., 2019; Phelps et al., 2017; Phillips et al., 2013, 2016). Experimental infection of hamsters with WEEV demonstrates that, similar to the other equine encephalitides, neurons are the preferred cell for infection in the CNS (Julander et al., 2007), though infection of microglia has also been demonstrated in an aerosol challenge model of WEEV in cynomolgus macaques (Reed et al., 2005).

The immune response has been shown to play a major role in the pathogenesis of VEEV infection. Following subcutaneous inoculation with VEEV, mice show upregulation of proinflammatory genes *Ifng, Il6, Il12, Il17,* and *Tnf,* chemokine genes *Cxcl9, Cxcl10, Cxcl11, Cxcl13, Ccl2, Ccl3, Ccl5,* and *Ccl12,* and genes associated with TLR signaling, including *Tlr1, Tlr2, Tlr3, Tlr7,* and *Tlr9,* correlating with blood brain barrier compromise and infiltration of neutrophils and mononuclear cells into the brain (Grieder et al., 1997; Gupta et al., 2017; Sharma et al., 2008; Sharma and Maheshwari, 2009). Infection of multiple strains of mice with the V3000 strain of VEEV results in induction of proinflammatory cytokines, including inducible nitric oxide synthase, IL-1β, IL-6, IL-12, and TNF-α (Schoneboom et al., 1999, 2000a, 2000b). Mice administered magnesium or cadmium demonstrate more rapid development and severe histopathological CNS lesions concurrent with upregulation of proinflammatory cytokines following infection with both VEEV and SFV (Seth et al., 2003). Treatment of BALB/c mice with IFN-α prior to VEEV challenge results in improved survival, reduced TNF-α production by macrophages, and delayed activation of the adaptive immune response (Lukaszewski and Brooks, 2000). Inhibition of GSK-3β, a serine/threonine protein kinase with immunomodulatory capabilities, protects C3H mice from VEEV TC83-induced encephalitis (Kehn-Hall et al., 2012).

The adaptive immune response also contributes to pathological changes in the CNS during VEEV infection. Severe combined immunodeficiency (SCID) mice, which lack both T cells and B cells, show improved survival times compared to wild-type mice despite demonstrating higher viral titers (Charles et al., 2001). Nude mice, which lack T cells, infected with VEEV do not develop the inflammatory demyelinating spinal cord lesions seen in wild-type (WT) counterparts, and mice administered anti-thymocyte serum, which depletes T cells, have delayed mortality, indicating T cells play a role in inducing VEEV pathology (Dal Canto and Rabinowitz, 1981; Woodman et al., 1975). However, the role and contributions of CD4+ versus CD8+ T cells or other immune cells, such as macrophages, to immune-mediated damage during VEEV infection has not yet been elucidated.

As inflammation in the brain is not a key aspect of brain pathology during EEEV infection, less is understood about the role the immune system plays in the pathogenesis of EEE. However, mice infected via the intranasal or aerosol route with EEEV do demonstrate upregulation of proinflammatory cytokines IFN-γ, CCL4, CCL5, CXCL9, and G-CSF (Honnold et al., 2015a). While few articles describing the local immune response to WEEV have been published, increases in pro-inflammatory cytokines such as IFN-γ, TNF-α, IL-12, CCL2, and CXCL10 have also been demonstrated at 3 DPI in the brains of WEEV MCM strain-infected CD-1 mice before death at 4–5 DPI (Logue et al., 2010; Phillips et al., 2013). Significant differences in clinical outcomes among various inbred mouse strains following WEEV infection, with C57BL/6 and BALB/c mouse strains showing high susceptibility and BALB/cBy and DBA/2 mouse strains being more resistant to severe disease, suggest that some of these disparities in neurovirulence due to host genetics are likely due to differences in the immune response (Blakely et al., 2015).

6.2 Sindbis virus (SINV)

Numerous extensive studies have been conducted using the naturally arthritogenic alphavirus SINV to study the neuropathogenesis of alphavirus infection. SINV is neurotropic in mice but, unlike most strains of the equine encephalitides, can be handled under biosafety level 2 conditions and therefore represents a more accessible model of alphavirus encephalomyelitis. Disease severity is dependent on the strain of SINV used; for example, while both viruses effectively disseminate throughout the CNS, AR339, the prototypic strain of SINV, is fatal in neonatal mice but nonvirulent in weanling mice, while neuroadapted SINV (NSV) induces ascending paralysis and 100% mortality in susceptible mice of all ages (Griffin, 1976; Griffin and Johnson, 1977; Jackson et al., 1987; Taylor et al., 1955). Similar to VEEV and WEEV, neuroinvasion occurs via transaxonal transport from the neuroepithelium to the olfactory bulb, after which SINV disseminates caudally throughout the brain and to the brainstem and spinal cord (Lee et al., 2013; Passoni et al., 2017; Thach et al., 2000). Ascending paralysis seen with SINV NSV infection is due to infection and degeneration of motor neurons in the lumbar spinal cord (Havert et al., 2000; Jackson et al., 1988). In brains of C57BL/6 mice infected with SINV, marked loss of neurons, particularly in the CA regions of hippocampus, is a key feature (Baxter et al., 2018; Kimura and Griffin, 2003). Perivascular cuffing and parenchymal infiltration of mononuclear cells is also considerable (Baxter et al., 2018; Irani and Griffin, 1991; Kimura and Griffin, 2003; Moench and Griffin, 1984; Stohlman et al., 1998). Hydrocephalus and dilation of the lateral ventricles may also be seen (Baxter and Griffin, 2020; Kimura and Griffin, 2003). Mice infected with the nonfatal TE strain of SINV demonstrate clinical disease characterized by gait deficits and exhibit behavioral changes concurrent to the development of these pathological changes in the brain (Baxter and Griffin, 2016; Baxter et al., 2018; Potter et al., 2015). In particular, hippocampal-dependent memory changes persist beyond the period of recovery from clinical disease and virus clearance (Potter et al., 2015). The mechanism of neuronal damage is dependent on the type of neuron, with evidence of both apoptosis and necrosis demonstrated (Havert et al., 2000; Kerr et al., 2002; Lewis et al., 1996; Nargi-Aizenman and Griffin, 2001).

Studies have shown that the immune system, particularly T cells, play the predominant role in inducing CNS pathology during SINV infection. During nonfatal SINV infection, inflammatory cells begin to infiltrate the CNS around 3 DPI, peaking at 7–10 DPI (Moench and Griffin, 1984). While NK cells peak early in infection, at 7 DPI, CD8+ T cells and macrophages predominate, though numbers of CD4+ T cells rapidly increase, becoming the main immune cell type by 10 DPI (Baxter and Griffin, 2020; Irani and Griffin, 1991, 1996; Metcalf and Griffin, 2011; Moench and Griffin, 1984). Onset of neurological disease and mortality in mice infected with SINV coincides with clearance of infectious virus and infiltration of immune cells into the brain (Griffin and Hardwick, 1997; Jackson et al., 1987; Kulcsar et al., 2014). SCID mice, which are deficient in both mature T and B cells, show 100% survival and minimal neuropathology following SINV NSV infection and do not develop neurological disease following SINV AR339 infection despite supporting high levels of virus replication (Levine et al., 1991; Wesselingh et al., 1994). Impaired proliferation and infiltration of lymphocytes into the brain of SINV-infected mice treated with 6-diazo-5-oxo-l-norleucine, a glutamine antagonist that arrests lymphocytes in the G1 phase of the cell cycle (Wang et al., 2011), leads to reduced mortality, weight loss, and development of neurological disease and cognitive deficits (Baxter et al., 2017; Manivannan et al., 2016; Potter et al., 2015). Mice deficient in multiple components of the cellular adaptive immune response, including TCRα, TCRβδ, β2m, TAP1, and CD4, but not CD8, show reduced mortality following SINV NSV infection, indicating T cells, especially CD4+ T cells, play a major role in inducing CNS pathology (Kimura and Griffin, 2000; Rowell and Griffin, 2002). In particular, Th17 cells expressing granzyme B, IL-22, and GM-CSF infiltrate the CNS during the development of SINV NSV-induced paralysis and fatal neurological disease (Kulcsar et al., 2014). These pathogenic Th17 cells are regulated by IL-10, with *Il10*$^{-/-}$ mice developing more severe weight loss and neurological disease and increased mortality concurrent with more CD4 + T cells skewed towards a Th1 and Th17 response but away from a Th2 and Treg response in the brain (Kulcsar et al., 2014; Kulcsar and Griffin, 2016; Martin and Griffin, 2018). BALB/c mice, which are more resistant to fatal encephalomyelitis than C57BL/6 mice, demonstrate fewer Th17 cells and more IL-10-dependent Tregs in the brain following SINV NSV infection (Kulcsar et al., 2015). These studies together demonstrate the importance of the T cell response in inducing neuropathology in the mouse model of alphavirus encephalomyelitis.

Proinflammatory cytokines have been shown to play a role in the immunopathogenesis in the SINV mouse model of alphavirus encephalomyelitis. Following infection with multiple strains of SINV, mouse brains show upregulation of proinflammatory cytokine gene expression, including *Tnf, Csf2, Il1b, Il4, Il6, Il12b, Il17a, Il23, Lif,* and *Ifng* (Baxter et al., 2018; Baxter and Griffin, 2016; Kulcsar et al., 2015; Lee et al., 2013; Martin and Griffin, 2018; Rowell and Griffin, 1999; Wesselingh et al., 1994). Mice deficient in interferon regulatory factors 3 and 7, who show increased disease severity and mortality respectively, have upregulated brain expression of multiple proinflammatory cytokines that correlate with increased severity of inflammation (*Irf3*$^{-/-}$ mice), numbers of CD4+ (*Irf3*$^{-/-}$ mice) and CD8+ (*Irf7*$^{-/-}$ mice) T cells, and TUNEL-positive cells (*Irf7*$^{-/-}$ mice) in the brain and spinal cord compared to WT mice following infection with the generally nonfatal TE strain of SINV (Schultz et al., 2019). IL-1β deficient mice are less susceptible to fatal encephalitis induced by SINV NSV, despite showing comparable histopathology and levels of apoptosis as WT mice (Liang et al., 1999). TNF-α deficient mice infected with SINV NSV show improved survival and reduced neuronal death compared to WT mice (Carmen et al., 2009), and TNF-receptor-1-deficient mice show reduced mortality and delayed development of paralysis compared to WT mice infected with the neurovirulent SVNI strain of SINV (Sarid et al., 2001). IFN-γ, the sole type II IFN, is known to contribute to the clearance of SINV in cooperation with anti-SINV antibody (Baxter and Griffin, 2016; Binder and Griffin, 2001; Burdeinick-Kerr et al., 2007; Lee et al., 2013) but is also involved in inducing pathology. WT mice lose more weight following SINV TE infection compared to *Ifng*$^{-/-}$ and *Ifngr1*$^{-/-}$ mice due to changes in feed intake indirectly related to levels of TNF-α in the brain (Baxter and Griffin, 2016). During the height of clinical disease at 7 DPI, *Ifng*$^{-/-}$ and *Ifngr1*$^{-/}$ mice show reduced inflammation in the brain and spinal cord and are less likely to demonstrate dilatation of the lateral ventricles compared to WT mice (Baxter and Griffin, 2020). Mice with intact IFN-γ signaling have more macrophages and degranulating NK cells in the brain but fewer CD8+ T cells producing granzyme B and perforin and show delayed initiation of viral RNA clearance compared to *Ifng*$^{-/-}$ and *Ifngr1*$^{-/-}$ mice (Baxter and Griffin, 2020; Lee et al., 2013). Upregulation of proinflammatory cytokines induced during alphavirus encephalomyelitis likely promotes the infiltration of inflammatory cells that then induce pathological changes and neuronal damage in the CNS.

Another indirect mechanism by which neurons may die during alphavirus infection of the CNS is excitotoxicity. In neuronal excitotoxicity, excess amounts of the neurotransmitter glutamate accumulate in the synaptic cleft, leading to overstimulating of a-amino-3-hydroxy-5-methyl-4-isoxazole propionic acid (AMPA), *N*-methyl-D-aspartate, and kainate-type receptors (reviewed in Sattler and Tymianski, 2001). Influx of metal ions, particularly Ca^{2+}, into the neuron through channels gated by these receptors leads to metabolic destruction of the cell (Lee et al., 1999; Nicotera and Orrenius, 1998). Glutamate excitotoxicity has been shown to play a role in the pathogenesis of SINV infection, with neuronal infection by multiple strains of SINV resulting in downregulation of glutamate transporter 1, a receptor primarily expressed on astrocytes responsible for preventing glutamate excitotoxicity by taking up the majority of glutamate from the synaptic cleft (Baxter et al., 2017; Carmen et al., 2009; Darman et al., 2004; Nargi-Aizenman and Griffin, 2001; Prow and Irani, 2007). Treatment of mice with either an *N*-methyl-D-aspartate receptor antagonist or AMPA receptor antagonist results in fewer necrotic and apoptotic neurons in the hippocampus following infection with SINV NSV (Nargi-Aizenman et al., 2004). Furthermore, spinal cord neuronal damage and paralysis is prevented, and mortality is reduced in mice treated with the AMPA receptor antagonist in a viral replication-independent manner (Greene et al., 2008; Nargi-Aizenman et al., 2004). However, inhibition of glutamate excitotoxicity either by AMPA receptor or glutamine antagonist treatment also results in reduced proliferation and impaired infiltration of immune cells in the draining lymph nodes and brain, respectively, during SINV infection (Baxter et al., 2017; Greene et al., 2008; Manivannan et al., 2016; Potter et al., 2015). Furthermore, mice deficient in TNF-α or IL-1β do not demonstrate glutamate transporter 1 downregulation in the brain following SINV NSV infection like WT mice do (Carmen et al., 2009; Prow and Irani, 2007). These findings indicate that multiple pathways involved in excitotoxicity also affect the immune response and emphasize the interconnectedness of different mechanisms known to induce neuronal damage during alphavirus encephalomyelitis.

6.3 Semliki forest virus (SFV)

SFV, an Old World alphavirus antigenically related to RRV and MAYV, is generally considered to be non-pathogenic in humans, only occasionally causing fever, myalgia, and arthralgia (Mathiot et al., 1990). However, a case of fatal encephalitis due to laboratory acquired SFV infection has been reported (Willems et al., 1979), and infection of mice with multiple strains of SFV by both direct and indirect routes consistently results in demyelinating meningoencephalomyelitis (Gates et al., 1984; Kelly et al., 1982). While some SFV strains produce acute fatal encephalitis, the well-studied avirulent A7(74) strain of SFV induces a transient viral infection that is cleared by 8 DPI in susceptible immunocompetent strains, such as BALB/c mice (Amor et al., 1996; Berger, 1980; Kelly et al., 1982; Morris et al., 1997). Perivascular cuffs composed of CD8+ T cells, B cells, and macrophages are found throughout the brain starting during the first week of infection, but beginning around 10–15 DPI, mice develop demyelinating lesions in the white matter of the cerebellum, corpus callosum, and brainstem (Amor et al., 1996; Berger, 1980; Kelly et al., 1982; Morris et al., 1997; Subak-Sharpe et al., 1993). Following infection and subsequent SFV clearance, focal areas of vacuolation in the myelin are filled with necrotic cells and surrounded by mononuclear cells, particularly lymphocytes extending cytoplasmic projections to astrocytes and macrophages (Kelly et al., 1982; Sheahan et al., 1983).

Similar to EEEV, following peripheral inoculation, SFV primarily enters the CNS via hematogenous spread by infecting vascular endothelial cells (Erälinna et al., 1996; Pathak and Webb, 1974; Soilu-Hänninen et al., 1994), though intranasal infection has shown that neuroinvasion via the olfactory bulb is also possible (Keogh et al., 2003; Oliver and Fazakerley, 1998). Upregulation of intercellular adhesion molecule 1 and fibrinogen in the CNS during SFV infection indicates that blood brain barrier integrity is also compromised (Erälinna et al., 1996; Soilu-Hänninen et al., 1994). While neurons are susceptible to SFV infection in an host age-dependent manner similar to SINV (Fazakerley et al., 1993; Fragkoudis et al., 2009; Oliver et al., 1997), unlike SINV and the other New World alphaviruses, oligodendrocytes represent the primary targeted cell in the CNS (Fazakerley et al., 2006; Fragkoudis et al., 2009; Sheahan et al., 1983). In optic nerves of SFV-infected BALB/c mice, oligodendrocytes develop swellings and vacuolations along internodal myelin sheaths, appearing more like immature cells (Butt et al., 1996); this suggests SFV infection induces the oligodendrocytes to dedifferentiate rather than die, and therefore the cells are capable of remyelination.

Numerous studies have shown that demyelination following SFV infection is primarily immune mediated. Mice lacking both T and B cells through genetics (SCID mice), irradiation, or cyclophosphamide administration show reduced and/or delayed pathology despite robust viral replication (Amor et al., 1996; Berger, 1980; Fazakerley and Webb, 1987a). Nude mice do not develop the perivascular inflammation or demyelinating lesions of immunocompetent mice (Amor et al., 1996; Fazakerley et al., 1988, 2006; Gates et al., 1984), and transfer of T cells, but not hyperimmune serum, back into nude or SCID mice results in demyelination (Amor et al., 1996; Fazakerley et al., 1983; Fazakerley and Webb, 1987a, 1987b). Depletion of CD8+ T cells, but not CD4+ T cells, reduces inflammation and eliminates demyelinating lesions, suggesting that CD8+ T cells are primarily responsible for the CNS damage (Subak-Sharpe et al., 1993). Production of MHC class I, which is expressed on oligodendrocytes and makes them a target for cytotoxic T cells (Fazakerley et al., 1993), is upregulated early in SFV infection (M. M. Morris et al., 1997), and interactions between lymphocytes and oligodendrocytes have been observed by electron microscopy at 7 DPI before demyelinating lesions are seen starting at 10 DPI (Pathak et al., 1983). While CNS expression of cytokines IL-1α, IL-1β, IL-10, and TGFβ increases early in infection, TNF-α, IL-2, and GM-CSF expression increases when cellular infiltrates are at their peak; IFN-γ and IL-6 first become apparent at 10 DPI immediately prior to the demyelination phase of disease, suggesting these cytokines may play a role in induction of CNS pathology (Blackman and Morris, 1984; McKimmie et al., 2005; Morris et al., 1997). Mice that lack the IFN-γ receptor, and therefore have deficient type II IFN signaling, demonstrate higher levels of neuronal necrosis associated with increased viral antigen and more macrophages and B cells in the brain, but comparable amounts of demyelination, following SFV infection (Keogh et al., 2003). Nude mice, which have impaired antibody class switching and therefore produce only antiviral IgM, μMT mice, which lack the ability to produce antibody, and SCID mice all demonstrate delayed viral clearance and/or viral persistence (Amor et al., 1996; Fazakerley et al., 2006; Fazakerley and Webb, 1987b; Fragkoudis et al., 2008). However, administration of polyclonal or monoclonal SFV antibody quickly reduces infectious virus to undetectable levels in SCID mice, indicating that, similar to SINV, anti-SFV antibody plays a major role in SFV clearance in the CNS (Amor et al., 1996; Fragkoudis et al., 2018).

SFV A7(74) has been shown to react immunologically with immune sera against several antigens in the CNS, including galactocerebroside (found on oligodendrocytes), glucocerebroside, total ganglioside, and GT1b ganglioside (found on neurons), indicating the possibility of SFV infection creating an autoimmune state against cells present in the CNS (Webb et al., 1984). Administering anti-SFV hyperimmune serum, but not normal mouse serum, results in CNS tissue degeneration, but no demyelination or inflammatory response, in both SFV-infected and mock-infected control mice, suggesting autoantibodies play a role in pathology (Fazakerley and Webb, 1987b). C57BL/6 mice immunized with SFV proteins show lymphocyte proliferation against peptides on the SFV E2 glycoprotein and myelin oligodendrocyte glycoprotein, inducing SFV-like demyelination; this suggests that SFV infection creates a molecular mimicry response against myelin oligodendrocyte glycoprotein, leading to autoimmune demyelination (Mokhtarian et al., 1999). Because of this, SFV has been used in mouse models of experimental autoimmune encephalomyelitis to evaluate the pathogenesis of autoimmune conditions such as multiple sclerosis (Wu et al., 1988). In C57BL/6 mice infected with SFV A7(74), CNS inflammation and virus clearance occur from 6 to 10 DPI, demyelination occurs from 15 to 21 DPI, and remyelination follows, with brains showing almost no lesions by 35 DPI (Mokhtarian et al., 2003). In this model, CD8+ T cell numbers peak between 8 and 15 DPI, concurrent with initial myelin injury, but B cells and microglia represent the predominant inflammatory cells present during the late period of demyelination at 21–28 DPI. γδ T cells and antibody against the SFV epitope, E2 T_h peptide$_2$, have been shown to contribute to the remyelination process (Mokhtarian et al., 2012; Safavi et al., 2011) via an unknown mechanism. The C57BL/6 experimental autoimmune encephalomyelitis model using SFV highlights the multifaceted role the immune response plays in inducing pathology and the challenges that must be overcome in understanding and addressing immune-mediated damage following alphavirus infection.

7. Conclusion and future directions

The increase in geographic expansion and frequency of outbreaks of disease induced by alphaviruses necessitates the development of effective therapies that not only curb the initial course of infection but address and/or prevent the development of chronic disease. Thanks to information assembled from reports of human cases and experimental infection of animals, we now know the immune system is responsible for a significant amount of clinical disease following alphavirus infection. Infection by alphaviruses promotes a proinflammatory state both systemically and in the tissues targeted by the viruses, and development of clinical disease coincides with the infiltration of immune cells, particularly T cells. Furthermore, blockage or inhibition of the inflammatory response helps mitigate the development and severity of clinical disease following alphavirus infection. However, the immune response is also responsible for restriction of viral replication, clearance of infectious virus and viral RNA, and prevention of persistent viral RNA reactivation and relapse of clinical disease. Some components of the immune response, such as TNF-α, IFN-γ, and CCL2, have been reported to be both pathogenic and protective during alphavirus infection. This double-edged sword highlights the complicated balance the immune response must navigate when responding to infection by an alphavirus, and understanding this interplay between pathology and protection is critical when considering potential treatments that modulate the immune response.

Many aspects of the role the immune response plays in development of clinical disease remain to be understood. For example, the pathomechanism of chronic disease following alphavirus infection, particularly with CHIKV, is currently unknown. Despite reports of debilitating pain, swelling, and stiffness in patients suffering from chronic chikungunya arthritis, damage to affected joints is not usually severe, raising the question of what unseen factors are mediating clinical symptoms. The recent increase in detection of CHIKV-associated GBS cases, especially in light of similar co-circulating arboviruses in affected regions, warrants further evaluation of the role of molecular mimicry in development of autoimmune disease following alphavirus infection. Understanding the effect of host genetics on the immune response and the role played in the range of disease severity and development of atypical manifestations is also in its infancy. Disparate manifestations of CHIKV-induced neurological disease in various geographical regions suggest that the diverse immune systems of the human populations among affected areas are likely responding to infection and mediating the course of disease in different manners. Robust animal models that effectively replicate the course of infection and development of disease and pathology, particularly chronic or atypical disease manifestations, will greatly assist in answering these questions, and critical evaluation, characterization, and modification of these models should be ongoing. Further illumination of the immunopathogenesis of alphavirus infection will help lead to therapies that effectively target the specific components of the immune response liable for tissue damage and pathology, leading to better disease outcomes and improved quality of life in affected individuals.

Acknowledgments

V.K.B. is supported by the National Institutes of Health (NIH) grants K01 OD026529 and U19 AI100625. M.T.H. is supported by NIH grants U19 AI100625 and U19 AI142759.

References

Acevedo, N., Waggoner, J., Rodriguez, M., Rivera, L., Landivar, J., Pinsky, B., Zambrano, H., 2017. Zika virus, Chikungunya virus, and Dengue virus in cerebrospinal fluid from adults with neurological manifestations, Guayaquil, Ecuador. Front. Microbiol. 8, 42. https://doi.org/10.3389/fmicb.2017.00042.

Adams, A.P., Aronson, J.F., Tardif, S.D., Patterson, J.L., Brasky, K.M., Geiger, R., Garza, l., de, M., Carrion, R., Weaver, S.C., 2008. Common marmosets (Callithrix jacchus) as a nonhuman primate model to assess the virulence of eastern equine encephalitis virus strains. J. Virol. 82, 9035–9042. https://doi.org/10.1128/JVI.00674-08.

Aguilar, M.J., 1970. Pathological changes in brain and other target organs of infant and weanling mice after infection with non-neuroadapted Western equine encephalitis virus. Infect. Immun. 2, 533–542.

Aletaha, D., Neogi, T., Silman, A.J., Funovits, J., Felson, D.T., Bingham, C.O., Birnbaum, N.S., Burmester, G.R., Bykerk, V.P., Cohen, M.D., Combe, B., Costenbader, K.H., Dougados, M., Emery, P., Ferraccioli, G., Hazes, J.M.W., Hobbs, K., Huizinga, T.W.J., Kavanaugh, A., Kay, J., Kvien, T.K., Laing, T., Mease, P., Ménard, H.A., Moreland, L.W., Naden, R.L., Pincus, T., Smolen, J.S., Stanislawska-Biernat, E., Symmons, D., Tak, P.P., Upchurch, K.S., Vencovský, J., Wolfe, F., Hawker, G., 2010. 2010 Rheumatoid arthritis classification criteria: an American College of Rheumatology/European League Against Rheumatism collaborative initiative. Arthritis Rheum. 62, 2569–2581. https://doi.org/10.1002/art.27584.

Alla, S.A.O., Combe, B., 2011. Arthritis after infection with chikungunya virus. Best Pract. Res. Cl. Rh. 25, 337–346. https://doi.org/10.1016/j.berh.2011.03.005.

Amaral, J.K., Schoen, R.T., 2018. A case report of chikungunya fever, rheumatoid arthritis, and Felty's Syndrome. Rheumatol. Ther. 5, 293–301. https://doi.org/10.1007/s40744-018-0103-9.

Amaral, J.K., Bilsborrow, J.B., Schoen, R.T., 2020a. Chronic Chikungunya Arthritis and Rheumatoid Arthritis: what they have in common. Am. J. Med. 133, e91–e97. https://doi.org/10.1016/j.amjmed.2019.10.005.

Amaral, J.K., Bingham, C.O., Schoen, R.T., 2020b. Successful methotrexate treatment of chronic chikungunya arthritis. J. Clin. Rheumatol. 26, 119–124. https://doi.org/10.1097/RHU.0000000000000943.

Amor, S., Scallan, M.F., Morris, M.M., Dyson, H., Fazakerley, J.K., 1996. Role of immune responses in protection and pathogenesis during Semliki Forest virus encephalitis. J. Gen. Virol. 77, 281–291. https://doi.org/10.1099/0022-1317-77-2-281.

Anand, K.S., Agrawal, A.K., Garg, J., Dhamija, R.K., Mahajan, R.K., 2019. Spectrum of neurological complications in chikungunya fever: experience at a tertiary care centre and review of literature. Trop. Doct. 49, 79–84. https://doi.org/10.1177/0049475518825219.

Anaya, J.-M., Rodríguez, Y., Monsalve, D.M., Vega, D., Ojeda, E., González-Bravo, D., Rodríguez-Jiménez, M., Pinto-Díaz, C.A., Chaparro, P., Gunturiz, M.L., Ansari, A.A.,

Gershwin, M.E., Molano-González, N., Ramírez-Santana, C., Acosta-Ampudia, Y., 2017. A comprehensive analysis and immunobiology of autoimmune neurological syndromes during the Zika virus outbreak in Cúcuta, Colombia. J. Autoimmun. 77, 123–138. https://doi.org/10.1016/j.jaut.2016.12.007.

Anderson, C.R., Downs, W.G., Wattley, G.H., Ahin, N.W., Reese, A.A., 1957. Mayaro virus: a new human disease agent. II. Isolation from blood of patients in Trinidad, B.W.I. Am. J. Trop. Med. Hyg. 6, 1012–1016.

Anfasa, F., Lim, S.M., Fekken, S., Wever, R., Osterhaus, A.D.M.E., Martina, B.E.E., 2019. Characterization of antibody response in patients with acute and chronic chikungunya virus disease. J. Clin. Virol. 117, 68–72. https://doi.org/10.1016/j.jcv.2019.06.001.

Arrigo, N.C., Adams, A.P., Weaver, S.C., 2010. Evolutionary patterns of eastern equine encephalitis virus in North versus South America suggest ecological differences and taxonomic revision. J. Virol. 84, 1014–1025. https://doi.org/10.1128/JVI.01586-09.

Aubry, M., Cao-Lormeau, V.M., 2019. History of arthropod-borne virus infections in French Polynesia. New Microbes New Infect. 29, 100513. https://doi.org/10.1016/j.nmni.2019.01.009.

de Azevedo, M.B., Coutinho, M.S.C., Silva, M.A.D., Arduini, D.B., Lima, J.D.V., Monteiro, R., Mendes, B.N.B., Lemos, M.C.F., Noronha, C.P., Saraceni, V., 2018. Neurologic manifestations in emerging arboviral diseases in Rio de Janeiro City, Brazil, 2015-2016. Rev. Soc. Bras. Med. Trop. 51, 347–351. https://doi.org/10.1590/0037-8682-0327-2017.

Azevedo, R.D.S.D.S., Oliveira, C.S., Vasconcelos, P.F.D.C., 2015. Chikungunya risk for Brazil. Rev. Saude Publica 49, 58. https://doi.org/10.1590/S0034-8910.2015049006219.

Azevedo, R.S.S., Silva, E.V.P., Carvalho, V.L., Rodrigues, S.G., Nunes-Neto, J.P., Monteiro, H., Peixoto, V.S., Chiang, J.O., Nunes, M.R.T., Vasconcelos, P.F.C., 2009. Mayaro fever virus, Brazilian Amazon. Emerg. Infect. Dis. 15, 1830–1832. https://doi.org/10.3201/eid1511.090461.

Badawi, A., Ryoo, S.G., Vasileva, D., Yaghoubi, S., 2018. Prevalence of chronic comorbidities in chikungunya: a systematic review and meta-analysis. Int. J. Infect. Dis. 67, 107–113. https://doi.org/10.1016/j.ijid.2017.12.018.

Balavoine, S., Pircher, M., Hoen, B., Herrmann-Storck, C., Najioullah, F., Madeux, B., Signate, A., Valentino, R., Lannuzel, A., Saint Louis, M., Cassadou, S., Cabie, A., Schepers, K., 2017. Guillain-Barré syndrome and chikungunya: description of all cases diagnosed during the 2014 outbreak in the french West Indies. Am. J. Trop. Med. Hyg. 97, 356–360. https://doi.org/10.4269/ajtmh.15-0753.

Bantle, C.M., Phillips, A.T., Smeyne, R.J., Rocha, S.M., Olson, K.E., Tjalkens, R.B., 2019. Infection with mosquito-borne alphavirus induces selective loss of dopaminergic neurons, neuroinflammation and widespread protein aggregation. NPJ Parkinsons Dis. 5, 20. https://doi.org/10.1038/s41531-019-0090-8.

Barr, K.L., Khan, E., Farooqi, J.Q., Imtiaz, K., Prakoso, D., Malik, F., Lednicky, J.A., Long, M.T., 2018. Evidence of chikungunya virus disease in Pakistan since 2015 with patients demonstrating involvement of the central nervous system. Front. Public Health 6, 186. https://doi.org/10.3389/fpubh.2018.00186.

Bartelloni, P.J., McKinney, R.W., Duffy, T.P., Cole, F.E., 1970. An inactivated eastern equine encephalomyelitis vaccine propagated in chick-embryo cell culture. II. Clinical and serologic responses in man. Am. J. Trop. Med. Hyg. 19, 123–126. https://doi.org/10.4269/ajtmh.1970.19.123.

Baxter, V.K., Glowinski, R., Braxton, A.M., Potter, M.C., Slusher, B.S., Griffin, D.E., 2017. Glutamine antagonist-mediated immune suppression decreases pathology but delays virus clearance in mice during nonfatal alphavirus encephalomyelitis. Virology 508, 134–149. https://doi.org/10.1016/j.virol.2017.05.013.

Baxter, V.K., Griffin, D.E., 2016. Interferon gamma modulation of disease manifestation and the local antibody response to alphavirus encephalomyelitis. J. Gen. Virol. 97, 2908–2925. https://doi.org/10.1099/jgv.0.000613.

Baxter, V.K., Griffin, D.E., 2020. Interferon-gamma modulation of the local T cell response to alphavirus encephalomyelitis. Viruses 12, 113–125. https://doi.org/10.3390/v12010113.

Baxter, V.K., Troisi, E.M., Pate, N.M., Zhao, J.N., Griffin, D.E., 2018. Death and gastrointestinal bleeding complicate encephalomyelitis in mice with delayed appearance of CNS IgM after intranasal alphavirus infection. J. Gen. Virol. 99, 309–320. https://doi.org/10.1099/jgv.0.001005.

Berger, M.L., 1980. Humoral and cell-mediated immune mechanisms in the production of pathology in avirulent Semliki Forest virus encephalitis. Infect. Immun. 30, 244–253.

Binder, G.K., Griffin, D.E., 2001. Interferon-gamma-mediated site-specific clearance of alphavirus from CNS neurons. Science 293, 303–306. https://doi.org/10.1126/science.1059742.

Blackman, M.J., Morris, A.G., 1984. Gamma interferon production and cytotoxicity of spleen cells from mice infected with Semliki Forest virus. J. Gen. Virol. 65, 955–961. https://doi.org/10.1099/0022-1317-65-5-955.

Blakely, P.K., Delekta, P.C., Miller, D.J., Irani, D.N., 2015. Manipulation of host factors optimizes the pathogenesis of western equine encephalitis virus infections in mice for antiviral drug development. J. Neurovirol. 21, 43–55. https://doi.org/10.1007/s13365-014-0297-8.

Bocan, T.M., Stafford, R.G., Brown, J.L., Akuoku Frimpong, J., Basuli, F., Hollidge, B.S., Zhang, X., Raju, N., Swenson, R.E., Smith, D.R., 2019. Characterization of brain inflammation, apoptosis, hypoxia, blood-brain barrier integrity and metabolism in venezuelan Equine Encephalitis Virus (VEEV TC-83) exposed mice by in vivo positron emission tomography imaging. Viruses 11, 1052. https://doi.org/10.3390/v11111052.

Bonifay, T., Lienne, J.-F., Bagoée, C., Santa, F., Vesin, G., Walter, G., Nacher, M., Vaserman, N., Djossou, F., Epelboin, L., 2018. Prevalence and risk factors of post chikungunya rheumatic musculoskeletal disorders: a prospective follow-up study in French Guiana. Eur. J. Clin. Microbiol. Infect. Dis. 37, 2159–2164. https://doi.org/10.1007/s10096-018-3353-0.

Booss, J., Esiri, M.M., 2003. The Arboviruses. In: Viral Encephalitis in Humans. American Society for Microbiology, pp. 163–189.

Borgherini, G., Poubeau, P., Jossaume, A., Gouix, A., Cotte, L., Michault, A., Arvin Berod, C., Paganin, F., 2008. Persistent arthralgia associated with chikungunya virus: a study of 88 adult patients on reunion island. Clin. Infect. Dis. 47, 469–475. https://doi.org/10.1086/590003.

Bouquillard, É., Combe, B., 2009. A report of 21 cases of rheumatoid arthritis following Chikungunya fever. A mean follow-up of two years. Joint Bone Spine 76, 654–657. https://doi.org/10.1016/j.jbspin.2009.08.005.

Bourque, P.R., Chardon, J.W., Massie, R., 2015. Autoimmune peripheral neuropathies. Clin. Chim. Acta 449, 37–42. https://doi.org/10.1016/j.cca.2015.02.039.

Bowen, G.S., Fashinell, T.R., Dean, P.B., Gregg, M.B., 1976. Clinical aspects of human Venezuelan equine encephalitis in Texas. Bull. Pan Am. Health Organ. 10, 46–57.

Braga, J.U., Bressan, C., Dalvi, A.P.R., Calvet, G.A., Daumas, R.P., Rodrigues, N., Wakimoto, M., Nogueira, R.M.R., Nielsen-Saines, K., Brito, C., Bispo de Filippis, A.M., Brasil, P., 2017. Accuracy of Zika virus disease case definition during simultaneous Dengue and Chikungunya epidemics. PLoS One 12, e0179725. https://doi.org/10.1371/journal.pone.0179725.

Broeckel, R.M., Haese, N., Ando, T., Dmitriev, I., Kreklywich, C.N., Powers, J., Denton, M., Smith, P., Morrison, T.E., Heise, M., DeFilippis, V., Messaoudi, I., Curiel, D.T., Streblow, D.N., 2019. Vaccine-induced skewing of T cell responses protects against chikungunya virus disease. Front. Immunol. 10, 2563. https://doi.org/10.3389/fimmu.2019.02563.

Brooke, C.B., Deming, D.J., Whitmore, A.C., White, L.J., Johnston, R.E., 2010. T cells facilitate recovery from Venezuelan equine encephalitis virus-induced encephalomyelitis in the absence of antibody. J. Virol. 84, 4556–4568. https://doi.org/10.1128/JVI.02545-09.

Brooke, C.B., Schäfer, A., Matsushima, G.K., White, L.J., Johnston, R.E., 2012. Early activation of the host complement system is required to restrict central nervous system invasion and limit neuropathology during Venezuelan equine encephalitis virus infection. J. Gen. Virol. 93, 797–806. https://doi.org/10.1099/vir.0.038281-0.

Brummer-Korvenkontio, M., Vapalahti, O., Kuusisto, P., Saikku, P., Manni, T., Koskela, P., Nygren, T., Brummer-Korvenkontio, H., Vaheri, A., 2002. Epidemiology of Sindbis virus infections in Finland 1981-96: possible factors explaining a peculiar disease pattern. Epidemiol. Infect. 129, 335–345. https://doi.org/10.1017/s0950268802007409.

Bruyn, H.B., Lennette, E.H., 1953. Western equine encephalitis in infants; a report on three cases with sequelae. Calif. Med. 79, 362–366.

Burdeinick-Kerr, R., Wind, J., Griffin, D.E., 2007. Synergistic roles of antibody and interferon in noncytolytic clearance of Sindbis virus from different regions of the central nervous system. J. Virol. 81, 5628–5636. https://doi.org/10.1128/JVI.01152-06.

Burrack, K.S., Montgomery, S.A., Homann, D., Morrison, T.E., 2015. CD8+ T cells control Ross River virus infection in musculoskeletal tissues of infected mice. J. Immunol. 194, 678–689. https://doi.org/10.4049/jimmunol.1401833.

Burt, F.J., Chen, W., Miner, J.J., Lenschow, D.J., Merits, A., Schnettler, E., Kohl, A., Rudd, P.A., Taylor, A., Herrero, L.J., Zaid, A., Ng, L.F.P., Mahalingam, S., 2017. Chikungunya virus: an update on the biology and pathogenesis of this emerging pathogen. Lancet Infect. Dis. 17, e107–e117. https://doi.org/10.1016/S1473-3099(16)30385-1.

Butt, A.M., Tutton, M.G., Kirvell, S.L., Amor, S., Jenkins, H.G., 1996. Morphology of oligodendrocytes during demyelination in optic nerves of mice infected with Semliki Forest virus. Neuropathol. Appl. Neurobiol. 22, 540–547. https://doi.org/10.1111/j.1365-2990.1996.tb01133.x.

Byrnes, A.P., Durbin, J.E., Griffin, D.E., 2000. Control of Sindbis virus infection by antibody in interferon-deficient mice. J. Virol. 74, 3905–3908.

Cain, M.D., Salimi, H., Gong, Y., Yang, L., Hamilton, S.L., Heffernan, J.R., Hou, J., Miller, M.J., Klein, R.S., 2017. Virus entry and replication in the brain precedes blood-brain barrier disruption during intranasal alphavirus infection. J. Neuroimmunol. 308, 118–130. https://doi.org/10.1016/j.jneuroim.2017.04.008.

Calisher, C.H., 1994. Medically important arboviruses of the United States and Canada. Clin. Microbiol. Rev. 7, 89–116.

Capasso, A., Ompad, D.C., Vieira, D.L., Wilder-Smith, A., Tozan, Y., 2019. Incidence of Guillain-Barré Syndrome (GBS) in Latin America and the Caribbean before and during the 2015-2016 Zika virus epidemic: A systematic review and meta-analysis. PLoS Negl. Trop. Dis. 13, e0007622. https://doi.org/10.1371/journal.pntd.0007622.

Carmen, J., Rothstein, J.D., Kerr, D.A., 2009. Tumor necrosis factor-α modulates glutamate transport in the CNS and is a critical determinant of outcome from viral encephalomyelitis. Brain Res. 1263, 143–154. https://doi.org/10.1016/j.brainres.2009.01.040.

Carrera, J.-P., Forrester, N., Wang, E., Vittor, A.Y., Haddow, A.D., López-Vergès, S., Abadía, I., Castaño, E., Sosa, N., Báez, C., Estripeaut, D., Díaz, Y., Beltrán, D., Cisneros, J., Cedeño, H.G., Travassos da Rosa, A.P., Hernandez, H., Martínez-Torres, A.O., Tesh, R.B., Weaver, S.C., 2013. Eastern Equine Encephalitis in Latin America. N. Engl. J. Med. 369, 732–744. https://doi.org/10.1056/NEJMoa1212628.

Carvalho, K., Biancardi, A.L., Provenzano, G., Moraes, H., 2020. Acute disseminated encephalomyelitis (ADEM) associated with mosquito-borne diseases: chikungunya virus X yellow fever immunization. Rev. Soc. Bras. Med. Trop. 53, e20190160. https://doi.org/10.1590/0037-8682-0160-2019.

Centers for Disease Control and Prevention, National Center for Emerging Zoonotic Infectious Diseases, Division of Vector-Borne Diseases, 2019. Eastern Equine Encephalitis [WWW Document]. Centers for Disease Control and Prevention. https://www.cdc.gov/easternequineencephalitis/tech/epi.html. accessed 2.14.20.

Cerny, T., Schwarz, M., Schwarz, U., Lemant, J., Gérardin, P., Keller, E., 2017. The range of neurological complications in Chikungunya fever. Neurocrit. Care 27, 447–457. https://doi.org/10.1007/s12028-017-0413-8.

Chaaitanya, I.K., Muruganandam, N., Sundaram, S.G., Kawalekar, O., Sugunan, A.P., Manimunda, S.P., Ghosal, S.R., Muthumani, K., Vijayachari, P., 2011. Role of proinflammatory cytokines and chemokines in chronic arthropathy in CHIKV infection. Viral Immunol. 24, 265–271. https://doi.org/10.1089/vim.2010.0123.

Chanana, B., Azad, R.V., Nair, S., 2007. Bilateral macular choroiditis following Chikungunya virus infection. Eye 21, 1020–1021. https://doi.org/10.1038/sj.eye.6702863.

Chang, A.Y., Encinales, L., Porras, A., Pachecho, N., Reid, S.P., Martins, K.A.O., Pacheco, S., Bravo, E., Navarno, M., Rico, A., Amdur, R., Kamalopathy, P., Firestein, G.S., Bethony, J.M., Simon, G.L., 2017. Frequency of chronic joint pain following chikungunya infection: a colombian cohort study. Arthritis Rheumatol. 70, 578–584. https://doi.org/10.1002/art.40384.

Chang, A.Y., Martins, K.A.O., Encinales, L., Reid, S.P., Acuña, M., Encinales, C., Matranga, C.B., Pacheco, N., Cure, C., Shukla, B., Ruiz Arteta, T., Amdur, R., Cazares, L.H., Gregory, M., Ward, M.D., Porras, A., Rico-Mendoza, A., Dong, L., Kenny, T., Brueggemann, E., Downey, L.G., Kamalapathy, P., Lichtenberger, P., Falls, O., Simon, G.L., Bethony, J.M., Firestein, G.S., 2018. Chikungunya arthritis mechanisms in the Americas. Arthritis Rheumatol. 70, 585–593. https://doi.org/10.1002/art.40383.

Charles, P.C., Trgovcich, J., Davis, N.L., Johnston, R.E., 2001. Immunopathogenesis and immune modulation of Venezuelan equine encephalitis virus-induced disease in the mouse. Virology 284, 190–202. https://doi.org/10.1006/viro.2001.0878.

Charles, P.C., Walters, E., Margolis, F., Johnston, R.E., 1995. Mechanism of neuroinvasion of Venezuelan equine encephalitis virus in the mouse. Virology 208, 662–671. https://doi.org/10.1006/viro.1995.1197.

Chen, W., Foo, S.-S., Li, R.W., Smith, P.N., Mahalingam, S., 2014. Osteoblasts from osteoarthritis patients show enhanced susceptibility to Ross River virus infection associated with delayed type I interferon responses. Virol. J. 11, 189. https://doi.org/10.1186/s12985-014-0189-9.

Chen, W., Foo, S.-S., Taylor, A., Lulla, A., Merits, A., Hueston, L., Forwood, M.R., Walsh, N.C., Sims, N.A., Herrero, L.J., Mahalingam, S., 2015. Bindarit, an inhibitor of monocyte chemotactic protein synthesis, protects against bone loss induced by chikungunya virus infection. J. Virol. 89, 581–593. https://doi.org/10.1128/JVI.02034-14.

Chen, W., Foo, S.-S., Zaid, A., Teng, T.-S., Herrero, L.J., Wolf, S., Tharmarajah, K., Vu, L.D., van Vreden, C., Taylor, A., Freitas, J.R., Li, R.W., Woodruff, T.M., Gordon, R., Ojcius, D.M., Nakaya, H.I., Kanneganti, T.-D., O'Neill, L.A.J., Robertson, A.A.B., King, N.J., Suhrbier, A., Cooper, M.A., Ng, L.F.P., Mahalingam, S., 2017. Specific inhibition of NLRP3 in chikungunya disease reveals a role for inflammasomes in alphavirus-induced inflammation. Nat. Microbiol. 2, 1435–1445. https://doi.org/10.1038/s41564-017-0015-4.

Chirathaworn, C., Poovorawan, Y., Lertmaharit, S., Wuttirattanakowit, N., 2013. Cytokine levels in patients with chikungunya virus infection. Asian Pac. J. Trop. Med. 6, 631–634. https://doi.org/10.1016/S1995-7645(13)60108-X.

Chirathaworn, C., Rianthavorn, P., Wuttirattanakowit, N., Poovorawan, Y., 2010. Serum IL-18 and IL-18BP levels in patients with Chikungunya virus infection. Viral Immunol. 23, 113–117. https://doi.org/10.1089/vim.2009.0077.

Chopra, A., Saluja, M., Venugopalan, A., 2014. Effectiveness of chloroquine and inflammatory cytokine response in patients with early persistent musculoskeletal pain and arthritis following chikungunya virus infection. Arthritis Rheumatol. 66, 319–326. https://doi.org/10.1002/art.38221.

Choudhary, N., Makhija, P., Puri, V., Khwaja, G.A., Duggal, A., 2016. An unusual case of myelitis with myositis. J. Clin. Diagn. Res. 10, 19–20. https://doi.org/10.7860/JCDR/2016/13095.7777.

Chow, A., Her, Z., Ong, E.K.S., Chen, J.-M., Dimatatac, F., Kwek, D.J.C., Barkham, T., Yang, H., Renia, L., Leo, Y.-S., Ng, L.F.P., 2011. Persistent arthralgia induced by Chikungunya virus infection is associated with interleukin-6 and granulocyte macrophage colony-stimulating factor. J. Infect. Dis. 203, 149–157. https://doi.org/10.1093/infdis/jiq042.

Chua, H.H., Abdul Rashid, K., Law, W.C., Hamizah, A., Chem, Y.K., Khairul, A.H., Chua, K.B., 2010. A fatal case of chikungunya virus infection with liver involvement. Med. J. Malaysia 65, 83–84.

Clarris, B.J., Doherty, R.L., Fraser, J.R.E., French, E.L., Muirden, K.D., 1975. Epidemic polyarthritis: a cytological, virological, and immunochemical study. Aust. N. Z. J. Med. 5, 450–457.

Contopoulos-Ioannidis, D., Newman-Lindsay, S., Chow, C., Labeaud, A.D., 2018. Mother-to-child transmission of Chikungunya virus: a systematic review and meta-analysis. PLoS Negl. Trop. Dis. 12, e0006510. https://doi.org/10.1371/journal.pntd.0006510.

Couderc, T., Chrétien, F., Schilte, C., Disson, O., Brigitte, M., Guivel-Benhassine, F., Touret, Y., Barau, G., Cayet, N., Schuffenecker, I., Desprès, P., Arenzana-Seisdedos, F., Michault, A., Albert, M.L., Lecuit, M., 2008. A mouse model for Chikungunya: young age and inefficient type-I interferon signaling are risk factors for severe disease. PLoS Pathog. 4, e29. https://doi.org/10.1371/journal.ppat.0040029.

Couzigou, B., Criquet-Hayot, A., Javelle, E., Tignac, S., Mota, E., Rigaud, F., Alain, A., Troisgros, O., Pierre-Francois, S., Abel, S., Banydeen, R., Cabie, A., 2018. Occurrence of chronic stage chikungunya in the general population of martinique during the first 2014 epidemic: a prospective epidemiological study. Am. J. Trop. Med. Hyg. 99, 182–190. https://doi.org/10.4269/ajtmh.17-0543.

Crosby, L., Perreau, C., Madeux, B., Cossic, J., Armand, C., Herrmann-Storke, C., Najioullah, F., Valentino, R., Thiéry, G., 2016. Severe manifestations of chikungunya virus in critically ill patients during the 2013-2014 Caribbean outbreak. Int. J. Infect. Dis. 48, 78–80. https://doi.org/10.1016/j.ijid.2016.05.010.

Dal Canto, M.C., Rabinowitz, S.G., 1981. Central nervous system demyelination in Venezuelan equine encephalomyelitis infection. J. Neurol. Sci. 49, 397–418. https://doi.org/10.1016/0022-510x(81)90030-7.

Danes, L., Rychterova, V., Kufner, J., Hruskova, J., 1973. The role of the olfactory route on infection of the respiratory tract with Venezuelan equine encephalomyelitis virus in normal and operated Macaca rhesus monkeys. II. Results of histological examination. Acta Virol. 17, 57–60.

Darman, J., Backovic, S., Dike, S., Maragakis, N.J., Krishnan, C., Rothstein, J.D., Irani, D.N., Kerr, D.A., 2004. Viral-induced spinal motor neuron death is non-cell-autonomous and involves glutamate excitotoxicity. J. Neurosci. 24, 7566–7575. https://doi.org/10.1523/JNEUROSCI.2002-04.2004.

Davis, L.E., DeBiasi, R., Goade, D.E., Haaland, K.Y., Harrington, J.A., Harnar, J.B., Pergam, S.A., King, M.K., DeMasters, B.K., Tyler, K.L., 2006. West Nile virus neuroinvasive disease. Ann. Neurol. 60, 286–300. https://doi.org/10.1002/ana.20959.

Davis, N.L., Grieder, F.B., Smith, J.F., Greenwald, G.F., Valenski, M.L., Sellon, D.C., Charles, P.C., Johnston, R.E., 1994. A molecular genetic approach to the study of Venezuelan equine encephalitis virus pathogenesis. Arch. Virol. Suppl. 9, 99–109. https://doi.org/10.1007/978-3-7091-9326-6_11.

de Castro-Jorge, L.A., de Carvalho, R.V.H., Klein, T.M., Hiroki, C.H., Lopes, A.H., Guimarães, R.M., Fumagalli, M.J., Floriano, V.G., Agostinho, M.R., Slhessarenko, R.D., Ramalho, F.S., Cunha, T.M., Cunha, F.Q., da Fonseca, B.A.L., Zamboni, D.S., 2019. The NLRP3 inflammasome is involved with the pathogenesis of Mayaro virus. PLoS Pathog. 15, e1007934. https://doi.org/10.1371/journal.ppat.1007934.

de Souza, C.G., da Costa, J.F., de Sousa Dantas, D., de Abreu Freitas, R.P., Lopes, J.M., Okano, A.H., 2019. Evaluation of pain, functional capacity and kinesiophobia in women in the chronic stage of chikungunya virus infection: a cross-sectional study in northeastern Brazil. Acta Trop. 199, 104853. https://doi.org/10.1016/j.actatropica.2018.12.008.

Deaton, A.V., Craft, S., Skenazy, J., 1986. Enduring psychiatric and neuropsychologic sequelae in the post-encephalitis patient. Int. J. Psychiatry Med. 16, 275–280. https://doi.org/10.2190/rv2q-ukfw-uy9l-5d8h.

Deeba, I.M., Hasan, M.M., Mosabbir, A.A., Siam, M.H.B., Islam, M.S., Raheem, E., Hossain, M.S., 2019. Manifestations of atypical symptoms of chikungunya during the Dhaka outbreak (2017) in Bangladesh. Am. J. Trop. Med. Hyg. 100, 1545–1548. https://doi.org/10.4269/ajtmh.19-0122.

Del Carpio-Orantes, L., Pola-Ramirez, M.R., Garcia-Mendez, S., Mata-Miranda, M.P., Perfecto-Arroyo, M.A., Solis-Sanchez, I., Trujillo-Ortega, B.M., Gonzalez-Flores, E.E., 2018. The most frequent causative agents of Guillain-Barre syndrome in a Mexican general hospital. Rev. Neurol. 67, 203–209.

Del Piero, F., Wilkins, P.A., Dubovi, E.J., Biolatti, B., Cantile, C., 2001. Clinical, pathologic, immunohistochemical, and virologic findings of eastern equine encephalomyelitis in two horses. Vet. Pathol. 38, 451–456.

Delgado Enciso, I., Paz Michel, B., Melnikov, V., Guzmán-Esquivel, J., Espinoza Gomez, F., Soriano-Hernandez, A.D., Rodriguez-Sanchez, I.P., Martinez-Fierro, M.L., Ceja Espiritu, G., Olmedo-Buenrostro, B.A., Galvan-Salazar, H.R., Delgado-Enciso, O.G., Delgado Enciso, J., Lopez-Lemus, U.A., Montes-Galindo, D.A., 2018. Smoking and female sex as key risk factors associated with severe arthralgia in acute and chronic phases of Chikungunya virus infection. Exp. Ther. Med. 15, 2634–2642. https://doi.org/10.3892/etm.2017.5668.

Deresiewicz, R.L., Thaler, S.J., Hsu, L., Zamani, A.A., 1997. Clinical and neuroradiographic manifestations of eastern equine encephalitis. N. Engl. J. Med. 336, 1867–1874. https://doi.org/10.1056/NEJM199706263362604.

Dias, C.N.d.S., Gois, B.M., Lima, V.S., Guerra-Gomes, I.C., Araújo, J.M.G., Gomes, J.d.A.S., Araújo, D.A.M., Medeiros, I.A., de Azevedo, F.d.L.A.A., Veras, R.C., Janebro, D.I., Amaral, I.P.G.D., Keesen, T.S.L., 2018. Human CD8 T-cell activation in acute and chronic chikungunya infection. Immunology 155, 499–504. https://doi.org/10.1111/imm.12992.

Dietz, W.H., Alvarez, O., Martin, D.H., Walton, T.E., Ackerman, L.J., Johnson, K.M., 1978. Enzootic and epizootic Venezuelan equine encephalomyelitis virus in horses infected by peripheral and intrathecal routes. J. Infect. Dis. 137, 227–237. https://doi.org/10.1093/infdis/137.3.227.

Dirlikov, E., Major, C.G., Mayshack, M., Medina, N., Matos, D., Ryff, K.R., Torres-Aponte, J., Alkis, R., Munoz-Jordan, J., Colon-Sanchez, C., Salinas, J.L., Pastula, D.M., Garcia, M., Segarra, M.O., Malave, G., Thomas, D.L., Rodríguez-Vega, G.M., Luciano, C.A., Sejvar, J., Sharp, T.M., Rivera-Garcia, B., 2016. Guillain-Barré Syndrome during ongoing Zika Virus Transmission - Puerto Rico, January 1-July 31, 2016. Morb. Mortal. Wkly. Rep. 65, 910–914. https://doi.org/10.15585/mmwr.mm6534e1.

Duvignaud, A., Fianu, A., Bertolotti, A., Jaubert, J., Michault, A., Poubeau, P., Fred, A., Méchain, M., Gauzere, B.A., Favier, F., Malvy, D., Gérardin, P., 2018. Rheumatism and chronic fatigue, the two facets of post-chikungunya disease: the TELECHIK cohort study on Reunion island. Epidemiol. Infect. 146, 633–641. https://doi.org/10.1017/S0950268818000031.

Earnest, M.P., Goolishian, H.A., Calverley, J.R., Hayes, R.O., Hill, H.R., 1971. Neurologic, intellectual, and psychologic sequelae following western encephalitis. A follow-up study of 35 cases. Neurology 21, 969–974.

Economopoulou, A., Dominguez, M., Helynck, B., Sissoko, D., Wichmann, O., Quenel, P., Germonneau, P., Quatresous, I., 2009. Atypical Chikungunya virus infections: clinical manifestations, mortality and risk factors for severe disease during the 2005-2006 outbreak on Réunion. Epidemiol. Infect. 137, 534–541. https://doi.org/10.1017/S0950268808001167.

Elsinga, J., Gerstenbluth, I., van der Ploeg, S., Halabi, Y., Lourents, N.T., Burgerhof, J.G., van der Veen, H.T., Bailey, A., Grobusch, M.P., Tami, A., 2017. Long-term chikungunya sequelae in curaçao: burden, determinants, and a novel classification tool. J. Infect. Dis. 216, 573–581. https://doi.org/10.1093/infdis/jix312.

Elsinga, J., Halabi, Y., Gerstenbluth, I., Tami, A., Grobusch, M.P., 2018. Consequences of a recent past dengue infection for acute and long-term chikungunya outcome: a retrospective cohort study in Curaçao. Travel Med. Infect. Dis. 23, 34–43. https://doi.org/10.1016/j.tmaid.2018.03.008.

Endy, T.P., 2020. Viral febrile illnesses and emerging pathogens. In: Ryan, E.T., Hill, D.R., Solomon, T., Endy, T.P., Aronson, N. (Eds.), Hunters Tropical Medicine and Emerging Infectious Disease, pp. 325–360. https://doi.org/10.1016/B978-0-323-55512-8.00036-3.

Erälinna, J.P., Soilu-Hänninen, M., Röyttä, M., Hukkanen, V., Salmi, A.A., Salonen, R., 1996. Blood-brain barrier breakdown and increased intercellular adhesion molecule (ICAM-1/CD54) expression after Semliki Forest (A7) virus infection facilitates the development of experimental allergic encephalomyelitis. J. Neuroimmunol. 66, 103–114. https://doi.org/10.1016/0165-5728(96)00031-8.

Farias, L.A.B.G., Ferragut, J.M., Pires Neto, R.D.J., 2018. Encephalitis and transverse myelitis in dengue and chikungunya coinfection. Rev. Soc. Bras. Med. Trop. 51, 403. https://doi.org/10.1590/0037-8682-0259-2017.

Fazakerley, J.K., Amor, S., Webb, H.E., 1983. Reconstitution of Semliki forest virus infected mice, induces immune mediated pathological changes in the CNS. Clin. Exp. Immunol. 52, 115–120.

Fazakerley, J.K., Cotterill, C.L., Lee, G., Graham, A., 2006. Virus tropism, distribution, persistence and pathology in the corpus callosum of the Semliki Forest virus-infected mouse brain: a novel system to study virus-oligodendrocyte interactions. Neuropathol. Appl. Neurobiol. 32, 397–409. https://doi.org/10.1111/j.1365-2990.2006.00739.x.

Fazakerley, J.K., Khalili-Shirazi, A., Webb, H.E., 1988. Semliki Forest virus (A7[74]) infection of adult mice induces an immune-mediated demyelinating encephalomyelitis. Ann. N. Y. Acad. Sci. 540, 672–673. https://doi.org/10.1111/j.1749-6632.1988.tb27208.x.

Fazakerley, J.K., Pathak, S., Scallan, M., Amor, S., Dyson, H., 1993. Replication of the A7(74) strain of Semliki Forest virus is restricted in neurons. Virology 195, 627–637. https://doi.org/10.1006/viro.1993.1414.

Fazakerley, J.K., Webb, H.E., 1987a. Semliki Forest virus induced, immune mediated demyelination: the effect of irradiation. Br. J. Exp. Pathol. 68, 101–113.

Fazakerley, J.K., Webb, H.E., 1987b. Semliki Forest virus-induced, immune-mediated demyelination: adoptive transfer studies and viral persistence in nude mice. J. Gen. Virol. 68, 377–385. https://doi.org/10.1099/0022-1317-68-2-377.

Fine, D.L., Roberts, B.A., Terpening, S.J., Mott, J., Vasconcelos, D., House, R.V., 2008. Neurovirulence evaluation of Venezuelan equine encephalitis (VEE) vaccine candidate V3526 in nonhuman primates. Vaccine 26, 3497–3506. https://doi.org/10.1016/j.vaccine.2008.04.044.

Finley, K.H., Longshore, W.A., Palmer, R.J., Cook, R.E., Riggs, N., 1955. Western equine and St. Louis encephalitis; preliminary report of a clinical follow-up study in California. Neurology 5, 223–235.

Forechi, L., Silveira-Nunes, G., Barbosa, M.A., Barbosa, É.G., Santos, D.L.D., Vieira, E.R., Barbosa, A.C., 2018. Pain, balance, grip strength and gait parameters of older adults with and without post-chikungunya chronic arthralgia. Trop. Med. Int. Health 23, 1394–1400. https://doi.org/10.1111/tmi.13154.

Forrester, N.L., Palacios, G., Tesh, R.B., Savji, N., Guzman, H., Sherman, M., Weaver, S.C., Lipkin, W.I., 2012. Genome-scale phylogeny of the alphavirus genus suggests a marine origin. J. Virol. 86, 2729–2738. https://doi.org/10.1128/JVI.05591-11.

Fragkoudis, R., Ballany, C.M., Boyd, A., Fazakerley, J.K., 2008. In Semliki Forest virus encephalitis, antibody rapidly clears infectious virus and is required to eliminate viral material from the brain, but is not required to generate lesions of demyelination. J. Gen. Virol. 89, 2565–2568. https://doi.org/10.1099/vir.0.2008/002238-0.

Fragkoudis, R., Dixon-Ballany, C.M., Zagrajek, A.K., Kedzierski, L., Fazakerley, J.K., 2018. Following acute encephalitis, Semliki Forest virus is undetectable in the brain by infectivity assays but functional virus RNA capable of generating infectious virus persists for life. Viruses 10, 273. https://doi.org/10.3390/v10050273.

Fragkoudis, R., Tamberg, N., Siu, R., Kiiver, K., Kohl, A., Merits, A., Fazakerley, J.K., 2009. Neurons and oligodendrocytes in the mouse brain differ in their ability to replicate Semliki Forest virus. J. Neurovirol. 15, 57–70. https://doi.org/10.1080/13550280802482583.

Franklin, R.P., Kinde, H., Jay, M.T., Kramer, L.D., Green, E.-G.N., Chiles, R.E., Ostlund, E., Husted, S., Smith, J., Parker, M.D., 2002. Eastern equine encephalomyelitis virus infection in a horse from California. Emerg. Infect. Dis. 8, 283–288. https://doi.org/10.3201/eid0803.010199.

Fraser, J.R., Becker, G.J., 1984. Mononuclear cell types in chronic synovial effusions of Ross River virus disease. Aust. N. Z. J. Med. 14, 505–506. https://doi.org/10.1111/j.1445-5994.1984.tb03629.x.

Fraser, J.R., Cunningham, A.L., Clarris, B.J., Aaskov, J.G., Leach, R., 1981. Cytology of synovial effusions in epidemic polyarthritis. Aust. N. Z. J. Med. 11, 168–173.

Fulton, J.S., Burton, A.N., 1953. After effects of western equine encephalomyelitis infection in man. Can. Med. Assoc. J. 69, 268–272.

Ganesan, K., Diwan, A., Shankar, S.K., Desai, S.B., Sainani, G.S., Katrak, S.M., 2008. Chikungunya encephalomyeloradiculitis: report of 2 cases with neuroimaging and 1 case with autopsy findings. Am. J. Neuroradiol. 29, 1636–1637. https://doi.org/10.3174/ajnr.A1133.

Ganjian, N., Riviere-Cinnamond, A., 2020. Mayaro virus in Latin America and the Caribbean. Rev. Panam. Salud Publica 44, e14. https://doi.org/10.26633/RPSP.2020.14.

Ganu, M.A., Ganu, A.S., 2011. Post-chikungunya chronic arthritis—our experience with DMARDs over two year follow up. J. Assoc. Physicians India 59, 83–86.

Gardner, C.L., Burke, C.W., Higgs, S.T., Klimstra, W.B., Ryman, K.D., 2012. Interferon-alpha/beta deficiency greatly exacerbates arthritogenic disease in mice infected with wild-type chikungunya virus but not with the cell culture-adapted live-attenuated 181/25 vaccine candidate. Virology 425, 103–112. https://doi.org/10.1016/j.virol.2011.12.020.

Gardner, C.L., Burke, C.W., Tesfay, M.Z., Glass, P.J., Klimstra, W.B., Ryman, K.D., 2008. Eastern and Venezuelan equine encephalitis viruses differ in their ability to infect dendritic cells and macrophages: impact of altered cell tropism on pathogenesis. J. Virol. 82, 10634–10646. https://doi.org/10.1128/JVI.01323-08.

Gardner, J., Anraku, I., Le, T.T., Larcher, T., Major, L., Roques, P., Schroder, W.A., Higgs, S., Suhrbier, A., 2010. Chikungunya virus Arthritis in adult wild-type mice. J. Virol. 84, 8021–8032. https://doi.org/10.1128/JVI.02603-09.

Gates, M.C., Sheahan, B.J., Atkins, G.J., 1984. The pathogenicity of the M9 mutant of Semliki Forest virus in immune-compromised mice. J. Gen. Virol. 65, 73–80. https://doi.org/10.1099/0022-1317-65-1-73.

Gauri, L.A., Thaned, A., Fatima, Q., Yadav, H., Singh, A., Jaipal, H.P., Chaudhary, A., 2016. Clinical spectrum of chikungunya in Bikaner (North Western India) in 2006 and follow up of patients for five years. J. Assoc. Physicians India 64, 22–25.

Gérardin, P., Barau, G., Michault, A., Bintner, M., Randrianaivo, H., Choker, G., Lenglet, Y., Touret, Y., Bouveret, A., Grivard, P., Le Roux, K., Blanc, S., Schuffenecker, I., Couderc, T., Arenzana-Seisdedos, F., Lecuit, M., Robillard, P.-Y., 2008. Multidisciplinary prospective study of mother-to-child chikungunya virus infections on the island of La Réunion. PLoS Med. 5, e60. https://doi.org/10.1371/journal.pmed.0050060.

Gérardin, P., Fianu, A., Malvy, D., Mussard, C., Boussaïd, K., Rollot, O., Michault, A., Gauzere, B.A., Bréart, G., Favier, F., 2011. Perceived morbidity and community burden after a Chikungunya outbreak: the TELECHIK survey, a population-based cohort study. BMC Med. 9, 5. https://doi.org/10.1186/1741-7015-9-5.

Gleiser, C.A., Gochenour, W.S., Berge, T.O., TIGERTT, W.D., 1962. The comparative pathology of experimental Venezuelan equine encephalomyelitis infection in different animal hosts. J. Infect. Dis. 110, 80–97. https://doi.org/10.1093/infdis/110.1.80.

Go, Y.Y., Balasuriya, U.B.R., Lee, C.-K., 2014. Zoonotic encephalitides caused by arboviruses: transmission and epidemiology of alphaviruses and flaviviruses. Clin. Exp. Vaccine Res. 3, 58–77. https://doi.org/10.7774/cevr.2014.3.1.58.

González-Galván, L.M., Arredondo-Ruiz, P., Chac-Lezama, G., Molina-Villena, A.A., Saenz-Castro, M., Perez-Garcia, J.A., Alonso-Martinez, D., Ramirez-Valadez, C., Pena-Perez, C.A., 2017. Transverse myelitis associated to chikungunya virus infection. Med. Int. Mex. 33, 109–115.

Gorelkin, L., Jahrling, P.B., 1975. Virus-initiated septic shock. Acute death of Venezuelan encephalitis virus-infected hamsters. Lab. Invest. 32, 78–85.

Gray, M.P., Gorelick, M.H., 2016. Acute disseminated encephalomyelitis. Pediatr. Emerg. Care 32, 395–400. https://doi.org/10.1097/PEC.0000000000000825.

Green, M.J., 2003. Serum MMP-3 and MMP-1 and progression of joint damage in early rheumatoid arthritis. Rheumatology 42, 83–88. https://doi.org/10.1093/rheumatology/keg037.

Greene, I.P., Lee, E.-Y., Prow, N., Ngwang, B., Griffin, D.E., 2008. Protection from fatal viral encephalomyelitis: AMPA receptor antagonists have a direct effect on the inflammatory response to infection. Proc. Natl. Acad. Sci. U. S. A. 105, 3575–3580. https://doi.org/10.1073/pnas.0712390105.

Greenlee, J.E., 2014. The equine encephalitides. Handb. Clin. Neurol. 123, 417–432. https://doi.org/10.1016/B978-0-444-53488-0.00019-5.

Grieder, F.B., Davis, B.K., Zhou, X.D., Chen, S.J., Finkelman, F.D., Gause, W.C., 1997. Kinetics of cytokine expression and regulation of host protection following infection with molecularly cloned Venezuelan equine encephalitis virus. Virology 233, 302–312. https://doi.org/10.1006/viro.1997.8617.

Griffin, D.E., 1976. Role of the immune response in age-dependent resistance of mice to encephalitis due to Sindbis virus. J. Infect. Dis. 133, 456–464.

Griffin, D.E., 2013. Alphaviruses. In: Knipe, D.M., Howley, P.M. (Eds.), Fields Virology. Lippincott Williams & Wilkins, Philadelphia, pp. 652–686.

Griffin, D.E., Hardwick, J.M., 1997. Regulators of apoptosis on the road to persistent alphavirus infection. Annu. Rev. Microbiol. 51, 565–592. https://doi.org/10.1146/annurev.micro.51.1.565.

Griffin, D.E., Johnson, R.T., 1977. Role of the immune response in recovery from Sindbis virus encephalitis in mice. J. Immunol. 118, 1070–1075.

Gualberto Cavalcanti, N., MeloVilar, K., Branco Pinto Duarte, A.L., de Melo Rêgo, M.J.B., Cristiny Pereira, M., da Rocha Pitta, I., Diniz Lopes Marques, C., Galdino da Rocha Pitta, M., 2019. IL-27 in patients with Chikungunya fever: a possible chronicity biomarker? Acta Trop. 196, 48–51. https://doi.org/10.1016/j.actatropica.2019.05.005.

Gunn, B.M., Jones, J.E., Shabman, R.S., Whitmore, A.C., Sarkar, S., Blevins, L.K., Morrison, T.E., Heise, M.T., 2018. Ross River virus envelope glycans contribute to disease through activation of the host complement system. Virology 515, 250–260. https://doi.org/10.1016/j.virol.2017.12.022.

Gunn, B.M., Morrison, T.E., Whitmore, A.C., Blevins, L.K., Hueston, L., Fraser, R.J., Herrero, L.J., Ramirez, R., Smith, P.N., Mahalingam, S., Heise, M.T., 2012. Mannose binding lectin is required for alphavirus-induced arthritis/myositis. PLoS Pathog. 8, e1002586. https://doi.org/10.1371/journal.ppat.1002586.

Gupta, P., Sharma, A., Han, J., Yang, A., Bhomia, M., Knollmann-Ritschel, B., Puri, R.K., Maheshwari, R.K., 2017. Differential host gene responses from infection with neurovirulent and partially-neurovirulent strains of Venezuelan equine encephalitis virus. BMC Infect. Dis. 17, 309. https://doi.org/10.1186/s12879-017-2355-3.

Haese, N.N., Broeckel, R.M., Hawman, D.W., Heise, M.T., Morrison, T.E., Streblow, D.N., 2016. Animal models of chikungunya virus infection and disease. J. Infect. Dis. 214, S482–S487. https://doi.org/10.1093/infdis/jiw284.

Hahn, C.S., Lustig, S., Strauss, E.G., Strauss, J.H., 1988. Western equine encephalitis virus is a recombinant virus. Proc. Natl. Acad. Sci. U. S. A. 85, 5997–6001. https://doi.org/10.1073/pnas.85.16.5997.

Hameed, S., Memon, M., Imtiaz, H., Kanwar, D., 2019. Longitudinally extensive transverse myelitis with seropositive chikungunya. BMJ Case Rep. 12. https://doi.org/10.1136/bcr-2019-231745.

Harley, D., Sleigh, A., Ritchie, S., 2001. Ross River virus transmission, infection, and disease: a cross-disciplinary review. Clin. Microbiol. Rev. 14, 909–932. https://doi.org/10.1128/CMR.14.4.909-932.2001.

Havert, M.B., Schofield, B., Griffin, D.E., Irani, D.N., 2000. Activation of divergent neuronal cell death pathways in different target cell populations during neuroadapted sindbis virus infection of mice. J. Virol. 74, 5352–5356. https://doi.org/10.1128/JVI.74.11.5352-5356.2000.

Hawman, D.W., Stoermer, K.A., Montgomery, S.A., Pal, P., Oko, L., Diamond, M.S., Morrison, T.E., 2013. Chronic joint disease caused by persistent chikungunya virus infection is controlled by the adaptive immune response. J. Virol. 87, 13878–13888. https://doi.org/10.1128/JVI.02666-13.

Heath, C.J., Lowther, J., Noël, T.P., Mark-George, I., Boothroyd, D.B., Mitchell, G., MacPherson, C., Desiree LaBeaud, A., 2018. The identification of risk factors for chronic Chikungunya Arthralgia in Grenada, West Indies: a cross-sectional cohort study. Open Forum Infect. Dis. 5, ofx234. https://doi.org/10.1093/ofid/ofx234.

Her, Z., Teng, T.-S., Tan, J.J.L., Teo, T.-H., Kam, Y.-W., Lum, F.-M., Lee, W.W.L., Gabriel, C., Melchiotti, R., Andiappan, A.K., Lulla, V., Lulla, A., Win, M.K., Chow, A., Biswas, S.K., Leo, Y.-S., Lecuit, M., Merits, A., Renia, L., Ng, L.F.P., 2015. Loss of TLR3 aggravates CHIKV replication and pathology due to an altered virus-specific neutralizing antibody response. EMBO Mol. Med. 7, 24–41. https://doi.org/10.15252/emmm.201404459.

Herzon, H., Shelton, J.T., Bruyn, H.B., 1957. Sequelae of western equine and other arthropod-borne encephalitides. Neurology 7, 535–548. https://doi.org/10.1212/wnl.7.8.535.

Hirsch, R.L., Griffin, D.E., Winkelstein, J.A., 1980. Role of complement in viral infections: participation of terminal complement components (C5 to C9) in recovery of mice from Sindbis virus infection. Infect. Immun. 30, 899–901.

Hoarau, J.J., Gay, F., Pellé, O., Samri, A., Jaffar-Bandjee, M.C., Gasque, P., Autran, B., 2013. Identical strength of the T cell responses against E2, nsP1 and capsid CHIKV proteins in recovered and chronic patients after the epidemics of 2005-2006 in La Reunion Island. PLoS One 8, e84695. https://doi.org/10.1371/journal.pone.0084695.

Hoarau, J.J., Jaffar-Bandjee, M.C., Krejbich-Trotot, P., Das, T., Li-Pat-Yuen, G., Dassa, B., Denizot, M., Guichard, E., Ribera, A., Henni, T., Tallet, F., Moiton, M.P., Gauzere, B.A., Bruniquet, S., Jaffar Bandjee, Z., Morbidelli, P., Martigny, G., Jolivet, M., Gay, F., Grandadam, M., Tolou, H., Vieillard, V., Debré, P., Autran, B., Gasque, P., 2010. Persistent chronic inflammation and infection by Chikungunya arthritogenic alphavirus in spite of a robust host immune response. J. Immunol. 184, 5914–5927. https://doi.org/10.4049/jimmunol.0900255.

Hollidge, B.S., González-Scarano, F., Soldan, S.S., 2010. Arboviral encephalitides: transmission, emergence, and pathogenesis. J. Neuroimmune Pharmacol. 5, 428–442. https://doi.org/10.1007/s11481-010-9234-7.

Honnold, S.P., Mossel, E.C., Bakken, R.R., Fisher, D., Lind, C.M., Cohen, J.W., Eccleston, L.T., Spurgers, K.B., Erwin-Cohen, R., Bradfute, S.B., Maheshwari, R.K., Glass, P.J., 2015a. Eastern equine encephalitis virus in mice I: clinical course and outcome are dependent on route of exposure. Virol. J. 12, 152. https://doi.org/10.1186/s12985-015-0386-1.

Honnold, S.P., Mossel, E.C., Bakken, R.R., Lind, C.M., Cohen, J.W., Eccleston, L.T., Spurgers, K.B., Erwin-Cohen, R., Glass, P.J., Maheshwari, R.K., 2015b. Eastern equine encephalitis virus in mice II: pathogenesis is dependent on route of exposure. Virol. J. 12, 154. https://doi.org/10.1186/s12985-015-0385-2.

Hoz, J.M.d.L., Bayona, B., Viloria, S., Accini, J.L., Juan-Vergara, H.S., Viasus, D., 2015. Fatal cases of Chikungunya virus infection in Colombia: diagnostic and treatment challenges. J. Clin. Virol. 69, 27–29. https://doi.org/10.1016/j.jcv.2015.05.021.

Hughes, R.A., Cornblath, D.R., 2005. Guillain-Barré syndrome. Lancet 366, 1653–1666. https://doi.org/10.1016/S0140-6736(05)67665-9.

Huits, R., De Kort, J., Van Den Berg, R., Chong, L., Tsoumanis, A., Eggermont, K., Bartholomeeusen, K., Ariën, K.K., Jacobs, J., Van Esbroeck, M., Bottieau, E., Cnops, L., 2018. Chikungunya virus infection in Aruba: diagnosis, clinical features and predictors of post-chikungunya chronic polyarthralgia. PLoS One 13, e0196630. https://doi.org/10.1371/journal.pone.0196630.

Irani, D.N., Griffin, D.E., 1991. Isolation of brain parenchymal lymphocytes for flow cytometric analysis. Application to acute viral encephalitis. J. Immunol. Methods 139, 223–231.

Irani, D.N., Griffin, D.E., 1996. Regulation of lymphocyte homing into the brain during viral encephalitis at various stages of infection. J. Immunol. 156, 3850–3857.

Jackson, A.C., Moench, T.R., Griffin, D.E., Johnson, R.T., 1987. The pathogenesis of spinal cord involvement in the encephalomyelitis of mice caused by neuroadapted Sindbis virus infection. Lab. Invest. 56, 418–423.

Jackson, A.C., Moench, T.R., Trapp, B.D., Griffin, D.E., 1988. Basis of neurovirulence in Sindbis virus encephalomyelitis of mice. Lab. Invest. 58, 503–509.

Jackson, A.C., Rossiter, J.P., 1997. Apoptotic cell death is an important cause of neuronal injury in experimental Venezuelan equine encephalitis virus infection of mice. Acta Neuropathol. 93, 349–353. https://doi.org/10.1007/s004010050626.

Jackson, A.C., SenGupta, S.K., Smith, J.F., 1991. Pathogenesis of Venezuelan equine encephalitis virus infection in mice and hamsters. Vet. Pathol. 28, 410–418. https://doi.org/10.1177/030098589102800509.

Jacobs, B.C., Rothbarth, P.H., van der Meché, F.G.A., Herbrink, P., Schmitz, P.I.M., de Klerk, M.A., van Doorn, P.A., 1998. The spectrum of antecedent infections in Guillain-Barré syndrome: a case-control study. Neurology 51, 1110–1115. https://doi.org/10.1212/WNL.51.4.1110.

Jahrling, P.B., Scherer, W.F., 1973. Histopathology and distribution of viral antigens in hamsters infected with virulent and benign Venezuelan encephalitis viruses. Am. J. Pathol. 72, 25–38.

Julander, J.G., Siddharthan, V., Blatt, L.M., Schafer, K., Sidwell, R.W., Morrey, J.D., 2007. Effect of exogenous interferon and an interferon inducer on western equine encephalitis virus disease in a hamster model. Virology 360, 454–460. https://doi.org/10.1016/j.virol.2006.10.031.

Kam, Y.-W., Simarmata, D., Chow, A., Her, Z., Teng, T.-S., Ong, E.K.S., Renia, L., Leo, Y.-S., Ng, L.F.P., 2012. Early appearance of neutralizing immunoglobulin G3 antibodies is associated with chikungunya virus clearance and long-term clinical protection. J. Infect. Dis. 205, 1147–1154. https://doi.org/10.1093/infdis/jis033.

Kashyap, R.S., Morey, S., Bhullar, S., Baheti, N., Chandak, N., Purohit, H., Taori, G., Daginawala, H., 2014. Determination of Toll-like receptor-induced cytokine profiles in the blood and cerebrospinal fluid of Chikungunya patients. Neuroimmunomodulation 21, 338–346. https://doi.org/10.1159/000358240.

Kehn-Hall, K., Narayanan, A., Lundberg, L., Sampey, G., Pinkham, C., Guendel, I., Van Duyne, R., Senina, S., Schultz, K.L., Stavale, E., Aman, M.J., Bailey, C., Kashanchi, F., 2012. Modulation of GSK-3β activity in Venezuelan equine encephalitis virus infection. PLoS One 7, e34761. https://doi.org/10.1371/journal.pone.0034761.

Kelly, W.R., Blakemore, W.F., Jagelman, S., Webb, H.E., 1982. Demyelination induced in mice by avirulent Semliki Forest virus. II. An ultrastructural study of focal demyelination in the brain. Neuropathol. Appl. Neurobiol. 8, 43–53. https://doi.org/10.1111/j.1365-2990.1982.tb00256.x.

Kelvin, A.A., Banner, D., Silvi, G., Moro, M.L., Spataro, N., Gaibani, P., Cavrini, F., Pierro, A., Rossini, G., Cameron, M.J., Bermejo-Martin, J.F., Paquette, S.G., Xu, L., Danesh, A., Farooqui, A., Borghetto, I., Kelvin, D.J., Sambri, V., Rubino, S., 2011. Inflammatory cytokine expression is associated with chikungunya virus resolution and symptom severity. PLoS Negl. Trop. Dis. 5, e1279. https://doi.org/10.1371/journal.pntd.0001279.

Keogh, B., Atkins, G.J., Mills, K., Sheahan, B.J., 2003. Role of interferon-γ and nitric oxide in the neuropathogenesis of avirulent Semliki Forest virus infection. Neuropathol. Appl. Neurobiol. 29, 553–562. https://doi.org/10.1046/j.0305-1846.2003.00492.x.

Kerr, D.A., Larsen, T., Cook, S.H., Fannjiang, Y.-R., Choi, E., Griffin, D.E., Hardwick, J.M., Irani, D.N., 2002. BCL-2 and BAX protect adult mice from lethal Sindbis virus infection but do not protect spinal cord motor neurons or prevent paralysis. J. Virol. 76, 10393–10400. https://doi.org/10.1128/jvi.76.20.10393-10400.2002.

Kimura, T., Griffin, D.E., 2000. The role of CD8(+) T cells and major histocompatibility complex class I expression in the central nervous system of mice infected with neurovirulent Sindbis virus. J. Virol. 74, 6117–6125.

Kimura, T., Griffin, D.E., 2003. Extensive immune-mediated hippocampal damage in mice surviving infection with neuroadapted Sindbis virus. Virology 311, 28–39. https://doi.org/10.1016/S0042-6822(03)00110-7.

Kindhauser, M.K., Allen, T., Frank, V., Santhana, R.S., Dye, C., 2016. Zika: the origin and spread of a mosquito-borne virus. Bull. World Health Organ. 94, 675–686C. https://doi.org/10.2471/BLT.16.171082.

Koeltz, A., Lastère, S., Jean-Baptiste, S., 2018. Intensive care admissions for severe chikungunya virus infection, french Polynesia. Emerg. Infect. Dis. 24, 794–796. https://doi.org/10.3201/eid2404.161536.

Krishnan, C., Kaplin, A.I., Pardo, C.A., Kerr, D.A., Keswani, S.C., 2006. Demyelinating disorders: update on transverse myelitis. Curr. Neurol. Neurosci. Rep. 6, 236–243. https://doi.org/10.1007/s11910-006-0011-1.

Kulcsar, K.A., Baxter, V.K., Abraham, R., Nelson, A., Griffin, D.E., 2015. Distinct immune responses in resistant and susceptible strains of mice during neurovirulent alphavirus encephalomyelitis. J. Virol. 89, 8280–8291. https://doi.org/10.1128/JVI.00173-15.

Kulcsar, K.A., Baxter, V.K., Greene, I.P., Griffin, D.E., 2014. Interleukin 10 modulation of pathogenic Th17 cells during fatal alphavirus encephalomyelitis. Proc. Natl. Acad. Sci. U. S. A. 111, 16053–16058. https://doi.org/10.1073/pnas.1418966111.

Kulcsar, K.A., Griffin, D.E., 2016. T cell-derived interleukin-10 is an important regulator of the Th17 response during lethal alphavirus encephalomyelitis. J. Neuroimmunol. 295-296, 60–67. https://doi.org/10.1016/j.jneuroim.2016.04.010.

Kumar, R., Rajvanshi, P., Khosla, H., Arora, S., 2019. Neuro-chikungunya: acute transverse myelopathy associated with chikungunya virus infection. J. Assoc. Physicians India 67, 84–85.

Labrada, L., Liang, X.H., Zheng, W., Johnston, C., Levine, B., 2002. Age-dependent resistance to lethal alphavirus encephalitis in mice: analysis of gene expression in the central nervous system and identification of a novel interferon-inducible protective gene, mouse ISG12. J. Virol. 76, 11688–11703. https://doi.org/10.1128/jvi.76.22.11688-11703.2002.

Laine, M., Luukkainen, R., Toivanen, A., 2004. Sindbis viruses and other alphaviruses as cause of human arthritic disease. J. Intern. Med. 256, 457–471. https://doi.org/10.1111/j.1365-2796.2004.01413.x.

Lalitha, P., Rathinam, S., Banushree, K., Maheshkumar, S., Vijayakumar, R., Sathe, P., 2007. Ocular involvement associated with an epidemic outbreak of chikungunya virus infection. Am. J. Ophthalmol. 144, 552–556. https://doi.org/10.1016/j.ajo.2007.06.002.

Lebrun, G., Chadda, K., Reboux, A.-H., Martinet, O., Gauzere, B.A., 2009. Guillain-Barré syndrome after chikungunya infection. Emerg. Infect. Dis. 15, 495–496. https://doi.org/10.3201/eid1503.071482.

Lee, E.-Y., Schultz, K.L.W., Griffin, D.E., 2013. Mice deficient in interferon-gamma or interferon-gamma receptor 1 have distinct inflammatory responses to acute viral encephalomyelitis. PLoS One 8, e76412. https://doi.org/10.1371/journal.pone.0076412.g010.

Lee, J.M., Zipfel, G.J., Choi, D.W., 1999. The changing landscape of ischaemic brain injury mechanisms. Nature 399, A7–14.

Lee, W.W.L., Teo, T.-H., Her, Z., Lum, F.-M., Kam, Y.-W., Haase, D., Renia, L., Rötzschke, O., Ng, L.F.P., 2015. Expanding regulatory T cells alleviates chikungunya virus-induced pathology in mice. J. Virol. 89, 7893–7904. https://doi.org/10.1128/JVI.00998-15.

Levine, B., Griffin, D.E., 1992. Persistence of viral RNA in mouse brains after recovery from acute alphavirus encephalitis. J. Virol. 66, 6429–6435.

Levine, B., Hardwick, J.M., Trapp, B.D., Crawford, T.O., Bollinger, R.C., Griffin, D.E., 1991. Antibody-mediated clearance of alphavirus infection from neurons. Science 254, 856–860.

Levitt, N.H., Miller, H.V., Edelman, R., 1979. Interaction of alphaviruses with human peripheral leukocytes: in vitro replication of Venezuelan equine encephalomyelitis virus in monocyte cultures. Infect. Immun. 24, 642–646.

Lewis, J., Wesselingh, S.L., Griffin, D.E., Hardwick, J.M., 1996. Alphavirus-induced apoptosis in mouse brains correlates with neurovirulence. J. Virol. 70, 1828–1835.

Lémant, J., Boisson, V., Winer, A., Thibault, L., André, H., Tixier, F., Lemercier, M., Antok, E., Cresta, M.P., Grivard, P., Besnard, M., Rollot, O., Favier, F., Huerre, M., Campinos, J.L., Michault, A., 2008. Serious acute chikungunya virus infection requiring intensive care during the Reunion Island outbreak in 2005-2006. Crit. Care Med. 36, 2536–2541. https://doi.org/10.1097/CCM.0b013e318183f2d2.

León, C.A., 1975. Sequelae of Venezuelan equine encephalitis in humans: a four year follow-up. Int. J. Epidemiol. 4, 131–140.

Liang, X.H., Goldman, J.E., Jiang, H.H., Levine, B., 1999. Resistance of interleukin-1beta-deficient mice to fatal Sindbis virus encephalitis. J. Virol. 73, 2563–2567.

Lidbury, B.A., Simeonovic, C., Maxwell, G.E., Marshall, I.D., Hapel, A.J., 2000. Macrophage-induced muscle pathology results in morbidity and mortality for Ross River virus-infected mice. J. Infect. Dis. 181, 27–34. https://doi.org/10.1086/315164.

Lim, E.X.Y., Lee, W.S., Madzokere, E.T., Herrero, L.J., 2018. Mosquitoes as suitable vectors for alphaviruses. Viruses 10, 84. https://doi.org/10.3390/v10020084.

Lima, M.E.d.S., Bachur, T.P.R., Aragão, G.F., 2019. Guillain-Barre syndrome and its correlation with dengue, Zika and chikungunya viruses infection based on a literature review of reported cases in Brazil. Acta Trop. 197, 105064. https://doi.org/10.1016/j.actatropica.2019.105064.

Lin, J., Chen, R.W.S., Hazan, A., Weiss, M., 2018. Chikungunya virus infection manifesting as intermediate uveitis. Ocul. Immunol. Inflamm. 26, 680–682. https://doi.org/10.1080/09273948.2016.1263340.

Lindenberg, R., Haymaker, W., 1982. Tissue reactions in the gray matter of the central nervous system. In: Haymaker, W., Adams, R.D. (Eds.), Histology and Histopathology of the Nervous System. Charles C Thomas, Springfield, IL, pp. 973–1275.

Llagonne-Barets, M., Icard, V., Leparc-Goffart, I., Prat, C., Perpoint, T., André, P., Ramière, C., 2016. A case of Mayaro virus infection imported from French Guiana. J. Clin. Virol. 77, 66–68. https://doi.org/10.1016/j.jcv.2016.02.013.

Logue, C.H., Bosio, C.F., Welte, T., Keene, K.M., Ledermann, J.P., Phillips, A., Sheahan, B.J., Pierro, D.J., Marlenee, N., Brault, A.C., Bosio, C.M., Singh, A.J., Powers, A.M., Olson, K.E., 2009. Virulence variation among isolates of western equine encephalitis virus in an outbred mouse model. J. Gen. Virol. 90, 1848–1858. https://doi.org/10.1099/vir.0.008656-0.

Logue, C.H., Phillips, A.T., Mossel, E.C., Ledermann, J.P., Welte, T., Dow, S.W., Olson, K.E., Powers, A.M., 2010. Treatment with cationic liposome-DNA complexes (CLDCs) protects mice from lethal Western equine encephalitis virus (WEEV) challenge. Antiviral Res. 87, 195–203. https://doi.org/10.1016/j.antiviral.2010.04.013.

Lohachanakul, J., Phuklia, W., Thannagith, M., Thonsakulprasert, T., Ubol, S., 2012. High concentrations of circulating interleukin-6 and monocyte chemotactic protein-1 with low concentrations of interleukin-8 were associated with severe chikungunya fever during the 2009-2010 outbreak in Thailand. Microbiol. Immunol. 56, 134–138. https://doi.org/10.1111/j.1348-0421.2011.00417.x.

Long, K.M., Ferris, M.T., Whitmore, A.C., Montgomery, S.A., Thurlow, L.R., McGee, C.E., Rodriguez, C.A., Lim, J.K., Heise, M.T., 2015. γδ T cells play a protective role in chikungunya virus-induced disease. J. Virol. 90, 433–443. https://doi.org/10.1128/JVI.02159-15.

Lubelczyk, C., Mutebi, J.P., Robinson, S., Elias, S.P., Smith, L.B., Juris, S.A., Foss, K., Lichtenwalner, A., Shively, K.J., Hoenig, D.E., Webber, L., Sears, S., Smith, R.P., 2013. An epizootic of eastern equine encephalitis virus, maine, USA in 2009: outbreak description and entomological studies. Am. J. Trop. Med. Hyg. 88, 95–102. https://doi.org/10.4269/ajtmh.2012.11-0358.

Luciani, K., Abadia, I., Martinez-Torres, A.O., Cisneros, J., Guerra, I., Garcia, M., Estripeaut, D., Carrera, J.P., 2015. Madariaga virus infection associated with a case of acute disseminated encephalomyelitis. Am. J. Trop. Med. Hyg. 92, 1130–1132. https://doi.org/10.4269/ajtmh.14-0845.

Lukaszewski, R.A., Brooks, T.J., 2000. Pegylated alpha interferon is an effective treatment for virulent venezuelan equine encephalitis virus and has profound effects on the host immune response to infection. J. Virol. 74, 5006–5015. https://doi.org/10.1128/jvi.74.11.5006-5015.2000.

Lum, F.-M., Teo, T.-H., Lee, W.W.L., Kam, Y.-W., Renia, L., Ng, L.F.P., 2013. An essential role of antibodies in the control of Chikungunya virus infection. J. Immunol. 190, 6295–6302. https://doi.org/10.4049/jimmunol.1300304.

Lundström, J.O., Pfeffer, M., 2010. Phylogeographic structure and evolutionary history of Sindbis virus. Vector Borne Zoonotic Dis. 10, 889–907. https://doi.org/10.1089/vbz.2009.0069.

MacDonald, G.H., Johnston, R.E., 2000. Role of dendritic cell targeting in Venezuelan equine encephalitis virus pathogenesis. J. Virol. 74, 914–922. https://doi.org/10.1128/jvi.74.2.914-922.2000.

MacGregor, A., Ollier, W., Thomson, W., Jawaheer, D., Silman, A., 1995. HLA-DRB1*0401/0404 genotype and rheumatoid arthritis: increased association in men, young age at onset, and disease severity. J. Rheumatol. 22, 1032–1036.

Mahendradas, P., Ranganna, S.K., Shetty, R., Balu, R., Narayana, K.M., Babu, R.B., Shetty, B.K., 2008. Ocular manifestations associated with chikungunya. Ophthalmology 115, 287–291. https://doi.org/10.1016/j.ophtha.2007.03.085.

Mahendradas, P., Shetty, R., Malathi, J., Madhavan, H.N., 2010. Chikungunya virus iridocyclitis in Fuchs' heterochromic iridocyclitis. Indian J. Ophthalmol. 58, 545–547. https://doi.org/10.4103/0301-4738.71707.

Mahesh, G., Giridhar, A., Shedbele, A., Kumar, R., Saikumar, S.J., 2009. A case of bilateral presumed chikungunya neuroretinitis. Indian J. Ophthalmol. 57, 148–150. https://doi.org/10.4103/0301-4738.45508.

Maity, P., Roy, P., Basu, A., Das, B., Ghosh, U.S., 2014. A case of ADEM following Chikungunya fever. J. Assoc. Physicians India 62, 441–442.

Mallilankaraman, K., Shedlock, D.J., Bao, H., Kawalekar, O.U., Fagone, P., Ramanathan, A.A., Ferraro, B., Stabenow, J., Vijayachari, P., Sundaram, S.G., Muruganandam, N., Sarangan, G., Srikanth, P., Khan, A.S., Lewis, M.G., Kim, J.J., Sardesai, N.Y., Muthumani, K., Weiner, D.B., 2011. A DNA vaccine against chikungunya virus is protective in mice and induces neutralizing antibodies in mice and nonhuman primates. PLoS Negl. Trop. Dis. 5, e928. https://doi.org/10.1371/journal.pntd.0000928.

Malta, J.M.A.S., Vargas, A., Leite, P.L.E., Percio, J., Coelho, G.E., Ferraro, A.H.A., Cordeiro, T.M.d.O., Dias, J.d.S., Saad, E., 2017. Guillain-Barré syndrome and other neurological manifestations possibly related to Zika virus infection in municipalities from Bahia, Brazil, 2015. Epidemiol. Serv. Saude 26, 9–18. https://doi.org/10.5123/S1679-49742017000100002.

Malvy, D., Ezzedine, K., Mamani-Matsuda, M., Autran, B., Tolou, H., Receveur, M.-C., Pistone, T., Rambert, J., Moynet, D., Mossalayi, D., 2009. Destructive arthritis in a patient with chikungunya virus infection with persistent specific IgM antibodies. BMC Infect. Dis. 9, 200. https://doi.org/10.1186/1471-2334-9-200.

Mancera-Páez, O., Román, G.C., Pardo-Turriago, R., Rodríguez, Y., Anaya, J.-M., 2018. Concurrent Guillain-Barré syndrome, transverse myelitis and encephalitis post-Zika: A case report and review of the pathogenic role of multiple arboviral immunity. J. Neurol. Sci. 395, 47–53. https://doi.org/10.1016/j.jns.2018.09.028.

Manivannan, S., Baxter, V.K., Schultz, K.L.W., Slusher, B.S., Griffin, D.E., 2016. Protective effects of glutamine antagonist 6-Diazo-5-Oxo-l-norleucine in mice with alphavirus encephalomyelitis. J. Virol. 90, 9251–9262. https://doi.org/10.1128/JVI.01045-16.

Martin, N.M., Griffin, D.E., 2018. Interleukin-10 modulation of virus clearance and disease in mice with alphaviral encephalomyelitis. J. Virol. 92, 1002. https://doi.org/10.1128/JVI.01517-17.

Mathiot, C.C., Grimaud, G., Garry, P., Bouquety, J.C., Mada, A., Daguisy, A.M., Georges, A.J., 1990. An outbreak of human Semliki Forest virus infections in Central African Republic. Am. J. Trop. Med. Hyg. 42, 386–393. https://doi.org/10.4269/ajtmh.1990.42.386.

McKimmie, C.S., Johnson, N., Fooks, A.R., Fazakerley, J.K., 2005. Viruses selectively upregulate Toll-like receptors in the central nervous system. Biochem. Biophys. Res. Commun. 336, 925–933. https://doi.org/10.1016/j.bbrc.2005.08.209.

Mehta, R., Soares, C.N., Medialdea-Carrera, R., Ellul, M., da Silva, M.T.T., Rosala-Hallas, A., Jardim, M.R., Burnside, G., Pamplona, L., Bhojak, M., Manohar, R., da Silva, G.A.M., Adriano, M.V., Brasil, P., Nogueira, R.M.R., Santos, D.C.C., Turtle, L., de Sequeira, P.C., Brown, D.W., Griffiths, M.J., de Filippis, A.M.B., Solomon, T., 2018. The spectrum of neurological disease associated with Zika and chikungunya viruses in adults in Rio de Janeiro, Brazil: A case series. PLoS Negl. Trop. Dis. 12, e0006212. https://doi.org/10.1371/journal.pntd.0006212.

Mercado, M., Acosta-Reyes, J., Parra, E., Guzmán, L., Beltrán, M., Gasque, P., Mejía-García, C., Viasus, D., 2018. Renal involvement in fatal cases of chikungunya virus infection. J. Clin. Virol. 103, 16–18. https://doi.org/10.1016/j.jcv.2018.03.009.

Mercado, M., Acosta-Reyes, J., Parra, E., Pardo, L., Rico, A., Campo, A., Navarro, E., Viasus, D., 2016. Clinical and histopathological features of fatal cases with dengue and chikungunya virus co-infection in Colombia, 2014 to 2015. Euro Surveill. 21https://doi.org/10.2807/1560-7917.ES.2016.21.22.30244.

Merle, H., Donnio, A., Jean-Charles, A., Guyomarch, J., Hage, R., Najioullah, F., Césaire, R., Cabié, A., 2018. Ocular manifestations of emerging arboviruses: dengue fever, Chikungunya, Zika virus, West Nile virus, and yellow fever. J. Fr. Ophtalmol. 41, e235–e243. https://doi.org/10.1016/j.jfo.2018.05.002.

Metcalf, T.U., Griffin, D.E., 2011. Alphavirus-induced encephalomyelitis: antibody-secreting cells and viral clearance from the nervous system. J. Virol. 85, 11490–11501. https://doi.org/10.1128/JVI.05379-11.

Miner, J.J., Aw Yeang, H.X., Fox, J.M., Taffner, S., Malkova, O.N., Oh, S.T., Kim, A.H.J., Diamond, M.S., Lenschow, D.J., Yokoyama, W.M., 2015. Chikungunya viral arthritis in the United States: a mimic of seronegative rheumatoid arthritis. Arthritis Rheumatol. 67, 1214–1220. https://doi.org/10.1002/art.39027.

Miner, J.J., Cook, L.E., Hong, J.P., Smith, A.M., Richner, J.M., Shimak, R.M., Young, A.R., Monte, K., Poddar, S., Crowe, J.E., Lenschow, D.J., Diamond, M.S., 2017. Therapy with CTLA4-Ig and an antiviral monoclonal antibody controls chikungunya virus arthritis. Sci. Transl. Med. 9. https://doi.org/10.1126/scitranslmed.aah3438.

Mittal, A., Mittal, S., Bharati, M.J., Ramakrishnan, R., Saravanan, S., Sathe, P.S., 2007. Optic neuritis associated with chikungunya virus infection in South India. Arch. Ophthalmol. 125, 1381–1386. https://doi.org/10.1001/archopht.125.10.1381.

Moench, T.R., Griffin, D.E., 1984. Immunocytochemical identification and quantitation of the mononuclear cells in the cerebrospinal fluid, meninges, and brain during acute viral meningoencephalitis. J. Exp. Med. 159, 77–88. https://doi.org/10.1084/jem.159.1.77.

Mogami, R., Pereira Vaz, J.L., de Fátima Barcelos Chagas, Y., de Abreu, M.M., Torezani, R.S., de Almeida Vieira, A., Junqueira Filho, E.A., Barbosa, Y.B., Carvalho, A.C.P., Lopes, A.J., 2017. Ultrasonography of hands and wrists in the diagnosis of complications of chikungunya fever. J. Ultrasound Med. 37, 511–520. https://doi.org/10.1002/jum.14344.

Mokhtarian, F., Huan, C.-M., Roman, C., Raine, C.S., 2003. Semliki Forest virus-induced demyelination and remyelination—involvement of B cells and anti-myelin antibodies. J. Neuroimmunol. 137, 19–31. https://doi.org/10.1016/s0165-5728(03)00039-0.

Mokhtarian, F., Safavi, F., Sarafraz-Yazdi, E., 2012. Immunization with a peptide of Semliki Forest virus promotes remyelination in experimental autoimmune encephalomyelitis. Brain Res. 1488, 92–103. https://doi.org/10.1016/j.brainres.2012.09.038.

Mokhtarian, F., Zhang, Z., Shi, Y., Gonzales, E., Sobel, R.A., 1999. Molecular mimicry between a viral peptide and a myelin oligodendrocyte glycoprotein peptide induces autoimmune demyelinating disease in mice. J. Neuroimmunol. 95, 43–54. https://doi.org/10.1016/s0165-5728(98)00254-9.

Molaei, G., Armstrong, P.M., Graham, A.C., Kramer, L.D., Andreadis, T.G., 2015. Insights into the recent emergence and expansion of eastern equine encephalitis virus in a new focus in the Northern New England USA. Parasit. Vectors 8, 516. https://doi.org/10.1186/s13071-015-1145-2.

Monlux, W.S., Luedke, A.J., 1973. Brain and spinal cord lesions in horses inoculated with Venezuelan equine encephalomyelitis virus (epidemic American and Trinidad strains). Am. J. Vet. Res. 34, 465–473.

Morris, M.M., Dyson, H., Baker, D., Harbige, L.S., Fazakerley, J.K., Amor, S., 1997. Characterization of the cellular and cytokine response in the central nervous system following Semliki Forest virus infection. J. Neuroimmunol. 74, 185–197. https://doi.org/10.1016/s0165-5728(96)00786-2.

Morrison, T.E., Fraser, R.J., Smith, P.N., Mahalingam, S., Heise, M.T., 2007. Complement contributes to inflammatory tissue destruction in a mouse model of Ross River virus-induced disease. J. Virol. 81, 5132–5143. https://doi.org/10.1128/JVI.02799-06.

Morrison, T.E., Oko, L., Montgomery, S.A., Whitmore, A.C., Lotstein, A.R., Gunn, B.M., Elmore, S.A., Heise, M.T., 2011. A mouse model of chikungunya virus-induced musculoskeletal inflammatory disease: evidence of arthritis, tenosynovitis, myositis, and persistence. Am. J. Pathol. 178, 32–40. https://doi.org/10.1016/j.ajpath.2010.11.018.

Morrison, T.E., Simmons, J.D., Heise, M.T., 2008. Complement receptor 3 promotes severe ross river virus-induced disease. J. Virol. 82, 11263–11272. https://doi.org/10.1128/JVI.01352-08.

Mulder, D.W., Parrott, M., Thaler, M., 1951. Sequelae of western equine encephalitis. Neurology 1, 318–327. https://doi.org/10.1212/wnl.1.7-8.318.

Murillo-Zamora, E., Mendoza-Cano, O., Trujillo-Hernández, B., Guzmán-Esquivel, J., Higareda-Almaraz, E., Higareda-Almaraz, M.A., Sánchez-Piña, R.A., Lugo-Radillo, A., 2018. Persistent arthralgia and related risks factors: a cohort study at 12 months from laboratory-confirmed chikungunya infection. Arch. Med. Res. 49, 65–73. https://doi.org/10.1016/j.arcmed.2018.04.008.

Murthy, K.R., Venkataraman, N., Satish, V., Babu, K., 2008. Bilateral retinitis following chikungunya fever. Indian J. Ophthalmol. 56, 329–331.

Musthafa, A.K.M., Abdurahiman, P., Jose, J., 2008. Case of ADEM following Chikungunya fever. J. Assoc. Physicians India 56, 473.

Mylonas, A.D., Brown, A.M., Carthew, T.L., McGrath, B., Purdie, D.M., Pandeya, N., Vecchio, P.C., Collins, L.G., Gardner, I.D., de Looze, F.J., Reymond, E.J., Suhrbier, A., 2002. Natural history of Ross River virus-induced epidemic polyarthritis. Med. J. Aust. 177, 356–360.

Nair, A.G., Biswas, J., Bhende, M.P., 2012. A case of bilateral Chikungunya neuroretinitis. J. Ophthalmic Inflamm. Infect. 2, 39–40. https://doi.org/10.1007/s12348-011-0038-6.

Nargi-Aizenman, J.L., Griffin, D.E., 2001. Sindbis virus-induced neuronal death is both necrotic and apoptotic and is ameliorated by N-methyl-D-aspartate receptor antagonists. J. Virol. 75, 7114–7121. https://doi.org/10.1128/JVI.75.15.7114-7121.2001.

Nargi-Aizenman, J.L., Havert, M.B., Zhang, M., Irani, D.N., Rothstein, J.D., Griffin, D.E., 2004. Glutamate receptor antagonists protect from virus-induced neural degeneration. Ann. Neurol. 55, 541–549. https://doi.org/10.1002/ana.20033.

Nathanson, N., Stolley, P.D., Boolukos, P.J., 1969. Eastern equine encephalitis. Distribution of central nervous system lesions in man and Rhesus monkey. J. Comp. Pathol. 79, 109–115. https://doi.org/10.1016/0021-9975(69)90034-6.

Neighbours, L.M., Long, K., Whitmore, A.C., Heise, M.T., 2012. Myd88-dependent toll-like receptor 7 signaling mediates protection from severe Ross River virus-induced disease in mice. J. Virol. 86, 10675–10685. https://doi.org/10.1128/JVI.00601-12.

Neri, V., Bacelar, O., Farinhas, J.G., França, L.C., Dantas, F.F., Leon, S.A., 2018. Optic neuritis and acute transverse myelitis after chikungunya virus infection. Int. J. Infect. Dis. 73, S179. https://doi.org/10.1016/j.ijid.2018.04.3820.

Ng, L.F.P., Chow, A., Sun, Y.-J., Kwek, D.J.C., Lim, P.-L., Dimatatac, F., Ng, L.-C., Ooi, E.-E., Choo, K.-H., Her, Z., Kourilsky, P., Leo, Y.-S., 2009. IL-1beta, IL-6, and RANTES as biomarkers of Chikungunya severity. PLoS One 4, e4261. https://doi.org/10.1371/journal.pone.0004261.

Nhan, T.-X., Claverie, A., Roche, C., Teissier, A., Colleuil, M., Baudet, J.-M., Cao-Lormeau, V.-M., Musso, D., 2014. Chikungunya virus imported into French Polynesia, 2014. Emerg. Infect. Dis. 20, 1773–1774. https://doi.org/10.3201/eid2010.141060.

Nickerson, J.P., Kannabiran, S., Burbank, H.N., 2016. MRI findings in eastern equine encephalitis: the "parenthesis" sign. Clin. Imaging 40, 222–223. https://doi.org/10.1016/j.clinimag.2015.10.012.

Nicotera, P., Orrenius, S., 1998. The role of calcium in apoptosis. Cell Calcium 23, 173–180.

Nilaratanakul, V., Chen, J., Tran, O., Baxter, V.K., Troisi, E.M., Yeh, J.X., Griffin, D.E., 2018. Germline IgM is sufficient, but not required, for antibody-mediated alphavirus clearance from the central nervous system. J. Virol. 92, e02081-17. https://doi.org/10.1128/JVI.02081-17. JVI.02081-17.

Ninla-aesong, P., Mitarnun, W., Noipha, K., 2019. Proinflammatory cytokines and chemokines as biomarkers of persistent arthralgia and severe disease after chikungunya virus infection: a 5-year follow-up study in Southern Thailand. Viral Immunol. 32, 442–452. https://doi.org/10.1089/vim.2019.0064.

Ninla-aesong, P., Mitarnun, W., Noipha, K., 2020. Long-term persistence of chikungunya virus-associated manifestations and anti-chikungunya virus antibody in southern Thailand: 5 years after an outbreak in 2008-2009. Viral Immunol. 33, 86–93. https://doi.org/10.1089/vim.2019.0168.

Oehler, E., Fournier, E., Leparc-Goffart, I., Larre, P., Cubizolle, S., Sookhareea, C., Lastère, S., Ghawche, F., 2015. Increase in cases of Guillain-Barré syndrome during a Chikungunya outbreak, French Polynesia, 2014 to 2015. Euro Surveill. 20, 30079. https://doi.org/10.2807/1560-7917.ES.2015.20.48.30079.

Oliver, K.R., Fazakerley, J.K., 1998. Transneuronal spread of Semliki Forest virus in the developing mouse olfactory system is determined by neuronal maturity. Neuroscience 82, 867–877. https://doi.org/10.1016/s0306-4522(97)00309-6.

Oliver, K.R., Scallan, M.F., Dyson, H., Fazakerley, J.K., 1997. Susceptibility to a neurotropic virus and its changing distribution in the developing brain is a function of CNS maturity. J. Neurovirol. 3, 38–48. https://doi.org/10.3109/13550289709015791.

Ozden, S., Huerre, M., Riviere, J.P., Coffey, L.L., Afonso, P.V., Mouly, V., de Monredon, J., Roger, J.-C., El Amrani, M., Yvin, J.-L., Jaffar, M.-C., Frenkiel, M.-P., Sourisseau, M., Schwartz, O., Gulter-Browne, G., Despres, P., Gessain, A., Ceccaldi, P.-E., 2007. Human muscle satellite cells as targets of chikungunya virus infection. PLoS One 2, e527. https://doi.org/10.1371/journal.pone.0000527.

Paessler, S., Aguilar, P., Anishchenko, M., Wang, H.-Q., Aronson, J., Campbell, G., Cararra, A.-S., Weaver, S.C., 2004. The hamster as an animal model for eastern equine encephalitis—and its use in studies of virus entrance into the brain. J. Infect. Dis. 189, 2072–2076. https://doi.org/10.1086/383246.

Palmer, R.J., Finley, K.H., 1956. Sequelae of encephalitis; report of a study after the California epidemic. Calif. Med. 84, 98–100.

Panato, C.S., Figueredo, E.D., Bassi, D., Felipe, I.M.A., Firmo, W.D.C.A., Rêgo, A.S., Silva, F.d.M.A.M., 2019. Evaluation of functional disability after Chikungunya infection. Rev. Soc. Bras. Med. Trop. 52, e20190112. https://doi.org/10.1590/0037-8682-0112-2019.

Partidos, C.D., Weger, J., Brewoo, J., Seymour, R., Borland, E.M., Ledermann, J.P., Powers, A.M., Weaver, S.C., Stinchcomb, D.T., Osorio, J.E., 2011. Probing the attenuation and protective efficacy of a candidate chikungunya virus vaccine in mice with compromised interferon (IFN) signaling. Vaccine 29, 3067–3073. https://doi.org/10.1016/j.vaccine.2011.01.076.

Passoni, G., Langevin, C., Palha, N., Mounce, B.C., Briolat, V., Affaticati, P., De Job, E., Joly, J.-S., Vignuzzi, M., Saleh, M.-C., Herbomel, P., Boudinot, P., Levraud, J.-P., 2017. Imaging of viral neuroinvasion in the zebrafish reveals that Sindbis and chikungunya viruses favour different entry routes. Dis. Model. Mech. 10, 847–857. https://doi.org/10.1242/dmm.029231.

Patel, D.M., Patel, M.V., Sharma, K.H., Patel, G.R., Patel, M.B., Shah, V.V., 2019. Post chikungunya chronic arthritis: systemic inflammatory status triggering acute coronary syndrome. Curr. Rheumatol. Rev. 15, 229–233. https://doi.org/10.2174/1573397114666180816112948.

Pathak, S., Illavia, S.J., Webb, H.E., 1983. The identification and role of cells involved in CNS demyelination in mice after Semliki Forest virus infection: an ultrastructural study. Prog. Brain Res. 59, 237–254. https://doi.org/10.1016/S0079-6123(08)63869-8.

Pathak, S., Webb, H.E., 1974. Possible mechanisms for the transport of Semliki forest virus into and within mouse brain. An electron-microscopic study. J. Neurol. Sci. 23, 175–184. https://doi.org/10.1016/0022-510x(74)90221-4.

Paul, B., Pariyapurath, R., 2018. Risk factor assessment of rheumatoid arthritis in North Kerala. Eur. J. Rheumatol. 5, 184–190. https://doi.org/10.5152/eurjrheum.2018.17111.

Phan, T.G., Somerville, E., Hueston, L., Dwyer, D.E., 1998. Probable association between Barmah Forest virus infection and Guillain-Barré syndrome. Med. J. Aust. 168, 422–423.

Phelps, A.L., O'Brien, L.M., Eastaugh, L.S., Davies, C., Lever, M.S., Ennis, J., Zeitlin, L., Nunez, A., Ulaeto, D.O., 2017. Susceptibility and lethality of Western equine encephalitis virus in Balb/c mice when infected by the aerosol route. Viruses 9, 163. https://doi.org/10.3390/v9070163.

Phelps, A.L., O'Brien, L.M., Eastaugh, L.S., Davies, C., Lever, M.S., Ennis, J., Zeitlin, L., Nunez, A., Ulaeto, D.O., 2019. Aerosol infection of Balb/c mice with eastern equine encephalitis virus; susceptibility and lethality. Virol. J. 16, 2. https://doi.org/10.1186/s12985-018-1103-7.

Phillips, A.T., Rico, A.B., Stauft, C.B., Hammond, S.L., Aboellail, T.A., Tjalkens, R.B., Olson, K.E., 2016. Entry Sites of Venezuelan and Western Equine Encephalitis Viruses in the mouse central nervous system following peripheral infection. J. Virol. 90, 5785–5796. https://doi.org/10.1128/JVI.03219-15.

Phillips, A.T., Stauft, C.B., Aboellail, T.A., Toth, A.M., Jarvis, D.L., Powers, A.M., Olson, K.E., 2013. Bioluminescent imaging and histopathologic characterization of WEEV neuroinvasion in outbred CD-1 mice. PLoS One 8, e53462. https://doi.org/10.1371/journal.pone.0053462.

Pinheiro, T.J., Guimarães, L.F., Silva, M.T.T., Soares, C.N., 2016. Neurological manifestations of Chikungunya and Zika infections. Arq. Neuropsiquiatr. 74, 937–943. https://doi.org/10.1590/0004-282X20160138.

Poo, Y.S., Nakaya, H., Gardner, J., Larcher, T., Schroder, W.A., Le, T.T., Major, L.D., Suhrbier, A., 2014. CCR2 deficiency promotes exacerbated chronic erosive neutrophil-dominated chikungunya virus arthritis. J. Virol. 88, 6862–6872. https://doi.org/10.1128/JVI.03364-13.

Poon, A.N., Simon, G.L., Chang, A.Y., 2019. Treatment of chronic chikungunya with methotrexate. J. Clin. Rheumatol. 00, 1–2. https://doi.org/10.1097/RHU.0000000000000998.

Porter, A.I., Erwin-Cohen, R.A., Twenhafel, N., Chance, T., Yee, S.B., Kern, S.J., Norwood, D., Hartman, L.J., Parker, M.D., Glass, P.J., DaSilva, L., 2017. Characterization and pathogenesis of aerosolized eastern equine encephalitis in the common marmoset (Callithrix jacchus). Virol. J. 14, 25. https://doi.org/10.1186/s12985-017-0687-7.

Potter, M.C., Baxter, V.K., Mathey, R.W., Alt, J., Rojas, C., Griffin, D.E., Slusher, B.S., 2015. Neurological sequelae induced by alphavirus infection of the CNS are attenuated by treatment with the glutamine antagonist 6-diazo-5-oxo-l-norleucine. J. Neurovirol. 21, 159–173. https://doi.org/10.1007/s13365-015-0314-6.

Potter, M.E., Currier, R.W., Pearson, J.E., Harris, J.C., Parker, R.L., 1977. Western equine encephalomyelitis in horses in the Northern Red River Valley, 1975. J. Am. Vet. Med. Assoc. 170, 1396–1399.

Powers, A.M., 2018. Vaccine and therapeutic options to control chikungunya virus. Clin. Microbiol. Rev. 31, e00104-16. https://doi.org/10.1128/CMR.00104-16.

Prow, N.A., Irani, D.N., 2007. The opioid receptor antagonist, naloxone, protects spinal motor neurons in a murine model of alphavirus encephalomyelitis. Exp. Neurol. 205, 461–470. https://doi.org/10.1016/j.expneurol.2007.03.013.

Przelomski, M.M., O'Rourke, E., Grady, G.F., Berardi, V.P., Markley, H.G., 1988. Eastern equine encephalitis in Massachusetts: a report of 16 cases, 1970-1984. Neurology 38, 736–739. https://doi.org/10.1212/wnl.38.5.736.

Reddy, V., Mani, R.S., Desai, A., Ravi, V., 2014. Correlation of plasma viral loads and presence of Chikungunya IgM antibodies with cytokine/chemokine levels during acute Chikungunya virus infection. J. Med. Virol. 86, 1393–1401. https://doi.org/10.1002/jmv.23875.

Reed, D.S., Lackemeyer, M.G., Garza, N.L., Norris, S., Gamble, S., Sullivan, L.J., Lind, C.M., Raymond, J.L., 2007. Severe encephalitis in cynomolgus macaques exposed to aerosolized Eastern equine encephalitis virus. J. Infect.Dis. 196, 441–450. https://doi.org/10.1086/519391.

Reed, D.S., Larsen, T., Sullivan, L.J., Lind, C.M., Lackemeyer, M.G., Pratt, W.D., Parker, M.D., 2005. Aerosol exposure to western equine encephalitis virus causes fever and encephalitis in cynomolgus macaques. J. Infect. Dis. 192, 1173–1182. https://doi.org/10.1086/444397.

Reilly, J.M., Xing, W., Levicky, V., Souccar, S., Rogers, C., Sathyavagiswaran, L., 2020. Postmortem chikungunya diagnosis: a case report and literature review. Am. J. Forensic Med. Pathol. 41, 48–51. https://doi.org/10.1097/PAF.0000000000000519.

Rezza, G., Chen, R., Weaver, S.C., 2017. O'nyong-nyong fever: a neglected mosquito-borne viral disease. Pathog. Glob. Health 111, 271–275. https://doi.org/10.1080/20477724.2017.1355431.

Rivas, F., Diaz, L.A., Cardenas, V.M., Daza, E., Bruzon, L., Alcala, A., Hoz, l., De, O., Caceres, F.M., Aristizabal, G., Martinez, J.W., Revelo, D., Hoz, l., De, F., Boshell, J., Camacho, T., Calderon, L., Olano, V.A., Villarreal, L.I., Roselli, D., Alvarez, G., Ludwig, G., Tsai, T., 1997. Epidemic Venezuelan equine encephalitis in La Guajira, Colombia, 1995. J. Infect. Dis. 175, 828–832.

Riyaz, N., Riyaz, A., Rahima, A., Latheef, E.N., Anitha, P.M., Aravindan, K.P., Nair, A.S., Shameera, P., 2010. Cutaneous manifestations of chikungunya during a recent epidemic in Calicut, north Kerala, south India. Indian J. Dermatol. Venereol. Leprol. 76, 671–676. https://doi.org/10.4103/0378-6323.72466.

Roberts, E.D., Sanmartin, C., Payán, J., Mackenzie, R.B., 1970. Neuropathologic changes in 15 horses with naturally occurring Venezuelan equine encephalomyelitis. Am. J. Vet. Res. 31, 1223–1229.

Robinson, M.C., 1955. An epidemic of virus disease in Southern Province, Tanganyika territory, in 1952–1953. Trans. R. Soc. Trop. Med. Hyg. 49, 28–32.

Rocha, V.F.D., de Oliveira, A.H.P., Bandeira, A.C., Sardi, S.I., Garcia, R.F., Magalhães, S. d.A., Sampaio, C.A., Campos Soares, G., 2018. Chikungunya virus infection associated with encephalitis and anterior uveitis. Ocul. Immunol. Inflamm. 26, 677–679. https://doi.org/10.1080/09273948.2017.1358378.

Rose, N., Anoop, T.M., John, A.P., Jabbar, P.K., George, K.C., 2011. Acute optic neuritis following infection with chikungunya virus in southern rural India. Int. J. Infect. Dis. 15, e147–e150. https://doi.org/10.1016/j.ijid.2010.10.004.

Ross, R.W., 1956. The Newala epidemic. III. The virus: isolation, pathogenic properties and relationship to the epidemic. J. Hyg. (Lond) 54, 177–191. https://doi.org/10.1017/s0022172400044442.

Rowell, J.F., Griffin, D.E., 1999. The inflammatory response to nonfatal Sindbis virus infection of the nervous system is more severe in SJL than in BALB/c mice and is associated with low levels of IL-4 mRNA and high levels of IL-10-producing CD4+ T cells. J. Immunol. 162, 1624–1632.

Rowell, J.F., Griffin, D.E., 2002. Contribution of T cells to mortality in neurovirulent Sindbis virus encephalomyelitis. J. Neuroimmunol. 127, 106–114.

Roy, C.J., Adams, A.P., Wang, E., Leal, G., Seymour, R.L., Sivasubramani, S.K., Mega, W., Frolov, I., Didier, P.J., Weaver, S.C., 2013. A chimeric Sindbis-based vaccine protects cynomolgus macaques against a lethal aerosol challenge of eastern equine encephalitis virus. Vaccine 31, 1464–1470. https://doi.org/10.1016/j.vaccine.2013.01.014.

Roy, C.J., Reed, D.S., Wilhelmsen, C.L., Hartings, J., Norris, S., Steele, K.E., 2009. Pathogenesis of aerosolized Eastern Equine Encephalitis virus infection in guinea pigs. Virol. J. 6, 170. https://doi.org/10.1186/1743-422X-6-170.

Rozdilsky, B., Robertson, H.E., Chorney, J., 1968. Western encephalitis: report of eight fatal cases. Saskatchewan epidemic, 1965. Can. Med. Assoc. J. 98, 79–86.

Rudd, P.A., Wilson, J., Gardner, J., Larcher, T., Babarit, C., Le, T.T., Anraku, I., Kumagai, Y., Loo, Y.-M., Gale, M., Akira, S., Khromykh, A.A., Suhrbier, A., 2012. Interferon response factors 3 and 7 protect against Chikungunya virus hemorrhagic fever and shock. J. Virol. 86, 9888–9898. https://doi.org/10.1128/JVI.00956-12.

Rulli, N.E., Guglielmotti, A., Mangano, G., Rolph, M.S., Apicella, C., Zaid, A., Suhrbier, A., Mahalingam, S., 2009. Amelioration of alphavirus-induced arthritis and myositis in a mouse model by treatment with bindarit, an inhibitor of monocyte chemotactic proteins. Arthritis Rheum. 60, 2513–2523. https://doi.org/10.1002/art.24682.

Ryman, K.D., Klimstra, W.B., Nguyen, K.B., Biron, C.A., 2000. Alpha/beta interferon protects adult mice from fatal Sindbis virus infection and is an important determinant of cell and tissue tropism. J. Virol. 74, 3366–3378.

Ryzhikov, A.B., Tkacheva, N.V., Sergeev, A.N., Ryabchikova, E.I., 1991. Venezuelan equine encephalitis virus propagation in the olfactory tract of normal and immunized mice. Biomed. Sci. 2, 607–614.

Safavi, F., Feliberti, J.P., Raine, C.S., Mokhtarian, F., 2011. Role of γδ T cells in antibody production and recovery from SFV demyelinating disease. J. Neuroimmunol. 235, 18–26. https://doi.org/10.1016/j.jneuroim.2011.02.013.

Sahu, R., Verma, R., Jain, A., Garg, R.K., Singh, M.K., Malhotra, H.S., Sharma, P.K., Parihar, A., 2014. Neurologic complications in dengue virus infection: a prospective cohort study. Neurology 83, 1601–1609. https://doi.org/10.1212/WNL.0000000000000935.

Sahu, S.P., Pedersen, D.D., Jenny, A.L., Schmitt, B.J., Alstad, A.D., 2003. Pathogenicity of a Venezuelan equine encephalomyelitis serotype IE virus isolate for ponies. Am. J. Trop. Med. Hyg. 68, 485–494. https://doi.org/10.4269/ajtmh.2003.68.485.

Salceanu, S.O., Raman, V., 2018. Recurrent chikungunya retinitis. BMJ Case Rep. 2018, 1–4. https://doi.org/10.1136/bcr-2017-222864.

Sane, J., Kurkela, S., Lokki, M.-L., Miettinen, A., Helve, T., Vaheri, A., Vapalahti, O., 2012. Clinical Sindbis alphavirus infection is associated with HLA-DRB1*01 allele and production of autoantibodies. Clin. Infect. Dis. 55, 358–363. https://doi.org/10.1093/cid/cis405.

Santiago, F.W., Halsey, E.S., Siles, C., Vilcarromero, S., Guevara, C., Silvas, J.A., Ramal, C., Ampuero, J.S., Aguilar, P.V., 2015. Long-term arthralgia after mayaro virus infection correlates with sustained pro-inflammatory cytokine response. PLoS Negl. Trop. Dis. 9, e0004104. https://doi.org/10.1371/journal.pntd.0004104.

Sarid, R., Ben-Moshe, T., Kazimirsky, G., Weisberg, S., Appel, E., Kobiler, D., Lustig, S., Brodie, C., 2001. vFLIP protects PC-12 cells from apoptosis induced by Sindbis virus: implications for the role of TNF-α. Cell Death Differ. 8, 1224–1231. https://doi.org/10.1038/sj.cdd.4400926.

Sattler, R., Tymianski, M., 2001. Molecular mechanisms of glutamate receptor-mediated excitotoxic neuronal cell death. Mol. Neurobiol. 24, 107–129. https://doi.org/10.1385/MN:24:1-3:107.

Sánchez, J.S., Cañón, A.M., Lombo, J.C., 2019. Subacute and chronic symptoms of chikungunya fever in a group of adults in Colombia. Biomedica 39, 587–594. https://doi.org/10.7705/biomedica.4350.

Schilte, C., Couderc, T., Chrétien, F., Sourisseau, M., Gangneux, N., Guivel-Benhassine, F., Kraxner, A., Tschopp, J., Higgs, S., Michault, A., Arenzana-Seisdedos, F., Colonna, M., Peduto, L., Schwartz, O., Lecuit, M., Albert, M.L., 2010. Type I IFN controls chikungunya virus via its action on nonhematopoietic cells. J. Exp. Med. 207, 429–442. https://doi.org/10.1084/jem.20090851.

Schoneboom, B.A., Catlin, K.M., Marty, A.M., Grieder, F.B., 2000a. Inflammation is a component of neurodegeneration in response to Venezuelan equine encephalitis virus infection in mice. J. Neuroimmunol. 109, 132–146. https://doi.org/10.1016/s0165-5728(00)00290-3.

Schoneboom, B.A., Fultz, M.J., Miller, T.H., McKinney, L.C., Grieder, F.B., 1999. Astrocytes as targets for Venezuelan equine encephalitis virus infection. J. Neurovirol. 5, 342–354. https://doi.org/10.3109/13550289909029475.

Schoneboom, B.A., Lee, J.S., Grieder, F.B., 2000b. Early expression of IFN-alpha/beta and iNOS in the brains of Venezuelan equine encephalitis virus-infected mice. J. Interferon Cytokine Res. 20, 205–215. https://doi.org/10.1089/107999000312621.

Schultz, K.L.W., Troisi, E.M., Baxter, V.K., Glowinski, R., Griffin, D.E., 2019. Interferon regulatory factors 3 and 7 have distinct roles in the pathogenesis of alphavirus encephalomyelitis. J. Gen. Virol. 100, 46–62. https://doi.org/10.1099/jgv.0.001174.

Scripsema, N.K., Sharifi, E., Samson, C.M., Kedhar, S., Rosen, R.B., 2015. Chikungunya-associated uveitis and exudative retinal detachment: A case report. Retin Cases Brief Rep. 9, 352–356. https://doi.org/10.1097/ICB.0000000000000232.

Sepúlveda-Delgado, J., Vera-Lastra, O.L., Trujillo-Murillo, K., Canseco-Ávila, L.M., Sánchez-González, R.A., Gómez-Cruz, O., Lugo-Trampe, A., Fernández-Salas, I., Danis-Lozano, R., Contreras-Contreras, A., Mendoza-Torres, A., Domínguez-Arrevillaga, S., Mena-Vela, B.A., Ocaña-Sibilla, M., Ramirez-Valdespino, J.C., Jara, L.J., 2017. Inflammatory biomarkers, disease activity index, and self-reported disability may be predictors of chronic arthritis after chikungunya infection: brief report. Clin. Rheumatol. 36, 695–699. https://doi.org/10.1007/s10067-016-3419-2.

Seth, P., Husain, M.M., Gupta, P., Schoneboom, A., Grieder, B.F.B., Mani, H., Maheshwari, R.K., 2003. Early onset of virus infection and up-regulation of cytokines in mice treated with cadmium and manganese. Biometals 16, 359–368. https://doi.org/10.1023/a:1020682716212.

Seymour, R.L., Adams, A.P., Leal, G., Alcorn, M.D.H., Weaver, S.C., 2015. A rodent model of chikungunya virus infection in RAG1 -/- mice, with features of persistence, for vaccine safety evaluation. PLoS Negl. Trop. Dis. 9, e0003800. https://doi.org/10.1371/journal.pntd.0003800.

Sharma, A., Bhattacharya, B., Puri, R.K., Maheshwari, R.K., 2008. Venezuelan equine encephalitis virus infection causes modulation of inflammatory and immune response genes in mouse brain. BMC Genomics 9, 289. https://doi.org/10.1186/1471-2164-9-289.

Sharma, A., Maheshwari, R.K., 2009. Oligonucleotide array analysis of Toll-like receptors and associated signaling genes in Venezuelan equine encephalitis virus-infected mouse brain. J. Gen. Virol. 90, 1836–1847. https://doi.org/10.1099/vir.0.010280-0.

Sheahan, B.J., Gates, M.C., Caffrey, J.F., Atkins, G.J., 1983. Oligodendrocyte infection and demyelination produced in mice by the M9 mutant of Semliki Forest virus. Acta Neuropathol. 60, 257–265. https://doi.org/10.1007/bf00691874.

Silverman, M.A., Misasi, J., Smole, S., Feldman, H.A., Cohen, A.B., Santagata, S., McManus, M., Ahmed, A.A., 2013. Eastern equine encephalitis in children, Massachusetts and New Hampshire, USA, 1970–2010. Emerg. Infect. Dis. 19, 194–201. https://doi.org/10.3201/eid1902.120039.

Sissoko, D., Malvy, D., Ezzedine, K., Renault, P., Moscetti, F., Ledrans, M., Pierre, V., 2009. Post-epidemic Chikungunya disease on reunion Island: course of rheumatic manifestations and associated factors over a 15-month period. PLoS Negl. Trop. Dis. 3, e389. https://doi.org/10.1371/journal.pntd.0000389.

Soilu-Hänninen, M., Erälinna, J.P., Hukkanen, V., Röyttä, M., Salmi, A.A., Salonen, R., 1994. Semliki Forest virus infects mouse brain endothelial cells and causes blood-brain barrier damage. J. Virol. 68, 6291–6298.

Sosa-Martínez, M.d.J., Orea-Flores, M., Vázquez-Cruz, I., Palacios-Castillo, V., Juanico-Morales, G., Pérez-Mijangos, L., 2018. Characterization of chronic clinical manifestations in patients with chikungunya fever. Rev. Med. Inst. Mex. Seguro Soc. 56, 239–245.

Steele, K.E., Davis, K.J., Stephan, K., Kell, W., Vogel, P., Hart, M.K., 1998. Comparative neurovirulence and tissue tropism of wild-type and attenuated strains of Venezuelan equine encephalitis virus administered by aerosol in C3H/HeN and BALB/c mice. Vet. Pathol. 35, 386–397. https://doi.org/10.1177/030098589803500508.

Steele, K.E., Reed, D.S., Glass, P.J., Hart, M.K., Ludwig, G.V., Pratt, W.D., Parker, M.D., Smith, J.F., 2007. Alphavirus encephalitides. In: Dembek, Z.F. (Ed.), Medical Aspects of Biological Warfare. Government Printing Office, Washington, DC, pp. 241–270.

Steele, K.E., Seth, P., Catlin-Lebaron, K.M.K., Schoneboom, B.A., Husain, M.M., Grieder, F., Maheshwari, R.K., 2006. Tunicamycin enhances neuroinvasion and encephalitis in mice infected with Venezuelan equine encephalitis virus. Vet. Pathol. 43, 904–913. https://doi.org/10.1354/vp.43-6-904.

Steele, K.E., Twenhafel, N.A., 2010. Pathology of animal models of alphavirus encephalitis. Vet. Pathol. 47, 790–805. https://doi.org/10.1177/0300985810372508.

Stegmann-Planchard, S., Gallian, P., Tressières, B., Leparc-Goffart, I., Lannuzel, A., Signate, A., Laouénan, C., Cabie, A., Hoen, B., 2019. Chikungunya, a risk factor for Guillain-Barré syndrome. Clin. Infect. Dis. 70, 1233–1235. https://doi.org/10.1093/cid/ciz625. ciz625.

Stohlman, S.A., Bergmann, C.C., Lin, M.T., Cua, D.J., Hinton, D.R., 1998. CTL effector function within the central nervous system requires CD4+ T cells. J. Immunol. 160, 2896–2904.

Storm, N., Weyer, J., Markotter, W., Kemp, A., Leman, P.A., Dermaux-Msimang, V., Nel, L.H., Paweska, J.T., 2014. Human cases of Sindbis fever in South Africa, 2006-2010. Epidemiol. Infect. 142, 234–238. https://doi.org/10.1017/S0950268813000964.

Styczynski, A.R., Malta, J.M.A.S., Krow-Lucal, E.R., Percio, J., Nóbrega, M.E., Vargas, A., Lanzieri, T.M., Leite, P.L., Staples, J.E., Fischer, M.X., Powers, A.M., Chang, G.-J.J., Burns, P.L., Borland, E.M., Ledermann, J.P., Mossel, E.C., Schonberger, L.B., Belay, E.B., Salinas, J.L., Badaro, R.D., Sejvar, J.J., Coelho, G.E., 2017. Increased rates of Guillain-Barré syndrome associated with Zika virus outbreak in the Salvador metropolitan area, Brazil. PLoS Negl. Trop. Dis. 11, e0005869. https://doi.org/10.1371/journal.pntd.0005869.

Subak-Sharpe, I., Dyson, H., Fazakerley, J., 1993. In vivo depletion of CD8+ T cells prevents lesions of demyelination in Semliki Forest virus infection. J. Virol. 67, 7629–7633.

Suhrbier, A., Jaffar-Bandjee, M.C., Gasque, P., 2012. Arthritogenic alphaviruses—an overview. Nat. Rev. Rheumatol. 8, 420–429. https://doi.org/10.1038/nrrheum.2012.64.

Tandale, B.V., Sathe, P.S., Arankalle, V.A., Wadia, R.S., Kulkarni, R., Shah, S.V., Shah, S.K., Sheth, J.K., Sudeep, A.B., Tripathy, A.S., Mishra, A.C., 2009. Systemic involvements and fatalities during Chikungunya epidemic in India, 2006. J. Clin. Virol. 46, 145–149. https://doi.org/10.1016/j.jcv.2009.06.027.

Tappe, D., Pérez-Girón, J.V., Just-Nübling, G., Schuster, G., Gómez-Medina, S., Günther, S., Muñoz-Fontela, C., Schmidt-Chanasit, J., 2016. Sustained elevated cytokine levels during recovery phase of Mayaro virus infection. Emerg. Infect. Dis. 22, 750–752. https://doi.org/10.3201/eid2204.151502.

Taraphdar, D., Roy, B.K., Chatterjee, S., 2015. Chikungunya virus infection amongst the acute encephalitis syndrome cases in West Bengal, India. Indian J. Med. Microbiol. 33, 153–156. https://doi.org/10.4103/0255-0857.150946. Suppl.

Taylor, R.M., Hurlbut, H.S., Work, T.H., Kingston, J.R., Frothingham, T.E., 1955. Sindbis virus: a newly recognized arthropod transmitted virus. Am. J. Trop. Med. Hyg. 4, 844–862.

Teixeira, A.A.R., Ferreira, L.F., Pires Neto, R.D.J., 2019. Acute disseminated encephalomyelitis after Chikungunya infection. JAMA Neurol. 76, 619. https://doi.org/10.1001/jamaneurol.2019.0113.

Teng, T.-S., Kam, Y.-W., Lee, B., Hapuarachchi, H.C., Wimal, A., Ng, L.-C., Ng, L.F.P., 2015. A systematic meta-analysis of immune signatures in patients with acute Chikungunya Virus infection. J. Infect. Dis. 211, 1925–1935. https://doi.org/10.1093/infdis/jiv049.

Teo, T.-H., Lum, F.-M., Claser, C., Lulla, V., Lulla, A., Merits, A., Renia, L., Ng, L.F.P., 2013. A pathogenic role for CD4+ T cells during Chikungunya virus infection in mice. J. Immunol. 190, 259–269. https://doi.org/10.4049/jimmunol.1202177.

Thach, D.C., Kimura, T., Griffin, D.E., 2000. Differences between C57BL/6 and BALB/cBy mice in mortality and virus replication after intranasal infection with neuroadapted Sindbis virus. J. Virol. 74, 6156–6161.

Thanapati, S., Ganu, M., Giri, P., Kulkarni, S., Sharma, M., Babar, P., Ganu, A., Tripathy, A.S., 2017. Impaired NK cell functionality and increased TNF-α production as biomarkers of chronic chikungunya arthritis and rheumatoid arthritis. Hum. Immunol. 78, 370–374. https://doi.org/10.1016/j.humimm.2017.02.006.

Theilacker, C., Held, J., Allering, L., Emmerich, P., Schmidt-Chanasit, J., Kern, W.V., Panning, M., 2013. Prolonged polyarthralgia in a German traveller with Mayaro virus infection without inflammatory correlates. BMC Infect. Dis. 13, 369. https://doi.org/10.1186/1471-2334-13-369.

Touret, Y., Randrianaivo, H., Michault, A., Schuffenecker, I., Kauffmann, E., Lenglet, Y., Barau, G., Fourmaintraux, A., 2006. Early maternal-fetal transmission of the Chikungunya virus. Presse Med. 35, 1656–1658. https://doi.org/10.1016/S0755-4982(06)74874-6.

Tournebize, P., Charlin, C., Lagrange, M., 2009. Neurological manifestations in Chikungunya: about 23 cases collected in Reunion Island. Rev. Neurol. (Paris) 165, 48–51. https://doi.org/10.1016/j.neurol.2008.06.009.

Tritsch, S.R., Encinales, L., Pacheco, N., Cadena, A., Cure, C., McMahon, E., Watson, H., Porras-Ramirez, A., Mendoza, A.R., Li, G., Khurana, K., Jaller-Raad, J.J., Mejia Castillo, S., Barrios Taborda, O., Jaller-Char, J.J., Avendaño Echavez, L., Jiménez, D., Gonzalez Coba, A., Alarcon Gomez, M., Ariza Orozco, D., Bravo, E., Martinez, V., Guerra, B., Simon, G., Firestein, G.S., Chang, A.Y., 2019. Chronic joint pain 3 years after Chikungunya virus infection largely characterized by relapsing-remitting symptohms. J. Rheumatol. 46, 1–8. https://doi.org/10.3899/jrheum.190162.

Tsetsarkin, K.A., Vanlandingham, D.L., McGee, C.E., Higgs, S., 2007. A single mutation in Chikungunya virus affects vector specificity and epidemic potential. PLoS Pathog. 3, e201–e212. https://doi.org/10.1371/journal.ppat.0030201.

Ulloa-Padilla, J.P., Dávila, P.J., Izquierdo, N.J., García-Rodríguez, O., Jiménez, I.Z., 2018. Ocular symptoms and signs of Chikungunya fever in Puerto Rico. P. R. Health Sci. J. 37, 83–87.

Venigalla, S.S.K., Premakumar, S., Janakiraman, V., 2020. A possible role for autoimmunity through molecular mimicry in alphavirus mediated arthritis. Sci. Rep. 10, 938. https://doi.org/10.1038/s41598-019-55730-6.

Venugopalan, A., Ghorpade, R.P., Chopra, A., 2014. Cytokines in acute chikungunya. PLoS One 9, e111305. https://doi.org/10.1371/journal.pone.0111305.

Vieira, M.A.D.C.E.S., Costa, C.H.N., Linhares, A.D.C., Borba, A.d.S., Henriques, D.F., Silva, E.V.P.D., Tavares, F.N., Batista, F.M.d.A., Guimarães, H.C.L., Martins, L.C., Monteiro, T.A.F., Cruz, A.C.R., Azevedo, R.D.S.D.S., Vasconcelos, P.F.D.C., 2018. Potential role of dengue virus, chikungunya virus and Zika virus in neurological diseases. Mem. Inst. Oswaldo Cruz 113, e170538. https://doi.org/10.1590/0074-02760170538.

Vijayakumar, K.P., Nair Anish, T.S., George, B., Lawrence, T., Muthukkutty, S.C., Ramachandran, R., 2011. Clinical profile of Chikungunya patients during the epidemic of 2007 in Kerala, India. J. Glob. Infect. 3, 221–226. https://doi.org/10.4103/0974-777X.83526.

Villamil-Gómez, W., Silvera, L.A., Páez-Castellanos, J., Rodriguez-Morales, A.J., 2016. Guillain-Barré syndrome after Chikungunya infection: a case in Colombia. Enferm. Infecc. Microbiol. Clin. 34, 140–141. https://doi.org/10.1016/j.eimc.2015.05.012.

Villari, P., Spielman, A., Komar, N., McDowell, M., Timperi, R.J., 1995. The economic burden imposed by a residual case of eastern encephalitis. Am. J. Trop. Med. Hyg. 52, 8–13.

Vogel, P., Abplanalp, D., Kell, W., Ibrahim, M.S., Downs, M.B., Pratt, W.D., Davis, K.J., 1996. Venezuelan equine encephalitis in BALB/c mice: kinetic analysis of central nervous system infection following aerosol or subcutaneous inoculation. Arch. Pathol. Lab. Med. 120, 164–172.

Vogel, P., Kell, W.M., Fritz, D.L., Parker, M.D., Schoepp, R.J., 2005. Early events in the pathogenesis of eastern equine encephalitis virus in mice. Am. J. Pathol. 166, 159–171. https://doi.org/10.1016/S0002-9440(10)62241-9.

Wachira, V.K., Peixoto, H.M., de Oliveira, M.R.F., 2019. Systematic review of factors associated with the development of Guillain-Barré syndrome 2007-2017: what has changed? Trop. Med. Int. Health 24, 132–142. https://doi.org/10.1111/tmi.13181.

Walker, D.H., Harrison, A., Murphy, K., Flemister, M., Murphy, F.A., 1976. Lymphoreticular and myeloid pathogenesis of Venezuelan equine encephalitis in hamsters. Am. J. Pathol. 84, 351–370.

Walton, T.E., Alvarez Jr., O., Buckwalter, R.M., Johnson, K.M., 1973. Experimental infection of horses with enzootic and epizootic strains of Venezuelan equine encephalomyelitis virus. J. Infect. Dis. 128, 271–282.

Wang, R., Dillon, C.P., Shi, L.Z., Milasta, S., Carter, R., Finkelstein, D., McCormick, L.L., Fitzgerald, P., Chi, H., Munger, J., Green, D.R., 2011. The transcription factor myc controls metabolic reprogramming upon T Lymphocyte activation. Immunity 35, 871–882. https://doi.org/10.1016/j.immuni.2011.09.021.

Watson, H., Tritsch, S.R., Encinales, L., Cadena, A., Cure, C., Ramirez, A.P., Mendoza, A.R., Chang, A.Y., 2020. Stiffness, pain, and joint counts in chronic chikungunya disease: relevance to disability and quality of life. Clin. Rheumatol. 372, 1231–1238. https://doi.org/10.1007/s10067-019-04919-1.

Wauquier, N., Becquart, P., Nkoghe, D., Padilla, C., Ndjoyi-Mbiguino, A., Leroy, E.M., 2011. The acute phase of Chikungunya virus infection in humans is associated with strong innate immunity and T CD8 cell activation. J. Infect. Dis. 204, 115–123. https://doi.org/10.1093/infdis/jiq006.

Weaver, S.C., Ferro, C., Barrera, R., Boshell, J., Navarro, J.-C., 2004. Venezuelan equine encephalitis. Annu. Rev. Entomol. 49, 141–174. https://doi.org/10.1146/annurev.ento.49.061802.123422.

Weaver, S.C., Lecuit, M., 2015. Chikungunya virus and the global spread of a mosquito-borne disease. N. Engl. J. Med. 372, 1231–1239. https://doi.org/10.1056/NEJMra1406035.

Weaver, S.C., Salas, R., Rico-Hesse, R., Ludwig, G.V., Oberste, M.S., Boshell, J., Tesh, R.B., 1996. Re-emergence of epidemic Venezuelan equine encephalomyelitis in South America. VEE Study Group. Lancet 348, 436–440.

Webb, H.E., Mehta, S., Gregson, N.A., Leibowitz, S., 1984. Immunological reaction of the demyelinating Semliki Forest virus with immune serum to glycolipids and its possible importance to central nervous system viral auto-immune disease. Neuropathol. Appl. Neurobiol. 10, 77–84.

Wesselingh, S.L., Levine, B., Fox, R.J., Choi, S., Griffin, D.E., 1994. Intracerebral cytokine mRNA expression during fatal and nonfatal alphavirus encephalitis suggests a predominant type 2 T cell response. J. Immunol. 152, 1289–1297.

Weyand, C.M., Hicok, K.C., Conn, D.L., Goronzy, J.J., 1992. The influence of HLA-DRB1 genes on disease severity in rheumatoid arthritis. Ann. Intern. Med. 117, 801–806. https://doi.org/10.7326/0003-4819-117-10-801.

Wielanek, A.C., Monredon, J.D., Amrani, M.E., Roger, J.C., Serveaux, J.P., 2007. Guillain-Barré syndrome complicating a Chikungunya virus infection. Neurology 69, 2105–2107. https://doi.org/10.1212/01.wnl.0000277267.07220.88.

Willems, W.R., Kaluza, G., Boschek, C.B., Bauer, H., Hager, H., Schütz, H.J., Feistner, H., 1979. Semliki forest virus: cause of a fatal case of human encephalitis. Science 203, 1127–1129. https://doi.org/10.1126/science.424742.

Willison, H.J., Jacobs, B.C., van Doorn, P.A., 2016. Guillain-Barré syndrome. Lancet 388, 717–727. https://doi.org/10.1016/S0140-6736(16)00339-1.

Wollish, A.C., Ferris, M.T., Blevins, L.K., Loo, Y.-M., Gale, M., Heise, M.T., 2013. An attenuating mutation in a neurovirulent Sindbis virus strain interacts with the IPS-1 signaling pathway in vivo. Virology 435, 269–280. https://doi.org/10.1016/j.virol.2012.09.008.

Woodman, D.R., McMANUS, A.T., Eddy, G.A., 1975. Extension of the mean time to death of mice with a lethal infection of Venezuelan equine encephalomyelitis virus by antithymocyte serum treatment. Infect. Immun. 12, 1006–1011.

Wu, L.X., Mäkelä, M.J., Röyttä, M., Salmi, A., 1988. Effect of viral infection on experimental allergic encephalomyelitis in mice. J. Neuroimmunol. 18, 139–153. https://doi.org/10.1016/0165-5728(88)90062-8.

Yactayo, S., Staples, J.E., Millot, V., Cibrelus, L., Ramon-Pardo, P., 2016. Epidemiology of Chikungunya in the Americas. J. Infect. Dis. 214, S441–S445. https://doi.org/10.1093/infdis/jiw390.

Zaid, A., Gérardin, P., Taylor, A., Mostafavi, H., Malvy, D., Mahalingam, S., 2018. Chikungunya arthritis: implications of acute and chronic inflammation mechanisms on disease management. Arthritis Rheumatol. 70, 484–495. https://doi.org/10.1002/art.40403.

Zambrano, H., Waggoner, J.J., Almeida, C., Rivera, L., Benjamin, J.Q., Pinsky, B.A., 2016. Zika virus and Chikungunya virus coinfections: a series of three cases from a single center in Ecuador. Am. J. Trop. Med. Hyg. 95, 894–896. https://doi.org/10.4269/ajtmh.16-0323.

Zeller, H., Van Bortel, W., Sudre, B., 2016. Chikungunya: its history in Africa and asia and its spread to new regions in 2013-2014. J. Infect. Dis. 214, S436–S440. https://doi.org/10.1093/infdis/jiw391.

CHAPTER TEN

Reverse genetic systems: Rational design of coronavirus live attenuated vaccines with immune sequelae

Zhiqian Ma[†], Zhiwei Li[†], Linfang Dong, Ting Yang, Shuqi Xiao*

College of Veterinary Medicine, Northwest A&F University, Yangling, Shaanxi, China

*Corresponding author: e-mail address: xiaoshuqi@nwsuaf.edu.cn

Contents

Abstract

Since the end of 2019, the global COVID-19 outbreak has once again made coronaviruses a hot topic. Vaccines are hoped to be an effective way to stop the spread of the virus. However, there are no clinically approved vaccines available for coronavirus infections. Reverse genetics technology can realize the operation of RNA virus genomes at the DNA level and provide new ideas and strategies for the development of new vaccines.

[†] These authors contributed equally to this work.

Advances in Virus Research, Volume 107
ISSN 0065-3527
https://doi.org/10.1016/bs.aivir.2020.06.003

In this review, we systematically describe the role of reverse genetics technology in studying the effects of coronavirus proteins on viral virulence and innate immunity, cell and tissue tropism and antiviral drug screening. An efficient reverse genetics platform is useful for obtaining the ideal attenuated strain to prepare an attenuated live vaccine.

1. Introduction

Coronaviruses are large, enveloped, positive-sense, single-stranded RNA viruses with genome sizes ranging from 26 to 32 kb that are distributed broadly among humans, other mammals, and birds and cause respiratory, enteric, hepatic, and neurologic diseases (Coronaviridae Study Group of the International Committee on Taxonomy of Viruses, 2020; Cui et al., 2019). *Coronaviridae* belongs to the order *Nidovirales* along with three other families (*Arteriviridae*, *Mesoniviridae*, and *Roniviridae*). *Coronaviridae* is further classified into four genera (Fig. 1A) based on phylogenetic analyses and genomic structures, namely, *Alphacoronavirus*, *Betacoronavirus*, *Gammacoronavirus*, and *Deltacoronavirus* (Wu et al., 2020). The alphacoronaviruses and betacoronaviruses infect only mammals; gammacoronaviruses and deltacoronaviruses infect birds, but some of them can also infect mammals. The genome of coronaviruses is arranged in the order of the 5′ untranslated region (5′ UTR), open reading frame 1a/b (orf1ab), spike (S) protein, envelope (E) protein, membrane (M) protein, nucleoprotein (N) protein, 3′ UTR, and the poly (A) tail, with regions encoding accessory proteins,

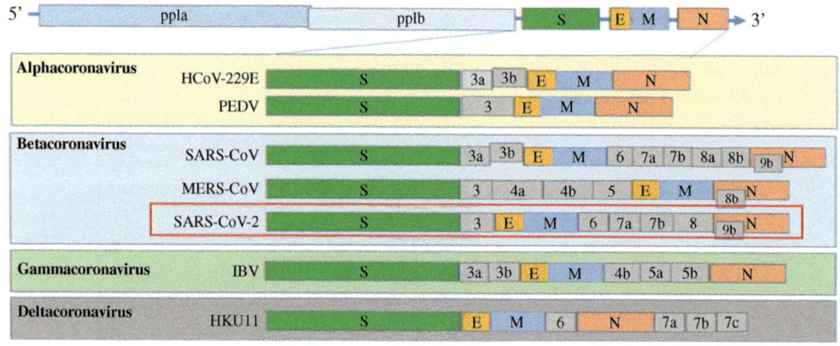

Fig. 1 The genomic structure of coronaviruses. (A) Classification of coronavirus, with the new coronavirus SARS-CoV-2 highlighted in red. (B) The genome structure of four genera of coronaviruses. Pp1a and pp1b represent the 2 long polypeptides that are processed into 16 nonstructural proteins. S, E, M, and N indicate the four structural proteins spike, envelope, membrane, and nucleocapsid proteins.

including orf3, 6, 7a, 7b, 8, and 9b, between those encoding structural proteins (Fig. 1B). The large replicase polyproteins pp1a and pp1ab, encoded by the partially overlapping 5′-terminal orf1a/b within the 5′ two-thirds of the genome, are proteolytically cleaved into 16 putative nonstructural proteins (nsps; Fig. 1B).

In early December 2019, a cluster of cases of pneumonia caused by a novel coronavirus named Sudden Acute Respiratory Syndrome Coronavirus 2 (SARS-CoV-2) resulted in tremendous challenges to China's public health and clinical treatment (Munster et al., 2020; Yan et al., 2020), and now it has been confirmed in more than 211 other countries and territories, causing a major global public health crisis (Day, 2020; Jernigan and Team, 2020; Peeri et al., 2020). SARS-CoV-2 belongs to the *Betacoronavirus* genus in the family *Coronaviridae*, and is closely related to severe acute respiratory syndrome coronavirus (SARS-CoV) and Middle East respiratory syndrome coronavirus (MERS-CoV) (Drosten et al., 2003; Wassenaar and Zou, 2020; Zaki et al., 2012). SARS-CoV, MERS-CoV and SARS-CoV-2 are three highly transmissible and pathogenic viruses that have emerged in humans in the past 2 decades and bats are considered their natural hosts (Zhou et al., 2020). They mainly infect the lower respiratory tract and cause severe pneumonia (Hotez et al., 2020), sometimes leading to fatal acute lung injury and acute respiratory distress syndrome, resulting in high morbidity and mortality. Cytokines and chemokines, which are triggered by the innate immune system during virus infections have long been thought to play an important role in immunity, but dysregulated and excessive immune responses may cause immunopathology (Channappanavar and Perlman, 2017). Research shows that robust SARS-CoV replication accompanied by delayed IFN-I signaling promotes the accumulation of pathogenic inflammatory monocyte-macrophages, resulting in elevated lung cytokine/chemokine levels, vascular leakage, and impaired virus-specific T cell responses (Channappanavar et al., 2016). In addition to SARS-CoV and MERS-CoV, SARS-CoV-2 infection also can induce proinflammatory cytokine responses, including interferon gamma (IFN-γ), tumor necrosis factor alpha (TNFα), interleukin-1 (IL-1), IL-2, IL-6 and IL-8 (Kadkhoda, 2020). In patients with severe COVID-19, lymphopenia is a common feature, with drastically reduced numbers of $CD4^+$ T cells, $CD8^+$ T cells, B cells and natural killer (NK) cells, as well as a reduced percentage of monocytes, eosinophils and basophils. Furthermore, SARS-CoV-2 may have an antibody-dependent enhancement effect (Cao, 2020; Xu et al., 2020).

Although there have been outbreaks of coronavirus diseases, such as SARS and MERS, when the novel coronavirus SARS-CoV-2 outbreak occurred, there were no effective vaccines or drugs available to prevent or cure the disease. Clinical development of multiple vaccines and screening of specific antiviral drugs were initiated in parallel with attempts to contain the outbreak. Currently, more than 115 candidate vaccines for SARS-CoV-2 are being developed globally. Five advanced candidates have recently moved into clinical development, including mRNA-1273 from Moderna, Ad5-nCoV from CanSino Biologicals, INO-4800 from Inovio, LV-SMENP-DC and pathogen-specific aAPC from Shenzhen Geno-Immune Medical Institute (Thanh Le et al., 2020).

It became apparent that coronaviruses can cross the species barrier and cause life-threatening infections in humans (Ji et al., 2020; Nishiura et al., 2020). Therefore, further attention needs to be paid to these new coronaviruses. In addition, coronaviruses causing high pathogenicity and high mortality in animals have attracted increasing attention. For example, highly virulent porcine epidemic diarrhea virus (PEDV), which appeared in China in 2010 and the United States in 2013, caused large global economic losses to the pig industry due to the lack of effective vaccines and drugs (Kong et al., 2019; Li et al., 2012; Stevenson et al., 2013). Another coronavirus, swine acute diarrhea syndrome coronavirus (SADS-CoV), which is a novel bat-HKU2-like coronavirus, was responsible for a large-scale fatal outbreak of in pigs in China that caused the death of more than 24,000 piglets across four farms (Zhou et al., 2018). Obviously, the role of vaccines is an effective way not only to enhance the ability of individuals to fight the virus but also to stop the spread of the virus. To be better prepared for future outbreaks of unknown human coronavirus pathogens, platform technologies to accelerate vaccine development should be employed.

2. Coronavirus reverse genetics systems

Reverse genetics systems are highly valuable research tools for RNA virus vaccine development, which may be more efficient than conventional approaches with live attenuation through passing (forward genetics) or inactivation (Stobart and Moore, 2014). Reverse genetics systems are useful tools for studying the modifications of viral genomes and for generating recombinant viruses to better understand their fundamental biology, develop novel vaccine candidates, and test antiviral therapeutics. Recently, reverse genetics techniques, including targeted RNA recombination, *in vitro* ligation and bacterial artificial chromosome systems, vaccinia virus vectors and transformation associated recombination (TAR) cloning, have been successfully used to manipulate the genome of coronaviruses (Fig. 2).

2.1 Reverse genetic system using targeted RNA recombination

Targeted RNA recombination was the first reverse genetic system devised for coronaviruses at a time when it was not clear whether a full-length clone could be successfully constructed (Masters and Rottier, 2005). Constructing an infectious clone of a coronavirus using targeted RNA recombination requires two steps: first, constructing a chimeric coronavirus that carries the mouse hepatitis virus (MHV) S gene and has a strict mouse cell affinity and second, rescuing the targeted recombinant coronavirus on a specific cell dependent on the virus (Fig. 3A). Targeted RNA recombination was the first reverse genetic system devised for MHV. Subsequently, the method was applied to the construction of full-length cDNA in other coronaviruses, including PEDV and feline infectious peritonitis virus (FIPV) (Haijema et al., 2003; Li et al., 2013). Targeted RNA recombination presents clear limitations; for example, it does not operate on parts of the genome that code for replicating enzymes.

2.2 Reverse genetic system using *in vitro* ligation

In vitro ligation uses unique type II restriction endonucleases (e.g., *Bgl*I, *Sap*I, and *Bsa*I) that cleave several bases away from their recognition site, allowing for the reassembly of authentic coronavirus genomes from smaller fragments. Detailed steps can be found in the published literature (Cockrell et al., 2017). The assembled full-length cDNA, containing a T7 RNA polymerase promoter at the 5′ end and a poly (A) tail at the 3′ end, is transcribed *in vitro* to generate capped full-length transcripts that are used together with capped N gene transcripts to efficiently rescue infectious virus after the transfection of susceptible cells (Fig. 3B). The first coronavirus full-length infectious cDNA clone was also generated for TGEV using *in vitro* ligation (Yount et al., 2000). In addition, SARS-CoV, MERS-CoV, PEDV and SARS-CoV-2 were constructed using *in vitro* ligation (Scobey et al., 2013; Xie et al., 2020; Yount et al., 2003; Zaki et al., 2012). This method can avoid the instability and virulence of viral cDNA in bacteria, as it depends on the T7 RNA polymerase.

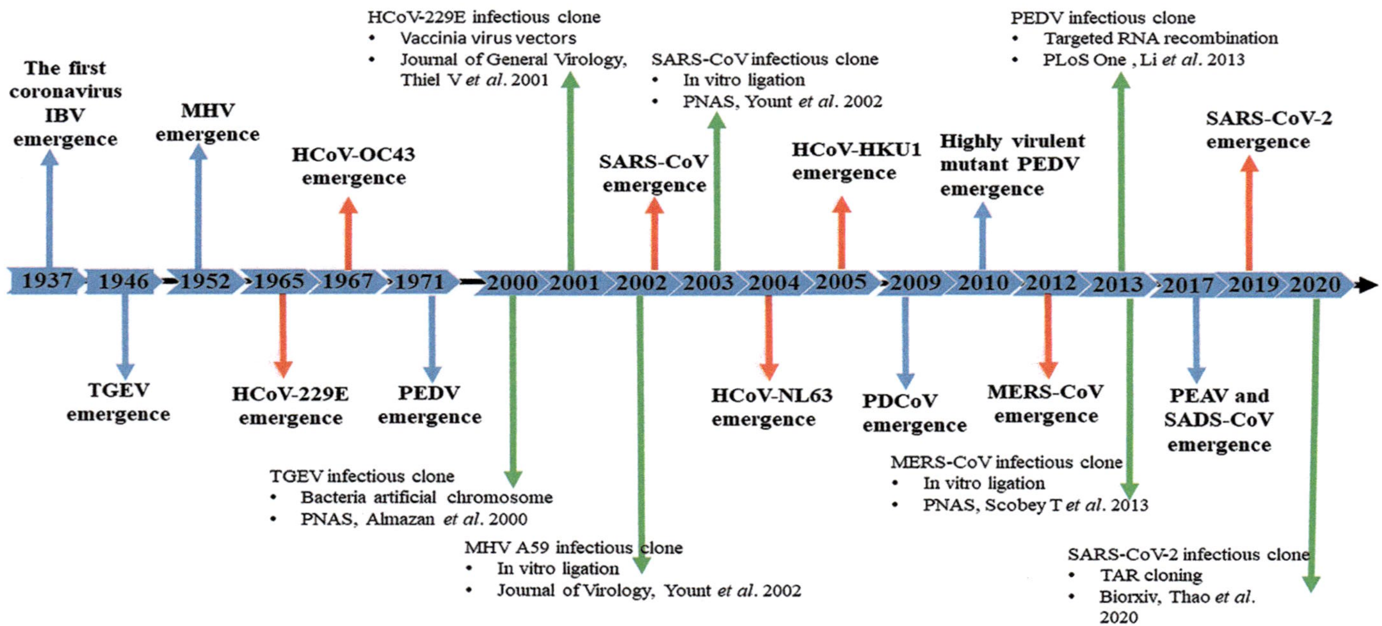

Fig. 2 Timeline of emerging representative coronavirus events and their infectious clones generated using reverse genetics systems. The timeline spans from the first coronavirus in history to the emergence of 2019-nCoV in 2019. Red arrows indicate human coronavirus outbreaks and identification events. Blue arrows indicate animal coronavirus outbreaks and identification events. Green arrows indicate the publication of infectious clones using different reverse genetics methods (Almazan et al., 2000; Li et al., 2013; Thao et al., 2020; Yount et al., 2002).

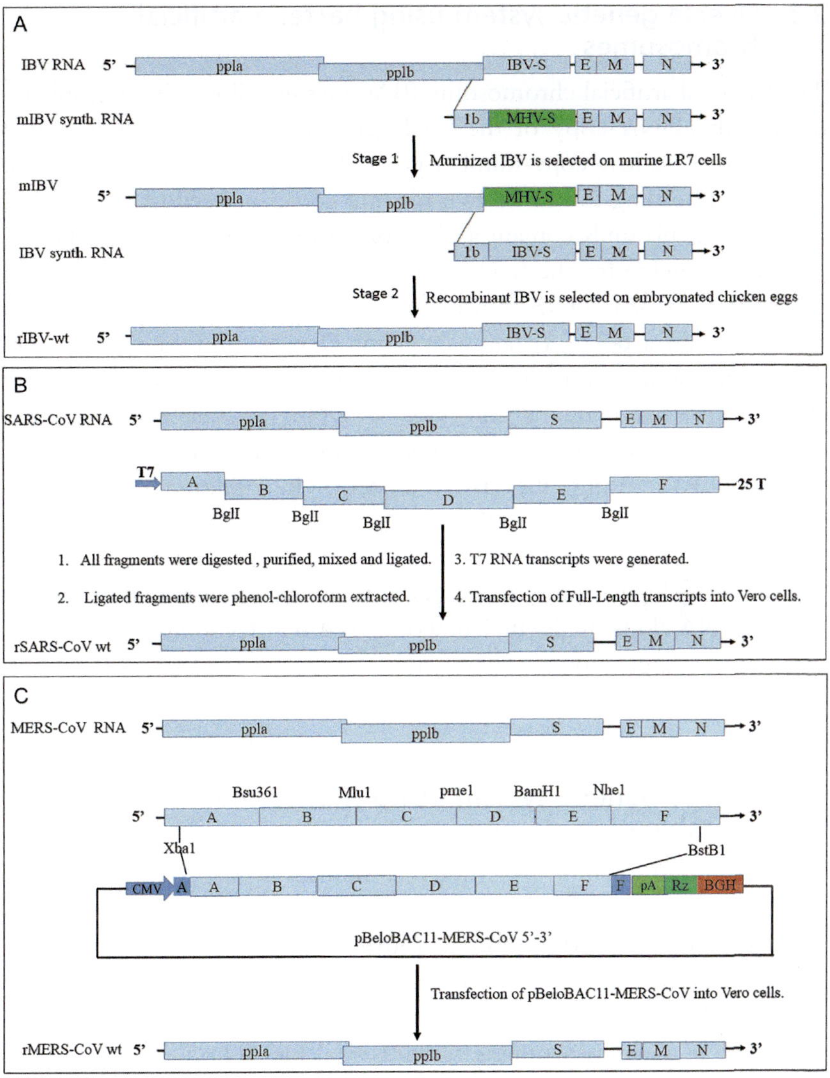

Fig. 3 Flowchart of three methods for constructing coronavirus infectious cloning. (A) Targeted RNA recombination scheme used to make interspecies chimeric viruses: mIBV (Stage 1) and recombinant IBV (Stage 2). (B) *In vitro* ligation. The SARS-CoV full-length genome is divided into seven segments, named A-F, with type II restriction endonucleases Bgl1 at both ends. (C) Bacterial artificial chromosome system. Restriction enzyme sites in the MERS-CoV genome were employed to ligate the full-length MERS-CoV genome into pBeloBAC11.

2.3 Reverse genetic system using bacteria artificial chromosomes

The bacterial artificial chromosome (BAC) system allows the assembly of a full-length cDNA copy of the viral genome in BAC plasmids such as pBeloBAC11, a low-copy-number plasmid that presents strictly controlled replication, leading to one or two plasmid copies per cell (Almazan et al., 2008). BAC plasmids contain a 5′ cytomegalovirus promoter of the viral genome, allowing for the transcription of the viral genome following the transfection of BAC DNA into mammalian cells. In addition, coronavirus BACs contain a poly (A) tail, a hepatitis D virus ribozyme, and bovine growth hormone termination and polyadenylation signals to create genomic RNA with an authentic 3′ end (Fig. 3C). The first full-length infectious coronavirus cDNA clone was generated for transmissible gastroenteritis virus (TGEV) using the BAC system. In addition, SARS-CoV, MERS-CoV, FIPV and PEDV have been successfully developed (Balint et al., 2012; Li et al., 2017). The BAC system presents several advantages, including the high stability of exogenous sequences, unlimited production of the cDNA clone, high efficiency of cDNA transfection into mammalian cells, easy manipulation for genes and intracellular expression of the viral RNA (Fehr, 2020).

2.4 Reverse genetic system using transformation associated recombination cloning

Recently, transformation associated recombination (TAR) cloning has been used to construct infectious clones of coronaviruses, including MHV, MERS-CoV and SARS-CoV-2 (https://www.biorxiv.org/content/10.1101/2020.02.21.959817v1). The overall strategy is as follows (Fig. 4A): viral RNA is prepared from several overlapping DNA fragments by RT-PCR or chemical synthesis (no virus available). The 5′- and 3′-terminal DNA fragments contain the T7-RNA polymerase promoter upstream of the coronavirus at the 5′ end, a restriction endonuclease cleavage site downstream of the poly(A) sequence, and overlapping sequences with the TAR plasmid pVC604. Then, all the fragments are used to transform *Saccharomyces cerevisiae* (strain VL6-48N), and the positive colonies are screened for the correct assembly of the yeast artificial chromosome (YAC) containing virus genome and purified the YAC. The plasmid is linearized using restriction endonuclease and subjected to T7 RNA polymerase-based *in vitro* transcription to generate transcripts. The RNA transcripts together with capped

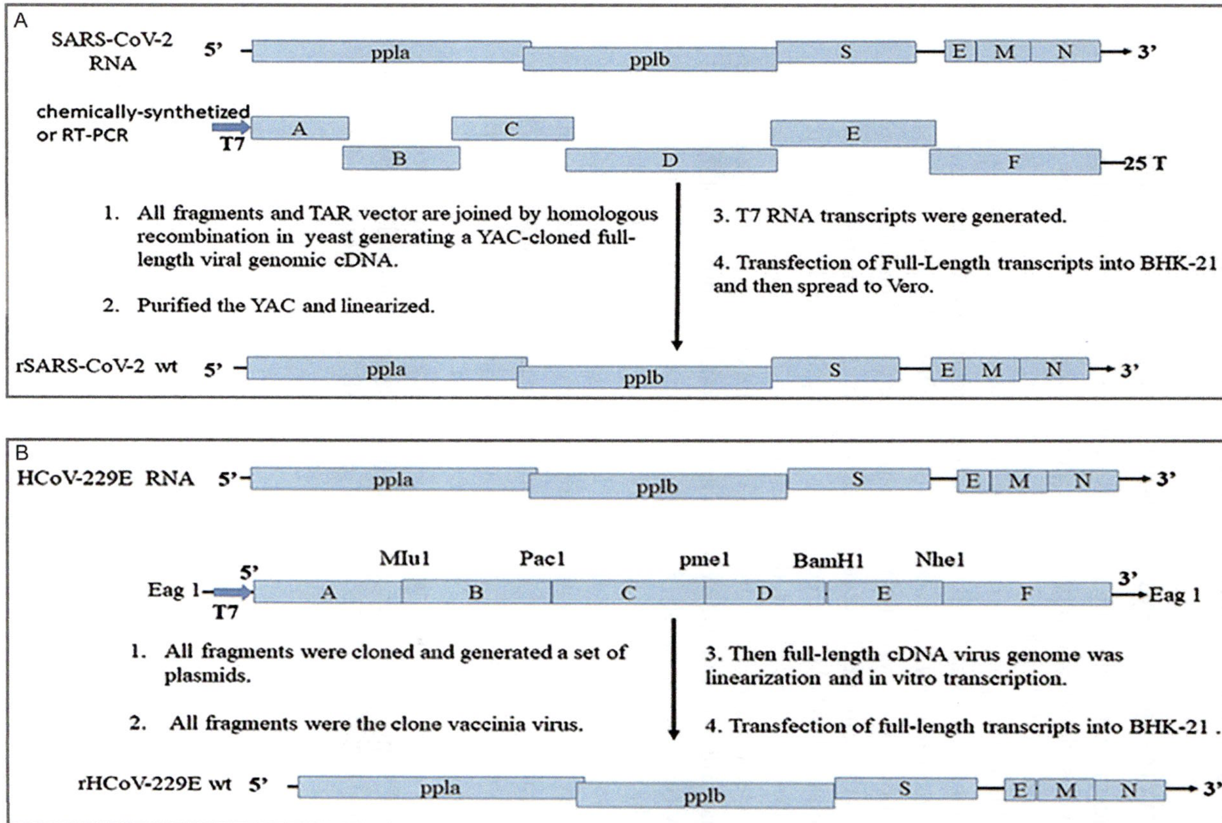

Fig. 4 Flowchart of two methods for constructing coronavirus infectious clones. (A) Transformation-associated recombination (TAR) cloning. A schematic representation of the general workflow of TAR cloning for SARS-CoV-2 infectious clones. (B) Vaccinia virus vectors. The general workflow of vaccinia virus vectors for constructing HCoV-229E infectious clones.

N gene transcripts are transfected into BHK cells to obtain the corresponding virions. The TAR cloning system can rapidly reassemble at least 14 fragments with remarkable efficacy (usually >90% of the clones are correct).

2.5 Reverse genetic system using vaccinia virus vectors

The vaccinia virus cloning vector also can be used to construct infectious coronavirus clones and represents a generic approach to reverse genetics (Thiel and Siddell, 2005). Vaccinia virus vectors allow large fragments (26–31 kb) of foreign genes to be inserted without affecting virus replication, and vaccinia viruses can easily produce high titer strains. The process of cloning coronavirus cDNA in vaccinia virus is as follows (Fig. 4B). First of all, the virus genome is analyzed for useful naturally encoded endonuclease restriction sites that can later be used to ligate cloned cDNA inserts and generate a set of plasmid DNAs that cover the full-length virus genomic sequence. Upstream of the viral 5′ end, there should be an *Eag*I or a Bsp120I restriction endonuclease site to allow insertion of the cDNA into the *Not*I site of the vaccinia virus genomic DNA by *in vitro* ligation. Downstream of the 3′ end of the virus genome, a poly(A) tail and unique restriction endonuclease site should be cloned. Second, long cDNA fragments are assembled by *in vitro* ligation to obtain a full-length virus cDNA fragment. Third, the full-length fragment or several fragments are inserted into a vaccinia virus genome by *in vitro* ligation. Finally, the vaccinia virus containing the full-length cDNA virus genome is linearized and used as a template for *in vitro* transcription using bacteriophage T7 RNA polymerase and transfected into BHK cells to obtain the corresponding virions. Currently, vaccinia virus vectors are successfully applied to reverse genetic manipulation of several coronaviruses, including HCoV-229E, IBV, MHV and SARS-CoV (Thiel et al., 2001; Thiel and Siddell, 2005; van den Worm et al., 2012).

2.6 Rapid gene manipulation of coronavirus

Chemical synthesis of the viral genome is widely used for the construction of infectious clones. Using a chemical synthesis system, the complete genome of the virus can be synthesized to construct a full-length genomic cDNA clone, especially in the absence of a natural template. This method was recently used to construct an infectious clone of SARS-like CoV (Becker et al., 2008). In general, the ability to quickly and efficiently introduce virtually any type of genetic modification (point mutations, insertions, or deletions) into the coronavirus genome and recover recombinant virus is desirable. CRISPR/Cas9 technology is an efficient platform for manipulating the modification of full-length infectious cloning of coronaviruses, which can generate recombinant viruses within a week (Peng et al., 2020; Shen et al., 2019; Wang et al., 2019a). Moreover, a combination of bacterial artificial chromosome and lambda red recombination with the I-*Sce*I homing endonuclease method was used to engineer genetic alterations of the MERS-CoV (Fehr, 2020; Muth et al., 2017). Importantly, the optimized reverse genetics platform used in this study will simplify the construction of mutant infectious clones and help accelerate the progress in coronavirus research.

3. Rational design of vaccine candidates using reverse genetics technology

Vaccination is expected to be an efficacious strategy in preventing individuals and animals from suffering from coronavirus infections. Virus vaccines must be immunogenic and sufficiently stable, safe, and suitable to induce long-lasting immunity. To date, various kinds of candidate vaccines for coronaviruses have been developed, including live attenuated vaccines, subunit vaccines, DNA vaccines, inactivated vaccines, and recombinant vector vaccines. Each vaccine type has different advantages and disadvantages. For example, traditional inactivated vaccines can generally induce highly potent immune responses and/or protection, but they may induce the antibody-dependent enhancement effect. In addition, they exit recovery of virulence if incomplete inactivation of viruses, resulting in significant safety concerns (Luo et al., 2018). Although DNA vaccines maintain strong safety, they are generally less immunogenic than are whole-virus-based vaccines, usually requiring the optimization of sequences or immune routes; and adjuvant coordination is also often required to achieve immune effects. Among them, live attenuated vaccines are considered highly effective because of their ability to replicate within host cells, inducing high levels of antigenic stimulation and robust long-term immunological memory. However, a major safety concern with live attenuated vaccines is the possibility of reversion to a pathogenic form and oral or aerosol route vaccination that can induce mucosal immune responses should be considered.

3.1 Live attenuated vaccines

Virus-infected cells react quickly to invading viruses by producing type I interferons (IFN-α/β) and type III interferons (IFN-$\lambda 1/\lambda 2/\lambda 3/\lambda 4$) and establish an innate antiviral state, which provides a first line of defense against viral infection. Type III IFNs are mainly produced by epithelial cells and play a major role in diseases that cause mucosal immunity, such as PEDV and PDCoV. However, most coronaviruses commonly suppress interferon levels in infected cells. Reverse genetics technology provides a very useful platform in the study of virulence and host innate immune-related mutations and in the rational design of coronavirus live attenuated vaccine candidates. According to current research, the 5′ UTR, several structural proteins, nonstructural proteins and the 3′ UTR of coronaviruses all have more or less influence on virulence and the life cycle of the virus. Here, we summarize the coronavirus genes associated with virulence, replication and innate immunity (Table 1) and potential vaccines based on reverse genetics (Table 2).

It is well known that among four coronavirus genera, only alphacoronaviruses and betacoronaviruses produce the nsp1 protein, which is located at the N-terminal of the replicase polyprotein pp1a. In terms of both sequence identity and size, nsp1 of alphacoronaviruses shares low similarity with nsp1 of betacoronaviruses. To date, the crystal structure of nsp1 from several coronaviruses (SARS-CoV, TGEV, and PEDV) has been resolved, which is important to understand the role of nsp1 in the pathogenesis of viruses and its interaction with host cell proteins (Shen et al., 2018, 2019). The nsp1 protein of coronaviruses is a vital virulence factor widely demonstrated in both alphacoronaviruses and betacoronaviruses (Jimenez-Guardeno et al., 2015; Lokugamage et al., 2015; Shen et al., 2018, 2019; Zhang et al., 2015). Not only inhibiting the expression of host genes with different mechanisms, but also playing a significant regulatory role in evading the host's natural immune response, nsp1 mainly inhibits the production of IFNs (Zhang et al., 2016, 2018b). Additionally, some studies have shown that abolishing the anti-IFN function of nsp1 can attenuate viruses, which has been demonstrated in MHV and SARS-CoV (Jimenez-Guardeno et al., 2015; Lei et al., 2013).

The nsp2 proteins of some coronaviruses are dispensable for viral replication in cell culture, but those viruses lacking nsp2, such as MHV and SARS-CoV, show attenuated viral growth and RNA synthesis (Graham et al., 2005). However, the pathogenicity of these viruses in mice has not been further reported. In addition, amino acids 1–120 of TGEV nsp2 are involved in the regulation of inflammation via NF-κB activation (Wang et al., 2018b).

The multidomain in nsp3, an essential component of the replication/transcription complex, is the largest protein encoded by the coronavirus

Table 1 Summary of different proteins (nonstructural, structural, and accessory proteins) in coronavirus pathobiology.

Classification	Protein	Pathobiology	Reference
Nonstructural proteins	nsp1	A vital virulence factor; inhibits the expression of the host gene; inhibits IFN production	Shen et al. (2019) and Zhang et al. (2018b)
	nsp2	Related to viral replication; regulates inflammation *via* NF-κB activation	Graham et al. (2005) and Wang et al. (2018b)
	nsp3	An essential component of the replication/transcription complex; inhibits IFN production	Lei et al. (2018)
	nsp4	Related to cellular membrane rearrangements and viral replication	Beachboard et al. (2015)
	nsp5	Viral replication; inhibits IFN production; cleaves polypeptides	Chen et al. (2019)
	nsp6	Related to autophagy and cellular membrane rearrangements	Angelini et al. (2013) and Cottam et al. (2014)
	nsp7	Inhibits IFN production (PEDV)	Zhang et al. (2016)
	nsp8	Inhibits IFN production (PEDV)	Zhang et al. (2018b)
	nsp9	Dimerization and RNA binding	Egloff et al. (2004)
	nsp10	Interacts with both nsp14 and nsp16; regulates SARS-CoV replication function and replication fidelity	Bouvet et al. (2014) and Smith et al. (2015)
	nsp14	Has 3′-to-5′ exoribonuclease activity and N7-methyltransferase (N7-MTase) activities; vital for virulence; inhibits IFN production	Case et al. (2018) and Graepel et al. (2017)
	nsp15	Endoribonuclease; vital virulence factor	Deng et al. (2017) and Zhang et al. (2018a)
	nsp16	Has 2′-*O*-methyltransferase (MTase) activity; vital virulence factor; negatively regulates innate immunity	Hou et al. (2019b) and Yong et al. (2019)

Continued

Table 1 Summary of different proteins (nonstructural, structural, and accessory proteins) in coronavirus pathobiology.—cont'd

Classification	Protein	Pathobiology	Reference
Structural proteins	S	Receptor binding and membrane fusion; vital virulence factor; target protein of a subunit vaccine	Hou et al. (2019a) and Shang et al. (2020a)
	M	Morphogenesis or assembly virus; inhibits IFN production	Lui et al. (2016)
	E	Involved in the viral life cycle, including assembly, budding, and pathogenesis; vital virulence factor; ion channel	Schoeman and Fielding (2019) and Shang et al. (2020b)
	N	Regulates viral RNA synthesis, packaging of the viral RNA in helical nucleocapsids and in virion assembly; inhibits IFN production	Lu et al. (2011) and Veit et al. (2018)
Accessory proteins	ORF3	Virulence factor; dispensable; inhibits IFN production; ion channel	Menachery et al. (2017b) and Van Beurden et al. (2018)
	ORF4	Virulence factor; dispensable	Menachery et al. (2017b)
	ORF5	Virulence factor; dispensable	Menachery et al. (2017b) and Van Beurden et al. (2018)
	ORF6	Inhibits IFN production	Hu et al. (2017) and Kopecky-Bromberg et al. (2007)
	NS6	Virulence factor; dispensable; inhibits IFN production	Fang et al. (2018)
	ORF8a	Virulence factor; dispensable	Castano-Rodriguez et al. (2018)

Table 2 Summary of the application of reverse genetics in live attenuated vaccines of different members of *Coronaviridae*.

Coronaviridae	Virus	Mutation sites	Evaluation	Reference
Alphacoronavirus	PEDV	nsp15: H262A	Attenuation	Deng et al. (2019)
		nsp16: KDKE/AAAA	Attenuation and protective immunity against parental strain infection	Hou et al. (2019a)
		S: ΔYxxΦEKVHVQ of the S protein or Δ197aa	ΔYxxΦEKVHVQ: attenuation Δ197aa: attenuation, reduces the protection	Hou et al. (2017, 2019b)
		ORF3: ΔORF3	Attenuation	Beall et al. (2016)
Betacoronavirus	MERS-CoV	Deletion: ΔE,3, 4a, 4b, and 5	None	Almazan et al. (2013)
		nsp16: D130A	Attenuation and protection against a lethal MERS-CoV challenge	Menachery et al. (2017a)
	MHV	nsp1: Δ99 nts (829–927 nt) or Δ27 nts between nts 780–808 (LLRKxGxKG)	Δ99 nts: attenuation and protection of mice from homologous and heterologous viral infections Δ27 nts: attenuation and protection against challenge with WT MHV-A59 virus	Lei et al. (2013) and Zust et al. (2007)
		nsp2: Δnsp2	None	Graham et al. (2005)
		nsp3: V787S or N1347A	V787S: reduced pathogenesis and protection against challenge with WT virus N1347A: reductions in lethality and weight loss	Fehr et al. (2015) and Mielech et al. (2015)

Continued

Table 2 Summary of the application of reverse genetics in live attenuated vaccines of different members of *Coronaviridae*.—cont'd

Coronaviridae	Virus	Mutation sites	Evaluation	Reference
		nsp5: T26I/D65G	Attenuation	Deng et al. (2014)
		nsp15: H262A	Attenuation and protective immunity against WT virus infection	Deng et al. (2017)
	SARS-CoV	nsp1: ΔLLRKNGNKG (121–129aa)Δ EDYEQNWNTKH (154–164aa)	No significant weight loss and 100% survived the challenge	Jimenez-Guardeno et al. (2015)
		nsp2: Δnsp2	None	Graham et al. (2005)
		nsp3: N1040A	Attenuation	Fehr et al. (2016)
		nsp14: D90A/E92A	Reductions in weight loss and lung titer	Graham et al. (2012)
		nsp16: D130A	Attenuation in a variety of pathogenic outcomes and minimal weight loss	Menachery et al. (2014)
		E: ΔPBM of the E protein	Decreased lung pathology	Jimenez-Guardeno et al. (2014)
Gammacoronavirus	IBV	Δ3ab/5ab	Attenuation and protection against a homologous challenge	Van Beurden et al. (2018)
		Δ3ab/5ab of Beaudette and the S gene was replaced by the S gene from the M14 strain	Attenuation and protection against a M14 strain challenge	Van Beurden et al. (2018)
Deltacoronavirus	PDCoV	NS6:ΔNS6	Attenuation	Zhang et al. (2020)

genome, with an average molecular mass of approximately 200 kDa (Lei et al., 2018). In all coronaviruses, highly conserved functional domains within nsp3 are present: papain-like protease domains, ubiquitin-like domains, and ADP-ribose-phosphatase domains (Lei et al., 2018). Using a recombinant SARS-CoV strain with reduced nsp3 de-ADP-ribosylation activity showed that this mutant strain led to virus attenuation in mice but protected them from an otherwise lethal SARS-CoV infection and significantly enhanced the innate immune response, indicating that it is an important virulence factor for SARS-CoV (Fehr et al., 2016). V787S of nsp3 is thought to induce attenuation of MHV, and the mutant MHV strain elicits protective immunity (Mielech et al., 2015). N1347A of nsp3 is also thought to be associated with the virulence of MHV (Fehr et al., 2015). PEDV and TGEV (590–1215aa, via NF-κB) nsp3 proteins have been proven to evade the host immune system (Wang et al., 2019b). In addition, 102 and 61 amino acid substitutions were found in nsp2 and nsp3 of SARS-CoV-2 compared with SARS-CoV or SARS-like bat CoV, respectively, and these differences warrant further investigation (Wu et al., 2020).

The nsp4 genes of coronaviruses are significant in viral replication and pertain to cellular membrane rearrangements (Beachboard et al., 2015). Both H120 and F121 in SARS-CoV nsp4 play critical roles in viral replication (Sakai et al., 2017). However, whether nsp4 affects the virulence of the virus needs to be further verified by using reverse genetics.

The coronavirus nsp5 protease (3CLpro; Mpro) processes nsp proteins at 11 cleavage sites and is essential for virus replication, making it a high value target for the development of anti-coronavirus therapeutics (Jo et al., 2020; Tomar et al., 2015). It has been reported that coronavirus nsp5 cleaves NF-κB essential modulator and STAT2, inhibiting IFN production, which has been verified in FIPV, PEDV and PDCoV (Chen et al., 2019; Wang et al., 2016; Zhu et al., 2017). MHV nsp5 T26I/D65G, which confers resistance to a broad-spectrum coronavirus 3C-like protease inhibitor, also has a slight effect on the virulence and replication of MHV (Deng et al., 2014).

Coronavirus nsp6 is associated with autophagy (Cottam et al., 2014). Scientists used reverse genetics to verify that deletion of any of regions encoding nsp7, nsp8, nsp9 and nsp10 in MHV was lethal to virus (Deming et al., 2006). Furthermore, PEDV nsp7 was found to inhibit the IFN-β and IRF3 promoter activities (Zhang et al., 2016), and nsp8 can suppress type III IFN activities (Zhang et al., 2018b).

The 148-amino acid nsp10 subunit contains 2 zinc fingers and is known to interact with both nsp14 and nsp16, stimulating their respective 3′-5′

exoribonuclease and 2′-O-methyltransferase activities. A series of studies have shown that nsp10 is a major regulator of the SARS-CoV replication function (Bouvet et al., 2014). The single and double mutations R80A/E82A-ExoN(+) in MHV nsp10 disrupt the nsp10–nsp14 interaction, and compared to the wild-type virus ExoN(+), these mutations rendered the virus 5 and 10 times more sensitive to treatment with the RNA mutagen 5-fluorouracil, respectively. The results showed that nsp10 is important for CoV replication fidelity and support the hypothesis that nsp10 functions to regulate nsp14-ExoN activity during viral replication (Smith et al., 2015).

Coronavirus nsp14 has 3′-to-5′ exoribonuclease (ExoN) and N7-methyltransferase (N7-MTase) activities. The former is a proofreading function that is required for high-fidelity replication. The inactivation of MHV-CoV ExoN(−) activity by the nsp14 mutations D89A and E91A are relatively stable, even after passaging MHV-ExoN(−) 250 times, without reversion of the ExoN(−) mutation site. However, novel amino acid changes within the RNA-dependent RNA polymerase and nsp14 of MHV-ExoN(−) P250, indicate that multiple replicase proteins could compensate for the ExoN function during replication and will likely inform the design of countermeasures for endemic and emerging CoVs by defining novel common targets for stable virus attenuation or direct inhibition (Graepel et al., 2017). MHV-ExoN activity is required for resistance to the innate immune response (Case et al., 2018). The inactivation of SARS-CoV ExoN activity by the nsp14 mutations D90A and E92A was attenuated in both young and aged disease models compared to virulent WT, and it provided complete protection against lethal challenge in a susceptible, immunosenescent mouse model of the lethal SARS-CoV infection (Graham et al., 2012). The recombinant TGEV virus with a mutation in zinc finger 1 of the ExoN domain of nsp14 induced weak antiviral responses, including reduced expression of beta interferon (IFN-β), tumor necrosis factor (TNF), and interferon-stimulated genes. Therefore, coronavirus nsp14 plays a potential role in regulating innate immune responses (Becares et al., 2016). In addition, PEDV nsp14 was found to suppress type I and type III IFN activities (Zhang et al., 2016, 2018b).

Nsp15 encoded by coronavirus is a nidoviral uridylate-specific endoribonuclease (NendoU) that plays an essential role in the life cycle of the virus (Zhang et al., 2018a). MHV with the nsp15 H262A mutation that inactivates the EndoU activity of nsp15, resulted in greatly attenuated disease in mice and stimulated a protective immune response (Deng et al., 2017). PEDV EndoU activity is also a key virulence factor and coronavirus nsp15 is a target for generating live attenuated vaccine (Deng et al., 2019).

Nsp16 has 2′-*O*-methyltransferase (MTase) activity and it is conserved across the entire coronavirus family, increasing its appeal as an attenuation target (Yong et al., 2019). Several recent works have shown that although disruption of nsp16 activity in MHV (D129A), SARS-CoV (D130A), MERS-CoV (D130A) and PEDV (KDKE to AAAA) rendered an attenuated strain, it induced stronger type I and type III interferon responses and protected animals from the challenge (Hou et al., 2019a; Menachery et al., 2017a, 2018). However, without sufficiently attenuated virulence, the nsp16 mutant could be combined with another attenuating mutation in nsp1, nsp14, or S protein, which will produce a stable, attenuated virus capable of protection from heterologous challenge (Hou et al., 2019a; Menachery et al., 2018).

The coronavirus S protein, which is a class I fusion protein located at the surface of the virus, includes the receptor binding S1 subunit and the membrane fusion S2 subunit (Shang et al., 2020a; Walls et al., 2020). The coronavirus S protein is considered a major target for the development of subunit vaccines and major virulence genes. Two chimeric viruses with the reciprocally exchanged S gene were generated; one is a highly virulent strain, and the other is an avirulent strain, showing that the S gene is only one of the necessary determinants for virus virulence (Wang et al., 2018a). Another report examined the role of the S gene in PEDV pathogenesis with different generations of strains in the same way. Although the results showed that the S gene plays a role in virulence, other genes might also important in determining virulence (Kao and Chang, 2019). Three recombinant IBVs, BeauR-M41 (S1), BeauR-QX (S1) and BeauR-M41 (S), based on the BeauR backbone expressing a heterologous S1 or S from M41 or QX conferred incomplete protection against homologous challenge based on ciliary activity and clinical signs. However, the protection efficiency of BeauR-M41(S) was higher than that of BeauR-M41 (S1) (Ellis et al., 2018). Coronavirus virions assemble at the endoplasmic reticulum (ER)-Golgi intermediate compartment (ERGIC) and some of them, including SARS-CoV, TGEV, IBV, and PEDV, possess conserved motifs of YxxΦ and/or KxHxx/KKxx in the cytoplasmic tail of the S protein (Hou et al., 2019b; Ujike et al., 2016). A recombinant PEDV with deletions or a mutation in the two motifs (YxxΦ and/or KxHxx/KKxx) showed that the YxxΦ motif triggers endocytosis of S proteins and that the motif KVHVQ is involved in the retention of the S proteins in the ER-Golgi intermediate compartment; the loss of both motifs significantly enhances syncytia formation in Vero cells and reduces virulence in pigs (Hou et al., 2019b).

The NTD of the S protein can attenuate a highly virulent coronavirus, which was confirmed by a recombinant virus (icPC22A-S1Δ197aa) and (icTGEV-S1Δ224aa), but the virus may lose important epitopes for inducing robust protective immunity, which provides new insights for the development of a new attenuated vaccine (Hou et al., 2017; Wang et al., 2019a). In addition, 27 amino acid substitutions were found in the SARS-CoV S protein compared with SARS or SARS-like CoV, and whether these differences could affect the host tropism and transmission property is worthy of future investigation (Wu et al., 2020).

The coronavirus E protein is the smallest protein of the major structural proteins and is an integral membrane protein involved in several aspects of the virus life cycle, including assembly, budding, and pathogenesis (Schoeman and Fielding, 2019). Viruses lacking E protein form infectious virions, but their replication efficiency and virulence are significantly reduced. This seems to indicate that the E protein is an important virulence factor in coronavirus but is not essential for some coronaviruses (Castano-Rodriguez et al., 2018; Fett et al., 2013; Jimenez-Guardeno et al., 2015; Shang et al., 2020b). MERS-CoV and TGEV lacking the E gene are not successfully rescued, but they can spread from cell to cell via expression of the E protein (Almazan et al., 2013; Ortego et al., 2002).

Coronavirus M proteins are major structural proteins and are required for virion assembly. The S, E and M proteins can form virus-like particles (VLPs) with structural integrity (Wang et al., 2017). Both peptides (M1–31 and M132–161) in the SARS-CoV M protein are immunogenic (He et al., 2005), but the M protein of SARS-CoV has not been investigated for its protective efficacy against SARS-CoV infection. Several studies have shown that the coronavirus M protein also inhibits IFN production (Shokri et al., 2019; Siu et al., 2009). The MERS-CoV M protein inhibits type I IFN production through the inhibition of TBK1-dependent phosphorylation of IRF3 (Lui et al., 2016).

Coronavirus N proteins have been associated with multiple functions in the viral life cycle, including the regulation of viral RNA synthesis, the packaging of the viral RNA in helical nucleocapsids, and virion assembly through the interaction with the viral M protein (Mcbride et al., 2014). Expression of the N protein is also necessary for efficient recovery of the virus from infectious cDNA clones. Unlike the S protein, the N protein has no ability to elicit neutralizing antibodies, but it may induce specific antibody and cellular immune responses. The N protein plays an important role in viral pathogenesis since BALB/c mice immunized with recombinant virus MVA-MERS-N exhibit stronger T cell responses and anti-N monoclonal antibodies protect mice from lethal infection by MHV (Nakanaga et al., 1986; Veit et al., 2018). Immunodominant SARS-CoV-N regions N1 (1–422aa) and N3 (110–422aa) produce specific antigens in BALB/c mice and react with the serum of SARS patients; hence, they can be used as effective SARS DNA vaccines (Dutta et al., 2008). The SARS-CoV N protein is also an IFN antagonist that inhibits IFN synthesis by inhibiting IRF-3 and NF-κB, interfering with TRIM25-mediated RIG-I ubiquitination or attenuating PTAC-mediated RIG-I/MDA5 activation (Ding et al., 2017; Hu et al., 2017).

Coronavirus accessory proteins have been generally regarded as dispensable for *in vitro* viral replication. For example, PEDV with the ORF3 gene entirely deleted could replicate similarly to wild-type PEDV (Beall et al., 2016). Likewise, deletions of NS6 from PDCoV, ORF8a from SARS-CoV, ORF3, ORF4 and ORF5 from MERS-CoV and ORF3 and ORF5 from IBV do not affect replication, and the deletion of these accessory proteins could attenuate viruses, showing that accessory proteins are also related to the virulence of viruses (Almazan et al., 2013; Castano-Rodriguez et al., 2018; Menachery et al., 2017b; Zhang et al., 2020). The accessory proteins of coronaviruses are often used as target regions for the insertion of exogenous gene since they are dispensable. Notably, some accessory proteins have ion channels and are related to the innate immune response (Kint et al., 2016; Wong et al., 2018; Yue et al., 2018; Zeng et al., 2016). Some accessory proteins interact with host cells, potentially helping viruses evade the immune system and increase their virulence (Comar et al., 2019; Fang et al., 2018). ORF4a, ORF4b and ORF8b of MERS-CoV seem to be involved in viral evasion of NF-κB mediated host immune responses (Lee et al., 2019). Therefore, the mutant virus lacking some accessory proteins is potentially a safe and promising vaccine candidate to prevent coronavirus infection.

3.2 Recombinant vector vaccines

To date, viral vectors utilizing modified vaccinia virus Ankara, measles virus, RV vectors or adenovirus have been used to express the glycoprotein of coronaviruses, including SARS-CoV and SARS-CoV-2, and have been shown to be immunogenic according to *in vivo* assays (Kato et al., 2019; Thanh Le et al., 2020). To date, a recombinant novel coronavirus vaccine using an adenovirus type 5 vector that expresses the S protein has entered phase I clinical trials (Thanh Le et al., 2020). Researchers designed recombinant replication-deficient adenovirus-based vaccines expressing the MERS-CoV S protein that induced the highest neutralizing antibody titer and the strongest cytokine-induced T cell responses, and these may confer protection against MERS-CoV infection (Kim et al., 2019). Recombinant parainfluenza virus 5 (PIV5)-expressing the MERS-CoV S protein can induce neutralizing antibodies and robust T cells in a human DPP4 knock-in C57BL/6 congenic mouse model (hDPP4 KI) and protect mice from fatal MERS-CoV infection (Li et al., 2020b). An attenuated parainfluenza virus encoding the full-length S protein of SARS-CoV was used for the vaccination of African green monkeys, and monkeys with this vaccine were protected from subsequent homologous SARS-CoV infection (Bukreyev et al., 2004). Another novel recombinant influenza A virus (H1N 1pdm09) with MERS-CoV was generated, and the inactivated chimeric bivalent vaccine induced potent and specific neutralizing antibodies against MERS-CoV and H1N1pdm09 in BALB/c mice (Shehata et al., 2019). The use of reverse genetics to construct recombinant vaccine strains could achieve a single immunization and prevent two diseases.

4. Study the cell and tissue tropism of coronaviruses using reverse genetics

Coronaviruses generally exhibit restricted cell and tissue tropism, which is dependent on the S glycoprotein of individual coronavirus strains (Millet and Whittaker, 2015). Understanding the cell and tissue tropism of a virus is helpful to further research the pathogenesis of the virus and save vaccine production costs. Vaccines against IBV, both live attenuated and inactivated, are currently produced in embryonated hen eggs, which is a cumbersome and expensive process, because most IBV strains do not replicate in cultured cells. Previous studies have shown that the S2 subunit of the avirulent BeauR strain is responsible for its extended cellular tropism for Vero cells (Bickerton et al., 2018b). Later, recombinant IBVs with the immunogenic S1 subunit were derived from the IBV vaccine strain and the virulent field strain was generated using reverse genetics. The recombinant IBVs are able to replicate in both primary chicken kidney and Vero cells, which would enable IBV vaccines to be grown in cell lines rather than expensive embryonated hen eggs (Bickerton et al., 2018a). Moreover, recombinant IBV with a mutant S2′ site (furin S2′ site) led to neurotropism (Cheng et al., 2019). TGEV has both enteric and respiratory tropism. TGEV S219A in the S protein was required to confer enteric tropism, and a 6nt insertion at position 1124 could increase virus stability and virus titers after passage in cell cultures (Sanchez et al., 2019). The MHV amino acid lysine at 194 in the nsp1 protein can promote virus replication in the liver, and this study further confirmed that nsp1 is a betacoronavirus virulence factor (Zhang et al., 2015).

5. Screening antiviral drugs using reverse genetics

A recombinant coronavirus would efficiently express green fluorescent protein (GFP), which is commonly applied as a tool for high-throughput drug screens and neutralizing antibody assays. Feline coronavirus (FCoV) is one of the most significant coronaviruses. Tissue culture-adapted type I FCoV often loses pathogenicity, which complicates research on type I FCoV-induced feline infectious peritonitis (FIP). Therefore, researchers established a recombinant reporter C3663 virus carrying the nanoluciferase (Nluc) gene. They examined the inhibitory effect of 68 compounds on C3663 replication in Fcwf-4 cells and infectivity in a canine-derived cell line by using the reporter C3663 virus. Finally, they successfully screened a canine cell line, A72, that permitted FCoV replication but with low efficiency and aberrant viral gene expression (Terada et al., 2019).

6. Conclusion and future directions

Coronaviruses have repeatedly crossed species barriers, and some have emerged as important human pathogens. The best-known examples include SARS-CoV, MERS-CoV and SARS-CoV-2 (Chan et al., 2020), all of which pose a marked threat to human health. In addition, four other coronaviruses are known to infect humans, including HCoV-229E, HKU-NL63, HCoV-OC43 and HCoV-HKU1, which cause mild illnesses (Li et al., 2020a; Yan et al., 2020). Bats are generally thought to be the natural reservoir of a range of coronaviruses (Graham et al., 2013). Some bat coronaviruses also have the potential to emerge in human populations (Letko et al., 2020), and these include the SARS-like virus SHC014-CoV in horseshoe bats (Menachery et al., 2015). Human coronaviruses likely originated from bats and then jumped into another amplification mammalian host before crossing species barriers to infect humans. In addition to coronaviruses harming human health, the new emergence of animal coronaviruses has been a threat to animal health in the past decade. For example, porcine enteric alphacoronavirus, PEDV and SADS-CoV

(Tan et al., 2020; Zaki et al., 2012; Zhou et al., 2018). To date, there are no clinically approved vaccines or antiviral drugs available for these coronavirus infections. An ideal live attenuated vaccine should replicate effectively in the host, not causing diseases, but stimulating enough protective immune responses without reverting to a virulent phenotype.

One of the most promising applications of reverse genetics is to realize the development of live attenuated vaccines by mutating or deleting virulence-related genes. To produce the ideal coronavirus vaccine strain through reverse genetics, a deeper knowledge of the coronavirus-host interaction, host immune responses, and pathogen immune evasion strategies is needed. For example, host cell pathways affected during coronavirus infection need to be identified, and these proteins of coronavirus suppressing the host innate immunity and the specific signaling pathway also need to be identified. Combining reverse genetics with metagenomics and structural biology can help characterize pre-emergent coronavirus populations, allowing the field to make predictions about which zoonotic coronaviruses are likely to emerge, prepare for future outbreaks, and facilitate the development of therapeutic strategies against coronavirus infection (Johnson et al., 2018). In addition, effective broad-spectrum antiviral drugs are still needed. Hopefully, live attenuated vaccines based on reverse genetics will be available in the near future and will prove to be highly effective tools against coronaviruses.

Acknowledgments

Support for this work was received from the National Key R&D Project of China (2017YFD0501102, 2017YFD0500605), Key R&D Project in Shaanxi Province of China (2019NY-076) and The Youth Innovation Team of Shaanxi Universities.

References

Almazan, F., Galan, C., Enjuanes, L., 2008. Engineering infectious cDNAs of coronavirus as bacterial artificial chromosomes. Methods Mol. Biol. 454, 275.

Almazan, F., DeDiego, M.L., Sola, I., Zuniga, S., Nieto-Torres, J.L., Marquez-Jurado, S., Andres, G., Enjuanes, L., 2013. Engineering a replication-competent, propagation-defective Middle East respiratory syndrome coronavirus as a vaccine candidate. mBio 4, e00650.

Almazan, F., Gonzalez, J.M., Penzes, Z., Izeta, A., Calvo, r.E., Plana-Duran, J., Enjuanes, L., 2000. Engineering the largest RNA virus genome as an infectious bacterial artificial chromosome. Proc. Natl. Acad. Sci. U. S. A. 97, 5516.

Angelini, M.M., Akhlaghpour, M., Neuman, B.W., Buchmeier, M.J., 2013. Severe acute respiratory syndrome coronavirus nonstructural proteins 3, 4, and 6 induce double-membrane vesicles. mBio 4, e00524-13.

Balint, A., Farsang, A., Zadori, Z., Hornyak, A., Dencso, L., Almazan, F., Enjuanes, L., Belak, S., 2012. Molecular characterization of feline infectious peritonitis virus strain DF-2 and studies of the role of ORF3abc in viral cell tropism. J. Virol. 86, 6258.

Beachboard, D.C., Anderson-Daniels, J.M., Denison, M.R., 2015. Mutations across murine hepatitis virus nsp4 alter virus fitness and membrane modifications. J. Virol. 89, 2080.

Beall, A., Yount, B., Lin, C.M., Hou, Y., Wang, Q., Saif, L., Baric, R., 2016. Characterization of a pathogenic full-length cDNA clone and transmission model for porcine epidemic diarrhea virus strain PC22A. mBio 7, e01451.

Becares, M., Pascual-Iglesias, A., Nogales, A., Sola, I., Enjuanes, L., Zuniga, S., 2016. Mutagenesis of coronavirus nsp14 reveals its potential role in modulation of the innate immune response. J. Virol. 90, 5399.

Becker, M.M., Graham, R.L., Donaldson, E.F., Rockx, B., Sims, A.C., Sheahan, T., Pickles, R.J., Corti, D., Johnston, R.E., Baric, R.S., 2008. Synthetic recombinant bat Sars-like coronavirus is infectious in cultured cells and in mice. Proc. Natl. Acad. Sci. U. S. A. 105, 19944.

Bickerton, E., Dowgier, G., Britton, P., 2018a. Recombinant infectious bronchitis viruses expressing heterologous S1 subunits: potential for a new generation of vaccines that replicate in Vero cells. J. Gen. Virol. 99, 1681.

Bickerton, E., Maier, H.J., Stevenson-Leggett, P., Armesto, M., Britton, P., 2018b. The S2 subunit of infectious bronchitis virus beaudette is a determinant of cellular tropism. J. Virol. 92, e01044-18.

Bouvet, M., Lugari, A., Posthuma, C.C., Zevenhoven, J.C., Bernard, S., Betzi, S., Imbert, I., Canard, B., Guillemot, J.C., Lecine, P., Pfefferle, S., Drosten, C., Snijder, E.J., Decroly, E., Morelli, X., 2014. Coronavirus Nsp10, a critical co-factor for activation of multiple replicative enzymes. J. Biol. Chem. 289, 25783.

Bukreyev, A., Lamirande, E.W., Buchholz, U.J., Vogel, L.N., Elkins, W.R., St Claire, M., Murphy, B.R., Subbarao, K., Collins, P.L., 2004. Mucosal immunisation of African green monkeys (Cercopithecus aethiops) with an attenuated parainfluenza virus expressing the SARS coronavirus spike protein for the prevention of SARS. Lancet 363, 2122.

Cao, X., 2020. COVID-19: immunopathology and its implications for therapy. Nat. Rev. Immunol. 20, 269–270.

Case, J.B., Li, Y., Elliott, R., Lu, X., Graepel, K.W., Sexton, N.R., Smith, E.C., Weiss, S.R., Denison, M.R., 2018. Murine hepatitis virus nsp14 exoribonuclease activity is required for resistance to innate immunity. J. Virol. 92, e01531-17.

Castano-Rodriguez, C., Honrubia, J.M., Gutierrez-Alvarez, J., DeDiego, M.L., Nieto-Torres, J.L., Jimenez-Guardeno, J.M., Regla-Nava, J.A., Fernandez-Delgado, R., Verdia-Baguena, C., Queralt-Martin, M., Kochan, G., Perlman, S., Aguilella, V.M., Sola, I., Enjuanes, L., 2018. Role of severe acute respiratory syndrome coronavirus viroporins E, 3a, and 8a in replication and pathogenesis. mBio 9, e02325-17.

Chan, J.F., Kok, K.H., Zhu, Z., Chu, H., To, K.K., Yuan, S., Yuen, K.Y., 2020. Genomic characterization of the 2019 novel human-pathogenic coronavirus isolated from a patient with atypical pneumonia after visiting Wuhan. Emerg. Microbes Infect. 9, 221.

Channappanavar, R., Perlman, S., 2017. Pathogenic human coronavirus infections: causes and consequences of cytokine storm and immunopathology. Semin. Immunopathol. 39, 529.

Channappanavar, R., Fehr, A.R., Vijay, R., Mack, M., Zhao, J., Meyerholz, D.K., Perlman, S., 2016. Dysregulated type I interferon and inflammatory monocyte-macrophage responses cause lethal pneumonia in SARS-CoV-infected mice. Cell Host Microbe 19, 181.

Chen, S., Tian, J., Li, Z., Kang, H., Zhang, J., Huang, J., Yin, H., Hu, X., Qu, L., 2019. Feline infectious peritonitis virus Nsp5 inhibits type I interferon production by cleaving NEMO at multiple sites. Viruses 12, 43.

Cheng, J., Zhao, Y., Xu, G., Zhang, K., Jia, W., Sun, Y., Zhao, J., Xue, J., Hu, Y., Zhang, G., 2019. The S2 subunit of QX-type infectious bronchitis coronavirus spike protein is an essential determinant of neurotropism. Viruses 11, 972.

Cockrell, A.S., Beall, A., Yount, B., Baric, R., 2017. Efficient reverse genetic systems for rapid genetic manipulation of emergent and preemergent infectious coronaviruses. Methods Mol. Biol. 1602, 59.

Comar, C.E., Goldstein, S.A., Li, Y., Yount, B., Baric, R.S., Weiss, S.R., 2019. Antagonism of dsRNA-induced innate immune pathways by NS4a and NS4b accessory proteins during MERS coronavirus infection. mBio 10, e00319-19.

Coronaviridae Study Group of the International Committee on Taxonomy of Viruses, 2020. The species severe acute respiratory syndrome-related coronavirus: classifying 2019-nCoV and naming it SARS-CoV-2. Nat. Microbiol. 5, 536–544.

Cottam, E.M., Whelband, M.C., Wileman, T., 2014. Coronavirus NSP6 restricts autophagosome expansion. Autophagy 10, 1426.

Cui, J., Li, F., Shi, Z.L., 2019. Origin and evolution of pathogenic coronaviruses. Nat. Rev. Microbiol. 17, 181.

Day, M., 2020. Covid-19: Italy confirms 11 deaths as cases spread from north. BMJ 368, m757.

Deming, D.J., Graham, R.L., Denison, M.R., Baric, R.S., 2006. MHV-A59 ORF1a replicase protein nsp7-nsp10 processing in replication. Adv. Exp. Med. Biol. 581, 101.

Deng, X., StJohn, S.E., Osswald, H.L., O'Brien, A., Banach, B.S., Sleeman, K., Ghosh, A.K., Mesecar, A.D., Baker, S.C., 2014. Coronaviruses resistant to a 3C-like protease inhibitor are attenuated for replication and pathogenesis, revealing a low genetic barrier but high fitness cost of resistance. J. Virol. 88, 11886.

Deng, X., Hackbart, M., Mettelman, R.C., O'Brien, A., Mielech, A.M., Yi, G., Kao, C.C., Baker, S.C., 2017. Coronavirus nonstructural protein 15 mediates evasion of dsRNA sensors and limits apoptosis in macrophages. Proc. Natl. Acad. Sci. U. S. A. 114, E4251.

Deng, X., van Geelen, A., Buckley, A.C., O'Brien, A., Pillatzki, A., Lager, K.M., Faaberg, K.S., Baker, S.C., 2019. Coronavirus endoribonuclease activity in porcine epidemic diarrhea virus suppresses type I and type III interferon responses. J. Virol. 93, e02000-18.

Ding, Z., Fang, L., Yuan, S., Zhao, L., Wang, X., Long, S., Wang, M., Wang, D., Foda, M.F., Xiao, S., 2017. The nucleocapsid proteins of mouse hepatitis virus and severe acute respiratory syndrome coronavirus share the same IFN-β antagonizing mechanism: attenuation of PACT-mediated RIG-I/MDA5 activation. Oncotarget 8, 49655.

Drosten, C., Gunther, S., Preiser, W., van der Werf, S., Brodt, H.R., Becker, S., Rabenau, H., Panning, M., Kolesnikova, L., Fouchier, R.A., Berger, A., Burguiere, A.M., Cinatl, J., Eickmann, M., Escriou, N., Grywna, K., Kramme, S., Manuguerra, J.C., Muller, S., Rickerts, V., Sturmer, M., Vieth, S., Klenk, H.D., Osterhaus, A.D., Schmitz, H., Doerr, H.W., 2003. Identification of a novel coronavirus in patients with severe acute respiratory syndrome. N. Engl. J. Med. 348, 1967.

Dutta, N.K., Mazumdar, K., Lee, B.H., Baek, M.W., Kim, D.J., Na, Y.R., Park, S.H., Lee, H.K., Kariwa, H., Le, Q.M., 2008. Search for potential target site of nucleocapsid gene for the design of an epitope-based SARS DNA vaccine. Immunol. Lett. 118, 65.

Egloff, M.P., Ferron, F., Campanacci, V., Longhi, S., Rancurel, C., Dutartre, H., Snijder, E.J., Gorbalenya, A.E., Cambillau, C., Canard, B., 2004. The severe acute respiratory syndrome-coronavirus replicative protein nsp9 is a single-stranded RNA-binding subunit unique in the RNA virus world. Proc. Natl. Acad. Sci. U. S. A. 101, 3792.

Ellis, S., Keep, S., Britton, P., de Wit, S., Bickerton, E., Vervelde, L., 2018. Recombinant infectious bronchitis viruses expressing chimeric spike glycoproteins induce partial protective immunity against homologous challenge despite limited replication in vivo. J. Virol. 92, e01473-18.

Fang, P., Fang, L., Ren, J., Hong, Y., Liu, X., Zhao, Y., Wang, D., Peng, G., Xiao, S., 2018. Porcine deltacoronavirus accessory protein NS6 antagonizes interferon beta production by interfering with the binding of RIG-I/MDA5 to double-stranded RNA. J. Virol. 92, e00712-18.

Fehr, A.R., 2020. Bacterial artificial chromosome-based lambda red recombination with the I-SceI homing endonuclease for genetic alteration of MERS-CoV. Methods Mol. Biol. 2099, 53.

Fehr, A.R., Athmer, J., Channappanavar, R., Phillips, J.M., Meyerholz, D.K., Perlman, S., 2015. The nsp3 macrodomain promotes virulence in mice with coronavirus-induced encephalitis. J. Virol. 89, 1523.

Fehr, A.R., Channappanavar, R., Jankevicius, G., Fett, C., Zhao, J., Athmer, J., Meyerholz, D.K., Ahel, I., Perlman, S., 2016. The conserved coronavirus macrodomain promotes virulence and suppresses the innate immune response during severe acute respiratory syndrome coronavirus infection. mBio 7, e01721-16.

Fett, C., DeDiego, M.L., Regla-Nava, J.A., Enjuanes, L., Perlman, S., 2013. Complete protection against severe acute respiratory syndrome coronavirus-mediated lethal respiratory disease in aged mice by immunization with a mouse-adapted virus lacking E protein. J. Virol. 87, 6551.

Graepel, K.W., Lu, X.T., Case, J.B., Sexton, N.R., Smith, E.C., Denison, M.R., 2017. Proofreading-deficient coronaviruses adapt for increased fitness over long-term passage without reversion of exoribonuclease-inactivating mutations. mBio 8, e01503-17.

Graham, R.L., Sims, A.C., Brockway, S.M., Baric, R.S., Denison, M.R., 2005. The nsp2 replicase proteins of murine hepatitis virus and severe acute respiratory syndrome coronavirus are dispensable for viral replication. J. Virol. 79, 13399.

Graham, R.L., Becker, M.M., Eckerle, L.D., Bolles, M., Denison, M.R., Baric, R.S., 2012. A live, impaired-fidelity coronavirus vaccine protects in an aged, immunocompromised mouse model of lethal disease. Nat. Med. 18, 1820.

Graham, R.L., Donaldson, E.F., Baric, R.S., 2013. A decade after SARS: strategies for controlling emerging coronaviruses. Nat. Rev. Microbiol. 11, 836.

Haijema, B.J., Volders, H., Rottier, P.J., 2003. Switching species tropism: an effective way to manipulate the feline coronavirus genome. J. Virol. 77, 4528.

He, Y., Zhou, Y., Siddiqui, P., Niu, J., Jiang, S., 2005. Identification of immunodominant epitopes on the membrane protein of the severe acute respiratory syndrome-associated coronavirus. J. Clin. Microbiol. 43, 3718.

Hotez, P.J., Bottazzi, M.E., Corry, D.B., 2020. The potential role of Th17 immune responses in coronavirus immunopathology and vaccine-induced immune enhancement. Microbes Infect. 22, 165–167.

Hou, Y., Lin, C.M., Yokoyama, M., Yount, B.L., Marthaler, D., Douglas, A.L., Ghimire, S., Qin, Y., Baric, R.S., Saif, L.J., Wang, Q., 2017. Deletion of a 197-amino-acid region in the N-terminal domain of spike protein attenuates porcine epidemic diarrhea virus in piglets. J. Virol. 91, e00227-17.

Hou, Y., Ke, H., Kim, J., Yoo, D., Su, Y., Boley, P., Chepngeno, J., Vlasova, A.N., Saif, L.J., Wang, Q., 2019a. Engineering a live attenuated porcine epidemic diarrhea virus vaccine candidate via inactivation of the viral 2'-O-methyltransferase and the endocytosis signal of the spike protein. J. Virol. 93, e00406-19.

Hou, Y., Meulia, T., Gao, X., Saif, L.J., Wang, Q., 2019b. Deletion of both the tyrosine-based endocytosis signal and the endoplasmic reticulum retrieval signal in the cytoplasmic tail of spike protein attenuates porcine epidemic diarrhea virus in pigs. J. Virol. 93, e01758-18.

Hu, Y., Li, W., Gao, T., Cui, Y., Jin, Y., Li, P., Ma, Q., Liu, X., Cao, C., Perlman, S., 2017. The severe acute respiratory syndrome coronavirus nucleocapsid inhibits type I interferon production by interfering with TRIM25-mediated RIG-I ubiquitination. J. Virol. 91, e02143.

Jernigan, D.B., Team, C.C.-R., 2020. Update: public health response to the coronavirus disease 2019 outbreak—United States, February 24, 2020. Morb. Mortal. Wkly Rep. 69, 216.

Ji, W., Wang, W., Zhao, X., Zai, J., Li, X., 2020. Cross-species transmission of the newly identified coronavirus 2019-nCoV. J. Med. Virol. 92, 433.

Jimenez-Guardeno, J.M., Nieto-Torres, J.L., DeDiego, M.L., Regla-Nava, J.A., Fernandez-Delgado, R., Castano-Rodriguez, C., Enjuanes, L., 2014. The PDZ-binding motif of severe acute respiratory syndrome coronavirus envelope protein is a determinant of viral pathogenesis. PLoS Pathog. 10, e1004320.

Jimenez-Guardeno, J.M., Regla-Nava, J.A., Nieto-Torres, J.L., DeDiego, M.L., Castano-Rodriguez, C., Fernandez-Delgado, R., Perlman, S., Enjuanes, L., 2015. Identification of the mechanisms causing reversion to virulence in an attenuated SARS-CoV for the design of a genetically stable vaccine. PLoS Pathog. 11, e1005215.

Jo, S., Kim, S., Shin, D.H., Kim, M.S., 2020. Inhibition of SARS-CoV 3CL protease by flavonoids. J. Enzyme Inhib. Med. Chem. 35, 145.

Johnson, B.A., Graham, R.L., Menachery, V.D., 2018. Viral metagenomics, protein structure, and reverse genetics: key strategies for investigating coronaviruses. Virology 517, 30.

Kadkhoda, K., 2020. COVID-19: an immunopathological view. mSphere 5, e00344-20.

Kao, C.F., Chang, H.W., 2019. Investigation of the role of the spike protein in reversing the virulence of the highly virulent Taiwan porcine epidemic diarrhea virus Pintung 52 strains and its attenuated counterpart. Viruses 12, 41.

Kato, H., Takayama-Ito, M., Iizuka-Shiota, I., Fukushi, S., Posadas-Herrera, G., Horiya, M., Satoh, M., Yoshikawa, T., Yamada, S., Harada, S., Fujii, H., Shibamura, M., Inagaki, T., Morimoto, K., Saijo, M., Lim, C.K., 2019. Development of a recombinant replication-deficient rabies virus-based bivalent-vaccine against MERS-CoV and rabies virus and its humoral immunogenicity in mice. PLoS One 14, e0223684.

Kim, M.H., Kim, H.J., Chang, J., 2019. Superior immune responses induced by intranasal immunization with recombinant adenovirus-based vaccine expressing full-length spike protein of Middle East respiratory syndrome coronavirus. PLoS One 14, e0220196.

Kint, J., Langereis, M.A., Maier, H.J., Britton, P., van Kuppeveld, F.J., Koumans, J., Wiegertjes, G.F., Forlenza, M., 2016. Infectious bronchitis coronavirus limits interferon production by inducing a host shutoff that requires accessory protein 5b. J. Virol. 90, 7519.

Kong, N., Shan, T., Wang, H., Jiao, Y., Zuo, Y., Li, L., Tong, W., Yu, L., Jiang, Y., Zhou, Y., Li, G., Gao, F., Yu, H., Zheng, H., Tong, G., 2019. BST2 suppresses porcine epidemic diarrhea virus replication by targeting and degrading virus nucleocapsid protein with selective autophagy. Autophagy 1.

Kopecky-Bromberg, S.A., Martinez-Sobrido, L., Frieman, M., Baric, R.A., Palese, P., 2007. Severe acute respiratory syndrome coronavirus open reading frame (ORF) 3b, ORF 6, and nucleocapsid proteins function as interferon antagonists. J. Virol. 81, 548.

Lee, J.Y., Bae, S., Myoung, J., 2019. Middle East respiratory syndrome coronavirus-encoded accessory proteins impair MDA5-and TBK1-mediated activation of NF-kappaB. J. Microbiol. Biotechnol. 29, 1316.

Lei, L., Ying, S., Baojun, L., Yi, Y., Xiang, H., Wenli, S., Zounan, S., Deyin, G., Qingyu, Z., Jingmei, L., Guohui, C., 2013. Attenuation of mouse hepatitis virus by deletion of the LLRKxGxKG region of Nsp1. PLoS One 8, e61166.

Lei, J., Kusov, Y., Hilgenfeld, R., 2018. Nsp3 of coronaviruses: structures and functions of a large multi-domain protein. Antiviral Res. 149, 58.

Letko, M., Marzi, A., Munster, V., 2020. Functional assessment of cell entry and receptor usage for SARS-CoV-2 and other lineage B betacoronaviruses. Nat. Microbiol. 5, 562–569.

Li, W., Li, H., Liu, Y., Pan, Y., Deng, F., Song, Y., Tang, X., He, Q., 2012. New variants of porcine epidemic diarrhea virus, China, 2011. Emerg. Infect. Dis. 18, 1350.

Li, C., Li, Z., Zou, Y., Wicht, O., van Kuppeveld, F.J., Rottier, P.J., Bosch, B.J., 2013. Manipulation of the porcine epidemic diarrhea virus genome using targeted RNA recombination. PLoS One 8, e69997.

Li, J., Jin, Z., Gao, Y., Zhou, L., Ge, X., Guo, X., Han, J., Yang, H., 2017. Development of the full-length cDNA clones of two porcine epidemic diarrhea disease virus isolates with different virulence. PLoS One 12, e0173998.

Li, J.Y., You, Z., Wang, Q., Zhou, Z.J., Qiu, Y., Luo, R., Ge, X.Y., 2020a. The epidemic of 2019-novel-coronavirus (2019-nCoV) pneumonia and insights for emerging infectious diseases in the future. Microbes Infect. 22, 80–85.

Li, K., Li, Z., Wohlford-Lenane, C., Meyerholz, D.K., Channappanavar, R., An, D., Perlman, S., McCray Jr., P.B., He, B., 2020b. Single-dose, intranasal immunization with recombinant parainfluenza virus 5 expressing Middle East respiratory syndrome coronavirus (MERS-CoV) spike protein protects mice from fatal MERS-CoV infection. mBio 11, e00554-20.

Lokugamage, K.G., Narayanan, K., Nakagawa, K., Terasaki, K., Ramirez, S.I., Tseng, C.T., Makino, S., 2015. Middle East respiratory syndrome coronavirus nsp1 inhibits host gene expression by selectively targeting mRNAs transcribed in the nucleus while sparing mRNAs of cytoplasmic origin. J. Virol. 89, 10970.

Lu, X., Pan, J.A., Tao, J., Guo, D., 2011. SARS-CoV nucleocapsid protein antagonizes IFN-β response by targeting initial step of IFN-β induction pathway, and its C-terminal region is critical for the antagonism. Virus Genes 42, 37.

Lui, P.Y., Wong, L.Y.R., Fung, C.L., Siu, K.L., Yeung, M.L., Yuen, K.S., Chan, C.P., Woo, P.C.Y., Yuen, K.Y., Jin, D.Y., 2016. Middle East respiratory syndrome coronavirus M protein suppresses type I interferon expression through the inhibition of TBK1-dependent phosphorylation of IRF3. Emerg. Microbes Infect. 5, e39.

Luo, F., Liao, F.L., Wang, H., Tang, H.B., Yang, Z.Q., Hou, W., 2018. Evaluation of antibody-dependent enhancement of SARS-CoV infection in rhesus macaques immunized with an inactivated SARS-CoV vaccine. Virol. Sin. 33, 201.

Masters, P.S., Rottier, P.J., 2005. Coronavirus reverse genetics by targeted RNA recombination. Curr. Top. Microbiol. Immunol. 287, 133.

Mcbride, R., Van Zyl, M., Fielding, B., 2014. The coronavirus nucleocapsid is a multifunctional protein. Viruses 6, 2991.

Menachery, V.D., Yount, B.L., Josset, L., Gralinski, L.E., Scobey, T., Agnihothram, S., Katze, M.G., Baric, R.S., 2014. Attenuation and restoration of severe acute respiratory syndrome coronavirus mutant lacking 2'-O-methyltransferase activity. J. Virol. 88, 4251.

Menachery, V.D., Yount Jr., B.L., Debbink, K., Agnihothram, S., Gralinski, L.E., Plante, J.A., Graham, R.L., Scobey, T., Ge, X.Y., Donaldson, E.F., Randell, S.H., Lanzavecchia, A., Marasco, W.A., Shi, Z.L., Baric, R.S., 2015. A SARS-like cluster of circulating bat coronaviruses shows potential for human emergence. Nat. Med. 21, 1508.

Menachery, V.D., Gralinski, L.E., Mitchell, H.D., Dinnon 3rd, K.H., Leist, S.R., Yount Jr., B.L., Graham, R.L., McAnarney, E.T., Stratton, K.G., Cockrell, A.S., Debbink, K., Sims, A.C., Waters, K.M., Baric, R.S., 2017a. Middle East respiratory syndrome coronavirus nonstructural protein 16 is necessary for interferon resistance and viral pathogenesis. mSphere 2, e00346-17.

Menachery, V.D., Mitchell, H.D., Cockrell, A.S., Gralinski, L.E., Yount Jr., B.L., Graham, R.L., McAnarney, E.T., Douglas, M.G., Scobey, T., Beall, A., Dinnon 3rd, K., Kocher, J.F., Hale, A.E., Stratton, K.G., Waters, K.M., Baric, R.S., 2017b. MERS-CoV accessory ORFs play key role for infection and pathogenesis. mBio 8, e00665-17.

Menachery, V.D., Gralinski, L.E., Mitchell, H.D., Dinnon 3rd, K.H., Leist, S.R., Yount Jr., B.L., McAnarney, E.T., Graham, R.L., Waters, K.M., Baric, R.S., 2018. Combination attenuation offers strategy for live attenuated coronavirus vaccines. J. Virol. 92, e00710-18.

Mielech, A.M., Deng, X., Chen, Y., Kindler, E., Wheeler, D.L., Mesecar, A.D., Thiel, V., Perlman, S., Baker, S.C., 2015. Murine coronavirus ubiquitin-like domain is important for papain-like protease stability and viral pathogenesis. J. Virol. 89, 4907.

Millet, J.K., Whittaker, G.R., 2015. Host cell proteases: critical determinants of coronavirus tropism and pathogenesis. Virus Res. 202, 120.

Munster, V.J., Koopmans, M., van Doremalen, N., van Riel, D., de Wit, E., 2020. A novel coronavirus emerging in China—key questions for impact assessment. N. Engl. J. Med. 382, 692.

Muth, D., Meyer, B., Niemeyer, D., Schroeder, S., Osterrieder, N., Muller, M.A., Drosten, C., 2017. Transgene expression in the genome of Middle East respiratory syndrome coronavirus based on a novel reverse genetics system utilizing red-mediated recombination cloning. J. Gen. Virol. 98, 2461.

Nakanaga, K., Yamanouchi, K., Fujiwara, K., 1986. Protective effect of monoclonal antibodies on lethal mouse hepatitis virus infection in mice. J. Virol. 59, 168.

Nishiura, H., Linton, N.M., Akhmetzhanov, A.R., 2020. Initial cluster of novel coronavirus (2019-nCoV) infections in Wuhan, China is consistent with substantial human-to-human transmission. J. Clin. Med. 9, 488.

Ortego, J., Escors, D., Laude, H., Enjuanes, L., 2002. Generation of a replication-competent, propagation-deficient virus vector based on the transmissible gastroenteritis coronavirus genome. J. Virol. 76, 11518.

Peeri, N.C., Shrestha, N., Rahman, M.S., Zaki, R., Tan, Z., Bibi, S., Baghbanzadeh, M., Aghamohammadi, N., Zhang, W., Haque, U., 2020. The SARS, MERS and novel coronavirus (COVID-19) epidemics, the newest and biggest global health threats: what lessons have we learned? Int. J. Epidemiol.

Peng, Q., Fang, L., Ding, Z., Wang, D., Peng, G., Xiao, S., 2020. Rapid manipulation of the porcine epidemic diarrhea virus genome by CRISPR/Cas9 technology. J. Virol. Methods 276, 113772.

Sakai, Y., Kawachi, K., Terada, Y., Omori, H., Matsuura, Y., Kamitani, W., 2017. Two-amino acids change in the nsp4 of SARS coronavirus abolishes viral replication. Virology 510, 165.

Sanchez, C.M., Pascual-Iglesias, A., Sola, I., Zuniga, S., Enjuanes, L., 2019. Minimum determinants of transmissible gastroenteritis virus enteric tropism are located in the N-terminus of spike protein. Pathogens 9, 2.

Schoeman, D., Fielding, B.C., 2019. Coronavirus envelope protein: current knowledge. Virol. J. 16, 69.

Scobey, T., Yount, B.L., Sims, A.C., Donaldson, E.F., Agnihothram, S.S., Menachery, V.D., Graham, R.L., Swanstrom, J., Bove, P.F., Kim, J.D., Grego, S., Randell, S.H., Baric, R.S., 2013. Reverse genetics with a full-length infectious cDNA of the Middle East respiratory syndrome coronavirus. Proc. Natl. Acad. Sci. U. S. A. 110, 16157.

Shang, J., Wan, Y., Liu, C., Yount, B., Gully, K., Yang, Y., Auerbach, A., Peng, G., Baric, R., Li, F., 2020a. Structure of mouse coronavirus spike protein complexed with receptor reveals mechanism for viral entry. PLoS Pathog. 16, e1008392.

Shang, W., Yang, Y., Rao, Y., Rao, X., 2020b. The outbreak of SARS-CoV-2 pneumonia calls for viral vaccines. NPJ Vaccines 5, 18.

Shehata, M.M., Kandeil, A., Mostafa, A., Mahmoud, S.H., Gomaa, M.R., El-Shesheny, R., Webby, R., Kayali, G., Ali, M.A., 2019. A recombinant influenza a/H1N1 carrying a short immunogenic peptide of MERS-CoV as bivalent vaccine in BALB/c mice. Pathogens 8, 281.

Shen, Z., Ye, G., Deng, F., Wang, G., Cui, M., Fang, L., Xiao, S., Fu, Z.F., Peng, G., 2018. Structural basis for the inhibition of host gene expression by porcine epidemic diarrhea virus nsp1. J. Virol. 92, e01896-17.

Shen, Z., Wang, G., Yang, Y., Shi, J., Fang, L., Li, F., Xiao, S., Fu, Z.F., Peng, G., 2019. A conserved region of nonstructural protein 1 from alphacoronaviruses inhibits host gene expression and is critical for viral virulence. J. Biol. Chem. 294, 13606–13618.

Shokri, S., Mahmoudvand, S., Taherkhani, R., Farshadpour, F., 2019. Modulation of the immune response by Middle East respiratory syndrome coronavirus. J. Cell. Physiol. 234, 2143.

Siu, K.L., Kok, K.H., Ng, M.H.J., Poon, V.K.M., Yuen, K.Y., Zheng, B.J., Jin, D.Y., 2009. Severe acute respiratory syndrome coronavirus M protein inhibits type I interferon production by impeding the formation of TRAF3.TANK.TBK1/IKKepsilon complex. J. Biol. Chem. 284, 16202.

Smith, E.C., Case, J.B., Blanc, H., Isakov, O., Shomron, N., Vignuzzi, M., Denison, M.R., Perlman, S., 2015. Mutations in coronavirus nonstructural protein 10 decrease virus replication fidelity. J. Virol. 89, 6418.

Stevenson, G.W., Hoang, H., Schwartz, K.J., Burrough, E.R., Sun, D., Madson, D., Cooper, V.L., Pillatzki, A., Gauger, P., Schmitt, B.J., Koster, L.G., Killian, M.L., Yoon, K.J., 2013. Emergence of porcine epidemic diarrhea virus in the United States: clinical signs, lesions, and viral genomic sequences. J. Vet. Diagn. Invest. 25, 649.

Stobart, C.C., Moore, M.L., 2014. RNA virus reverse genetics and vaccine design. Viruses 6, 2531.

Tan, L., Li, Y., He, J., Hu, Y., Cai, X., Liu, W., Liu, T., Wang, J., Li, Z., Yuan, X., Zhan, Y., Yang, L., Deng, Z., Wang, N., Yang, Y., Wang, A., 2020. Epidemic and genetic characterization of porcine epidemic diarrhea virus strains circulating in the regions around Hunan, China, during 2017-2018. Arch. Virol. 165, 877–889.

Terada, Y., Kuroda, Y., Morikawa, S., Matsuura, Y., Maeda, K., Kamitani, W., 2019. Establishment of a virulent full-length cDNA clone for type I feline coronavirus strain C3663. J. Virol. 93, e01208-19.

Thanh Le, T., Andreadakis, Z., Kumar, A., Gomez Roman, R., Tollefsen, S., Saville, M., Mayhew, S., 2020. The COVID-19 vaccine development landscape. Nat. Rev. Drug Discov. 19, 305–306.

Thao, T.T.N., Labroussaa, F., Ebert, N., V'Kovski, P., Stalder, H., Portmann, J., Kelly, J., Steiner, S., Holwerda, M., Kratzel, A., Gultom, M., Schmied, K., Laloli, L., Husser, L., Wider, M., Pfaender, S., Hirt, D., Cippa, V., Crespo-Pomar, S., Schroder, S., Muth, D., Niemeyer, D., Corman, V., Muller, M.A., Drosten, C., Dijkman, R., Jores, J., Thiel, V., 2020. Rapid reconstruction of SARS-CoV-2 using a synthetic genomics platform. Nature.

Thiel, V., Siddell, S.G., 2005. Reverse genetics of coronaviruses using vaccinia virus vectors. Curr. Top. Microbiol. Immunol. 287, 199.

Thiel, V., Herold, J., Schelle, B., Siddell, S.G., 2001. Infectious RNA transcribed in vitro from a cDNA copy of the human coronavirus genome cloned in vaccinia virus. J. Gen. Virol. 82, 1273.

Tomar, S., Johnston, M.L., St John, S.E., Osswald, H.L., Nyalapatla, P.R., Paul, L.N., Ghosh, A.K., Denison, M.R., Mesecar, A.D., 2015. Ligand-induced dimerization of Middle East respiratory syndrome (MERS) coronavirus nsp5 protease (3CLpro): implications for nsp5 regulation and the development of antivirals. J. Biol. Chem. 290, 19403.

Ujike, M., Huang, C., Shirato, K., Makino, S., Taguchi, F., 2016. The contribution of the cytoplasmic retrieval signal of severe acute respiratory syndrome coronavirus to intracellular accumulation of S proteins and incorporation of S protein into virus-like particles. J. Gen. Virol. 97, 1853.

Van Beurden, S.J., Berends, A.J., Krämer-Kühl, A., Spekreijse, D., Chenard, G., Philipp, H.C., Mundt, E., Rottier, P.J.M., Verheije, M.H., 2018. Recombinant live attenuated avian coronavirus vaccines with deletions in the accessory genes 3ab and/or 5ab protect against infectious bronchitis in chickens. Vaccine 36, 1085.

van den Worm, S.H., Eriksson, K.K., Zevenhoven, J.C., Weber, F., Züst, R., Kuri, T., Dijkman, R., Chang, G., Siddell, S.G., Snijder, E.J., Thiel, V., 2012. Reverse genetics of SARS-related coronavirus using Vaccinia virus-based recombination. PLoS One 7, e32857.

Veit, S., Jany, S., Fux, R., Sutter, G., Volz, A., 2018. CD8+ T cells responding to the Middle East respiratory syndrome coronavirus nucleocapsid protein delivered by vaccinia virus MVA in mice. Viruses 10, 718.

Walls, A.C., Park, Y.J., Tortorici, M.A., Wall, A., McGuire, A.T., Veesler, D., 2020. Structure, function, and antigenicity of the SARS-CoV-2 spike glycoprotein. Cell 181, 281–292.e6.

Wang, D., Fang, L., Shi, Y., Zhang, H., Gao, L., Peng, G., Chen, H., Li, K., Xiao, S., 2016. Porcine epidemic diarrhea virus 3C-like protease regulates its interferon antagonism by cleaving NEMO. J. Virol. 90, 2090.

Wang, C., Zheng, X., Gai, W., Zhao, Y., Wang, H., Wang, H., Feng, N., Chi, H., Qiu, B., Li, N., Wang, T., Gao, Y., Yang, S., Xia, X., 2017. MERS-CoV virus-like particles produced in insect cells induce specific humoural and cellular imminity in rhesus macaques. Oncotarget 8, 12686.

Wang, D., Ge, X., Chen, D., Li, J., Cai, Y., Deng, J., Zhou, L., Guo, X., Han, J., Yang, H., 2018a. The S gene is necessary but not sufficient for the virulence of porcine epidemic diarrhea virus novel variant strain BJ2011C. J. Virol. 92, e00603-18.

Wang, L., Qiao, X., Zhang, S., Qin, Y., Guo, T., Hao, Z., Sun, L., Wang, X., Wang, Y., Jiang, Y., Tang, L., Xu, Y., Li, Y., 2018b. Porcine transmissible gastroenteritis virus nonstructural protein 2 contributes to inflammation via NF-kappaB activation. Virulence 9, 1685.

Wang, G., Liang, R., Liu, Z., Shen, Z., Shi, J., Shi, Y., Deng, F., Xiao, S., Fu, Z.F., Peng, G., 2019a. The N-terminal domain of spike protein is not the enteric tropism determinant for transmissible gastroenteritis virus in piglets. Viruses 11, 313.

Wang, Y., Sun, A., Sun, Y., Zhang, S., Xia, T., Guo, T., Hao, Z., Sun, L., Jiang, Y., Qiao, X., Cui, W., Tang, L., Xu, Y., Li, Y., Wang, L., 2019b. Porcine transmissible gastroenteritis virus inhibits NF-kappaB activity via nonstructural protein 3 to evade host immune system. Virol. J. 16, 97.

Wassenaar, T.M., Zou, Y., 2020. 2019_nCoV/SARS-CoV-2: rapid classification of betacoronaviruses and identification of traditional Chinese medicine as potential origin of zoonotic coronaviruses. Lett. Appl. Microbiol. 70, 342–348.

Wong, H.H., Fung, T.S., Fang, S., Huang, M., Le, M.T., Liu, D.X., 2018. Accessory proteins 8b and 8ab of severe acute respiratory syndrome coronavirus suppress the interferon signaling pathway by mediating ubiquitin-dependent rapid degradation of interferon regulatory factor 3. Virology 515, 165.

Wu, A., Peng, Y., Huang, B., Ding, X., Wang, X., Niu, P., Meng, J., Zhu, Z., Zhang, Z., Wang, J., Sheng, J., Quan, L., Xia, Z., Tan, W., Cheng, G., Jiang, T., 2020. Genome composition and divergence of the novel coronavirus (2019-nCoV) originating in China. Cell Host Microbe 27, 325–328.

Xie, X., Muruato, A., Lokugamage, K.G., Narayanan, K., Zhang, X., Zou, J., Liu, J., Schindewolf, C., Bopp, N.E., Aguilar, P.V., Plante, K.S., Weaver, S.C., Makino, S., LeDuc, J.W., Menachery, V.D., Shi, P.Y., 2020. An infectious cDNA clone of SARS-CoV-2. Cell Host Microbe 27, 841–848.e3.

Xu, Z., Shi, L., Wang, Y., Zhang, J., Huang, L., Zhang, C., Liu, S., Zhao, P., Liu, H., Zhu, L., Tai, Y., Bai, C., Gao, T., Song, J., Xia, P., Dong, J., Zhao, J., Wang, F.S., 2020. Pathological findings of COVID-19 associated with acute respiratory distress syndrome. Lancet Respir. Med. 8, 420.

Yan, J., Guan, Z.Y., Zhu, W.F., Zhong, L.Y., Qiu, Z.Q., Yue, P.F., Wu, W.T., Liu, J., Huang, X., 2020. Preparation of puerarin chitosan oral nanoparticles by ionic gelation method and its related kinetics. Pharmaceutics 12, 216.

Yong, C.Y., Ong, H.K., Yeap, S.K., Ho, K.L., Tan, W.S., 2019. Recent advances in the vaccine development against Middle East respiratory syndrome-coronavirus. Front. Microbiol. 10, 1781.

Yount, B., Curtis, K.M., Baric, R.S., 2000. Strategy for systematic assembly of large RNA and DNA genomes: transmissible gastroenteritis virus model. J. Virol. 74, 10600.

Yount, B., Curtis, K.M., Fritz, E.A., Hensley, L.E., Jahrling, P.B., Prentice, E., Denison, M.R., Geisbert, T.W., Baric, R.S., 2003. Reverse genetics with a full-length infectious cDNA of severe acute respiratory syndrome coronavirus. Proc. Natl. Acad. Sci. U. S. A. 100, 12995.

Yount, B., Denison, M.R., Weiss, S.R., Baric, R.S., 2002. Systematic assembly of a full-length infectious cDNA of mouse hepatitis virus strain A59. J Virol. 76, 11065.

Yue, Y., Nabar, N.R., Shi, C.S., Kamenyeva, O., Xiao, X., Hwang, I.Y., Wang, M., Kehrl, J.H., 2018. SARS-coronavirus open reading frame-3a drives multimodal necrotic cell death. Cell Death Dis. 9, 904.

Zaki, A.M., van Boheemen, S., Bestebroer, T.M., Osterhaus, A.D., Fouchier, R.A., 2012. Isolation of a novel coronavirus from a man with pneumonia in Saudi Arabia. N. Engl. J. Med. 367, 1814.

Zeng, L.P., Gao, Y.T., Ge, X.Y., Zhang, Q., Peng, C., Yang, X.L., Tan, B., Chen, J., Chmura, A.A., Daszak, P., Shi, Z.L., 2016. Bat severe acute respiratory syndrome-like coronavirus WIV1 encodes an extra accessory protein, ORFX, involved in modulation of the host immune response. J. Virol. 90, 6573.

Zhang, R., Li, Y., Cowley, T.J., Steinbrenner, A.D., Phillips, J.M., Yount, B.L., Baric, R.S., Weiss, S.R., 2015. The nsp1, nsp13, and M proteins contribute to the hepatotropism of murine coronavirus JHM.WU. J. Virol. 89, 3598.

Zhang, Q., Shi, K., Yoo, D., 2016. Suppression of type I interferon production by porcine epidemic diarrhea virus and degradation of CREB-binding protein by nsp1. Virology 489, 252.

Zhang, L., Li, L., Yan, L., Ming, Z., Jia, Z., Lou, Z., Rao, Z., 2018a. Structural and biochemical characterization of endoribonuclease Nsp15 encoded by Middle East respiratory syndrome coronavirus. J. Virol. 92, e00893-18.

Zhang, Q., Ke, H., Blikslager, A., Fujita, T., Yoo, D., 2018c. Type III interferon restriction by porcine epidemic diarrhea virus and the role of viral protein nsp1 in IRF1 signaling. J. Virol. 92, e01677-17.

Zhang, M., Li, W., Zhou, P., Liu, D., Luo, R., Jongkaewwattana, A., He, Q., 2020. Genetic manipulation of porcine deltacoronavirus reveals insights into NS6 and NS7 functions: a novel strategy for vaccine design. Emerg. Microbes Infect. 9, 20–31.

Zhou, P., Fan, H., Lan, T., Yang, X.L., Shi, W.F., Zhang, W., Zhu, Y., Zhang, Y.W., Xie, Q.M., Mani, S., Zheng, X.S., Li, B., Li, J.M., Guo, H., Pei, G.Q., An, X.P., Chen, J.W., Zhou, L., Mai, K.J., Wu, Z.X., Li, D., Anderson, D.E., Zhang, L.B., Li, S.Y., Mi, Z.Q., He, T.T., Cong, F., Guo, P.J., Huang, R., Luo, Y., Liu, X.L., Chen, J., Huang, Y., Sun, Q., Zhang, X.L., Wang, Y.Y., Xing, S.Z., Chen, Y.S., Sun, Y., Li, J., Daszak, P., Wang, L.F., Shi, Z.L., Tong, Y.G., Ma, J.Y., 2018. Fatal swine acute diarrhoea syndrome caused by an HKU2-related coronavirus of bat origin. Nature 556, 255.

Zhou, P., Yang, X.L., Wang, X.G., Hu, B., Zhang, L., Zhang, W., Si, H.R., Zhu, Y., Li, B., Huang, C.L., Chen, H.D., Chen, J., Luo, Y., Guo, H., Jiang, R.D., Liu, M.Q., Chen, Y., Shen, X.R., Wang, X., Zheng, X.S., Zhao, K., Chen, Q.J., Deng, F., Liu, L.L., Yan, B., Zhan, F.X., Wang, Y.Y., Xiao, G.F., Shi, Z.L., 2020.

A pneumonia outbreak associated with a new coronavirus of probable bat origin. Nature 579, 270–273.

Zhu, X., Wang, D., Zhou, J., Pan, T., Chen, J., Yang, Y., Lv, M., Ye, X., Peng, G., Fang, L., Xiao, S., 2017. Porcine deltacoronavirus nsp5 antagonizes type I interferon signaling by cleaving STAT2. J. Virol. 91, e00003-17.

Zust, R., Cervantes-Barragan, L., Kuri, T., Blakqori, G., Weber, F., Ludewig, B., Thiel, V., 2007. Coronavirus non-structural protein 1 is a major pathogenicity factor: implications for the rational design of coronavirus vaccines. PLoS Pathog. 3, 1062.

9780128215883